Nutrition, Exercise, and Behavior

AN INTEGRATED APPROACH TO WEIGHT MANAGEMENT

SECOND EDITION

LIANE M. SUMMERFIELD, PH.D.

Marymount University

W9-CCC-537

WADSWORTH
CENGAGE Learning™

Australia • Brazil • Japan • Korea • Mexico • Singapore • Spain • United Kingdom • United States

WADSWORTH
CENGAGE Learning

Nutrition, Exercise, and Behavior: An Integrated Approach to Weight Management, Second Edition
Liane M. Summerfield

Senior Acquisitions Editor: Peggy Williams

Assistant Editor: Shannon Holt

Senior Marketing Manager: Laura McGinn

Senior Marketing Communications Manager: Linda Yip

Content Project Management: PreMediaGlobal

Design Director: Rob Hugel

Art Director: John Walker

Print Buyer: Linda Hsu

Rights Acquisitions Specialist (Image & Text): Thomas McDonough

Production Service: PreMediaGlobal

Photo Researcher: Sara Golden

Text Researcher: Andrew Tremblay

Copy Editor: Pattie Stechschulte

Cover Designer: Riezebos Holzbaur/Tim Heraldo

Cover Image: Vegetable Image/Shutterstock, Royalty-Free; People with Resistance Bands (inset)/Shutterstock, Royalty-Free; Woman Stretching (inset)/Getty Images, Royalty-Free; Spinning Class (inset)/Shutterstock, Royalty-Free

Compositor: PreMediaGlobal

Library of Congress Control Number: 2011926914

ISBN-13: 978-0-8400-6924-5

ISBN-10: 0-8400-6924-3

Wadsworth
20 Davis Drive
Belmont, CA, 94002-3098
USA

Cengage Learning is a leading provider of customized learning solutions with office locations around the globe, including Singapore, the United Kingdom, Australia, Mexico, Brazil, and Japan. Locate your local office at **international.cengage.com/region**.

Cengage Learning products are represented in Canada by Nelson Education, Ltd.

To learn more about Wadsworth, visit **www.cengage.com/wadsworth**

Purchase any of our products at your local college store or at our preferred online store **www.cengagebrain.com**.

Printed in the United States of America
2 3 4 5 6 7 15 14 13 12

To the memory of my parents for all that they gave me.

Brief Contents

PART VI

Identification, Prevention, and Treatment of Eating Disorders and Childhood Obesity

Contents

PART II Metabolic and Physiological Aspects of Weight Management

CHAPTER 3
Energy Metabolism   81

PART III Nutrition for Health and Weight Management

PART IV Physical Activity for Health and Weight Management

CHAPTER 7
Physical Activity and Exercise: The Basics

CHAPTER 8

Physical Activity, Health, and Weight Management   271

PART V Approaches to Weight Management

CHAPTER 9

PART VI Identification, Prevention, and Treatment of Eating Disorders and Childhood Obesity

T en years have passed since the publication of the first edition of this text. Since that time, we have learned more about the incidence, prevalence, and consequences of obesity/overweight and eating disorders; we have seen updates to Dietary Guidelines, the food pyramid, and recommended nutrient intakes and have been given revised exercise guidelines. Additional physiological factors affecting weight have been discovered and treatment approaches evolved. And yet, obesity is still on the rise.

In the United States today, over two-thirds of the adult population and one-third of children and adolescents are overweight or obese. While the United States still leads the way, overweight and obesity rates have increased in all the developed nations and in many developing countries as well. For example, more than half of the population is overweight in the United Kingdom, Poland, and Chile, and just under half is overweight in Germany, Canada, and Australia. This excess weight, and particularly excess fat, is responsible for a rise in serious health conditions, an increase in health care costs by billions of dollars each year, and a reduction of quality of life. About 80% of obese adults have cardiovascular disease, hypertension, type 2 diabetes, cancer, osteoarthritis, complications during pregnancy, sleep apnea, and approximately 150,000 excess deaths in the United States alone have been attributed to obesity.

Furthermore, obesity is now seen at much earlier ages. Ten years ago, 14% of 6- to 11-year-olds and 12% of 12- to 19-year-olds were obese. Today, 20% of children and 18% of adolescents are obese. In the past ten years, rates of obesity among low income preschoolers increased from 12% to almost 15%. These rates are poised to have a profoundly negative effect on health and longevity, possibly lowering life expectancy of younger generations for the first time in decades.

Weight management is a complex topic, far more complicated than popular diet books, reality television programs, or exercise videos would suggest. This book looks at weight management holistically, considering the role of physiology, the environment, and human behavior to explain obesity and eating disorders. It offers in-depth coverage of important areas supported by current evidence with tables and figures to synthesize and summarize key points. An extensive reference list at the end of each chapter allows students to read original research.

WHY THIS TEXT WAS DEVELOPED

The first edition of this book was written to fill a gap in the textbook market: There were no comprehensive texts on weight management. Such a gap still exists, and the need for information has never been greater. Many people who want to lose weight will try almost anything that seems to offer a glimmer of success. This makes the promotion of evidence-based treatment approaches very important, and this text offers consideration of many of these. It also emphasizes the necessity of prevention. To address this, the text incorporates a public health approach to issues of weight management, where not only individual factors but societal, family, and environmental factors contributing to eating disorders and overweight/obesity are reviewed in each chapter.

This book is designed for students and professionals in many disciplines who are confronted—and confounded—by weight-management issues. While many people think of excess weight as an individual failure—not enough exercise and too much food—addressing this problem requires more than individual effort. Environmental factors so significantly influence people's individual behaviors that any interventions aimed at lowering rates of obesity and preventing overweight must take into consideration the individual, family, community, and broader environment. Without a multifaceted approach that involves health, fitness, and nutrition professionals, almost everyone could be overweight or obese within the next two decades.

BOOK DESIGN

This second edition reorders and updates information from the previous edition. It features twelve chapters organized in six sections: (1) An Introduction to Weight Management, (2) Metabolic and Physiological Aspects of Weight Management, (3) Nutrition for Health and Weight Management, (4) Physical Activity for Health and Weight Management, (5) Approaches to Weight Management, and (6) Identification, Prevention, and Treatment of Eating Disorders and Childhood Obesity.

Two introductory chapters lead off the text: an overview of overweight, underweight, and obesity; and methods of assessment relevant to weight management professionals—anthropometric, clinical, and biochemical measures and assessment of physical activity and diet. Chapters 3 and 4 review metabolic and physiological aspects of weight management. These chapters were moved closer to the front to introduce readers earlier to concepts of energy transformation in the body, the components of energy expenditure, and the roles of body systems and genetics in determining how and where we store fat.

Chapters 5 and 6 focus on energy nutrients, vitamins, minerals, and water. The chapters review the impact of each energy nutrient, specific vitamins and minerals on weight management. In addition, Chapter 5 contains a new section on dietary approaches to weight management.

Updated chapters on the role of physical activity in promoting health and weight management (Chapters 7 and 8) and behavioral and nonbehavioral approaches to weight management (Chapters 9 and 10) remain in this edition of the text. In addition, a new sixth section has been added: identification, prevention, and treatment of eating disorders (Chapter 11) and childhood obesity (Chapter 12). Because of rising rates of childhood obesity, Chapter 12 was completely rewritten to focus exclusively on primary, secondary, and tertiary prevention of childhood obesity.

As college pedagogy continues to evolve from teacher-centered approaches, where lecture predominates, to student-centered approaches, where case studies and other active learning techniques are used, so has the design of this text. Each chapter includes a "running" case study with multiple parts and discussion questions. Instructors can assign students to read a case or one section of a case before the class meets; initiate class with a brief lecture; and then use class time to discuss elements of the case in small groups. Research outside of class can also supplement each of the cases.

ACKNOWLEDGEMENTS

Many thanks to Dr. Carolyn Oxenford, professor and director of the Center for Teaching Excellence at Marymount University, for her help with developing case studies and to Sylvia Whitman, Marymount University Writing Specialist, who could always find just the right word when I could not. Reviewers of this second edition provided detailed comments, which were extremely helpful in revising the text. Many thanks to Susan Berkow (George Mason University), Jeffery Betts (Central Michigan University), Jeffery Harris (West Chester University), Cindy Marshall (Saddleback Community College), Kathy Munoz (Humboldt State University), and Susan Perry (Appalachian State University) for their thorough reviews and insights.

Overweight, Underweight, and Obesity

CHAPTER OUTLINE

R. Gino Santa Maria/Shutterstock.com

cappi thompson/Shutterstock.com

lorga Studio/Shutterstock.com

Over two-thirds of American adults and one-third of American children are at an unhealthy weight. By some estimates, three-quarters of the adult population could be overweight or obese by 2015[1] and 86% by 2030, with over half of those being obese.[2] The annual health care costs of this are approximately $147 billion, almost twice the cost of all cancers combined, according to the Secretary of Health and Human Services at the Centers for Disease Control's (CDC) July 2009 Weight of the Nation Conference. The U.S. Department of Health and Human Services set three goals related to overweight and obesity in its goals for the nation, *Healthy People 2010*: increase to 60% the proportion of adults who are at a healthy weight; reduce the proportion of adults who are obese to 15%; and reduce the proportion of children and adults who are overweight or obese to 5%. None of these goals has been met. In fact, we seem to move farther away from the goals every year. At this rate, the three goals of Healthy People 2010 will most likely be used again for *Healthy People 2020*.

For a variety of reasons—feeling better, looking better, fitting into clothes—most people do not wish to be overweight. Bookstores stock hundreds of "diet" books, and these books as well as special foods, weight-loss programs, and over-the-counter diet aids generate an estimated $35–$50 billion in sales each year. Our national preoccupation with weight has not only kept alive the weight-loss industry, but it has also contributed to a rise in eating disorders. In the quest for thinness, millions of individuals engage in severe caloric restriction, excessive exercise, and abuse of laxatives and other drugs.

Between 1% and 4% of the adult and adolescent population has an eating disorder. While the incidence of eating disorders is miniscule compared with obesity rates, many more individuals are afflicted with subclinical or partial eating disorders, which almost always begin with concern about weight and dieting behaviors.

fotoluminate, 2010/Shutterstock.com

Two-thirds of American adults and one-third of American children are at an unhealthy weight.

Losing weight is difficult. People who enter weight-loss programs can usually lose about 10% of their body weight; however, keeping off lost weight is even harder, and an unacceptably high proportion of those who lose weight regain it within 3–5 years. The psychological and physiological tolls of this repeated cycle of "failure" can be considerable. The individual may adopt extreme dieting and exercise behaviors, which become chronic. The impact can also be passed along to the individual's children, who may develop their own problems connected with food.

Who needs to lose weight? Who needs to gain weight? What health problems are associated with overweight? Is there a "best weight" for long life and good health? This chapter provides some answers to these questions and examines some of the problems associated with being over- and underweight.

AN OVERVIEW OF WEIGHT MANAGEMENT DEFINITIONS AND CONCERNS

What Do the Terms Overweight, Underweight, and Obese Mean?

Before looking more closely at the problems associated with over- and underweight, let's clarify the terminology used in weight management. Health professionals often use overweight and obesity interchangeably, and words such as plump, full-figured, large, and even fat are common euphemisms for both terms. Some individuals in the fat-acceptance movement believe that the term fat should be reclaimed as an adjective, like brunette, tall, or skinny, and not used as an expression of contempt, and they accept the use of terms like heavy, large, large-size, big, super-size, plus-size, and BBW for "big, beautiful women." However, obese women and men in a recent study did not like the terms fat, fatness, excess fat, obesity, large size, or heaviness.[3] Individuals in that study found the terms weight, excess weight, and even body mass index (BMI) to be more acceptable.

The point of this is that when professionals are talking among themselves, they should be clear about the meanings that they assign to the terms underweight, average weight, overweight, or obese. And when talking with clients, parents, or the public, they should avoid pejorative or censorious terms and adopt neutral terminology, such as weight and BMI. In this book, the terms obese and overweight will be used as characterized below, and when something applies to both, the term overweight/obese will be used.

BMI as an Indicator of Overweight, Underweight, and Obesity

Overweight has historically meant weight in excess of the average for a given individual's height, just as **underweight** refers to weight that is less than average. Determination of each required the measurement of height and weight. **Obesity** is defined as an excess of body fat, and its determination was not possible for the average person or even for a health care provider.

overweight
Weight in excess of the average for a given height, based on height–weight tables; or, in adults, a BMI of 25–29.9 kg/m^2.

underweight
Weight less than the average for a given height, based on height–weight tables; or, in adults, a BMI less than 18.5 kg/m^2.

obesity Excessive accumulation of body fat, generally considered to be 25% or more in men and 32% or more in women; sometimes defined as BMI 30 kg/m^2 or greater.

TABLE 1-1   Formulas for Calculating BMI

BMI = weight in kilograms (kg) ÷ (height in meters)^2

or

BMI = [weight in pounds (lb) × 703] ÷ (height in inches)^2

body mass index (BMI) *Indicator of overweight and obesity calculated by dividing the weight in kg by the height in meters squared.*

In the past 10 years, the **body mass index (BMI)** has come into widespread use as an indicator of overweight, underweight, and obesity. As illustrated in Table 1-1, BMI is calculated by dividing body weight in kilograms by height in meters squared (kg/m^2), or by using pounds and inches with a formula (or by using a table in which all calculations have already been done, such as the one in Appendix A, Table 1). Because height and weight are easy to obtain, BMI is a relatively simple method of assessment. In addition, a good correlation between BMI and body fat has been found, which means that most of the time people who have high BMIs are not only overweight but also obese.

In 1998, the National Heart, Lung, and Blood Institute (NHLBI), which is part of the National Institutes of Health (NIH), issued the first federal guidelines on the identification, evaluation, and treatment of overweight and obesity (guidelines are currently under revision and expected to be published in fall 2011). An expert panel defined overweight as a BMI of 25–29.9 kg/m^2 and obesity as a BMI of 30 kg/m^2 and above.[4] These cutoff points for designating overweight and obesity are also used internationally. NHLBI classification of underweight, overweight, and obesity is found in Table 1-2. The NHLBI classifies underweight as a BMI < 18.5 kg/m.^{2}

Obesity can also be determined by measuring body fat. Desirable body fat ranges are better agreed on than the ranges that constitute obesity. Men are considered to be in healthy fat ranges when between 10% and 20% of their body weight is composed of fat; for women, the range is 17%–25%. Men are said to be obese when 25% or more of their body weight is in the form of fat; women are considered obese at 32% fat or higher. In non-clinical settings, percent fat is usually determined by using skinfold measurements or bioelectrical impedance analysis. These techniques require a great deal more skill than measuring weight and are discussed at length in Chapter 2.

Defining obesity based on BMI is more difficult with children and adolescents because they have not yet reached their maximum height and their body composition is changing. When skinfold measurements are used, a fat percentage of 20% in boys and 25% in girls is considered moderately high. When BMI is used, gender- and age-specific charts must be consulted (these charts may be found in Appendix A). For ages 2 to 19 years, a BMI at or above the 85th percentile designates overweight and a BMI at or above the 95th percentile defines obesity.

Use of Height and Weight to Define Overweight, Underweight, and Obese

Because it is easier to measure body weight than body fat, and because bathroom scales are readily available, indicators of fatness based on weight are still used by the general public. The 1959 Metropolitan Desirable Weights Table and the 1983 Metropolitan Height and Weight Tables are examples of these and can be found in Appendix A, Tables A-2 and A-3. A person whose weight falls within the desirable range for height is said to be at ideal body weight (IBW). Sometimes health professionals refer to people at IBW as of "normal" weight, suggesting that anyone who either exceeds or weighs less than the IBW is somehow abnormal and in need of correction. A better way to describe an individual at IBW is at "expected" or "healthy" weight. And, conversely, a person whose weight falls outside that range might be considered at an unhealthy weight.

Wallenrock, 2010/Shutterstock.com

The oversize muscles of a bodybuilder.

Although 120% IBW in women and 124% IBW in men (based on the 1983 Metropolitan Height and Weight Tables) are sometimes used as indicators of obesity, body weight alone is not a reliable measure. Total body weight consists of fat, bone, water, muscle, organs and other tissues. Overweight individuals may have an excess of any or all of these components. Many overweight people have excessive body fat, but some, especially those at lower levels of overweight, may not. Bodybuilders, for example, carry a lot of muscle weight and may weigh more than their IBW, but most are not overfat (obese) or even unhealthy.

As people gain progressively more weight, it is increasingly likely that the excess weight is fat, not muscle. Individuals who weigh twice their IBW or who are more than 100 pounds over IBW are most certainly obese. So, you would be safe in saying that not everyone who is overweight is obese, but probably everyone who is obese is overweight.

Fat Distribution

The pattern of fat distribution and the characteristics of fat cells can also describe obesity. Fat stores may be distributed internally or subcutaneously (literally, under the skin). Internal fat consists of visceral and nonvisceral fat. **Internal abdominal visceral fat**, fat stored in the abdominopelvic region, is of particular interest in obesity. You may also hear it referred to as central, truncal, or intraabdominal fat. This pattern of fat distribution, often called upper-body or central obesity, is thought to be more harmful to health than the **subcutaneous** distribution of fat and is pictured in Figure 1-1. So-called love

internal fat *Fat stored deep in the abdomen, sometimes called visceral, central, truncal, or intraabdominal fat.*

subcutaneous fat *Fat stored under the skin throughout the body.*

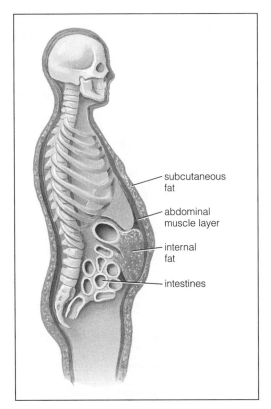

subcutaneous
fat

abdominal
muscle layer

internal
fat

intestines

FIGURE 1-1 Internal fat and subcutaneous fat. The fat lying deep within the body's abdominal cavity may pose an especially high risk to health.

handles are usually made up of subcutaneous, not visceral, fat. When fat is stored subcutaneously in the hips, buttocks, and thighs, fewer harmful effects are noted. In fact, some researchers believe that measures of central obesity are superior to the BMI in predicting the health risks of obesity.

Either waist circumference or a measure called the waist-to-hip ratio can be used to determine the pattern of fat distribution and is discussed further in Chapter 2. Recently obesity researchers have found the waist-to-height ratio as another useful indicator of central obesity and health risk. In adults a waist-to-height ratio of 0.5 or higher is predictive of cardiovascular and metabolic abnormalities.[5]

These concepts are discussed in detail in Chapter 2. Table 1-2 summarizes the ways that obesity can be classified: (1) by percentage of body weight made up of fat; (2) by location of fat stores; (3) by overweight; and (4) by BMI.

How Many People Are Overweight, Obese, or Underweight?

The National Center for Health Statistics has been collecting data on the diet behaviors and health status of U.S. residents since 1960 in a series of studies known as the National Health and Nutrition Examination Surveys (NHANES). Each survey has included thousands of adults and children ages 6 through 74 years. Data from NHANES I (1970–1974) suggested that a quarter of the adult population was overweight. When data were

TABLE I-2　Classification of Obesity

Obesity may be classified by

A. Level of fatness
　　32% fat in women　　　　　25% fat in men
B. Distribution of fat
　　Upper body, central, or abdominal obesity
　　　　Waist circumference > 40 inches in men; > 35 inches in women
　　　　Waist-to-height ratio ≥ 0.5
　　Lower body – subcutaneous fat deposited in the hips, buttocks, and thighs
　　　　Waist-to-hip ratio < 0.95 in men; < 0.8 in women
C. Body weight
　　120% of ideal body weight in women　　　　124% of ideal body weight in men
D. Body mass index (kg/m^2)
　　Underweight　≤ 18.5
　　Overweight　25–29.9
　　Obesity　30–34.9
　　Obesity (Grade 2)　35–39.9
　　Obesity (Grade 3)　≥ 40

Case Study. The Health Writer, Part I

Clara is a newly hired health writer for a company that produces educational materials on a variety of subjects. Her company houses its approximately 100 employees on several floors of a building in a large city. During new employee orientation, Clara learned that employees are provided free parking in the garage under the building. There is no cafeteria, but fast food and other restaurants abound in the 3–4 block area around the building. In addition, there are vending machines on the third floor that sell soda and chips, and employees have access to a fairly large television lounge where many people spend their 45-minute lunch period and take breaks.

While eating lunch in the lounge, Clara met Ginny, editor of an automotive e-publication. "Maybe you can help me understand this BMI that my doctor was talking about at my annual physical last week," Ginny says. "I don't understand where he got that number after the nurse took my height and weight." Ginny is willing to share her height and weight with you: she is 5'5" tall and weighs 160 lb. "I've put on about 20 lbs over the past two years, and my doctor says I need to lose some weight."

- Calculate Ginny's "old" and "new" BMI
- Use Table 1-2 to determine her BMI classification. Has it changed in two years?
- Besides the BMI, what other information would be useful for Ginny to understand her health risk.

examined from the first phase of NHANES III (1988–1994), researchers were startled to find that there had been a dramatic increase in the prevalence of overweight among U.S. adults in a relatively short time. Based on the findings from NHANES III, one-third (33.3%) of the adult population of the United States was estimated to be overweight—31% of men and 35% of women. Results from NHANES (2003–2006) indicated that, while rates of overweight (BMI 25–29.9 kg/m^2) held to about one-third of the population, rates of obesity (BMI ≥ 30 kg/m^2) continued to increase. NHANES data from 2007 to 2008 show some leveling off of obesity rates for women and small increases for men.[6]

Table 1-3 provides current estimates of adult overweight and obesity in the United States. Today, over two-thirds of the adult population (72% of men and 64% of women) is overweight or obese (having a BMI at or above 25), and 34% has a BMI ≥ 30 (32% of men and 35.5% of women).

The prevalence of overweight and obesity among children and adolescents has increased significantly as well. Approximately 34% of children and adolescents are overweight, with almost 20% of 6- to 11-year-olds and 18% of 12- to 19-year-olds at or above the 95th BMI percentile and considered obese.[7] Table 1-4 presents obesity prevalence data for children and adolescents. Childhood obesity is discussed in Chapter 12.

Contrast these figures with the prevalence of underweight. Just under 2% of the adult U.S. population is underweight (BMI < 18.5 kg/m^2). Most of these individuals (3%) are 20- to 39-years-old; 1% are 40 and older.[8] Among children and adolescents, about 3% are underweight (below the 5th age- and gender-specific BMI percentile)—3% of 2- to 11-year-olds and 4% of those over age 11.[9] As mentioned in the introduction to this chapter, 1–4% of the population has an eating disorder, and underweight may be

TABLE 1-3 Prevalence of Overweight and Obesity, U.S. Adults[1]

	Overweight (BMI ≥ 25–29.9 kg/m^2) (percentage of population)		Obese (BMI ≥ 30 kg/m^2) (percentage of population)		
	1976–80	2003–06	1976–80	2003–06	2007–08[2]
Both genders	32.2	32.8	15.1	34.1	33.9
Male	40.1	39.5	12.8	33.1	32.2
Female	24.9	26.0	17.1	35.2	35.5
White, not Hispanic or Latino					
Male	41.4	39.1	12.4	33.0	31.9
Female	23.3	24.9	15.4	32.5	33.0
Black or African American					
Male	34.8	35.7	16.5	36.3	37.3
Female	31.6	26.2	31.0	54.3	49.6
Mexican American					
Male	45.9	46.9	15.7	30.4	35.9
Female	43.0	31.8	26.6	42.6	45.1
All Hispanics[2]					
Male	—	—	—	—	34.3
Female	—	—	—	—	43.0

[1] Adults are ages 20–74 years
[2] Data in the 2007–08 column are from Flegal, K.M., Carroll, M.D., Ogden, C.L., & Curtin, L.R. (2010). Prevalence and trends in obesity among U.S. adults, 1999–2008. *JAMA, 303*(3), p. 236.
Source: Adapted from *Health 2009, United States*, pp. 320–321.

TABLE I-4   Prevalence of Obesity, U.S. Children

	BMI ≥ 95th percentile (age- and sex-specific) (percentage of population)		
	1976–88	2003–06	2007–08*
6–11 years of age			
Both genders	6.5	17.0	19.6
Boys	6.6	18.0	21.2
White, not Hispanic or Latino	6.1	15.5	20.5
Black or African American	6.8	18.6	17.7
Mexican American	13.3	27.5	27.1
Hispanic	—	—	28.3
Girls	6.4	15.8	18.0
White, not Hispanic or Latino	5.2	14.4	17.4
Black or African American	11.2	24.0	21.1
Mexican American	9.8	19.7	22.3
Hispanic	—	—	21.9
12–19 years of age			
Both genders	5.0	17.6	18.1
Boys	4.8	18.2	19.3
White, not Hispanic or Latino	3.8	17.3	16.7
Black or African American	6.1	18.5	19.8
Mexican American	7.7	22.1	26.8
Hispanic	—	—	25.5
Girls	5.3	16.8	16.8
White, not Hispanic or Latino	4.6	14.5	14.5
Black or African American	10.7	27.7	29.2
Mexican American	8.8	19.9	17.4
Hispanic	—	—	17.5

*Data in the 2007–08 column are from Ogden, C.L., Carroll, M.D., Curtin, L.R., Lamb, M.M., & Flegal, K.M., & (2010). Prevalence of high body mass index in U.S. children and adolescents, 2007–2008. *JAMA, 303*(3), 244–245.
Source: Adapted from *Health 2009, United States*, p. 305.

an indicator of anorexia nervosa, a very serious eating disorder that will be discussed in Chapter 11.

Gender and Racial Differences in Overweight and Obesity
The increase in overweight and obesity among Americans has spared no population group. Both males and females and people of all ages, races, education levels, and geographic regions have been affected. Racial-ethnic minorities, particularly women, have higher rates of obesity than whites, as Table 1-3 illustrates. Over three-quarters of

Mexican American men and women are overweight or obese, and almost half of African American women are obese. (NOTE: NHANES, from which these data were derived, was designed to study only people of Mexican origin and not all Hispanic groups in the United States. Only data for 2007–08 include a category for all Hispanics). These racial differences are apparent early in childhood. Notice that Asian Americans are not included in these tables. Asian Americans are the one minority group that has much lower rates of overweight and obesity than other groups in the U.S. population.

Racial and ethnic differences in overweight/obesity have several probable causes. Some of these are cultural/familial, while others are tied to socioeconomic status.

Cultural and Familial Factors

Attitudes about exercise, food preparation methods, and preferred foods are all responsive to cultural influences. In addition, parents who were raised in poverty may model the same dietary and activity behaviors that they were exposed to as children. When these behaviors are obesity-promoting, obesity is perpetuated.

Some cultures and some families have a greater acceptance of—and even a preference for—heavier body weight. A larger child may mean "healthier" in these families; a larger parent may be seen as "stronger."

Socioeconomic Status

Educational attainment, income, and occupation combine to produce socioeconomic status. In industrialized societies, obesity is more prevalent among people at low socioeconomic levels, regardless of race or ethnicity. In the United States, poverty rates are higher for African Americans and Hispanics than for whites, which is linked to obesity for several reasons:

food deserts
Areas, often in low-income communities and neighborhoods, where there is limited access to healthy foods.

- Many socioeconomically disadvantaged communities are said to exist in **food deserts**, areas with limited access to lower-fat foods, including fresh fruits and vegetables, whole grains, low-fat milk, and fresh fish. A study in Chicago found that residents of predominantly African American neighborhoods had to go twice as far to access a supermarket as a fast food restaurant.[10] Prices may be higher in these urban markets, and poor access to transportation creates an additional barrier.[11]
- When people must rely on government-surplus foods and donated foods from community agencies, they do not necessarily receive low-fat or reduced-calorie items.
- Lack of equipment and instruction may result in less use of healthy food preparation techniques, such as steaming and microwaving.
- Health clubs, gyms, and afterschool sports programs may not be affordable.
- Fears for personal safety may restrict walking or participation in community-based activity programs.
- Less access to health care prevents regular contact with health professionals.

social determinants of health *Factors within families, communities, and the environment that affect health, including income, education, employment, relationships with others, and access to health care.*

Many of the factors discussed above are known as **social determinants of health**, defined by the World Health Organization as "the conditions in which people are born, grow, live, work, and age, including the health system" (www.who.int/social_determinants/en/). Social determinants of health include standard of living, education, employment, the built and natural environments in which we live, social relationships, freedom from discrimination, and access to health care. You would be right if you concluded that social determinants of health are factors that lead to health disparities and contribute to higher rates of overweight and obesity among racial and ethnic minorities.

Global Perspectives on the Prevalence of Overweight and Obesity

Excess weight is not only a problem in the United States. The World Health Organization estimates that more than 400 million adults throughout the world are obese and 1.6 billion are overweight. These numbers may double by 2015. In fact, overweight and obesity are on the rise in all the developed nations and in many developing countries. As one obesity researcher notes: "Heavier people are getting heavier at a faster rate and thinner people are getting heavier at a slower rate."[12]

The World Health Organization's International Obesity Task Force maintains the Global Database on Body Mass Index (available at http://apps.who.int/bmi/index.jsp). Figures 1-2 and 1-3 include a sampling of nations and show that, while the United States leads the way, other countries are beginning to catch up. The rate of overweight in several countries—the United Kingdom, Poland, Hungary, Chile—has surpassed half the population, and in other countries—Germany, Canada, and Australia—includes just under half of the adult population.

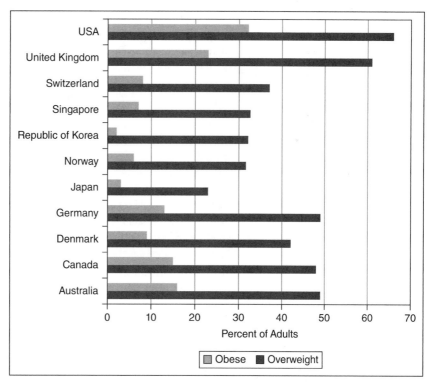

FIGURE 1-2 Global Prevalence of Adult Overweight and Obesity, Developed Countries.

Source: Data from the Global Database on Body Mass Index and Low, S., Chin, M.C., Deurenberg-Yap, M. (2009). Review on epidemic of obesity. *Ann Acad Med Singapore, 38,* p. 59.

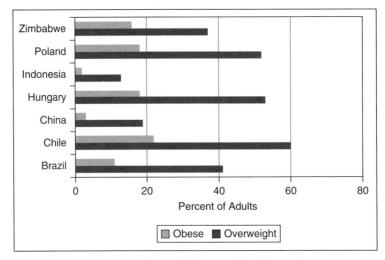

FIGURE 1-3   Global Prevalence of Obesity, Developing Countries.

Source: Adapted by the author from the Global Database on Body Mass Index and Low, S., Chin, M.C., Deurenberg-Yap, M. (2009). Review on epidemic of obesity. *Ann Acad Med Singapore, 38,* p. 59.

Why Is Weight Gain on the Rise in the United States and Elsewhere?

Obesity unquestionably has biological origins. Thousands of years ago, periodic food shortages and, in times of severe environmental challenge, outright food scarcity favored individuals who could store fat. This was particularly true for pregnant or nursing women, who needed enough stored energy for themselves as well as their developing fetus or a breast-feeding baby. Survivors passed on their fat-conserving or thrifty genotype to their children, giving future generations the ability to store more fat to adapt to food shortages.

Obesity almost certainly became socially advantageous. Early carvings and paintings from times when obesity was rare feature people with large hips and buttocks. So, to the extent to which food availability allowed it, weight gain was probably once considered to be desirable.

The dramatic increase in the prevalence of overweight and obesity today no doubt results from a combination of factors, including genetics. The increase cannot be explained solely by biologic factors, which presumably would always have been present in humans and would affect a predictable proportion of the population. Although heredity gives us the capacity for obesity, environment encourages us to store excessive quantities of fat. Most experts agree that two factors are primarily responsible for the fattening of the population—diet and inactivity.

Figure 1-4 shows the relationship between environmental factors, individual factors, and obesity.

Diet and Body Weight

Today, food is abundant for most residents of developed countries. For the average American, this has translated into consumption of about 300 **kilocalories (kcals)** more each day than in 1980. The reason caloric consumption has increased is no mystery:

kilocalorie (kcal)
Measure of the energy value of food or physical activity.

- Dining outside the home has increased;
- Eating larger portion sizes than 25 years ago;

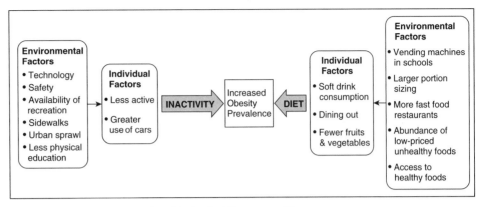

FIGURE 1-4 Environmental and Individual Factors in Obesity.

- Consuming energy-dense foods (those that are high in sugars, fats, and oils), particularly snack foods, and fewer fruits and vegetables. Snackers eat as much as 25% more in a typical day than non-snackers.[13]
- Consuming sweetened soft drink has never been higher. Whereas many people will reduce solid food consumption after they have had a high-calorie meal, fewer people compensate after ingesting excess liquid calories.
- Exporting inexpensive sweeteners and oils has increased globally, allowing even poor nations to supplement the diet with products that improve food flavor—but at greater caloric cost.[14]
- Eating more throughout the day. Researchers recently reported that the time between "eating occasions" decreased by an hour over the past 30 years.[15]

Various societal and environmental conditions contribute to increased calorie consumption:

- The number of single parent families and women working outside the home has increased, so there may be less time for meal preparation and greater reliance on convenience foods.
- Unhealthy foods are generally less expensive than healthy foods;
- There has been a proliferation of fast food restaurants (more than 240,000 in the United States alone).
- Whether at a fast food restaurant or big box store, foods and beverages are increasingly available in "supersized" portions, encouraging people to value quantity more than quality;
- Food advertising continues to be aimed at children, often using appealing cartoon-like characters;

Inactivity and Body Weight

There is some evidence that more active people weigh less and store less fat. Unfortunately, most Americans do not exercise. Over one-third of the U.S. adult population engages in no leisure-time physical activity, and 30% are only somewhat active.[16] Similar trends are being seen globally. The reasons for this are a combination of individual and environmental factors:

- Individuals walk less and drive more, often due to suburban sprawl in un-walkable communities.

Elena Elisseeva/Shutterstock

Standard portion sizes are larger today, contributing to obesity.

- The average household has three television sets,[17] and the average American watches just under 5 hours of television each day.[18] Not only are most people inactive while viewing television, they eat more.
- For many children, television viewing, computer use, and video games have replaced afterschool recreational physical activity.

Donna Day/Stone/Getty Images

Television watching influences children's eating habits and activity patterns.

- Some communities, particularly in rural and low-income areas, lack recreational areas.
- Fewer opportunities exist for physical activity in schools, including elimination of physical education and recess in some schools.
- Social networking popular with adolescents, such as Facebook and MySpace, allows us to socialize without ever leaving our homes.
- Throughout the world technology has reduced the caloric expenditure required to perform many household and occupational tasks.

The "Perfect Storm" of Inactivity and Excess Calories

The combination of lots of available food and lower caloric expenditure is a "perfect storm" for creating overweight and obesity. Addressing this problem will require more than individual effort. Environmental factors so significantly influence people's individual behaviors that any interventions aimed at lowering rates of obesity and preventing overweight must take into consideration the individual, family, community, and broader environment. Without this multifaceted approach, almost everyone could be overweight or obese within the next two decades. The public health effort to reduce tobacco use in the United States may offer some useful lessons for the prevention of obesity. Comprehensive policies on everything from the cost of a pack of cigarettes to cigarette advertising to smoking bans in public spaces resulted in a decrease in smoking rates from about 25% of the U.S. population in 1992 to 19% in 2006. Policies related to food and activity might have a similar effect on the prevalence of obesity. Table 1-5 illustrates parallel strategies for tobacco control and obesity prevention.

Chapters that follow consider diet and activity in more detail. Chapter 12 presents a discussion of policy changes that could help prevent obesity.

TABLE 1-5   Parallel Strategies for Preventing Smoking and Obesity

Strategy	Tobacco	Obesity
Cost	• Increase cigarette tax	• Reduce cost of fruits and vegetables • Tax unhealthy foods and earmark revenue for public health • Increase gasoline tax
Availability	• Anti-smoking laws	• Remove unhealthy foods from schools, hospitals, businesses
Image	• Restrictions on cigarette advertising on television • Prohibitions against using cartoon characters in cigarette advertising • Anti-smoking ads that show the true impact of smoking	• Restrict advertising of unhealthy foods, particularly during children's television programming • Develop advertisements that show the impact of harmful beverages and foods • Post calorie content of foods on fast food menus

An Overview of Weight Management Definitions and Concerns: Summary

In the United States, about two-thirds of the adult population is overweight or obese. While the prevalence of overweight has remained about the same for the past 30 years, obesity rates have more than doubled in that time period, with some racial and ethnic groups disproportionately affected. Globally, rates of overweight and obesity are increasing at a similar rate in developed nations, and developing nations are also seeing weight gain in their populations. Consumption of excess calories and the widespread lack of physical activity seen in our society are probably most to blame for Americans' weight gain. Reversing this trend will require not only individual effort, but significant attention to social determinants of health and environmental interventions.

THE IMPACT OF WEIGHT ON HEALTH

Both morbidity and mortality are increased with overweight and obesity. CDC researchers estimate that obesity was responsible for almost 170,000 excess deaths in 2004.[19] Obesity may be worse than smoking or heavy drinking in its impact on health.[20] The economic cost of this is high. RAND Corporation researchers point out that the elimination of one disease will not generally have a significant impact on health care costs, but obesity might be the exception.[21] The reason is that obese older adults do not have a dramatic reduction in life expectancy, so they may live as long and spend as many years in the Medicare system as nonobese older adults, but almost all of those years will be spent in disability. The obese 70-year-old will cost Medicare $38,000 more over his lifetime than the healthy-weight 70-year-old. Even in other developed nations, which have a lower prevalence of obesity than the United States, 2 to 4% of national health expenditures result from obesity-related causes.[22]

What Are Morbidity and Mortality?

morbidity *Illness.*

mortality *Death.*

The term **morbidity** refers to illness. A morbidity rate is calculated by dividing the rate of illness in a particular population by the number of people in that group. The term **mortality** refers to death. Mortality rates are usually expressed by number of deaths per thousand, ten thousand, or hundred thousand people in a population.

As future health professionals, you might ask, appropriately, "What is the best body weight for long life, good health, and happiness?" There is no single answer to this question. Although a great deal of research has been done to seek an answer, particularly to questions about morbidity and mortality, the relationship between weight and health is neither clear nor consistently agreed upon.

The best studies for drawing sweeping conclusions about the relationships between weight and morbidity or mortality use thousands of subjects who are examined over many years, often for decades. But performing such studies is very difficult. Regularly collecting data from a large group of subjects is time consuming and costly. People frequently move and are inaccessible for follow-up. In addition, people in these studies have many and varied habits that may affect their health and longevity independent of weight.

Members of the general public become frustrated when studies reach apparently contradictory conclusions. When reviewing the evidence on the effects of body weight on morbidity and mortality, keep the following in mind, which may explain some of the differences seen between studies:

- Measurement issues: In some large-scale studies, such as the NHANES, researchers actually train technicians to measure the height, weight, and other characteristics of the research subjects. In most studies, however, height and weight are not actually measured. Instead, researchers rely on research subjects' recollections of their weight at various stages of life. This is less reliable than having a trained technician take measurements. In addition, other factors affect accuracy of measurement, such as use of a calibrated balance scale or an electronic-load cell scale and clothing worn by research subjects when they are weighed.
- Body composition: When only weight is measured (or requested from subjects), the contribution of specific components of body composition is not determined. Weight may consist of greater or lesser proportions of fat and muscle, which should affect health differently. Patterns of fat storage also differ among people and may have a differential effect on health.
- Human factors: Several factors that exist independent of body fat can affect health and longevity, including activity level, smoking, diet, and drug and alcohol use. If these are not taken into consideration by researchers, then conclusions can be seriously flawed. For example, without controlling for detrimental health habits, the researcher cannot know if lower-weight people have higher death rates because low body weight is a health risk or because more of them smoke, a behavior often associated with lower body weight.

Which Health Problems Are Associated with Obesity?

comorbidities
Diseases that exist in clusters and contribute to overall morbidity.

(3) Obese people often have high blood cholesterol, high blood pressure, and poor control of blood sugar. These conditions are known as **comorbidities** because they tend to occur in clusters and contribute as a group to morbidity. But not all obese individuals have the same risk of disease. This is because, in addition to the human factors listed previously, five factors, which vary among people, affect the risk of health impairment in overweight or obese individuals. These are summarized in Table 1-6.

- Level of overweight: As weight increases, the incidence of cardiovascular disease, diabetes, and some cancers also increases.
- Location of stored fat: Some obesity researchers believe that location of body fat stores is at least as important and may be even more significant than total body fat in increasing the risk of comorbidities. The fatty acids released from internal visceral abdominal fat stores are rapidly taken up by the liver. As a result of this

TABLE 1-6   Factors that Affect the Risk of Disease in Overweight/Obese Individuals

- Human factors: Activity level, physical fitness, smoking, composition of the diet, drug and alcohol use
- Level of overweight: Higher weights associated with increased risk of disease
- Location of stored fat: Intraabdominal fat increases risk
- Age and duration of overweight and obesity: Higher risk when obesity begins at younger ages
- Race-ethnicity: African Americans and Hispanics at greater risk

abundant supply of fatty acids, the liver takes up less insulin, which leads to hyper-insulinemia as less insulin is cleared from the bloodstream (discussed later in this chapter) and release of triglycerides from the liver into the general circulation. In addition, visceral adipocytes produce various hormones that contribute to metabolic disorders. As a result, individuals with the central pattern of fat deposition have a higher prevalence of type 2 diabetes, higher blood pressure, and increased risk of cardiovascular diseases. An interesting study of Japanese sumo wrestlers illustrates this point. Young sumo wrestlers having an average BMI of 36 and consumption of 5,000–7,000 kcal per day were found to have a high proportion of fat stored subcu-taneously. Although they would be clinically defined as obese, their cholesterol, plasma glucose, and triglyceride levels were in low to normal ranges, possibly due to fat patterns and as a result of diet and training program.[22]

- Age and duration of overweight and obesity: Particularly in young adults, obesity and excess weight can have a powerful negative effect on health and mortality. For example, high blood pressure and high blood cholesterol levels are far more com-mon in overweight 20- to 24-year-olds than in average-weight people of the same ages. On the one hand, obesity beginning in childhood may also be associated with greater risk of ill health later in life, including heart disease in both males and females, gout in males, and arthritis in women. The specific level of fatness at which risk increases is not known. On the other hand, some experts believe that the longer people have remained overweight or obese without any health complica-tions, the less susceptible they are to the health impairments associated with obesity.

- Race-ethnicity: There is some evidence that African Americans have a lower health risk (and Asians a higher health risk) at a given BMI level, perhaps due to differ-ences in lean mass and fat mass with increasing BMIs.[6] Nevertheless, life expectancy rates among African Americans, Hispanics, and American Indians are generally lower than those of Caucasians. High rates of type 2 diabetes occur among Hispanics and American Indians, and hypertension rates are highest in African Americans. Whether obesity is a cause or an effect of these diseases and, indeed, the overall impact of obesity on disease among different racial-ethnic groups remains under investigation.

- Inactivity and unhealthy diet: Whether people are obese or not, these increase the risk of cardiovascular disease and other health problems. But when poor diet and inactivity are coupled with obesity, the risk of health problems is compounded.

(3) The diseases most commonly associated with excess fat are cardiovascular disease, hypertension, type 2 diabetes, cancer, and osteoarthritis. In addition, obesity may cause liver disease, complications during pregnancy, and sleep apnea. About 80% of obese adults have at least one of the morbidities discussed here.[23] The relationship between underweight and disease is less straightforward, although osteoporosis is more common in women who are underweight than overweight. The impact of weight on mortality is presented in the next section.

cardiovascular disease (CVD)
Disease group that includes diseases of the heart and blood vessels, such as coronary heart disease, myocardial infarction (heart attack), angina pectoris (chest pain on exertion), stroke, and hypertension.

Cardiovascular Disease

Diseases of the heart and blood vessels are collectively known as **cardiovascular diseases** (CVD). CVD is the leading cause of death in the United States and a major factor in loss of functional capacity. Researchers expect that within 10 years CVD will be the leading cause of death throughout the world, in large measure due to increasing rates of obesity.[24]

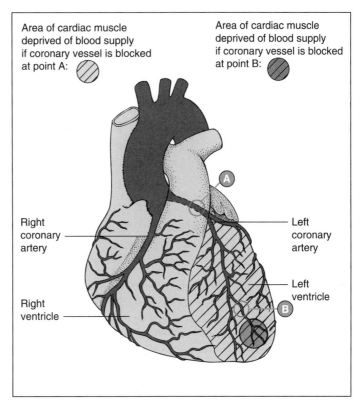

Area of cardiac muscle deprived of blood supply if coronary vessel is blocked at point A:

Area of cardiac muscle deprived of blood supply if coronary vessel is blocked at point B:

Right coronary artery

Left coronary artery

Left ventricle

Right ventricle

FIGURE I-5 The extent of damage to the heart depends on the size of the blocked artery.

One factor in obesity-related CVD is mechanical. Increased blood volume and cardiac output, characteristic of obese individuals, require their hearts to work harder. This can lead to hypertrophy of the left ventricle and left atrium. The result: increased risk of atrial fibrillation and congestive heart failure. Estimates suggest that obesity is the underlying cause of about 14% of heart failure cases in women and 11% in men.[25]

Most cases of CVD are due to coronary artery disease, also known as coronary heart disease, in which atherosclerotic plaque gradually narrows the coronary arteries, leading to disability, chest pain, and, in many cases, heart attack. Figure 1-5 illustrates the outcome of such a blockage.

Although much is still to be learned about the development of atherosclerosis, the following events are involved:

- The innermost layer of endothelial cells that line an artery—called the intima—becomes more susceptible to the uptake of cells that eventually evolve into plaque, by a number of possible scenarios. There may be actual injury to the endothelial cells from hypertension, a virus, or oxidation. Or the body's own immune response may make the cells rougher so that material passing through the bloodstream is more likely to stick to them. High levels of circulating glucose may also affect the endothelium. Places where coronary arteries branch into smaller arteries (called bifurcation points) seem particularly vulnerable.

cytokines
Chemicals (other than antibodies) secreted by white blood cells that are associated with inflammation. Adipokines are similar chemicals secreted by fat cells.

myocardial infarction *Lack of oxygen in the heart muscle due to narrowing or blockage of one or more arteries, which causes an interruption in normal heart functioning; also known as a heart attack.*

transient ischemic attack *Temporary lack of oxygen to a small area of the brain, which may precede a cerebrovascular accident, or stroke.*

- The resulting lesions on the intima attract white blood cells called monocytes, which differentiate into macrophages, cells that engulf and attempt to neutralize foreign substances as part of the body's immune system.
- Dietary cholesterol and cholesterol synthesized by the body are transported in particles called **low-density lipoprotein cholesterol (LDL)**, sometimes referred to as "bad" cholesterol. LDL particles may be oxidized when they come into contact with substances known as **free radicals**, produced at higher levels in smokers and people who eat a diet high in fat and low in fruits and vegetables (this is discussed in detail in Chapter 6). In the development of atherosclerosis, macrophages surround oxidized LDL particles, creating foam cells. LDLs also attract platelets to the area, which promotes clotting.
- Oxidized LDLs and macrophages stimulate inflammation. This makes the endothelial cells even more adhesive to passing cells and increases passage of foam cells into the smooth muscle layer of the artery—called the media—and plaque develops. As a result, specific markers of inflammation may be released into the blood stream, including **cytokines** and **C-reactive protein (CRP)**, for which there is a blood test.

Figure 1-6 shows you what plaque looks like, both in a drawing and in a photomicrograph of a blocked coronary artery. When a large area of the heart is affected, such as in condition A, it is deprived of needed oxygen, causing a **myocardial infarction**, or heart attack, which can be fatal. Blockage of an artery that brings oxygenated blood to the brain, such as the carotid artery, can cause a **transient ischemic attack** or a cerebrovascular accident (stroke). Arteries that serve the brain may be blocked by atherosclerotic plaque or a chunk of plaque that travels from a larger artery, or they may become weakened and rupture. Interestingly, not all plaque becomes dangerous, and scientists are just beginning to sort out the differences between high-risk and low-risk plaque.

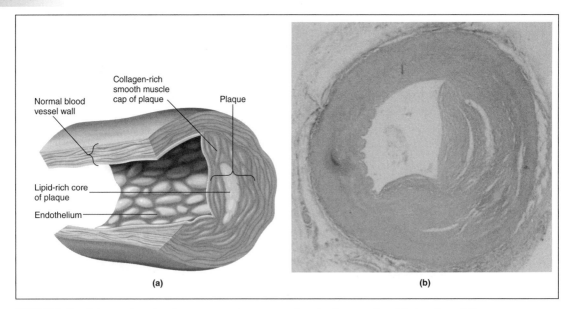

Normal blood vessel wall

Collagen-rich smooth muscle cap of plaque

Plaque

Lipid-rich core of plaque

Endothelium

(a)

(b)

FIGURE 1-6 Atherosclerotic plaque accumulates over time in the arteries. (a) drawing of the components of plaque; (b) photomicrograph of severe atherosclerosis in a coronary artery.

Storage fat itself produces inflammatory cytokines known as adipokines, including leptin and adiponectin, which are discussed further in Chapter 4. These increase the risk of CVD. In obesity, central fat deposition and insulin resistance CRP levels may be elevated. Thus, obesity is a major risk factor for CVD. Other major risk factors include family history, high blood cholesterol, hypertension, smoking, physical inactivity, and diabetes. Nonsmokers who are lean throughout life have the lowest risk of CVD, but obese nonsmokers who lose weight often improve both their cholesterol level and their blood pressure which, when accompanied by physical activity and healthy diet, can lower the risk of CVD.

Hypertension

hypertension
High blood pressure (140 or greater mm Hg systolic and/or 90 or greater mm Hg diastolic).

Approximately 30% of American adults and over 2 million children have **hypertension**. In adults, mild hypertension is indicated by a systolic blood pressure of 140 mm Hg or higher, a diastolic pressure of 90 mm Hg or higher, or a combination of both. Left untreated, high blood pressure is a potentially serious condition. It is a leading cause of kidney damage, stroke, and heart disease, and this is true for all racial and ethnic groups and both men and women.

Most people with hypertension are overweight, even children. Obese adults are six times more likely to develop high blood pressure than lean adults.[26] The prevalence of hypertension is particularly high among African Americans—over 40% have high blood pressure.

There are several reasons why obesity elevates blood pressure: increased blood volume, high fat diet, and inflammation are all factors. Drops in blood pressure generally

Monkey Business Images, 2010/Shutterstock.com

High blood pressure is the leading cause of kidney damage, stroke, and heart disease.

follow weight loss, especially when weight loss is accomplished by modifying diet and increasing physical activity. Even a small weight loss (3–5 kg, or 7–11 lbs) can have beneficial effects on blood pressure and reduce or eliminate the need for blood pressure–lowering drugs. Sometimes lifestyle changes will normalize blood pressure, even when weight remains the same. Lifestyle changes associated with blood pressure reduction include becoming more physically active, eating less salt, eating more fruits and vegetables, and stopping smoking.

Sleep Apnea

sleep apnea
Disorder characterized by episodes of slowed or stopped breathing during sleep.

Sleep-disordered breathing includes a spectrum of conditions ranging from snoring and snorting to periods during sleep when breathing actually stops. Obstructive **sleep apnea** is a potentially life-threatening condition in which episodes of slowed or stopped breathing occur during sleep. The condition is caused by collapsibility of the upper airway. Because breathing may be interrupted more than 100 times during the night, which usually causes the affected individual to wake abruptly, drowsiness during the day is common and may lead to automobile or other accidents.

About 1 in 5 adults has mild and 1 in 15 has moderate obstructive sleep apnea.[27] Obesity is a significant risk factor for sleep apnea, with estimates ranging from between 40% to 90% of moderately overweight or obese individuals having the condition. Individuals with central obesity are at greater risk, as are individuals with neck circumferences greater than 17.5 inches (44 cm).[28] Reduction of 10–15% of body weight is an effective treatment for sleep apnea.

type 2 diabetes mellitus *Preferred term for what was formerly called non–insulin-dependent diabetes mellitus (NIDDM); a metabolic disorder characterized by high blood glucose and cellular resistance to insulin. Overweight and obese individuals account for most cases.*

There is a relationship between sleep apnea and several comorbidities also associated with obesity. The combination of hypoxia (oxygen deficiency) and sleep deprivation affect glucose metabolism, increasing the risk of type 2 diabetes.[29] Elevated blood pressure may also be a consequence of sleep apnea.[30]

Type 2 Diabetes Mellitus

According to the CDC, the incidence of diabetes mellitus in this country has almost doubled in the past decade, from a rate of 4.8 per 1,000 to 9.0 per 1,000.[31] Today more than 24 million Americans have diabetes, and 90 to 95% of cases are type 2. Equally alarming, the number of cases of **type 2 diabetes** has increased almost 10-fold among children and adolescents, and almost one-third of diabetes cases diagnosed in 11- to 18-year-olds are type 2.[32] Rates of type 2 diabetes throughout the world are also on the increase creating what several researchers have called "one of the major threats to human health in the twenty-first century."[33] Diabetes rates in Egypt and Mexico equal U.S. rates, and India and China account for half of diabetes cases diagnosed every day.[14] Type 2 diabetes affects men and women about equally, but racial and ethnic minorities are disproportionately affected. The highest rates of diabetes in the world occur in American Indians and some Pacific Islander and aboriginal people. In the United States, Hispanics and African Americans have twice the prevalence of diabetes as Caucasians.[34]

hyperinsulinemia
Excessive secretion of insulin.

insulin resistance *Cellular resistance to the uptake of glucose in the presence of insulin.*

β-cells *Beta cells of the pancreas that secrete insulin.*

Type 2 diabetes, which used to be called non–insulin-dependent diabetes mellitus (NIDDM), is characterized by high blood glucose levels, high circulating insulin levels (also known as **hyperinsulinemia**), and **insulin resistance**. When glucose enters the bloodstream, insulin should be released from the **β-cells** of the pancreas to promote the uptake of glucose into the insulin-sensitive tissues—primarily muscle, fat, and liver.

Insulin also signals the uptake of fat into the fat cells. Sometimes, for reasons not fully understood, insulin is over-secreted. This may result from several factors:

- Chronic overconsumption of a diet very high in carbohydrates, especially sugars, requires secretion of increasing amounts of insulin;
- Internal abdominal visceral fat stores release high levels of fatty acids into the liver, which prevents insulin clearance from the bloodstream.

Insulin stimulates fat storage. As the fat cells become larger, they become increasingly resistant to the normal effects of insulin. This keeps even more insulin in circulation and hyperinsulinemia worsens. Insulin resistance can also occur first, with hyperinsulinemia being the result. The combination of insulin resistance and hyperinsulinemia leads to β-cell stress, and, when the β-cells can no longer compensate, impaired glucose tolerance, and type 2 diabetes.

Because enlarged fat cells reduce the capacity of insulin to bind to insulin receptors in adipose tissue, obesity is considered the biggest risk factor for type 2 diabetes. In addition, central fat may play a more significant role than lower body fat. Among the additional links between obesity and type 2 diabetes are:

- Consumption of energy-dense foods. These highly processed foods are high in fat and calories, high on the glycemic index (promoting greater release of glucose), and low in fiber. Insulin secretion rises in response to the need to store excess fat and glucose. A good example of the effect of diet on insulin sensitivity is the contrast between the Arizona and Mexican Pima Indians. Among the more obese Arizonan Pima, who consume a high-fat diet, 34% of men and 41% of women have type 2 diabetes. Among the Mexican Pima, who eat more traditional foods with a lower fat and higher fiber content, only 6% of men and 8.5% of women have the disease.[35]
- Fructose. Fructose metabolism is different than glucose metabolism, even though they are both simple sugars. Fructose is readily taken up by the liver, where it is quickly metabolized and increases the synthesis of triglycerides. Fructose is a component of the widely used high-fructose corn syrup (HFCS), a sweetener in soft drinks and some snack foods. While HFCS has not been directly connected to increased rates of obesity, there are indications that fructose intake is linked to the risk of type 2 diabetes.[36]
- Physical inactivity. Active people generally have better blood glucose and insulin profiles than **sedentary** people. In the United States and throughout the world, native populations going from active to inactive lifestyles have experienced an increase in type 2 diabetes. Inactive muscles store less carbohydrate, so blood glucose remains elevated, stimulating the release of more insulin. Active muscle cells, in contrast, take up glucose with less insulin, which keeps blood glucose and insulin levels in normal ranges. In addition, active people burn more fat than carbohydrate, which helps to reduce the amount of dietary fat available for storage in the fat cells.

sedentary *Inactive.*

Figure 1-7 summarizes the interaction of multiple factors in the development of type 2 diabetes mellitus.

When left untreated, hyperinsulinemia, poor blood glucose control, and type 2 diabetes create many health problems, including CVD, kidney failure, lower-extremity amputations, and eye disease. The risk of colorectal cancer may be higher in people, particularly men, with hyperinsulinemia.[37] Even children with hyperinsulinemia are at greater risk of CVD due to high blood lipids and hypertension. Unfortunately type 2 diabetes is often

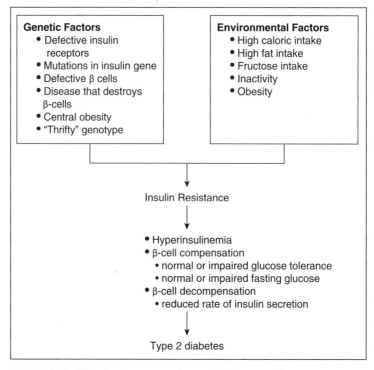

FIGURE 1-7   The development of type 2 diabetes mellitus.

Source: Adapted by the author from Olefsky, J.M., & Nolan, J.J. (1995). Insulin resistance and non-insulin-dependent diabetes mellitus: Cellular and molecular mechanisms. *American Journal of Clinical Nutrition, 61* (Suppl), p. 983S.

not diagnosed until one of these complications appears. People of color experience more complications from diabetes than Caucasians do, including higher rates of peripheral vascular disease, end-stage kidney disease, and blindness.[38] However, since not all obese individuals experience the same health complications, one researcher theorizes that the obese who have more adverse effects are those with insulin resistance.[39]

Fortunately, overweight type 2 diabetics who lose as little as 5–10% of their body weight can reduce the risk of disease and early mortality. Weight loss almost always improves blood glucose control and insulin sensitivity and prevents health-related complications that accompany diabetes. Lifestyle modification can even prevent progression of poor glucose control to diabetes. The Diabetes Prevention Program, which included 30 minutes of walking 5 days per week and dietary modifications that allowed obese participants to lose about 7% of their body weight, reduced the number of new cases of diabetes by 58%.[40] Other programs have produced similar results.

Metabolic Syndrome

Up to this point cardiovascular disease, hypertension, and diabetes are seen as separate health problems that afflict many overweight/obese individuals. The **metabolic syndrome** clusters these health problems under one heading. Although there have been many definitions of the metabolic syndrome over the years, most characterize it as a collection of conditions that includes central obesity, hypertension, abnormal blood lipids, and

metabolic syndrome *Cluster of metabolic abnormalities—hypertension, abnormal blood lipids, elevated fasting glucose, and central obesity—associated with increased risk of disease.*

TABLE 1-7 IDF Consensus Worldwide Definition of the Metabolic Syndrome

For a person to be defined as having the metabolic syndrome, he/she must have:	
• Central obesity	Defined as waist circumference ≥ 94 cm (37 in) for Europid men and ≥ 80 cm (31.5 in) for Europid women, with ethnicity specific values for other groups (see Table 1-8)
Plus any two of the following four factors:	
• Raised triglyceride level	≥ 150 mg/dL, or specific treatment for this lipid abnormality
• Reduced HDL cholesterol	< 40 mg/dL in males and < 50 mg/dl in females, or specific treatment for this lipid abnormality
• Raised blood pressure	Systolic BP > 130 or diastolic ≥ 85 mm Hg, or treatment of previously diagnosed hypertension
• Raised fasting plasma glucose	≥ 100 mg/dL, or previously diagnosed type 2 diabetes

Source: International Diabetes Federation. *The IDF consensus worldwide definition of the metabolic syndrome: Part I: Worldwide definition for use in clinical practice.* Available from http://www.idf.org/webdata/docs/MetSyndrome_FINAL.pdf; accessed June 29, 2010.

elevated fasting glucose. Table 1-7 provides the widely accepted International Diabetes Federation (IDF) definition.[41]

Increased waist circumference is one of the criteria for the metabolic syndrome. The IDF provides ethnic-specific values for waist circumference, which are provided in Table 1-8.

TABLE 1-8 Ethnic Specific Values for Waist Circumference to Determine Central Obesity

Country/Ethnic Group	Waist Circumference	
	Male	Female
Europids *NOTE: in the United States, ATP III values of 102 cm (40.2 in) for males and 88 cm (34.6 in) for females are likely to continue to be used for clinical purposes*	≥ 94 cm (37 in)	≥ 80 cm (31.5 in)
South Asians *(Based on a Chinese, Malay, and Asian-Indian population)*	≥ 90 cm (35.4 in)	≥ 80 cm (31.5 in)
Chinese	≥ 90 cm (35.4 in)	≥ 80 cm (31.5 in)
Japanese	≥ 85 cm	≥ 90 cm
Ethnic South and Central Americans	Use South Asian recommendations until more specific data are available	
Sub-Saharan Africans	Use European data until more specific data are available	
Eastern Mediterranean and Middle East (Arab) populations	Use European data until more specific data are available	

Source: International Diabetes Federation. *The IDF consensus worldwide definition of the metabolic syndrome: Part I: Worldwide definition for use in clinical practice.* Available from http://www.idf.org/webdata/docs/MetSyndrome_FINAL.pdf; accessed June 29, 2010.

Notice that the Europids (white people of European origin) have two waist circumference values—the IDF values ($\geq$ 94 cm for men and $\geq$ 80 cm for women) and the U.S. National Cholesterol Education Program Adult Treatment Panel III (ATP-III) values (> 102 cm for men and > 88 cm for women). Also notice that there are no values for three groups of people: ethnic South and Central Americans, Sub-Saharan Africans, and Eastern Mediterranean and Middle East populations. Research with these groups is ongoing. Recently researchers provided the first data available relating waist circumference specifically to insulin resistance in people of African descent. Their data suggest that the waist circumference values most diagnostic for this population are $\geq$ 102 cm for men (consistent with ATP-III guidelines) and $\geq$ 98 cm for women (higher than both the ATP-III and IDF values).[42]

Because definitions differ, prevalence estimates differ, but probably about 24% of U.S. adults have the metabolic syndrome.[43] The prevalence rises with age, from about 7% of 20-year-olds to 40% of 60-year-olds. There does not appear to be a single cause of metabolic syndrome, although many researchers theorize that insulin resistance is an important factor. The linkage to obesity is probably related to characteristics of adipose tissue: the fat cells themselves not only become insulin resistant but also produce adipokines that independently promote insulin resistance.

It is correct to assume that people with metabolic syndrome have a higher risk of coronary artery disease, heart attack, stroke, and diabetes. Metabolic syndrome can be reversed with weight loss. Metabolic syndrome has gone into remission with as little as a 10% weight loss among moderately obese adults, although morbidly obese adults require greater weight loss to see improvements.[44]

Cancer

Cancer is the second leading cause of death in the United States, and about 1.5 million new cases are diagnosed each year. The American Cancer Society estimates that one-third of the over 500,000 cancer deaths in 2010 will be related to overweight/obesity.[45] Several forms of cancer may be more common in obese than lean individuals. These include cancers of the breast in postmenopausal women, endometrium (the lining of the uterus), kidney, esophagus, pancreas, liver, and colon. People with BMIs at or greater than 40 kg/m^2 have the highest death rates from cancer, 52% higher in men and 62% higher in women.[46]

Cancers of the breast and endometrium are considered to be hormonally related. Body weight, weight gain, and location of fat deposits influence hormone levels and, therefore, may play a role in the development of these cancers. Gastroesophageal reflux, discussed below, probably plays a role in cancer of the esophagus. It is also possible that inactivity and unhealthy diet (one that is low in fruits, vegetables, and whole grains) interact with obesity to increase the risk of some cancers, such as pancreatic and colon cancer. Thus, an overweight individual who adopts a healthy diet and who is moderately active may have a lower cancer risk than an average-weight person who is inactive and has a poor diet. It goes without saying that anyone who smokes is at a higher risk of cancer.

osteoarthritis
Degenerative joint disease causing breakdown of cartilage in the affected joints and inflammation.

Osteoarthritis

Osteoarthritis is a painful and disabling condition characterized by breakdown and inflammation of the cartilage in the joints. The condition is most common in the hips, the knees, and the hands. Symptoms typically appear when people are in their 50s.

Factors that increase the likelihood of developing arthritis include injuries, inactivity, and overweight.

People with higher body weights even in early adulthood are at a greater risk of developing knee or hip osteoarthritis later in life. This risk is directly related to the stress of excess weight exerted on the joints of the lower body, which causes cartilage to break down. CDC estimates that 69% of people with doctor-diagnosed arthritis are overweight or obese.[47] Oddly enough, even osteoarthritis of the hand is more common in the obese. This may be due to overall higher bone mineral densities in heavier individuals, or it may be due to some systemic circulating factor that affects cartilage breakdown at various sites in the body.

For people who are already overweight, the risk for developing osteoarthritis of the knee increases by about 15% for each additional kg/m^2 increase in BMI.[48] As you might expect, the activity limitations imposed by arthritis of the knee could make weight loss difficult and further weight gain inevitable. Still, weight loss not only lowers the risk of developing arthritis but also helps minimize symptoms of arthritis.

Additional Health Issues Associated with Overweight/Obesity

preeclampsia *A condition that occurs in 7–10% of pregnancies in which the mother develops hypertension, edema, protein in the urine, and sometimes serious liver disease.*

thrombophlebitis *Painful inflammation of a vein.*

gastroesophageal reflux disease (GERD) *Caused when stomach acid backs up into the esophagus and produces inflammation.*

nonalcoholic fatty liver disease (NAFLD) *Results from accumulation of fat in the liver not associated with alcohol over-consumption, sometimes leading to fibrosis, cirrhosis, and liver failure.*

- Complications of pregnancy. About half of women of childbearing age (ages 18- to 44-years old) are overweight or obese.[49] Both overweight and obese women experience more complications during pregnancy, including **preeclampsia**, hypertension, **gestational diabetes mellitus**, **thrombophlebitis**, late delivery, and greater chance of caesarean delivery. Infants born to obese mothers have more congenital defects, such as heart and neural tube abnormalities, and are more likely to be at a lower birth weight; their risk of death within the first year after birth is twice the rate of infants born to lean mothers.[50]

- Polycystic ovary syndrome (PCOS). Between 6–10% of women of childbearing age have **polycystic ovary syndrome (PCOS)**.[51,52] PCOS is more common in obese women than lean, particularly those with central obesity. It is characterized by smaller than normal ovarian follicles, irregular or absent menstruation, increased secretion of androgenic hormones (those associated with maleness), acne and, sometimes, excessive hairiness. Insulin resistance, hypertension, and elevated blood lipids occur with PCOS, so the risk of CVD is increased. Almost half of women with PCOS also have metabolic syndrome.[52]

- **Gastroesophageal Reflux Disease (GERD).** GERD, sometimes called acid reflux, occurs when stomach liquid (which is acidic) backs up, or refluxes, into the esophagus. The esophagus does not have a protective lining like the stomach does, so it is damaged. The resulting inflammation is called esophagitis. Sometimes chronic, untreated esophagitis can lead to esophageal adenocarcinoma, a type of cancer. Overweight/obese individuals have a greater risk of GERD for several reasons. Accumulation of adipose tissue in the upper torso can increase pressure on the stomach and relax the sphincter between the esophagus and the stomach. In addition, anyone who eats a high-fat diet may experience reflux, particularly in the evening when lying down.

- Fatty Liver. **Nonalcoholic fatty liver disease** (NAFLD) occurs when large triglyceride droplets accumulate in the liver. There seems to be a relationship between obesity, insulin resistance, and elevated triglycerides, which contributes to NAFLD.[53]

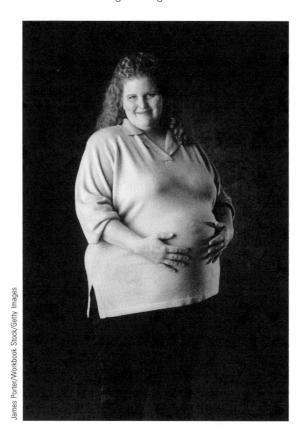

James Porter/Workbook Stock/Getty Images

High BMI during pregnancy can increase the health risk to mother and baby.

As excess fat accumulates in the liver, cytokines, which you have already learned to be associated with inflammation and CVD, increase. Untreated, NAFLD is a leading cause of fibrosis, cirrhosis, and liver failure.

Are There Harmful Health Effects from Being Underweight?

The relationship between underweight and disease is different from that between obesity and disease. Healthy underweight people (weighing less than average for height, or BMI under 18.5) do not have a higher risk of most chronic diseases than people of average weight, assuming they are consuming adequate amounts of food and nutrients and are not exercising to excess. Unhealthy, underweight people often do have higher morbidity rates, for two primary reasons:

- Both cigarette smoking and heavy alcohol use are associated with lower than average body weight. These behaviors also contribute to disease.
- Underlying diseases may cause low body weight. Cancer, heart failure, alcoholism, depression, and chronic lung disease are associated with low body weight as well as ill health.

A third reason that underweight may be associated with high morbidity and mortality occurs when the individual is severely underweight due to anorexia nervosa. Physiological conditions associated with anorexia are discussed in Chapter 11.

The key to determining whether there is a link between underweight and disease is in determining whether low body weight is voluntary or involuntary, explained or unexplained. An underweight person in poor health should be medically evaluated to rule out an eating disorder, drug or alcohol abuse, or an underlying disease before you presume that low body weight is the cause of ill health.

Osteoporosis and Low Body Weight

osteoporosis

Disease characterized by loss of bone mineral density that reduces bone mass and increases susceptibility to fractures.

The bones of people with **osteoporosis** absorb less calcium and gradually lose their density. The hip and spinal column are particularly affected, resulting in reduced height and greater susceptibility to fractures. Osteoporosis is more common in women than men and, in most cases, is due to a combination of low calcium intake, inactivity, and estrogen status. Pre-menopausal women who have cessation of menstruation, known as amenorrhea, are also at risk of early osteoporosis due to reduced estrogen. Smoking and alcohol use are additional risk factors for osteoporosis.

Bone mineral density is closely related to BMI. Weight may offer a type of protection that is very similar to the protection offered by physical activity—by regularly subjecting bone to mechanical stress (whether from exercise or from carrying extra weight in the form of muscle or fat), calcium uptake improves and bones become denser. This suggests that underweight people may have increased risk of osteoporosis. However, recent studies have shown that in both men and women, lean body mass rather than fat mass has a more profound influence on development of strong, dense bones.[54,55] So a person at a lower BMI who maintains a healthy lean body mass, does not smoke or consume immoderate amounts of alcohol, and eats a diet of calcium- and vitamin D-rich foods may be as able to preserve bone as a heavier person.

Estrogen can be produced from an enlarged fat mass, which may help to maintain bone in older women. However, we are just learning that the cytokines secreted by adipose tissue may have a negative effect on bone.

What Is the Best Weight for Long Life?

Life insurance companies rely on actuarial tables to predict length of life for people of various ages. Actuarial tables tell us, for example, that a 25-year-old male has a life

---

Case Study. The Health Writer, Part 2

Ginny typically runs next door for fast food at noon, but on days when she brings a lunch, Clara usually eats with her in the television lounge. One day she confides that her weight gain happened gradually as she got older. "I didn't really notice until I realized that my sweat pants were my most comfortable pants," Ginny said one day, "and now I feel like I can never go back to my college weight." She is concerned with the lab results from her physical exam, which show her with borderline high cholesterol, in addition to her increased blood pressure. So far her blood glucose is in the high-normal range. But Ginny's mother has diabetes, and Ginny is especially afraid of becoming diabetic.

- Why might Ginny be at greater risk of type 2 diabetes?
- What are some small steps that Ginny could take to begin minimizing her risk of type 2 diabetes?

expectancy of another 51 years, whereas if that same male lives to age 60, he has a life expectancy of 21 additional years. This section addresses the question of whether excess body weight, especially body weight associated with excess fat, reduces predicted lifespan. The quick answer to this question is yes, particularly for people at higher BMIs.

Links between Weight and Mortality

According to various estimates, middle-aged obese adults live between 2 and 10 years less than nonobese adults.[56,57] Many of the findings about mortality and weight have come from long-term, large-scale investigations that include hundreds of thousands of research subjects, such as NHANES,[19] the Nurses' Health Study,[58,59] the Physicians' Health Study,[60] the American Cancer Society's Cancer Prevention Study I and II,[46,61] the NIH-AARP Diet and Health Study,[62] and the smaller Framingham Heart Study.[56] These studies have consistently concluded that mortality is highest among obese adults—BMIs of 30 kg/m^2 and above—for both genders and all racial/ethnic groups. Most recent studies find overweight and non-overweight adults to have about the same death rates.[57]

A number of factors muddy the relationship between BMI and mortality: smoking (smokers tend to be lighter in weight and have high mortality rates); underlying diseases (these sometimes cause weight loss, and they certainly increase mortality); and comorbidities like type 2 diabetes and hypertension (these are seen more commonly in heavier people but are themselves linked to higher mortality rates). Researchers typically describe both a U-shaped and a J-shaped association between BMI and mortality (see Figure 1-8):

- In a U-shaped association, very low and very high weights are associated with high mortality, and weights between high and low are associated with low mortality. This shape may be most relevant in studies that include smokers.
- In a J-shaped association, very low weights are associated with a slight increase in mortality; then, as weights increase, there is a successively larger association, as shown in Figure 1-8. This relationship is most commonly described in studies of weight and mortality.

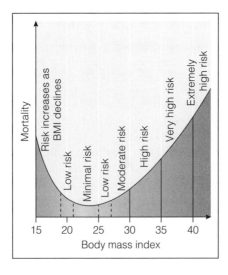

FIGURE 1-8   BMI and mortality.

Cardiovascular disease is the culprit in most obesity-related deaths. Given what you now know about the high prevalence of cardiovascular risk factors in obesity, this should not be a surprise. Cancer is the second leading cause of death among obese adults, with 50% to 60% higher mortality rates from cancer among individuals whose BMIs are 40 kg/m^2 or higher. As cancer death rates are even slightly higher among overweight persons, researchers estimate that 90,000 deaths per year from cancer might be avoided if all adults could maintain a BMI of 25 kg/m^2 or lower.[46] Death rates from diabetes are also higher among obese adults. Type 2 diabetes raises the risk of death from all causes, and particularly from CVD, five- to seven-fold.[59] Central obesity has been identified as an important factor explaining increased mortality in obesity. In one study the risk of death increased with waist circumference of about 95 cm (37 in) in men and 77 cm (30 in) in women.[63]

What about underweight adults? Smoking, heavy use of alcohol, muscle wasting, and pre-existing diseases (diabetes, chronic respiratory conditions, heart disease) are all factors associated with loss of weight that might make it appear as if underweight increases risk of death. However, the majority of research looking at the other end of the J-shaped association substantiates that healthy underweight adults—non-smokers with normal levels of lean body mass—do not have higher mortality rates than adults with BMIs from 18–29.9 kg/m^2.

Does Weight Loss Lengthen Life?

The previous sections present a strong case linking obesity with greater morbidity and mortality. If evidence suggests that higher BMIs are associated with higher mortality, then should overweight and obese people be advised to lose weight? And given that most people who lose weight eventually regain it, how does a lifetime of dieting and regaining (so-called weight cycling) affect mortality?

To answer these questions, it is necessary to make a distinction between voluntary (intentional) and involuntary (unintentional) weight loss. In other words, did people lose weight due to poor health or undiagnosed illness (involuntary) or because they were actively trying to achieve a lower body weight (voluntary)? For example, as type 2 diabetes progresses, some people lose weight due to kidney and heart disease. They die prematurely from the disease process, not because they lost weight.

weight cycling
Repeated episodes of weight gain and loss.

Let's look at intentional weight loss. Most evidence indicates that losing weight doesn't necessarily provide a survival advantage, except perhaps among people diagnosed with type 2 diabetes.[25,64] What about people who voluntarily lose weight, regain it, and keep cycling through this lose–gain process, called **weight cycling** or yo-yo dieting? Weight cycling does not seem to have an adverse effect on length of life. However, according to the Nurses' Health Study II, weight cycling may have the unintended effect of promoting greater weight gain and increased risk of binge eating.[65]

So, perhaps the best advice is for overweight and obese individuals who have health problems to undertake moderate weight loss using methods that will accomplish permanent and attainable lifestyle change. And for those who find weight loss impossible, a better strategy is to avoid additional weight gain.

Is There an Optimal Weight to Enhance Quality of Life?

This question is the most difficult to answer because quality of life encompasses lots of things that affect well-being—physical functioning, perception of health, income, education, housing, family relationships, job satisfaction. Most research conducted with overweight/obese participants has looked at health-related quality of life, which is best characterized as an individual's perception of his/her physical and mental health.

CDC added a health-related quality of life measure to its 1993 Behavioral Risk Factor Surveillance System. Data collected from the 2000 survey indicate that being obese—but not overweight—lowered respondents' health-related quality of life.[66] Compared to the nonobese, obese individuals reported more activity limitations and more days on which they were physically or mentally unhealthy. These findings are similar to those from the National Physical Activity and Weight Loss Survey, in which obese adults had lower self-rated health and more physically and mentally unhealthy days and days of activity limitation.[67] However, those at any BMI level who reported more physical activity had higher health-related quality of life.

An important question is whether obesity itself or the comorbidities of obesity result in lower quality of life. Comorbidities are likely in obesity, and morbidly obese adults are likely to have the most health problems, which perhaps explains why health-related quality of life decreases as BMI increases. Much of the research looking at the influence of comorbidities comes from medical centers that perform gastric bypass surgery. Obese individuals seeking surgery report more health problems than those who do not want surgery, and gastric bypass surgery candidates also report poorer health-related quality of life than other obese adults.[68] This result is not surprising. Heavier, sicker people certainly would be expected to have lower vitality, more pain, difficulty in bending and climbing stairs, and even depression and anxiety.

Another way to examine this question of quality of life is to look at people after weight loss and determine whether they are happier and feel better. Severely obese people who significantly reduce weight through gastric surgery experience greatly improved health-related quality of life, which is sustained over at least a year.[69] Reductions in pain and improved vitality and physical function have also been reported in overweight adults who lost smaller amounts of weight.[69] Particularly effective in improving health-related quality of life are weight management strategies that include dietary changes *and* exercise.[66]

Unfortunately, repeated cycles of diet-induced weight loss and regain can have detrimental psychological effects, including feelings of failure, unhappiness, and depression. The sense of deprivation that goes along with dieting may be particularly responsible for the poor psychological well-being seen in some people after weight loss, when compared with weight-stable people. The National Association to Advance Fat Acceptance (NAAFA) is among several groups that have attempted to promote an un-dieting movement and tried to move overweight individuals toward greater self-acceptance. At least one study has found improved mood and self-esteem in severely obese individuals who were counseled to "stop dieting and accept their bodies."[70]

Weight Stigmatism and Bias

During a decade in which the prevalence of obesity has dramatically increased, weight bias and discrimination has also risen. Much of this is fueled by people's belief that overweight and obesity are essentially personal problems—fat people are undisciplined, lazy, and unwilling to take responsibility for their weight—and shaming them into recognizing that they are too fat will motivate them to lose weight. These attitudes foster media bias and discrimination in schools, workplaces, and health care agencies. Lower health-related quality of life is directly related to weight stigmatism and bias.

- In the media, thin women dominate television and movie screens. Larger characters tend to be guest actors on television shows, rather than regular players, and are portrayed as more passive. They are more likely to be seen on comedy programs than dramas.[71]
- In the workplace, overweight/obese individuals are frequently discriminated against in hiring, they are often rated as less disciplined/competent than non-obese workers, and they earn less than non-overweight workers.[72]
- In educational settings, some teachers rate overweight students as sloppy; and classmates consider them to be lazy and undesirable playmates.[72]
- In health care settings, the environment may be unwelcome to overweight/obese patients—from small chairs, to blood pressure cuffs that don't fit a larger arm, to scales that only go up to 350 pounds. And provider attitudes may be confrontational and suggest that a lack of compliance is expected among obese clients.

There is no federal law that recognizes weight discrimination, although Michigan and several cities have such legislation. People who face weight bias may earn less, miss school days, avoid health care, and—alarmingly—eat more.[72] Given the difficulty in losing weight that most people encounter and the role of multiple social and environmental conditions in promoting weight gain, it is unreasonable to believe that blaming the "victim" will increase the likelihood of losing weight. More appropriately, NAAFA suggests promoting Health at Any Size. Their general principles are reprinted in Table 1-9.

TABLE 1-9   Health at Any Size

- Accepting and respecting the diversity of body shapes and sizes
- Recognizing that health and well-being are multi-dimensional and that they include physical, social, spiritual, occupational, emotional, and intellectual aspects
- Promoting all aspects of health and well-being for people of all sizes
- Promoting eating in a manner which balances individual nutritional needs, hunger, satiety, appetite, and pleasure
- Promoting individually appropriate, enjoyable, life-enhancing physical activity, rather than exercise that is focused on a goal of weight loss

Source: National Association to Advance Fat Acceptance (http://www.naafaonline.com/dev2/education/haes.html)

Impact of Weight on Health: Summary

As BMI increases, the risk of comorbidities, such as cardiovascular disease, high blood pressure, type 2 diabetes, metabolic syndrome, cancer, osteoarthritis, and sleep apnea, increase similarly. Particularly at risk of comorbidities and higher mortality rates are the obese. Underweight is not associated with high morbidity or mortality, unless low body weight is the result of smoking, anorexia nervosa, or a health problem that causes weight loss. Among obese individuals who have health problems, weight loss may increase life-span; among obese individuals who do not have health problems, weight loss may not convey any added years of life. There is no consensus about whether weight loss improves quality of life or makes people happier, although health problems that may affect life quality are more common in people whose BMI exceeds 30. Weight stigma and bias are prevalent in our society, adversely affecting quality of life.

WHO NEEDS WEIGHT-MANAGEMENT INFORMATION?

After reading this chapter, you probably have reached at least two conclusions: (1) Body weight and body fat do affect health and longevity, and (2) there is no one best weight for everybody. In some people, particular levels of overweight and under-weight are clearly associated with health risk. For example, the presence of glucose intolerance or hypertension may make at least a modicum of weight loss imperative, or a family history of osteoporosis may indicate a need for gain of lean body mass. In others, body weight itself may not be compromising health, but the use of unsafe methods of weight loss may present a hazard and potentially lead to an eating disorder.

So, there is no single answer to our question, "What is the best body weight for low risk of illness and disease, for long life, and for happiness or high quality of life?" Even among people who have the same BMI, there may be a wide disparity in health risk and longevity. Why? Because of differences in dietary practices, physical activity, and other lifestyle habits, like smoking; because of dieting and other weight-control practices that are more harmful than the excess weight itself, practices that sometimes even lead to eating disorders in susceptible people; and because of biological differences and limitations.

One valid conclusion that could be reached from reading this chapter is that not everyone needs to lose weight to live a long, healthy, quality life. Those whose BMI exceeds 30 and who have comorbidities will certainly benefit from an intervention. Some of the factors that are more likely to cause severe health consequences without weight reduction are:

- Obesity with onset before age 40
- Hypertension, poor glucose tolerance, and elevated blood lipids
- Other cardiac risk factors, such as smoking and inactivity
- Central fat distribution
- Sleep apnea

Additional factors that should be considered in evaluating the need for weight reduction are symptoms of knee osteoarthritis and social problems, including discrimination at school or the workplace.

Overweight individuals who have complicating factors or comorbidities such as those listed previously should alter diet and activity behaviors to prevent additional weight gain. Dramatic weight loss may not be needed to gain health benefits. Two weight-management experts suggest that obese individuals be counseled to attain a "reasonable" weight, which they define as the lowest weight a person has been able to maintain since age 21 without extreme diet or exercise adjusting for current age by adding 1 pound for each year over age 21.[73] If the weight loss needed to reach that weight would be extreme, a more modest goal of a loss of 10% of body weight is considered to be attainable, maintainable, and associated with improved health. In addition, when people are advised to consider a small weight loss as a success, their mood and self-esteem may improve.

When severe health consequences are not predicted, maintenance of a stable body weight might be an appropriate outcome. In most cases, this can be accomplished through improved diet and activity levels, which are discussed in detail later in this book. Such improvements, if attainable and maintainable, will surely improve health and might even result in healthy weight loss.

Garry Wade/Taxi/Getty Images

Physical activity is the best way to prevent obesity or minimize further weight gain.

Case Study. The Health Writer, Part 3

One day, after Clara had been on the job about a month, the executive director of the company called her into his office. He said, "I noticed that you have a health degree. Maybe you can help my employees to shape up. I can't stand seeing all these fat people working for me. They're killing our health insurance costs. Figure out some kind of a diet program to get them to lose weight." Clara was somewhat taken aback by his comments, but agreed to give this some thought.

A few days later Clara heard Ginny fuming that her request for a longer lunch was denied. "I asked if I could start work 15 minutes earlier and take a full hour for lunch," she said. "My carpool gets here early anyway, and if I could just have a few more minutes at lunchtime I could get some exercise then. But the boss said no." Her friend nods. "Remember when I wanted to use the conference room for a yoga class? He said no to that, too."

Clara gathered her thoughts and decided to share some ideas with her supervisor. He was enthusiastic about her approach. Clara's supervisor accompanied her when she went back to talk with the director. After a nudge from her supervisor, Clara took a deep breath and said, "Sir, I think I have a better idea than a diet program. Here it is."

- If you were Clara, what would you suggest to the director for helping to reduce his company health care costs?

REFERENCES

1. Wang, Y., & Beydoun, M. A. (2007). The obesity epidemic in the United States gender, age, socioeconomic, racial/ethnic, and geographic characteristics: A systematic review and meta-regression analysis. *Epidemiologic Reviews, 29*(1), 6–28.

2. Wang, Y., Beydoun, M. A., Liang, L., Caballero, B., & Kumanyika, S.K. (2008). Will all Americans become overweight or obese? Estimating the progression and cost of the US obesity epidemic. *Obesity, 16*(10), 2323–2330.

3. Wadden, T. A., & Didie, E. (2003). What's in a name? Patients' preferred terms for describing obesity. *Obesity Research, 11,* 1140–1146.

4. National Heart, Lung, and Blood Institute. (1998). *Clinical guidelines on the identification, evaluation, and treatment of overweight and obesity in adults: The evidence report.* Bethesda, MD: National Institutes of Health.

5. Maher, V., O'Dowd, M., Carey, M., Markham, C., Byrne, A., Hand, E., & McInerney, D. (2009). Association of central obesity with early carotid intima-media thickening is independent of that from other risk factors. *International Journal of Obesity, 33*(1), 136–143.

6. Flegal, K.M., Carroll, M.D., Ogden, C.L., & Curtin, L.R. (2010). Prevalence and trends in obesity among US adults, 1999–2008. *JAMA, 303*(3), 235–241.

7. Ogden, C.L., Carroll, M.D., Curtin, L.R., Lamb, M.M., & Flegal, K.M., & (2010). Prevalence of high body mass index in U.S. children and adolescents, 2007–2008. *JAMA, 303*(3), 242–249.

8. Fryar, C.D., & Ogden, C.L. (2009). Prevalence of underweight among adults: United States, 2003–2006. *NCHS Health E-Stat.* Hyattsville, MD: National Center for Health Statistics. Available from. http://www.cdc.gov/nchs/data/hestat/underweight/underweight_adults.htm. Accessed June 29, 2010

9. Fryar, C.D., & Ogden, C.L. (2009b). Prevalence of underweight among children and adolescents: United States, 2003–2006. *NCHS Health E-Stat.* Hyattsville, MD: National Center for Health Statistics. Available from http://www.cdc.gov/nchs/data/hestat/underweight/underweight_children.htm. Accessed June 29, 2010.

10. Mari Gallagher Research & Consulting Group. (2006). *Examining the impact of food deserts on public health in Chicago.* Chicago: Author. Available online at http://www.agr.state.il.us/marketing/ILOFFTaskForce/ChicagoFoodDesertReportFull.pdf. Accessed July 8, 2010.

11. Ver Ploeg, M., Breneman, V., Farrigan, T., Hamrick, K., Hopkins, D., Kaufman, P., et al. (2009). Access to affordable and nutritious food: Measuring and understanding food deserts and their consequences. *Economic*

Research Service (ERS) Report. Administrative Publication (AP-036). Available online at http://www.ers.usda.gov/publications/ap/ap036. Accessed July 8, 2010.

12. Bleich, S., Cutler, D., Murray, C., & Adams, A. (2008). Why is the developed world obese? *Annual Review of Public Health, 29,* 273–295.

13. McCrory, M. A., Suen, V. M. M., & Roberts, S. B. (2002). Biobehavioral influences on energy intake and adult weight gain. *Journal of Nutrition, 132*(12), 3830S-3834.

14. McLellan, F. (2002). Obesity rising to alarming levels around the world. *Lancet, 359,* 1412.

15. Popkin, B.M., & Duffey, K.J. (2010). Does hunger and satiety drive eating anymore? Increasing eating occasions and decreasing time between eating occasions in the United States. *American Journal of Clinical Nutrition, 91*(5), 1342–1347.

16. National health interview surveys: Summary health statistics for U.S. adults. (2009). Available online at http://www.cdc.gov/nchs/nhis.htm. Accessed June 29, 2010.

17. Kaiser Family Foundation. (2005) *Generation M: Media in the lives of 8–18 year-olds.* Menlo Park, CA: Kaiser Family Foundation.

18. Neilsen Media Research. (2009). *A2/M2 three screen report.* Available from http://enus.nielsen.com/main/insights/nielsen_a2m2_three. Accessed June 29, 2010.

19. Flegal, K. M., Graubard, B. I., Williamson, D. F., & Gail, M. H. (2007). Cause-specific excess deaths associated with underweight, overweight, and obesity. *JAMA, 298*(17), 2028–2037.

20. Sturm, R. (2002). The effects of obesity, smoking, and drinking on medical problems and costs. *Health Affairs, 21*(2), 245–253.

21. Goldman, D. P., Cutler, D. M., Shekelle, P. G., Bhattacharya, J., Shang, B., Joyce, G. F., et al. (2008). *Modeling the health and medical care spending of the future elderly* (No. RB-9324). Santa Monica, CA: RAND Corporation.

22. Matsuzawa, Y., Fujioka, S., Tokunaga, K., & Tarui, S. (1992). Classification of obesity with respect to morbidity. *Proceedings of the Society for Experimental Biology and Medicine, 200*(2), 197–201.

23. Koplan, J. P., & Dietz, W. H. (1999). Caloric imbalance and public health policy. *JAMA, 282*(16), 1579–1581.

24. Hansson, G. K. (2005). Inflammation, atherosclerosis, and coronary artery disease. *New England Journal of Medicine, 352*(16), 1685–1695.

25. Haslam, D.W., & James, W.P.T. (2005). Obesity. *Lancet, 366,* 1197–1209.

26. Poirier, P., Giles, T. D., Bray, G. A., Hong, Y., Stern, J. S., Pi-Sunyer, F. X., et al. (2006). Obesity and cardiovascular disease: Pathophysiology, evaluation, and effect of weight loss: An update of the 1997 American Heart Association scientific statement on obesity and heart disease from the Obesity Committee of the Council on Nutrition, Physical Activity, and Metabolism. *Circulation, 113*(6), 898–918.

27. Shamsuzzaman, A. S. M., Gersh, B. J., & Somers, V. K. (2003). Obstructive sleep apnea: Implications for cardiac and vascular disease. *JAMA, 290*(14), 1906–1914.

28. Mayo Clinic Staff. (2008). *Sleep apnea risk factors.* Available at http://www.mayoclinic.com/health/sleep-apnea/ds00148/dsection=risk-factors. Accessed July 15, 2009.

29. Tasali, E., Mokhlesi, B., & Van Cauter, E. (2008). Obstructive sleep apnea and type 2 diabetes: Interacting epidemics. *Chest, 133*(2), 496–506.

30. Schwartz, A. R., Patil, S. P., Laffan, A. M., Polotsky, V., Schneider, H., & Smith, P. L. (2008). Obesity and obstructive sleep apnea: Pathogenic mechanisms and therapeutic approaches. *Proceedings of the American Thoracic Society, 5*(2), 185–192.

31. State-specific incidence of diabetes among adults—participating states, 1995–1997 and 2005–2007. (2008). *Morbidity and Mortality Weekly Reports, 57*(43), 1169–1173.

32. Isganaitis, E., & Lustig, R. H. (2005). Fast food, central nervous system insulin resistance, and obesity. *Arteriosclerosis, Thrombosis, and Vascular Biology, 25*(12), 2451–2462.

33. Zimmet, P., Shaw, J., & Alberti, K. G. (2003). Preventing type 2 diabetes and the dysmetabolic syndrome in the real world: A realistic view. *Diabetic Medicine, 20*(9), 693–702.

34. Cowie, C. C., Rust, K. F., Ford, E. S., Eberhardt, M. S., Byrd-Holt, D. D., Li, C., et al. (2009). Full accounting of diabetes and pre-diabetes in the U.S. population in 1988–1994 and 2005–2006. *Diabetes Care, 32*(2), 287–294.

35. Schulz, L. O., Bennett, P. H., Ravussin, E., Kidd, J. R., Kidd, K. K., Esparza, J., et al. (2006). Effects of traditional and western environments on prevalence of type 2 diabetes in Pima Indians in Mexico and the United States. *Diabetes Care, 29*(8), 1866–1871.

36. Gross, L. S., Li, L., Ford, E. S., & Liu, S. (2004). Increased consumption of refined carbohydrates and the epidemic of type 2 diabetes in the United States:

An ecologic assessment. *American Journal of Clinical Nutrition, 79*(5), 774–779.

37. Kim, Y. I. (1998). Diet, lifestyle, and colorectal cancer: Is hyperinsulinemia the missing link? *Nutrition Reviews, 56*(9), 275–279.

38. Carter, J. S., Pugh, J. A., & Monterrosa, A. (1996). Non-insulin-dependent diabetes mellitus in minorities in the United States. *Annals of Internal Medicine, 125*(3), 221–232.

39. Reaven, G. (2005). All obese individuals are not created equal: Insulin resistance is the major determinant of cardiovascular disease in overweight/obese individuals. *Diabetes and Vascular Disease Research, 2*(3), 105–112.

40. Diabetes Prevention Program (DPP) Research Group. (2002). The diabetes prevention program (DPP): Description of lifestyle intervention. *Diabetes Care, 25*(12), 2165–2171.

41. International Diabetes Federation (IDF). *The IDF consensus worldwide definition of the metabolic syndrome: Part I: Worldwide definition for use in clinical practice.* Available from http://www.idf.org/webdata/docs/MetSyndrome_FINAL.pdf. Accessed June 29, 2010.

42. Sumner, A. E. (2008). The relationship of body fat to metabolic disease: Influence of sex and ethnicity. *Gender Medicine, 5*(4), 361–371.

43. Eckel, R. H., Grundy, S. M., & Zimmet, P. Z. (2005). The metabolic syndrome. *Lancet, 365*(9468), 1415–1428.

44. Lundgren, J. D., Malcolm, R., Binks, M., & O'Neil, P. M. (2009). Remission of metabolic syndrome following a 15-week low-calorie lifestyle change program for weight loss. *International Journal of Obesity, 33*(1), 144–150.

45. American Cancer Society. (2010). *Cancer facts & figures 2010.* Atlanta, GA: American Cancer Society. Available at http://www.cancer.org/docroot/STT/STT_0.asp. Accessed June 29, 2010.

46. Calle, E. E., Rodriguez, C., Walker-Thurmond, K., & Thun, M. J. (2003). Overweight, obesity, and mortality from cancer in a prospectively studied cohort of U.S. adults. *New England Journal of Medicine, 348*(17), 1625–1638.

47. Excess body weight among people with arthritis. (2008). NHIS Arthritis Surveillance. Accessed July 24, 2009. Available from http://www.cdc.gov/arthritis/data_statistics/national_nhis.htm. Accessed June 29, 2010.

48. Teichtahl, A. J., Wang, Y., Wluka, A. E., & Cicuttini, F. M. (2008). Obesity and knee osteoarthritis: New insights provided by body composition studies. *Obesity, 16*(2), 232–240.

49. Ogden, C. L., Carroll, M. D., Curtin, L. R., McDowell, M. A., Tabak, C. J., & Flegal, K. M. (2006). Prevalence of overweight and obesity in the United States, 1999–2004. *JAMA, 295*(13), 1549–1555.

50. Baeten, J. M., Bukusi, E. A., & Lambe, M. (2001). Pregnancy complications and outcomes among overweight and obese nulliparous women. *American Journal of Public Health, 91*(3), 436–440.

51. Vrbikova, J., & Hainer, V. (2009). Obesity and polycystic ovary syndrome. *Obesity Facts, 2,* 26–35.

52. Alexander, C.J., Tangchitnob, E.P., & Lepor, N.E. (2009). Polycystic ovary syndrome: A major unrecognized cardiovascular risk factor in women. *Reviews in Obstetrics and Gynecology, 2*(4), 232–239.

53. Schwimmer, J. B., Pardee, P. E., Lavine, J. E., Blumkin, A. K., & Cook, S. (2008). Cardiovascular risk factors and the metabolic syndrome in pediatric nonalcoholic fatty liver disease. *Circulation, 118*(3), 277–283.

54. Janicka, A., Wren, T. A., Sanchez, M. M., Dorey, F., Kim, P. S., Mittelman, S. D., et al. (2007). Fat mass is not beneficial to bone in adolescents and young adults. *Journal of Clinical Endocrinology and Metabolism, 92*(1), 143–147.

55. Edelstein, S. L., & Barrett-Connor, E. (1993). Relation between body size and bone mineral density in elderly men and women. *American Journal of Epidemiology, 138*(3), 160–169.

56. Peeters, A., Barendregt, J. J., Willekens, F., Mackenbach, J. P., Al Mamun, A., Bonneux, L., et al. (2003). Obesity in adulthood and its consequences for life expectancy: A life-table analysis. *Annals of Internal Medicine, 138*(1), 24–32.

57. Prospective Studies Collaboration, Whitlock, G., Lewington, S., Sherliker, P., Clarke, R., Emberson, J., et al. (2009). Body-mass index and cause-specific mortality in 900 000 adults: Collaborative analyses of 57 prospective studies. *Lancet, 373*(9669), 1083–1096.

58. Manson, J. E., Willett, W. C., Stampfer, M. J., Colditz, G. A., Hunter, D. J., Hankinson, S. E., et al. (1995). Body weight and mortality among women. *New England Journal of Medicine, 333*(11), 677–685.

59. Hu, F. B., Stampfer, M. J., Solomon, C. G., Liu, S., Willett, W. C., Speizer, F. E., et al. (2001). The impact of diabetes mellitus on mortality from all causes and coronary heart disease in women: 20 years of follow-up. *Archives of Internal Medicine, 161*(14), 1717–1723.

60. Ajani, U. A., Lotufo, P. A., Gaziano, J. M., Lee, I. M., Spelsberg, A., Buring, J. E., et al. (2004). Body mass index and mortality among US male physicians. *Annals of Epidemiology, 14*(10), 731–739.

61. Stevens, J., Cai, J., Pamuk, E. R., Williamson, D. F., Thun, M. J., & Wood, J. L. (1998). The effect of age on the association between body-mass index and mortality. *New England Journal of Medicine, 338*(1), 1–7.

62. Adams, K. F., Schatzkin, A., Harris, T. B., Kipnis, V., Mouw, T., Ballard-Barbash, R., et al. (2006). Overweight, obesity, and mortality in a large prospective cohort of persons 50 to 71 years old. *New England Journal of Medicine, 355*(8), 763–778.

63. Pischon, T., Boeing, H., Hoffmann, K., Bergmann, M., Schulze, M. B., Overvad, K., et al. (2008). General and abdominal adiposity and risk of death in Europe. *New England Journal of Medicine, 359*(20), 2105–2120.

64. Williamson, D. F., Pamuk, E., Thun, M., Flanders, D., Byers, T., & Heath, C. (1999). Prospective study of intentional weight loss and mortality in overweight white men aged 40–64 years. *American Journal of Epidemiology, 149*(6), 491–503.

65. Field, A. E., Manson, J. E., Taylor, C. B., Willett, W. C., & Colditz, G. A. (2004). Association of weight change, weight control practices, and weight cycling among women in the Nurses' Health Study II. *International Journal of Obesity and Related Metabolic Disorders, 28*(9), 1134–1142.

66. Hassan, M. K., Joshi, A. V., Madhavan, S. S., & Amonkar, M. M. (2003). Obesity and health-related quality of life: A cross-sectional analysis of the US population. *International Journal of Obesity and Related Metabolic Disorders, 27*(10), 1227–1232.

67. Kruger, J., Bowles, H. R., Jones, D. A., Ainsworth, B. E., & Kohl, H. W., 3rd. (2007). Health-related quality of life, BMI and physical activity among US adults (>/=18 years): National physical activity and weight loss survey, 2002. *International Journal of Obesity, 31*(2), 321–327.

68. Kolotkin, R. L., Crosby, R. D., Pendleton, R., Strong, M., Gress, R. E., & Adams, T. (2003). Health-related quality of life in patients seeking gastric bypass surgery vs non-treatment-seeking controls. *Obesity Surgery, 13*(3), 371–377.

69. Fontaine, K. R., & Barofsky, I. (2001). Obesity and health-related quality of life. *Obesity Reviews, 2*(3), 173–182.

70. Polivy, J., & Herman, C. P. (1992). Undieting: A program to help people stop dieting. *International Journal of Eating Disorders, 11*(3), 261–268.

71. Greenberg, B. S., Eastin, M., Hofschire, L., Lachlan, K., & Brownell, K. D. (2003). Portrayals of overweight and obese individuals on commercial television. *American Journal of Public Health, 93*(8), 1342–1348.

72. Friedman, R. R. (2008). *Weight bias: The need for public policy.* New Haven, CT: Rudd Center for Food Policy and Obesity.

73. Brownell, K. D., & Wadden, T. A. (1992). Etiology and treatment of obesity: Understanding a serious, prevalent, and refractory disorder. *Journal of Consulting and Clinical Psychology, 60*(4), 505–517.

Assessment Approaches in Weight Management

CHAPTER OUTLINE

R. Gino Santa Maria/Shutterstock.com

cappi thompson/Shutterstock.com

Iorga Studio/Shutterstock.com

Weight-management assessment has three broad purposes for the client and professional:

- To provide baseline information about body composition, health status, dietary practices, and activity habits
- To help devise realistic goals for change and develop an individualized treatment program based on those goals
- To document changes that occur as a result of treatment/behavior change

nutrition assessment *Assessment that involves obtaining and interpreting anthropometric, bio-chemical, clinical, dietary, and physical activity information.*

An additional purpose of assessment is research—to document population trends in weight that can guide broad policy needs.

Most are probably familiar with the term **nutrition assessment**, which includes gathering and interpreting data from a variety of sources—anthropometric, dietary, biochemical, and clinical—to produce a picture of an individual's health status. This chapter describes the elements of nutrition assessment that are most applicable to individuals in weight management programs, including physical activity assessment, which is not usually discussed in nutritional assessment textbooks.

ANTHROPOMETRIC ASSESSMENT: BODY SIZE, SHAPE, AND COMPOSITION

anthropometrics *Measures of body size and proportion (height, weight, circumferences, and skinfolds).*

The first thing that usually comes to mind when you are working with overweight or underweight clients is the need to determine height, weight, and level of fatness. Measures of body size and proportions—height, weight, circumferences, and skinfolds—are known as **anthropometrics**. From these measures, we can calculate the body mass index (BMI), pattern of fat distribution, and percent fat. This information, when used as part of an overall health evaluation, can be extremely helpful in assessing health risks associated with fat weight, fat distribution, or low body weight or muscle mass.

A word of caution: Some overweight and obese individuals may not significantly reduce weight or fatness in a weight-management program and may still be technically classified as overweight and/or obese after an intervention. Nevertheless, changes in diet or activity may have improved their health. Weight-management professionals should avoid overemphasizing anthropometric changes as markers of their clients' success.

fat mass *Proportion of body weight made up of storage and essential fat.*

fat-free mass (FFM) *Proportion of body weight made up of bone and other dense connective tissue, body water, and the fat-free portion of the body's cells.*

What Are We Made of?

Two components of the human body are particularly relevant to weight management: **fat mass** and **fat-free mass**. Fat mass includes storage fat as well as small amounts of essential fat found in the nervous system and cell membranes. Fat-free mass (FFM) consists of bone and other dense connective tissue, body water, and the protein-rich, fat-free portion of the cells that comprises organs, muscles, and the immune system (collectively known as the body cell mass). Figures 2-1A and B illustrates the proportions of fat and FFM in a "typical" adult male and female and in children of various ages. These proportions can vary significantly, however, as a result of physical training, aging, obesity, and anorexia nervosa.

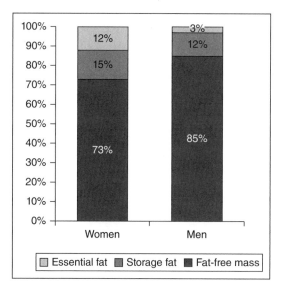

FIGURE 2-1A   What are we made of? Adults.

Source: Adapted from Behnke, A.R., & Wilmore, J.H. (1974). *Evaluation and regulation of body build and composition.* Englewood Cliffs, NJ: Prentice-Hall.

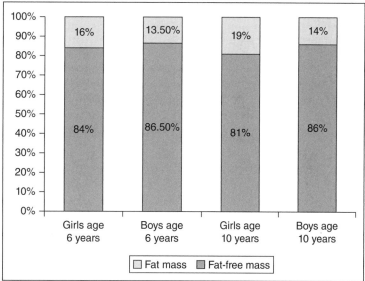

FIGURE 2-1B   What are we made of? Children.

Source: Adapted from Fomon, S.J., Haschke, F., Ziegler, E.E., & Nelson, S.E. (1982). Body composition of reference children from birth to age 10 years. *American Journal of Clinical Nutrition, 35,* 1169–1175.

Fat Mass and FFM

essential fat
Minimum percentage of body fat needed to support physiological functions.

Notice in Figure 2-1A that fat is classified as essential or storage fat. **Essential fat** is the minimum percentage of fat needed to maintain health. Females need more essential fat than males to support reproductive functions (12% versus 3%). Storage fat is fat accumulated over and above physiological need. Although storage fat is not required for physiological functions, it will support health during times of famine and is harmful only when accumulated in excessive amounts.

FFM comprises the largest portion of body weight. FFM is mostly water, about 16–19% protein and 5–8% mineral. There are age, gender, and ethnic differences in the composition of the FFM. For example, children have a lower mineral content than adults, more water in their fat-free tissue, and more variability in the FFM.[1] Because of these variations, mathematical equations used to convert measures of body size into estimates of percent fat or percent FFM are population specific. (This concept will be further explained in the section on skinfold measurement.)

Laboratory Determination of Fat and FFM

Several sophisticated techniques that permit accurate assessment of body composition are summarized in Table 2-1: underwater weighing, whole-body air-displacement

TABLE 2-1   Laboratory Methods for Assessment of Body Composition

Assessment methods	Description	Limitations
Underwater weighing (hydrodensitometry; hydrostatic weighing)	Person seated in chair suspended in water tank exhales maximally and then submerges. Accounting for residual lung volume, body density is calculated by dividing weight in air by loss of weight in water.	Cannot be used with people unable to submerge. Density of fat mass and FFM have not been validated by cadaver studies in populations other than Caucasian males.
Whole-body air-displacement plethysmography (ADP; Bod Pod)	Person sits in small chamber in which body volume is determined by subtracting air volume in empty chamber from volume while subject is in chamber.	Few limitations other than expense of device and training of operator. Compares well with DXA and underwater weighing. Can be used with wide range of subjects but not validated with subjects under 40 kg. Research is new.
Computed tomography (CT)	X-ray produces high-resolution body image based on differences in density of tissues. Examination of cross-sectional areas permits regional determination of fat, muscle, and bone and calculation of body composition.	Exposes subjects to ionizing radiation. Cannot be used in children, pregnancy. Costly and trained technician required.
Nuclear magnetic resonance imaging (NMR; MRI)	Electromagnetic radiation causes cell nuclei to spin and absorb energy. When nuclei realign, released energy creates an image. Examination of cross-sectional areas permits regional determination of fat, muscle, and bone and calculation of body composition.	No known health risks. Costly and time-consuming. Access to equipment may be limited and trained technician required.
Dual-energy x-ray absorptiometry (DXA; DEXA)	Low-dose x-ray scans produce image of bone and other tissues. Fat and FFM can be calculated. Regional fat can also be assessed.	Costly. Can be used with a wide range of subjects, although not in pregnancy. People taller or larger than the scan area cannot be measured.

plethysmography (ADP; Bod Pod), computed tomography (CT), nuclear magnetic resonance imaging (NMR, MRI), and dual energy x-ray absorptiometry (DXA, DEXA). These techniques are costly, complex, and reserved for the research laboratory, but they serve as the foundations for the body composition assessments that health professionals use in the field. Books and articles about body composition analysis often refer to these methods, so it is a good idea to become familiar with them.

What Are Common Measures of Body Size?

Even though measures of body size do not directly address body composition, the relationship between height and weight has been used as a standard indicator of health since the 1950s. Until a decade ago, height–weight tables were the primary tools for this. Today experts recommend using the BMI to determine a person's healthy weight. To calculate a person's BMI, height and weight must be determined.

Determination of Stature

stature *Standing height.*

stadiometer *Sturdy wall-mounted or free-standing device used to measure stature.*

Standing height, or **stature**, is most accurately measured using a sturdy, wall-mounted **stadiometer**. A stadiometer is a vertical measuring board attached to a wall with a sliding headpiece affixed to the board at a right angle. Alternatively, a nonflexible measuring tape can be affixed to a wall, and a piece of wood or other nonflexible object can be placed at a right angle to the wall. The flexible measurement rod that is part of a balance-beam scale should not be used to determine height because it is usually too loose and floppy to provide an accurate measurement.

Measure the individual as follows, which is illustrated in Figure 2-2:

Frankfort horizontal plane *In the measurement of stature, a head position that aligns the lowest margin of the orbit of the eye with the tragion of the ear. The tragion is the deepest part of the notch just above the cartilage projection of the ear known as the tragus.*

- With shoes off, the client should stand as straight as possible with the shoulder blades, buttocks, and heels touching the stadiometer. The back of the head may or may not be touching the stadiometer.
- The client should look straight ahead, holding the head in the **Frankfort horizontal plane**. Most people assume this position naturally. (See Figure 2-3 for an illustration of this position.)
- The client's arms should be at the sides with palms facing the thighs.
- Lower the stadiometer headpiece (or block of wood) to the crown of the client's head as he or she straightens the spine as much as possible while taking a deep breath.
- Record the measurement to the nearest millimeter or quarter-inch. When the reading falls directly between two millimeters or quarter-inch, use the lower number.[2]

supine *Lying on one's back.*

goniometer *Device that measures the angle of a joint.*

Knee Height

Knee height may be used to estimate stature in people who cannot stand or stand erect. To measure knee height, have the client lie **supine** on a flat surface and bend the left knee 90°. Check the angle of the knee with a **goniometer**. Using a large caliper, measure

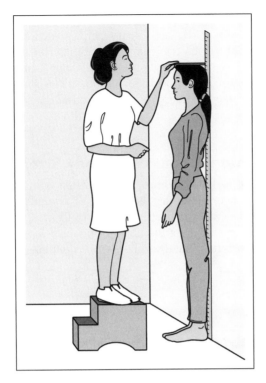

FIGURE 2-2 Measurement of height.

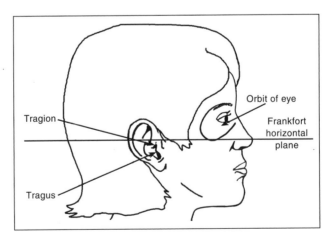

FIGURE 2-3 Frankfort Horizontal Plane aligns the lowest margin of the orbit of the eye with the tragion of the ear. The tragion is the deepest part of the notch just above the cartilage projection of the ear known as the tragus.

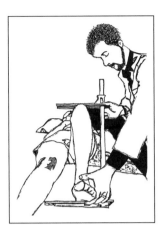

FIGURE 2-4 Knee height.

from the heel to the anterior surface of the thigh, just above the patella (knee cap). Height can be derived from the following equations:[3]

Male height (cm) = (2.02 × knee hgt [cm]) − (0.04 × age [yr]) + 64.19
Female height (cm) = (1.83 × knee hgt [cm]) − (0.24 × age [yr]) + 84.88

Weight

Weight should be measured on the most accurate scale available. Bathroom-type scales, although found in most homes, are not acceptable for professional use. A balance-beam scale (sometimes called a physician's scale) or, preferably, an electronic (digital) scale are used for determining the weight of children and adults.

For measurement of weight, the client should wear minimal clothing and remove shoes. While the client stands in the center of the scale's platform, record weight to the nearest 0.1 kg or 0.25 lb. Scales should be calibrated regularly. Calibrating a scale is not difficult. Simply weigh yourself, then step back onto the scale while holding a 10-lb calibrated weight. Your weight should increase accordingly.

How Should the BMI Be Used in Assessment?

The BMI is an indicator of overweight and obesity that is derived from height and weight measures. For an accurate BMI, careful measures of height and weight are required. The BMI is calculated by dividing weight in kilograms by height in meters squared (kg/m^2). The BMI may also be figured by multiplying weight in pounds by 703 and dividing the total by height in inches squared (e.g., [weight in lbs × 703] divided by height in inches squared). Alternatively, you may use the BMI calculator in Appendix A Table 1.

The BMI is a very simple and inexpensive method for classifying people as overweight or obese. A single formula works for children, adolescents, and adult men and women. It is not as good an indicator of fatness as some of the sophisticated methods of analyzing body composition summarized in Table 2-1; but a BMI may be just as good as percent fat in predicting health risks associated with excess weight and fat, including diabetes, hypertension, and coronary artery disease. In addition, a BMI of 17.5 or less is one diagnostic criterion for significant underweight in people with anorexia nervosa.

Limitations of the BMI

A person's body weight includes not just fat but also bone, water, and muscle. Any measure of obesity that relies primarily on weight, an indirect estimate of body fat, has limitations. While you can be confident that a BMI ≥ 30 indicates obesity, BMIs between 25 and 29 may be seen in some people who are not overfat, such as athletes who have a substantial lean body mass.

Consider the following when using the BMI:

edema *Fluid accumulation in the tissues, which may result from congestive heart failure, kidney disease, corticosteriod therapy, liver diseases, or other metabolic conditions.*

- The BMI will overestimate overweight/obesity and health risk in lean, muscular athletes or people with high bone densities.
- Individuals with **edema**, in which fluid accumulates in the tissues, may gain enough weight to increase their BMI to levels suggesting obesity.
- Postmenopausal women who have lost height due to osteoporosis will have higher BMIs but not necessarily excess fat.
- In general, BMI may be less accurate in adults who are under 5 ft tall.[4] Researchers do not find BMI as predictive of health risk among short-statured adults as among adults of average height.[5]
- The BMI typically underestimates body fatness in older adults, because fat mass increases with age while lean body mass decreases, yet weight may not increase enough to significantly increase BMI.
- The relationship between BMI and fatness probably differs among various racial and ethnic groups, yet there are not currently ethnic- or racially-specific BMI tables.
- The BMI does not provide information about fat distribution, which may have a more profound effect on health than weight alone.

Because the BMI cannot account for differences in body build, level of fatness, or fat distribution, it needs to be interpreted carefully with all other available health data.

Interpretation of BMI

Table 2-2 provides the BMIs that classify adults as underweight, overweight, obese, or extremely obese. An adult BMI below 18.5 indicates underweight, and a BMI of 25 or greater indicates overweight. Notice also that the table relates disease risk to both BMI and waist circumference, which is discussed in the next section. BMI values that suggest health risk in adults should not be applied to children and adolescents. CDC provides age- and gender-specific BMI charts for 2- to 20-year-olds, which may be found in Tables 8 and 9 in Appendix A. Using these charts, an Expert Committee of the American Academy of Pediatrics defines obesity in children and adolescents as a BMI at or above the 95th percentile and overweight as at or above the 85th percentile. Chapter 12 discusses the use of BMI with children and adolescents.

Little is served by simply telling someone that they are "underweight" or "obese" based on the BMI. It is more important to inform individuals about health risks associated with their BMI and to conduct further screening to assess fat distribution, fat mass and FFM, activity level, dietary patterns, cardiovascular risk factors, and risk of other health problems. While these additional assessments are not as simple as height and weight, they can help to overcome some of the limitations of the BMI.

Should We Use Height–Weight Tables?

The Metropolitan Life Insurance Company published its table of Desirable Weights for Men and Women in 1959. This particular table was derived from data collected on 4.5 million life insurance policies issued to 25- to 59-year-olds from 1935 to 1954. Later, data from 4.2 million additional policies issued from 1950 to 1972 were analyzed to create a revised table called the 1983 Metropolitan Height and Weight Tables for Men and Women. In the Metropolitan tables, the weight ranges given for each height correspond with lowest death rates among the company's life insurance holders. Both the 1959 and 1983 tables may be found in Appendix A, Tables 2 and 3.

The general public may be more familiar with height–weight tables than with the BMI. However, height–weight tables are problematic for many reasons and are not as useful as the BMI in predicting health risks from excess weight in relationship to height:

- Because the tables were originally created using data from 25- to 59-year-old purchasers of personal life insurance, the data are not representative of the population as a whole and underrepresent older adults, minorities, and people in lower socioeconomic groups.
- Information used to develop the height–weight tables is of uneven quality. Some people self-reported their heights and weights; people were weighed at different times of the year, wearing varying amounts of clothing; and there was an overall lack of standardization in collecting the data.
- Weights included in the tables are weights obtained when the person purchased insurance. Changes in weight that might have occurred before death are not reflected.
- There is an overall presumption that weight is equated with fat. In fact, a lean person may weigh more than the desirable weight yet have a low percentage of fat. In contrast, an overfat person may be close to healthy weight.
- The inclusion of smokers in the population used to construct tables complicates interpretation of the relationship between weight and health. Cigarette smokers tend to be lower in weight than nonsmokers, yet smokers have higher morbidity and mortality. This correlation might give the appearance that lighter weights are less healthy and heavier weights more healthy. Some experts suggest that the cumulative effect of cigarette smoking and the impact of smoking on weight invalidate tables published after 1959.[6]

What Do Circumference Measures Tell Us About Body Composition?

Circumferences are used to estimate FFM, fat mass, and fat distribution. Circumference measures may be preferable to skinfolds in some situations. They are certainly easier to obtain than skinfolds, and there is better agreement about the precise location of circumference sites than skinfolds.[7] Circumferences are more sensitive than skinfolds or bioelectrical impedance analysis (BIA) to changes in body composition that result from weight loss.[8]

The client being measured should wear underwear, a bathing suit, or a lightweight gown. Two measurements at each site are taken to the nearest 0.1 cm or 0.25 inch using a nonstretch, flexible tape. When the two measurements are within 5 mm of each other, they can be averaged. Circumferences most commonly used in calculating body density are of the abdomen, hip, thigh, calf, and mid-arm.

Prediction Equations

Equations are available to convert circumference measures into estimated percent fat. The only equations that should be used are those derived from studies of people similar to those being assessed. Age, gender, and race are important characteristics to consider when choosing equations. Prediction equations for calculating body density from circumference measures are given in Table 7 of Appendix A for older women, obese individuals, and female athletes.

Limitations of Circumferences in Determining Body Composition

subcutaneous fat
Fat stored directly under the skin, which is measured with skinfolds.

The biggest limitation of circumference measures is that girth represents not only **subcutaneous fat** but also muscle, bone, blood vessels, nerves, and internal fat. People with large musculature may be erroneously classified as obese. In addition, clinical edema and even simple fluid retention after sitting or standing for a prolonged period increases girth measurements. However, circumferences are of great value in determining the pattern of fat distribution.

Use of Circumferences in Determining Body Fat Distribution

Fat is stored both subcutaneously (under the skin) and internally. Two people with the same BMI might store fat quite differently. Because of the high association between internal fat and health problems, determination of fat distribution is an important component of assessment. Magnetic resonance imaging (MRI) and computed tomography are precise methods for determining body fat distribution. Not only can these techniques provide information about the quantity of internal fat, but they can also distinguish between subcutaneous abdominal and internal visceral abdominal fat (e.g., fat stored deep in the abdo-minopelvic region). However, computed tomography involves irradiation, and both are expensive diagnostic procedures, making them unlikely to become screening tools for fat distribution.

The waist-to-hip ratio (WHR) has been used for many years as an indicator of fat distribution. The WHR is calculated by dividing waist circumference by hip circumference. Upper body fat deposition is indicated as the ratio gets closer to 1.00. Health risk is increased in young men with WHR > 0.95; in young women with WHR > 0.86; in men ages 60–69 with WHR > 1.03; and in women ages 60–69 with WHR > 0.90.[9]

While there continues to be some debate on this point, many experts believe that waist circumference (WC) alone is an effective indicator of visceral abdominal fat. In addition, WC requires only one measurement, simplifying assessment.

Measuring Waist Girth

iliac crest *Upper margin of the part of the pelvis known as the ilium, which can be readily felt by most people; an anatomical landmark for several anthropometric measurements.*

Waist circumference requires only one measurement—the waist. It seems straightforward, but what is your "waist"? Most recommendations call for the measurement to be taken at the narrowest point on the torso, below the bottom of the rib cage and just above the lateral border of the **iliac crest**, while exhaling normally.[10,11] Measure to the nearest 0.1 cm or 0.25 in. Ideally, the client will wear no more than lightweight clothing around the waist while the measurement is taken.

Limitations of Waist Circumference

There are some limitations to using the WC as an indicator of health risks associated with visceral adiposity. Most important, the waist measurement gives only an approximation of visceral fat, and it is less accurate in children and adolescents than in adults. Although waist circumference correlates well with internal fat, it is by no means as precise as imaging techniques in distinguishing between subcutaneous and visceral abdominal fat. In addition, waist measurements indicating increased risk of disease may be less predictive in adults under 5 ft. tall.[4]

Interpretation of Waist Circumference

The waist circumference measure can be interpreted with BMI using Table 2-2. According to the National Heart, Lung, and Blood Institute, as waist circumference increases above 102 cm (40 in) in men and 88 cm (35 in) in women, so does the risk of developing type 2 diabetes, elevated blood LDL cholesterol and triglycerides, hypertension, and other markers of cardiovascular disease, such as elevated C-reactive proteins. This risk applies to men and women from all racial and ethnic groups that have been studied. Several researchers note that adults with a BMI below 25 kg/m^2 with a larger waist circumference may also have health risks. One study of over 500,000 men and women found that the risk of death was actually greater for people with a lower BMI and a higher waist circumference.[12] Analysis of NHANES III and Canadian Health Survey Data similarly found that even people with healthy BMIs experience greater risk of cardiovascular disease if their waist circumferences exceed specific values.[13] Table 2-3 provides waist circumferences at all BMIs that predict cardiovascular disease risk.

Among older adults, waist circumference may be more valuable than BMI in assessing disease risk.[14] Waist circumference is a good indicator of short-term changes in body composition resulting from improved diet, increased activity, and lifestyle behavior modification. Waist circumference interpretation in children and adolescents is discussed in Chapter 12.

TABLE 2-2 Classification of Adult Overweight and Obesity by BMI, Waist Circumference, and Associated Disease Risk

	BMI (kg/m²)	Obesity Class	Disease Risk* (Relative to Normal Weight and Waist Circumference)	
			Men ≤ 40 in (≤102 cm) Women ≤ 35 in (≤ 88 cm)	Men > 40 in (>102 cm) Women > 35 in (>88 cm)
Underweight	<18.5			
*Normal***	18.5–24.9			
Overweight	25.0–29.9		Increased	High
Obesity	30.0–34.9	I	High	Very High
	35.0–39.9	II	Very High	Very High
Extreme Obesity	≥40	III	Extremely High	Extremely High

*Disease risk for type 2 diabetes, hypertension, and cardiovascular disease.
**Increased waist circumference can be a marker for increased risk even in persons of normal weight.
Source: NHLBI. (2000). *The practical guide: identification, evaluation, and treatment of overweight and obesity in adults.* Washington, D.C.: U.S. Department of Health and Human Services, p. xvii.
Available online at http://www.nhlbi.nih.gov/guidelines/obesity/ob_gdlns.pdf

TABLE 2-3 Waist Circumferences Associated with Increased Health Risk in Adults at BMIs 18.5 to >35 kg/m^2

BMI	Optimal Waist Circumference (in cm) for Predicting CVD Risk	
	Females	**Males**
<18.5	—	—
18.5–24.9	79	87
25–29.9	92	98
30.0–34.9	103	109
≥35.0	115	124

Source: Adapted from data reported in Ardem, C.I., Janssen, I., Ross, R., & Katzmarzyk. (2004). Development of health-related waist circumference thresholds within BMI categories. *Obesity Research, 12*(7), 1094–1103.

Waist-to-Height Ratio

Recently several researchers have suggested that the relationship between waist circumference and height may be a useful indicator of visceral adiposity and health risk. A study of over 5,000 German Caucasian individuals found the waist-to-height ratio (WHtR) was better than any other anthropometric measure in predicting metabolic syndrome, abnormal blood lipids, and type 2 diabetes.[15] In adults, the waist circumference should be less than height. A waist-to-height ratio of 0.5 or higher is predictive of cardiovascular and metabolic abnormalities.[16]

How Are Skinfold Measures Used in Assessment?

As you learned in the last section, fat is stored both subcutaneously and internally. Skinfolds measure the amount of fat stored subcutaneously. Experts agree that certain skinfolds are good indicators of overall subcutaneous fat and that subcutaneous fat is a good predictor of internal fat. So, after measuring two or three skinfold sites, mathematical equations can be used to estimate total body fat.

Skinfold measurements are taken with metal or plastic calipers, as follows:

- Locate the appropriate site (abdomen, bicep, calf, chest, midaxilla, subscapula, suprailium, thigh, and/or tricep) and raise a fold of skin and fat, which can be pulled slightly away from the underlying tissue.
- While continuing to hold the fold with your fingers, place the tips of the caliper around the fold and take a reading.
- Record the reading to the nearest millimeter.
- Take at least two measurements at each site after a brief interval to let the skin and fat recoil. Successive measurements should be within 1–2 mm, or measurements should be repeated.

In the United States, skinfold measurements are usually taken on the right side of the body because most equations available for converting skinfolds to percent body fat were developed using the right side. In Great Britain, Australia, and Europe, skinfold measurements are usually taken on the left side. Most equations for

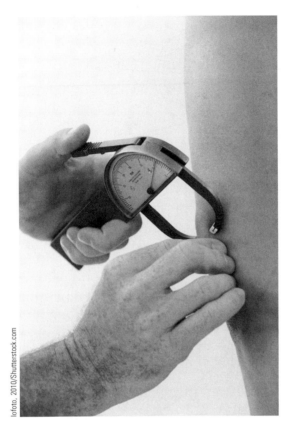

Person measuring skinfolds. Subcutaneous fat measured with a skinfold caliper is a good indicator of internal fat.

converting skinfold measurements to percent fat were developed with the Lange (United States) or Harpendon (Great Britain and Europe) calipers, which are considered to be quite accurate and can be calibrated. The inexpensive, plastic Slim Guide caliper is also considered to be acceptable.

Skinfolds can be very difficult to obtain on obese individuals. A skinfold should include only subcutaneous fat and skin as shown in Figure 2-5. If fat does not separate easily from underlying muscle, then the fatfold will be inaccurate. Also, in extreme obesity, even when fat and muscle can be separated, the size of the skinfold may exceed the opening of the caliper.

Various combinations of nine skinfold sites may be used in calculating percent fat in adults and children. These are abdominal, biceps, calf, chest, midaxillary, subscapular, suprailiac, thigh, and triceps (in children, the subscapular and triceps skinfolds are most useful). Consult Table 5 in Appendix A for detailed descriptions of these sites. Familiarity with anatomical landmarks is critical. For accuracy, beginners are encouraged to actually draw a mark on the client's skin before measuring the skinfold.

Skinfold Prediction Equations

The equations used to predict body density or percent fat from skinfold measures were developed using a statistical technique called multiple regression analysis. In this technique, several skinfolds are analyzed and combined mathematically to yield the

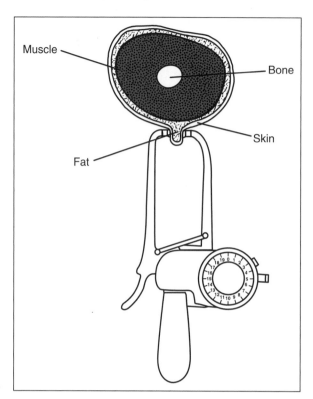

FIGURE 2-5 Anatomy of a skinfold.

combination that best predicts percent fat and/or body density. For example, researchers might measure nine skinfold sites in a group of people and use multiple regression analysis to determine which three skinfolds will best predict body density. Previously, body density would have been determined by a very accurate laboratory method, such as underwater weighing. A prediction equation would then be developed from the three skinfolds.

The only equations that clinicians should use are those derived from studies of people similar to the client being assessed. Age, gender, and race are particularly important characteristics to consider when choosing equations. Fortunately, many equations are available, so there should be no difficulty finding one that matches the characteristics of the client. Table 5 in Appendix A provides prediction equations for children, American Indian women, African American men and women, Hispanic women, Caucasian men and women, anorexic women, and male and female athletes.

Limitations of Skinfolds

Percent fat estimated by skinfolds might vary 3–5% from percent fat determined by underwater weighing or dual energy x-ray absorptiometry. When errors are made—and there are many opportunities for error when using skinfolds—percent fat may be miscalculated by more than 10%. The usual sources of error in the skinfold method are—:

- Poor technique: Technical skill is required to measure skinfolds accurately. The caliper must be placed correctly and read with accuracy. Proficiency is acquired through several hours of practice under the direction of a knowledgeable teacher.

- Incorrect caliper placement: Use of a reference source like the *Anthropometric Standardization Reference Manual* will help you develop consistency in locating skinfold sites.[17] Caliper placement is, however, also dictated by the equation being used to calculate body density. There are generalized equations that are widely used to measure sites slightly differently from the methods described in the *Anthropometric Standardization Reference Manual*.[18,19]
- Poor quality calipers: Technicians should always use precision calipers and periodically recalibrate calipers.
- The client: Even experienced technicians will have difficulty obtaining skinfolds from some people. Individuals with a lot of fat or muscle can be quite difficult to measure because fat and underlying muscle may not separate. As Figure 2-5 illustrates, the sides of the skinfold should be approximately parallel. If they are not, then the measure cannot be considered accurate. In addition, other client factors may invalidate skinfold measures: edema common in certain diseases, water retention that occurs during menstruation, and dilation of the blood vessels for a period of time after exercise.
- Failure to use equations specific to the client being measured: The relationship between subcutaneous fat and internal fat varies with age, gender, ethnicity, and physical activity (athletes versus nonathletes). Significant over- or underestimates of body fat result when prediction equations do not match clients.

Interpretation of Skinfolds

Determination of body composition brings additional information to the screening profile. When fat percentage is used with the BMI, it can help differentiate between people who are heavy because of muscularity versus fat. In the elderly, it can also help determine to what extent fat has replaced lean tissue.

Most important, percent fat predicts health risk. Although there are disagreements about what fat percentage constitutes obesity, there are also indications that fat percentages below 5% in men and 8% in women and at or above 25% in men and 32% in women convey increased risk of health complications. In children, fat percentages at or above 25% in boys and 30% in girls are associated with increased risk of hypertension and elevated blood lipids.[20] Health classifications of fat percentages in adults and children are presented in Tables 2-4 and 2-5.

TABLE 2-4 Interpretation of Percent Fat in Adults

Health risk category	Men	Women
Minimal weight	≤5%	≤8%
Below average	6–14%	9–22%
Average	15%	23%
Above average	16–24%	24–31%
At risk[†]	≥25%	≥32%

[†]At risk for diseases associated with obesity.

Source: Adapted from Lohman, T.G. (1992). Advances in body composition assessment: *Current issues in exercise science series (Monograph no. 3)*. Champaign, IL: Human Kinetics, p. 80.

TABLE 2-5 Interpretation of Percent Fat in Children 6–17 Years Old

Risk category	Boys	Girls
Very low	<6%	12%
Low	7–10%	12–15%
Optimal range	10–19%	15–25%
Moderately high	20–25%	25–30%
High	25–31%	30–35%
Very high	>31%	>35%

Source: Lohman, T.G. (1987). The use of skinfolds to estimate body fatness on children and youth. *Journal of Physical Education, Recreation, and Dance, 58*(9), 98–102. Adapted with permission.

Determination of percent fat can also help in setting weight goals for athletes and very lean individuals. Male athletes should maintain a fat percentage of 5% and female athletes 8%. Sometimes, athletes in sports with weight classifications, such as wrestling or judo, will attempt weight loss to move into a lower classification. The rationale for this is that it is better to be the heaviest athlete in a lower weight class than to be the lightest athlete in a higher weight class. Athletes at the minimum healthy fat percentage should be discouraged from losing additional weight.

 Using skinfolds to monitor short-term change in body composition is not recommended. Because skinfolds only directly measure subcutaneous fat, they are not sensitive to changes in internal fat. A person may experience a significant reduction in visceral abdominal fat after an exercise program but show little change in skinfolds. In addition, an increase in activity level and a decrease in body fat may change the density of the FFM and require the use of either a more sophisticated method of fat assessment or different equations.[7]

bioelectrical impedance analysis (BIA)

Method for determining fat-free mass, in which a weak electrical current passed through body tissues is impeded by tissues that contain little water (such as fat).

What Is Bioelectrical Impedance Analysis?

Bioelectrical impedance analysis (BIA) is a safe, cost-effective, technician-friendly, and client-friendly method for assessing body composition. It is widely used in health clubs, employee fitness centers, and other places where health screenings take place. Although BIA is not technically an anthropometric measure, it provides a safe, easy, and noninvasive way to estimate FFM quickly. A BIA analyzer delivers a weak (500–800 µA) electrical current at a 50 khz frequency. The magnitude of the current is enough to produce measurable voltage but not so high that the individual notices it. The current easily passes through tissues with high water and electrolyte content, such as skeletal muscle, but it is impeded by tissues with little water, such as bone and fat. The value obtained from this procedure gives an approximation of total body water. FFM can then be calculated and, from that figure, body fat can be estimated.

The procedure for BIA varies slightly with the analyzer model being used:

- The client either lies supine on a nonconductive surface, such as a padded examination table, or stands with feet apart.
- For models in which clients lie on their backs, the technician cleans the skin with alcohol and places two electrodes on the back of the right hand and wrist and two

electrodes on the right foot and ankle. There should be at least 5 cm between each electrode. For the standing model, the client holds the device in front, away from the body, and at a 90° angle to the torso.

- The client's arms should not be touching the trunk, and the legs should not be touching each other. When supine, the arms and legs should be abducted at a 30°–45° angle. Individuals who cannot separate the legs or arms sufficiently from the body, such as the very obese, should have dry clothing separating the limbs.

Prediction Equations

Like the skinfold method, BIA depends on good prediction equations to give an accurate estimate of total body water. Equations provided by BIA manufacturers should be closely examined to determine whether they match the client's gender, age, ethnicity, level of fatness, level of activity, and health status. Unfortunately, it is often difficult to determine exactly which data have been used by a given manufacturer. In fact, most equations were developed using Caucasian subjects, and these are less accurate for non-Hispanic blacks.[21] At the present stage of research, experts do not recommend using BIA to estimate FFM in African Americans or Hispanics. And even for Caucasians, BIA may be problematic. A recent study of overweight Caucasian women found handheld BIA analyzers to underestimate percent fat by almost 6%, suggesting that care in interpreting results should be taken.[22]

There are no prediction equations for children. Because of the variability in limb and trunk length in children, researchers have not found suitable equations to predict FFM from BIA—and perhaps they never will.[1]

Limitations of BIA

The ability of BIA to estimate FFM accurately is affected by several factors:

- The technician can make three mistakes that would significantly affect the result: incorrectly positioning the individual being measured (e.g., not on the back, or with legs or arms touching), placing electrodes improperly, or failing to obtain accurate height and weight measurements.
- BIA analyzers manufactured by different companies may produce results that vary considerably.[23] There is no way to overcome this limitation other than to use the same BIA analyzer when monitoring body composition changes in a given client.
- Anything that affects a person's hydration status will affect BIA. Consumption of food or beverages, skin temperature, recent physical activity, use of diuretics, and even stage of the menstrual cycle in women who experience weight fluctuations at different stages of the cycle can affect resistance measurements. To avoid errors, advise clients to avoid eating and drinking for several hours before the measurement but to be well-hydrated; to avoid strenuous exercise, caffeine, and alcohol the day before BIA; and to refrain from using diuretics for at least a week before the measurement. Women whose weight fluctuates during the menstrual cycle should not measured when they feel heavy or bloated. Because BIA is actually calculating body water, any disease, disorder, or activity that disrupts body water will reduce the validity of the prediction equations, which were developed using healthy people.
- BIA is based on the assumption that the body is symmetrical, so BIA should not be used to determine the body composition of clients with body asymmetry. For

example, people with amputations or hemiparesis (muscle weakness on one side) are not good candidates for BIA.

- Environmental factors may contribute to errors. The room temperature should be comfortable to prevent shivering or perspiration.
- Population-specific prediction equations must be used. There are no generalizable equations for estimating FFM from BIA. For groups for whom valid prediction equations are not yet available, other methods of body composition assessment should be used. Table 6 in Appendix A provides prediction equations for several population groups.

Interpretation of BIA

Once FFM has been calculated using a prediction equation, percent fat can be determined mathematically (generally, the analyzers do this for you):

Fat mass (kg) = body weight (kg) − FFM (kg)
% fat = fat mass ÷ total body weight

Table 2-4 may be consulted to determine the health risk associated with a particular percent fat. BIA is not a useful tool for measuring short-term changes in FFM or percent fat that occur when obese individuals lose weight or when individuals with eating disorders gain or lose weight.[24] BIA is also of no use in determining regional fat distribution, which may be more important than percent fat in predicting health risk.

Anthropometric Assessment: Summary

Anthropometric assessments that are part of a weight-management assessment include height, weight, skinfolds or BIA, and circumferences. From these measures, you can estimate BMI, FFM, percent fat, and fat distribution to initially determine health risk.

Case Study. Personal Assessment, Part I

You will be the subject of this case study. (If you prefer, and with permission of your instructor, you may select someone else to be your subject. If you select someone else, remember that all information collected for the case study is to be treated with confidentiality).

- Provide a brief description of the subject of this case (gender, age, occupation, and other relevant characteristics).
- Calculate your subject's BMI (or use Table I in Appendix A) and measure waist circumference. Use Table 2-2 to estimate the health risk associated with this waist circumference and BMI.
- For a person of your subject's gender and height, compare the recommended weights from the 1959 and 1983 Metropolitan height/weight tables in Appendix A.
- If an elbow caliper is available, measure your subject's elbow breadth and use Table 4 in Appendix A to determine frame size. How does measured frame size compare with what you estimate your subject's frame size to be, and how does this affect recommended weight?
- Comment on any apparent contradictions between recommended weight from height–weight tables and BMI health risk category.

CLINICAL ASSESSMENT

Components of a clinical assessment are the physical examination and health history. The clinical assessment has two primary purposes:

1. To identify risk of disease, presence of disease, or nutritional insufficiency
2. To assess the safety of making changes in diet, activity, or other behaviors

The clinical assessment may happen first and may prompt a client to seek advice about weight management. Or the clinical assessment may happen after a client contacts a weight-management professional.

What Does a Physical Examination Include?

A physician, physician's assistant, or nurse practitioner performs the physical examination component of the clinical assessment. Physical examinations always include:

- Vital signs—temperature, respiration rate, heart rate, and blood pressure
- Measurement of height and weight
- A thorough "head-to-toe" examination by inspection, palpation (touch), and auscultation (listening)

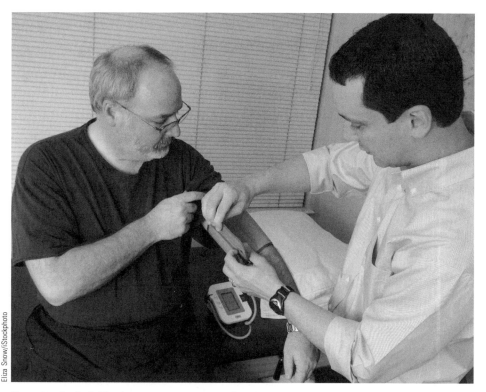

Screening for high blood pressure is one part of a physical examination.

TABLE 2-6 Physical Examination Components Relevant to Weight Management

Examination of the individual known or suspected to have an eating disorder should include special attention to:
- Condition of the hair, skin, and nails
- Signs of edema
- Bowel sounds
- Condition of the teeth and mouth
- Presence of lesions on the hands from self-induced vomiting
- Neck and salivary glands
- Cardiac abnormalities

Examination of the obese individual should include specific attention to:
- Signs of narrowing of the upper airway
- Arthritis, flat feet, or foot deformities
- Blood pressure
- Cardiac abnormalities
- Signs of edema
- Indications of diabetes

Source: From Dunn, S.A. (1998). *Mosby's primary care consultant.* St. Louis: Mosby.

Table 2-6 lists components of the "head-to-toe" examination that would be especially important in an individual known or suspected to have an eating disorder or in assessing the health risks associated with obesity.

What Information Is Needed to Construct a Health History?

The health professional who performs the physical examination will also take a comprehensive health history. The medical health history provides information about:

- Current illnesses or complaints
- Past illnesses, surgery, and hospitalizations
- Family medical history
- Allergies
- Medications
- Drug and alcohol use
- Preventive health behaviors

Dietetics professionals typically take an abbreviated health history from clients who seek assistance for problems related to weight. Fitness professionals may also take a modified health history to aid in developing a safe exercise program and to determine whether a medical referral is needed before increasing physical activity. Screening for exercise is described in Chapter 8.

Health History for the Underweight Client

Although most underweight clients do not have an eating disorder, one purpose of the health history for underweight individuals is to determine whether an eating disorder or another chronic disease is the cause of weight loss or failure to gain weight. Information should be obtained about:

- Weight history: Has the client always had a low body weight, or has weight change been relatively recent?
- Family history: Has anyone in the family had an eating disorder?
- Attitude about body weight: Does the client believe that a problem exists?
- Menstrual history: Is amenorrhea present and, if so, for how long?
- Psychoactive drug use, including alcohol: Could drug abuse account for weight loss?
- Exercise history, including sports participation: Is the client exercising excessively or involved in a sport where weight is a concern?
- Use of laxatives, emetics, or diuretics.
- Depression or suicidal thoughts.

Health History for the Overweight Client

Weight-management professionals should seek information that is helpful in determining any health risks of excess weight, diet and activity habits that may be contributing to health problems as well as factors that may make physical activity hazardous. Information should be obtained about:

- Weight history: Is excess weight a recent or life-long phenomenon?
- Dieting habits: Does the client restrict caloric intake? After dieting, does the client quickly regain weight?
- Activity habits: Is the individual accustomed to physical activity?
- Occupation: Are there opportunities for physical activity at work?
- Family weight history: Are other immediate family members obese, which might suggest a hereditary link?
- Medical problems: Is there anything in the patient's history that contraindicates exercise or diet modification?
- Motivation for weight loss: Why is the individual seeking help—to improve health, please a family member, or meet personal goals?

Given what we know about the influence of environment on diet and activity, the health history interview might also provide an opportunity to find out if the client has access to healthy foods, the length of car commuting time, and access to neighborhood venues for activity.

Although direct determination of the number and size of a client's fat cells is impractical for the clinician, it is possible to form an opinion about the potential for increased fat cell size and/or number based on the health history. This information can be useful in predicting the potential for weight loss and, perhaps, the likelihood of weight regain. In Chapter 4, you will learn that obese individuals who have a greater number of fat cells may find weight loss more difficult than those with enlarged fat cells. Points to consider in estimating the number and size of fat cells include:

hyperplasia

Excess number of fat cells.

- Onset of overweight: In general, earlier onset of obesity is associated with having more fat cells, called **hyperplasia**. In people of average weight throughout childhood and adolescence, weight gain during adulthood usually occurs through increased fat cell size, not number.[25] Still, significant weight gain in adulthood (more than 30% of the ideal body weight) could be accompanied by increased fat cell number.[26]

- Family history of obesity: Having one obese parent somewhat increases the likelihood that a child will become obese, and having two obese parents dramatically increases the probability of obesity in the child. Obese children with obese parents almost always have increased fat cell number.[26]
- Weight loss history: People who rapidly regain weight after weight loss may be more likely to have an excess number of fat cells. Hyperplasia may hinder the maintenance of weight loss.

Clinical Assessment: Summary

The physical examination and health history are valuable in identifying the risk or presence of disease or nutritional insufficiency and in predicting the safety of making changes in diet, activity, or other behaviors. Medical professionals conduct physical examinations. Nutritionists and exercise physiologists may take health histories to obtain information specific to weight-management programs, including environmental circumstances that might play a role in weight gain.

BIOCHEMICAL ASSESSMENT

Biochemical tests have several purposes in weight-management assessment:

- Screening for the presence of disease in a group of people, which can help determine program priorities
- Diagnosing disease in individuals and determining disease severity
- Evaluating the effectiveness of treatment

In anorexia nervosa or extreme weight loss, laboratory tests may identify nutritional deficiencies and metabolic abnormalities. In obesity, laboratory tests may identify abnormalities in glucose and blood lipids that suggest an increased risk for cardiovascular disease and diabetes. The next section looks briefly at biochemical tests that are typically indicated during weight-management assessment. For additional information about laboratory tests, refer to comprehensive texts on nutritional assessment.

What Are Biochemical Tests of General Nutritional Status?

hemoglobin *Iron-containing, oxygen-carrying protein found in red blood cells.*

Blood tests that measure hemoglobin, hematocrit, and serum albumin present a quick snapshot of nutritional status. Normal values for these tests are shown in Table 2-7.

- **Hemoglobin** (sometimes abbreviated Hb) is the iron-containing, oxygen-carrying protein that makes up red blood cells. Hemoglobin levels may drop in many diseases, including hyperthyroidism and anemia.

hematocrit *Percentage of packed red blood cells in whole blood.*

- The **hematocrit** (sometimes abbreviated Hct) represents the percentage of packed red blood cells in whole blood. Hematocrit drops in diseases like hyperthyroidism and anemia, and it becomes elevated in severe dehydration.

TABLE 2-7 Laboratory Tests of Nutritional Status

Laboratory Test	Acceptable Range
Hemoglobin (Hb)	Male: 13.5–17.5 g/dL Female: 12.0–16.0 g/dL
Hematocrit (Hct)	Male: 39–49% Female: 35–45%
Serum Albumin	3.4–4.8 g/dL
Glucose, fasting	74–106 mg/dL >60 yr: 80–115 mg/dL
Glycosylated hemoglobin (HbA$_{1c}$)	5.0–7.5% of Hb
Creatinine	Male: 0.7–1.3 mg/dL Female: 0.6–1.1 mg/dL

Source: Rolfes, S.R., Pinna. K., & Whitney, E. (2006). *Understanding normal and clinical nutrition*, 8th edition. Belmont, CA: Wadsworth (p. 603).

albumin *Protein involved in maintenance of osmotic pressure and transport of substances through the bloodstream that serves as a marker of adequate nutrition.*

- **Albumin** is a protein involved in maintenance of osmotic pressure and transport of various hormones, nutrients, and other substances. Serum albumin reflects long-term (e.g., over about 6 weeks) protein intake and protein metabolism. Low serum albumin levels indicate poor protein nutrition.

In addition, adults with diabetes should have serum creatinine measured as an indicator of renal function. Normal creatinine values may be found in Table 2-7.

Which Blood Glucose Tests Are Useful?

fasting plasma glucose (FPG) and 2-hour postload glucose(2-h PG) *Blood tests that measure blood glucose level, commonly used to diagnose diabetes.*

glycosylated hemoglobin (HbA$_{1c}$) *Blood test that measures glucose bound to hemoglobin during the previous several months to assess long-term blood glucose control.*

Several tests that measure blood glucose are important in the diagnosis of pre-diabetes and diabetes. Normal and abnormal values for these tests are summarized in Table 2-7. Suspicion of diabetes may be raised by a non-fasting (sometimes referred to as "casual") plasma glucose of 200 mg/dL or higher. The preferred test for children and non-pregnant adults is the **fasting plasma glucose** (sometimes abbreviated **FPG**), which requires an 8-hour overnight fast and one blood draw. An FPG between 100 and 125 mg/dL is a sign of pre-diabetes and typically referred to as impaired fasting glucose; and a value at or above 126 mg/dL indicates diabetes.

The 2-hour (75 gram) oral glucose tolerance test (OGTT) may also be helpful when the client has an impaired fasting glucose, but it is not routinely used in clinical practice. This test requires an overnight fast followed by ingestion of a 75 gram glucose load. When blood glucose measures 140 to 200 mg/dL two hours after taking the glucose load, the client is said to have impaired glucose tolerance.

Glycosylated hemoglobin is a nonfasting blood test that assesses glucose control during the preceding 2–3 months. Thus, it is not used as a screening test, but rather an assessment of compliance with treatment goals. Glucose binds to hemoglobin to create glycosylated hemoglobin. Once hemoglobin is glycosylated, it stays that way for the 120-day life of the red blood cell. The more glucose in the bloodstream, the more glycosylated hemoglobin. Hemoglobin A$_{1C}$ is a specific test of glycosylated hemoglobin that measures glucose binding at one site on hemoglobin. A normal HbA_{1C} indicates that under 6% of hemoglobin is made up of HbA_{1C}. A high HbA_{1C} (over 7%) indicates poor glucose control. When glucose control is poor, the HbA_{1C} test is generally repeated every 3 months.

The American Diabetes Association recommends that all people age 45 years or older be screened for diabetes, particularly if they have a BMI at or above 25.[27] People under 45 should be screened if they are overweight and have at least one of the following risk factors:

- Inactive
- First-degree relative with diabetes
- Member of high risk group (African American, Hispanic, Native American, Asian or Pacific Islander)
- Delivered a baby weighing more than 9 pounds, or were diagnosed with **gestational diabetes** during pregnancy
- Hypertensive
- Low HDL cholesterol (<35 mg/dL) and/or high triglycerides (>250 mg/dL)
- Have a history of vascular disease
- Have **polycystic ovary syndrome** or another condition associated with insulin resistance
- Have previously had impaired glucose tolerance

Children and adolescents who are at risk of type 2 diabetes should also be screened with fasting plasma glucose. Risk is indicated by overweight (≥85th BMI percentile) or obesity (≥95th BMI percentile) and any two of the following:[27]

- A first- or second-degree relative with type 2 diabetes
- Member of high risk group (African American, Hispanic, Native American, Asian or Pacific Islander)
- Signs of insulin resistance or a condition associated with insulin resistance (polycystic ovary syndrome, **acanthosis nigricans**, hypertension, abnormal blood lipids)
- Mother having a history of diabetes or gestational diabetes.

Which Laboratory Tests Comprise the Lipid Profile?

The lipid profile includes total **cholesterol**, **high-density lipoprotein cholesterol (HDL)**, **low-density lipoprotein cholesterol (LDL)**, and **triglycerides**. A fasting blood sample is preferred, but cholesterol can be measured in the nonfasting state.

Blood Cholesterol

Cholesterol is a sterol in the fat family that is found both in the diet (in animal products) and in the bloodstream (from dietary intake of animal products and the body's own cholesterol production). Cholesterol is transported through the bloodstream in the form of HDL and LDL cholesterol. Cholesterol carried in the tiny, cholesterol-rich LDL particles can be taken up by the blood vessels to form atherosclerotic plaque, the principal cause of cardiovascular disease. A high total cholesterol or LDL cholesterol level increases the risk of heart disease.

gestational diabetes *Diabetes that develops in the mother during pregnancy.*

polycystic ovary syndrome *A condition in which the ovaries are enlarged, ovulation is infrequent or absent, and high levels of circulating androgen hormones are present; menstruation is characteristically absent or erratic.*

acanthosis nigricans *A skin disease characterized by dark, velvety skin markings typically in body creases and folds, such as the neck, underarm area, and groin.*

cholesterol *Animal sterol in the fat family that is associated with CVD.*

high-density lipoprotein (HDL) cholesterol *Fat and cholesterol carrier involved in transporting cholesterol out of the tissues and back to the liver, where it can be disposed of; low levels are associated with CVD.*

TABLE 2-8 Blood Lipids: Desirable Values in Adults

Blood Test	Desirable (mg/dL)	Borderline or High Risk (mg/dL)
Total Cholesterol	<200	Borderline high: 200–239 High: ≥240
LDL Cholesterol	Optimal: <100 Near or above optimal: 100–129	Borderline high: 130–159 High: 160–189 Very high: ≥190
HDL Cholesterol	>40 High: ≥60	Low: <40
Triglycerides	<150	Borderline high: 150–199 High: 200–499 Very high: ≥500

Source: Adapted from Expert Panel on Detection, Evaluation, and Treatment of High Blood Cholesterol in Adults. (2001). Executive summary of the third report of the National Cholesterol Education Program (NCEP) Expert Panel on Detection, Evaluation, and Treatment of High Blood Cholesterol in Adults (Adult Treatment Panel III). *JAMA, 245*(19), 2486–2497.

low-density lipoprotein (LDL) cholesterol *Fat and cholesterol carrier that transports cholesterol out of the liver to the tissues; high levels are associated with an increased risk of CVD.*

triglycerides *Form of fat that makes up most of dietary and storage fat.*

LDL particles that are small and dense may increase cardiovascular risk to a greater extent than larger LDLs.[28] Measuring small, dense LDL cholesterol (SDLDL) is expensive but might be indicated in people with normal LDL but a strong family history of cardiovascular disease.

HDL cholesterol, in contrast, is involved in the transport of cholesterol back to the liver and out of the body. Therefore, high HDL levels are associated with a reduced risk of heart disease. The ratio of total cholesterol to HDL cholesterol is also considered an excellent indicator of cardiovascular disease risk. For people who have desirable cholesterol levels but HDL cholesterol at or below 40 mg/dL, modifications in diet and physical activity are indicated. Because of the connection between blood lipids and type 2 diabetes mellitus, the American Diabetes Association recommends a FPG for overweight adults and children who have a low HDL or high triglycerides. Normal and abnormal blood lipid values are shown in Table 2-8. Values for children and adolescents are presented in Table 12-9.

Triglycerides

Triglycerides make up over 90% of the dietary fats that we consume and store in our fat cells. The link between blood triglycerides and cardiovascular disease is complex. Generally, when blood triglyceride levels are elevated, particularly when total blood cholesterol is high, the risk of cardiovascular disease increases. Elevated triglycerides may be a sign that remnants of lipoproteins are abundant in the blood, which could promote atherosclerosis. Diabetes is also sometimes accompanied by high triglyceride levels. Alcohol restriction, increased physical activity, and weight loss (when appropriate) are often recommended for people with high triglycerides.

Biochemical Assessment: Summary

Laboratory tests can confirm the diagnosis of diseases associated with underweight, such as anemia, as well as conditions associated with overweight, such as diabetes and high blood lipids. When identifiable disease is not present, laboratory tests may indicate the individual's risk of developing chronic disease. Coupled with anthropometric data and information from clinical assessment, a health risk profile can be created for the client.

DIETARY AND PHYSICAL ACTIVITY ASSESSMENT

Weight-management professionals often find that assessment of dietary intake and energy expenditure provides useful information for weight-management planning. Although this is not clinical information, such as that garnered by anthropometric, physical, and biochemical assessments, it can prove useful in planning for change. One of the first steps in the behavior-change process, discussed in Chapter 9, is self-monitoring. Carefully completed dietary and activity assessments identify diet and activity patterns as well as help estimate caloric intake and expenditure to within 10–20% of "true" values. This section focuses on two methods: the 3-day diet record and the physical activity record.

What Is a 3-Day Diet Record?

A 3-day diet record requires the client to record all foods and beverages consumed during a designated 3-day period. The days chosen for examination should be typical, and at least one weekend day should be included. The 3-day diet record is advantageous because it is usually more representative of typical intake than a 1-day record, but it is not as burdensome as a 7-day record.

Recording Intake

The client uses a form to record as much detail as possible about dietary intake—time of day, place of eating, specific food or beverage consumed and its quantity (including alcohol, gum, candy, and water), brand names of specific foods, cooking methods used, and vitamin and mineral supplements. When combination foods are prepared (lasagna, for example), the individual should indicate all ingredients; combination foods eaten in restaurants should be described as fully as possible.

Accurately estimating the quantity of food eaten can be difficult. To minimize the problems associated with estimating portion sizes of foods typically consumed, the nutritionist should provide hands-on instruction with food models and standard household measures. The nutritionist might even demonstrate typical portions using glasses, cups, bowls, spoons, and plates from the client's home. Photographs of food in varying portion sizes may help to stimulate recall of the quantity consumed.

For clients unable to complete diet records without assistance, instructions should be given to a spouse, a caretaker, or a parent. A sample 3-day diet record is provided in Appendix B, Table 1.

Analysis of Food Records

Food records can be analyzed using composition of foods tables or computerized diet-analysis programs. Data that have been coded and put into a computerized nutrition analysis program can be easily analyzed for calories, fat, carbohydrate, protein, fiber, cholesterol, vitamins, and minerals. Computerized diet-analysis programs also indicate the proportion of calories from fat, carbohydrate, and protein and may even indicate number of servings from each food group. Computer analysis has the added advantage of quickly averaging intake from several days.

Good computerized diet-analysis programs are based on food composition data from the Nutrient Database for Standard Reference (NDSR), which is updated by the Agricultural Research Service of the U.S. Department of Agriculture (USDA). When selecting a computerized diet-analysis program, consider:

- Quality of food data: Numerous published and unpublished sources of data should have been used to construct the database.
- Number of food items in the database: At least 2,000 different food items should be included, and there should be many types of foods analyzed. For comparison, the NDSR contains over 7,000 foods.
- Quality of the vendor: The number of years in business and the support services offered should be considered.
- Frequency with which the database is updated. When you consider that almost 2,000 new products have come into the marketplace since 2004 just from the top 25 food manufacturers, you can see how difficult this is.[29]

Hundreds of diet-assessment programs are available. The International Nutrient Databank Directory lists organizations that provide software and databases for diet analysis (available at http://www.nutrientdataconf.org). In addition, nutrition journals, such as *Nutrition Today* and the *Journal of the American Dietetic Association*, regularly review diet-analysis programs.

MyPyramidTracker from the USDA (www.mypyramidtracker.gov) is a satisfactory tool for clients who have access to a computer. The online tracking system allows clients to enter food intake and physical activity information. Not all nutrients are analyzed (vitamin D, for example, is not included in the analysis), and users do not record non-quantitative information (such as meal timing, whether food is part of a meal or a snack, and place of eating).

Limitations of Diet Records

Determining how accurately 3-day diet records represent actual intake is very difficult without secretly observing people to verify that the foods and portions reported are the same as the foods actually consumed. In general, people tend to underreport their food intake, resulting in estimated caloric intakes that are lower than actual intake.[30-34] Among adults, obese individuals are more likely than average-weight people to underreport intake.[34-35] Underreporting has been noted among various racial and ethnic groups and people of all ages.[30,37,38] Whether this occurs due to forgetfulness, guilt about eating, or poor motivation, the tendency to underestimate food intake calls into question the accuracy of diet records.

In addition to client motivation and honesty, other factors limit the accuracy of diet records:

- Literacy of the client: He/she may not be able to read a food label to estimate quantities consumed and may have difficulty filling out a food record.
- Altered eating patterns: The client may eat atypically to conform to what he or she thinks you want to hear or to make recording intake easier.
- Errors in estimating portion sizes: This source of error could be avoided by having clients weigh and measure everything they eat. However, the inconvenience and time involved in weighing and measuring all foods and beverages may itself

introduce considerable error into diet assessment as people change their eating behavior to make reporting easier.

• Failure to represent habitual intake: Diet records represent only a snapshot of a client's eating patterns. People eat differently on weekends than weekdays, so one 24-hour recall may not be enough to fully assess intake.[39] Even a 7-day diet record may not fully represent typical intake. A person would need to keep diet records for at least a month to truly reflect customary food consumption.[40]

Interpretation of Diet Records

The nutritionist will need to interpret both quantitative information, which is obtained from computerized diet analysis, and non-quantitative information, which is obtained by "eyeballing" food records. Quantitative information is usually interpreted by comparing the individual's average intake with government standards and recommendations, such as the Dietary Guidelines for Americans and the Dietary Reference Intakes/Recommended Dietary Allowances. Of particular interest will be:

1. Total daily caloric intake
2. Percent of kilocalories from fats, carbohydrates, and protein
3. Intake of fiber (a good indicator of consumption of fruits, vegetables, and whole foods)
4. Cholesterol intake (an indicator of animal fats in the diet)

Non-quantitative data provide information about dietary practices. For example, the food record may reveal that most eating occurs at a particular time of day or in places other than dining areas. Snacks might be providing the bulk of calories. The pattern of food intake might indicate episodes of binge eating or, conversely, restrictive eating. And the record might suggest environmental factors that promote unhealthy eating, such as stopping into a fast food restaurant located in proximity to home. These kinds of observations may help the client to set goals related to selecting more healthful foods, developing healthier eating patterns, and overcoming environmental elements affecting food choice.

Individuals whose anthropometric measures, clinical assessment, and laboratory tests suggest that body weight presents a health risk but whose diet assessments report a nearly perfect diet need to be questioned further about dietary intake. Are there reporting errors? Were errors intentional or unintentional? Clients who report very low caloric intake (under 1,200 kcal/day) and whose body weight is unchanging may be assumed to be underreporting intake unless clinical assessment suggests extreme energy efficiency.[41] Underreporting, altering customary intake for the diet record, or deliberately falsifying information indicate a lack of readiness for self-examination and change. Confronting individuals about reporting errors has been demonstrated to improve reporting.[33] Nevertheless, if a client is unwilling to keep accurate diet records, then little is served by continuing this type of assessment. But, if the diet record seems to be largely accurate, then physical activity assessment may provide an explanation for weight loss or gain.

How Is Physical Activity Assessed?

Assessment of physical activity is an important component of weight management. Estimating caloric expenditure and identifying activity patterns will complement dietary data obtained through food records. Several tools are available:

- Doubly labeled water: This method is a very accurate and unobtrusive measure of energy expenditure, which is described in Chapter 3. It is an expensive and impractical method for assessing energy expenditure outside of research settings. In addition, doubly labeled water can only indicate that energy was expended, not what the energy was expended for. Therefore, even if this technique were used, additional tools would be needed to fully describe the quantity and the quality of physical activity.
- Activity questionnaires: Many activity questionnaires are available, including several from CDC, which oversees the National Health and Nutrition Examination Survey and the Behavioral Risk Factor Surveillance System. Seventeen activity questionnaires were reviewed in 1997 in the journal *Medicine and Science in Sports and Exercise*.[42]
- Electronic devices: Quantitative assessment of energy expenditure is now readily available with accelerometers and pedometers. Use of these devices is described later in this section.
- Physical activity records: Like diet records, physical activity records allow the client to self-report daily activity on a standard form. Because of the relative ease in using this method, it is described more fully here.

Physical Activity Records

Individuals may record activity as frequently as every minute, hourly, or in 10- or 15-minute increments. A form that uses 15-minute increments is not as tedious as one that uses 1-minute intervals and will provide more information than a form based on hourly intervals. A physical activity record based on 15-minute intervals may be found in Table 2 of Appendix B. You will have the opportunity to use this tool for the case study at the end of the chapter.

Individuals using an activity record are asked to record all of their activities, even sleeping and eating, for a 24-hour period. Although maintaining an activity record can be monotonous, individuals should examine more than one day, especially if activity varies considerably from one day to the next. If atypical days are examined, then activity records are of little use.

Once all activities have been recorded, the caloric value of each activity can be estimated. The easiest way to express the energy cost of activity is in METs. One **metabolic equivalent (MET)** is approximately equal to an individual's resting metabolic rate (RMR), which has a value of 3.5 ml oxygen consumption/kg body weight/minute, or 1 kcal/kg/hour. Although measuring RMR and precisely determining the oxygen and calorie value of 1 MET would certainly be more accurate, using a standard value for 1 MET allows quick calculation of energy expenditure based upon the individual's body weight. A range of activities—from sleeping to mountain climbing—can be categorized by their MET value.

metabolic equivalent (MET) *One MET is approximately equal to an individual's resting metabolic rate—3.5 ml oxygen consumption per kilogram body weight per minute, or about 1 kcal/kg/hour. Higher MET levels represent multiples of the metabolic rate.*

Analysis of Physical Activity Data

After all activities have been recorded for a 24-hour period, either the client or the weight-management professional determines the MET level of each 15-minute activity segment. If the client has been given instructions on MET classifications, then she may choose to do this herself. Complete instructions for calculating total daily caloric expenditure from activity data are provided with the physical activity record in Table 2 of Appendix B.

Limitations of Physical Activity Records

The limitations of activity records are similar to the limitations of food records: Both presume that individuals will be careful, conscientious, and candid in recording entries. Neither can be used by people with low literacy. A recent review of physical activity assessment methods found a prevalence of underestimates, overestimates, and errors in activity reporting using activity diaries and questionnaires.[43] Overweight people may be slightly more likely to overreport exercise.[44] And children, who tend to engage in activity intermittently, are least likely to maintain accurate physical activity records and are better served with more objective measures of activity, such as pedometers.[45]

Accuracy and compliance in adults can be improved by teaching them to use activity records. If clients are being asked to categorize their activities by MET level, then they will need help in determining the MET value of their typical activities. Otherwise, there is a tendency for less-fit people to overestimate the intensity of their activities.[44] Sedentary people rarely exceed 4 METs in daily activity, so the physical activity record in Appendix B has only one column for activities at 6 METs and higher. This can help reinforce the idea that activities at or above 6 METs are rarely engaged in by non-athletes. For very active people, the form can be adapted by adding separate columns for higher METs.

Interpretation of Physical Activity Records

A great deal of information is provided on a physical activity form. First, the assessment gives a broad picture of the kinds of activities in which the individual engages. A person who reports that he does not exercise because exercise makes him "feel bad" may be choosing the wrong kinds of activity or working at too high an intensity. The pattern of daily activity may identify potential problems, such as a tendency to sit and overeat in the evening. If a person's activity level drops to 1 MET at 6 P.M. due to passive television viewing while lying on a sofa, different activities during those hours may add to caloric expenditure and prevent boredom-related snacking. An activity record may also help in time management by assisting individuals to identify times of day, even in small segments, when some activity can be incorporated into their routines. It can highlight environmental issues that may be preventing the person from being active—and can suggest environmental modifications that may help promote activity.

In addition, by assigning the appropriate MET value to each activity, total caloric expenditure for the day(s) examined can be estimated. When activity records are kept for more than 1 day, caloric output total should be averaged. If the days examined are typical, then a comparison of caloric intake and caloric expenditure may help explain why weight change has occurred and what needs to be done to promote weight gain or loss. If clients are willing to use pedometers, described in the next section, it will not be necessary to use the MET portion of the activity record.

Individuals who appear to have expended considerable energy in physical activity but are having difficulty with weight loss should be questioned in further detail. Perhaps they are overestimating intensity of activity. Other assessments, such as a fitness test of cardiovascular endurance or a blood HDL level, can provide indirect confirmation of activity level. (More active people will have higher cardiovascular endurance and, usually, higher HDLs.) As with diet records, there is little benefit to be gained in keeping activity records if accuracy is questionable.

Pedometers and Accelerometers

These devices provide a more objective assessment of physical activity than a self-report. They are both small and light enough to remain unobtrusive to the wearer. Accelerometers are particularly useful because they can indicate intensity and duration of activity, but they are currently too expensive for general use, and they require some technical expertise to manage. Pedometers, on the other hand, can be purchased for as little as $10 and, while they do not provide as much information as accelerometers, they are accurate in assessing walking and running. Pedometers correlate well with accelerometers in evaluating distance traveled.[46] Used in combination with a physical activity record, pedometers can provide fairly accurate information about a client's ambulation.

A pedometer is worn clipped to the waistband and functions by counting steps. The user usually has to set stride length before the device can function accurately. Pedometers all provide directions for doing this and should be read carefully. Pedometers range in price from $10–55, the difference being their capacity to store data and perhaps even upload it to a computer, and where they can be worn. More expensive models are available that can be carried in a pocket or handbag and even clipped to a shoe. Both the Yamax Digi-Walker, which costs about $25, and the Omron HJ-151, which costs about $15, have been highly rated in scientific studies.[45,47–49]

Big Cheese Photo/Jupiter Images

Assessment of physical activity can provide useful information about contributors to weight management problems.

Interpretation of Pedometer Data

Most pedometers contain a spring-lever mechanism that detects up-and-down motion and displays it digitally as a step count. Thus, pedometers can be used for monitoring steps per day, accumulated either through walking or running on a treadmill or in one's daily activities. An individual can keep a daily record of steps taken and monitor this as an indicator of physical activity.

Recommendations for steps per day have been promoted by various groups as a method for maintaining weight, losing weight, or becoming more fit. For example, America on the Move (www.americaonthemove.org) suggests adding 2,000 extra steps to your daily routine. Shape Up America (www.shapeup.org) recommends gradually working up to 10,000 steps a day. Most people take between 1,300 and 2,000 steps when walking or running a mile.[45] A person who takes between 2,600 and 4,000 steps each day is probably close to CDC recommendations for daily physical activity, although a pedometer cannot indicate exercise intensity.

However, when individuals participate in activities during a delimited period, such as a 45-minute walk, a 30-minute aerobics class, or a 50-minute physical education class, steps taken can be converted to steps per minute and provide an indicator of activity intensity. Here is how that is done:

- Start with the assumption that 2,000 steps is equivalent to 1 mile
- Mrs. G. takes a 45-minute walk, and her pedometer registers 3,600 steps during that period
- Dividing 3600 steps taken by 2000 steps in a mile, we estimate that she walked 1.8 miles in 45 minutes (many pedometers make this calculation for you)
- 1.8 miles/45 minutes = 2.4 miles/60 minutes, so she walked about 2.4 mph
- Consulting Table 3 in Appendix B, you will see that walking 2.5 mph is equivalent to 3 METs, which is categorized as light intensity for a middle-aged person (the concept of intensity will be discussed further in Chapter 8)

Limitations of Pedometers

Because pedometers only detect vertical motion, their use is limited to walking and running. They will not detect household tasks, such as ironing, cooking, and weeding, which may expend energy but do not involve vertical acceleration. An activity record would have to be consulted to estimate the contribution of these lifestyle tasks to energy expenditure.

Spring-levered pedometers have been estimated to be accurate to within 3–25% of actual step count when walking between 2–4 mph.[50] They tend to overestimate steps in slow walking, which would be significant for obese individuals, who tend to walk slower.[47,50]

Other concerns in using pedometers echo those for diet and physical activity records: individuals will alter typical activity patterns because they know that they are being monitored; and users have to be honest in reporting steps taken (unless they wear a pedometer with a memory storage function).

Diet and Physical Activity Assessment: Summary

Three-day diet records and multiple-day physical activity records are helpful in estimating caloric intake and energy expenditure as well as patterns of intake and activity that contribute to weight-management problems. These records are subject to client error in estimating portion sizes and level of activity. In addition, client motivation, literacy, and candor may affect the accuracy of these records. For physical activity monitoring, commercially available pedometers can provide more objective data to supplement the physical activity record.

PUTTING ASSESSMENT COMPONENTS TOGETHER

Children and adults who are overweight and obese have a greater risk of health problems, particularly when obesity is coupled with risk factors such as diabetes, heart disease, elevated blood pressure and blood lipids, impaired glucose tolerance, inactivity, and a family history of these disorders. So, assessment may start with examination of body weight, body composition, and fat distribution, but other factors must be considered, too.

Individuals who are underweight (BMI under 18.5), anorexic, or bulimic may also have weight-related health problems. When weight loss has been severe or when weight loss practices have been extreme, the individual with an eating disorder is at risk of cardiovascular, endocrine, digestive, and neuromuscular complications. Even in the absence of an eating disorder, athletes endanger their health when they attempt to maintain a very low percentage of body fat or lose and regain weight rapidly.

Figure 2-6 offers an example of a recording form that can be used to consolidate the elements of a weight-management assessment. Once information about the client has been compiled and recorded, the weight-management professional can do a health risk factor analysis.

What Is a Health Risk Factor Analysis?

A health risk factor analysis looks at all of the available information about a client, not just weight and percent fat. The Expert Panel on the Identification, Evaluation, and Treatment of Overweight and Obesity in Adults suggests three categories of risk for individuals whose BMI is greater than 25 or whose waist circumference exceeds 88 cm (female) or 102 cm (male):[14]

1. Very high risk: Individuals in this category have one or more of the following conditions:

   - Established cardiovascular disease (previous heart attack, stable or unstable angina pectoris, coronary artery surgery, or angioplasty)
   - Other atherosclerotic disease (peripheral artery disease, abdominal aortic aneurysm, symptomatic carotid artery disease)
   - Type 2 diabetes mellitus (FPG at or over 126 mg/dL or 2-hour postprandial plasma glucose of 200 mg/dL or greater)
   - Sleep apnea

2. High risk: Individuals with three or more of the following are considered at high absolute risk for developing health complications:

   - Cigarette smoking
   - Hypertension (140 or higher systolic and/or 90 or higher diastolic), or taking blood pressure–lowering drugs

Client name _____ Date _____

Anthropometric Assessment

Height _____ (cm)(in) Weight _____ (kg)(lb) BMI _____

Percent fat _____ % Waist circumference _____ (cm)(in)

Disease risk based on anthropometrics _____

Clinical Assessment

Age _____ Blood pressure _____

Medical problems _____

Key findings from health history _____

Weight history _____

Family weight history _____

Exercise contraindications _____

Dietary contraindications _____

Biochemical Assessment

Hb _____ Hct _____ Serum albumin _____

FPG _____ 2-hr PG _____ HbA_{1c} (if diabetic) _____

Total cholesterol _____ HDL _____ LDL _____ Triglyceride _____

Total cholesterol /HDL cholesterol ratio _____

Disease risk based on anthropometric, clinical, and biochemical assessment _____
_____
_____

Dietary Assessment

Average daily caloric intake _____ kcal Cholesterol intake _____ mg

Percent kcal from fat _____ carbohydrate _____ protein _____

Indication of binge eating or restrictive eating _____

Eating pattern (time of day, place of eating, snacking) _____

Physical Activity Assessment

Average daily caloric expenditure _____ kcal Above 3 METs? _____

Activity pattern _____

FIGURE 2-6 A sample weight-management assessment record.

- LDL cholesterol of 160 mg/dL or higher, or a borderline high-risk LDL cholesterol (139 to 159 mg/dL) plus two or more additional risk factors
- HDL cholesterol of 35 mg/dL or lower
- Impaired fasting glucose (110–125 mg/dL)
- Family history of early cardiovascular disease (for example, heart attack or sudden death at or before age 55 in father or other male first-degree relative, or at or before age 65 in mother or other female first-degree relative). NOTE: first-degree relatives share about half their genes with an individual family member; these include biological parents, children, and siblings.
- Age (males 45 years or older; females 55 years or older, or postmenopausal)

3. Risk of obesity-related diseases that are not life-threatening: Individuals with these conditions are not at a high risk of mortality but may improve quality of life if they reduce weight:

- Gynecological abnormalities
- Osteoarthritis
- Gallstones
- Stress incontinence

Additional risk factors that suggest the need for weight reduction in obese individuals include inactivity and high triglycerides.

Figure 2-7 indicates how components of the anthropometric and clinical assessment may establish the health risks of excess body fat. A continuum of intervention approaches are determined by degree of health risk. For example, a person with a BMI of 40 who is at very high risk may require an extreme intervention, such as gastric surgery, to lose weight rapidly and restore health. A person with a lower BMI and no serious risk factors may benefit from diet and activity recommendations, whether or not weight loss occurs. Figure 2-8 illustrates the continuum of intervention approaches, based upon risk classification.

Individuals who are underweight or have anorexia or bulimia should have a health risk factor analysis that includes:

- Degree of underweight: The person whose BMI is near 19 is not at as great a level of risk as the one whose BMI is 17 or lower.
- Presence of diseases: Cardiovascular disease and diabetes, for example, could be made more severe by weight loss or extreme dietary practices.
- Psychological problems: Anxiety disorders, depression, and some other psychological problems may complicate treatment.

Putting Assessment Components Together: Summary

The process of assessment can help people realize that change is needed to support health. Assessment results may also suggest treatment approaches and lifestyle change strategies. Anthropometric data should be only one of several criteria used in assessing health risk.

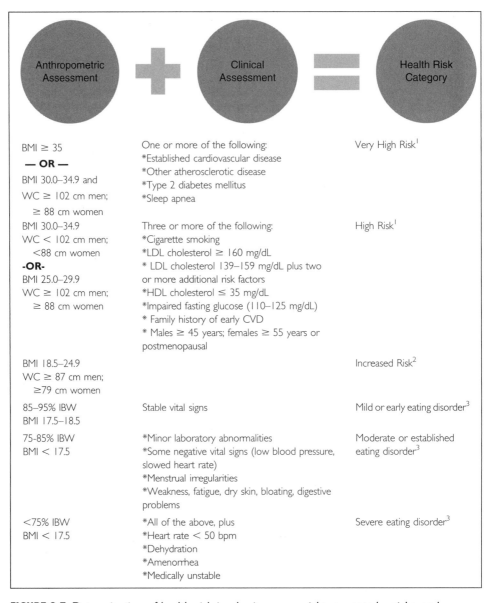

Anthropometric Assessment	Clinical Assessment	Health Risk Category
BMI ≥ 35 **— OR —** BMI 30.0–34.9 and WC ≥ 102 cm men; ≥ 88 cm women	One or more of the following: *Established cardiovascular disease *Other atherosclerotic disease *Type 2 diabetes mellitus *Sleep apnea	Very High Risk[1]
BMI 30.0–34.9 WC < 102 cm men; <88 cm women **-OR-** BMI 25.0–29.9 WC ≥ 102 cm men; ≥ 88 cm women	Three or more of the following: *Cigarette smoking *LDL cholesterol ≥ 160 mg/dL * LDL cholesterol 139–159 mg/dL plus two or more additional risk factors *HDL cholesterol ≤ 35 mg/dL *Impaired fasting glucose (110–125 mg/dL) * Family history of early CVD * Males ≥ 45 years; females ≥ 55 years or postmenopausal	High Risk[1]
BMI 18.5–24.9 WC ≥ 87 cm men; ≥79 cm women		Increased Risk[2]
85–95% IBW BMI 17.5–18.5	Stable vital signs	Mild or early eating disorder[3]
75-85% IBW BMI < 17.5	*Minor laboratory abnormalities *Some negative vital signs (low blood pressure, slowed heart rate) *Menstrual irregularities *Weakness, fatigue, dry skin, bloating, digestive problems	Moderate or established eating disorder[3]
<75% IBW BMI < 17.5	*All of the above, plus *Heart rate < 50 bpm *Dehydration *Amenorrhea *Medically unstable	Severe eating disorder[3]

FIGURE 2.7 Determination of health risk in obesity, overweight, expected weight, and underweight.

Source: [1]NHLBI. (2000). *The practical guide: Identification, evaluation, and treatment of overweight and obesity in adults.* Bethesda, MD: National Institutes of Health. [2]Ardern, C. I., et al.. (2004). Development of health-related waist circumference thresholds within BMI categories. *Obesity Research, 12*(7), 1094–1103. [3]Rome, E.S., et al., (2003). Children and adolescents with eating disorders: The state of the art. *Pediatrics, 111*(1), e98–e108.

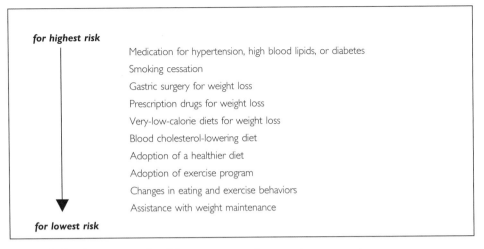

for highest risk

Medication for hypertension, high blood lipids, or diabetes

Smoking cessation

Gastric surgery for weight loss

Prescription drugs for weight loss

Very-low-calorie diets for weight loss

Blood cholesterol-lowering diet

Adoption of a healthier diet

Adoption of exercise program

Changes in eating and exercise behaviors

Assistance with weight maintenance

for lowest risk

FIGURE 2-8   Continuum of weight-management intervention approaches.

Case Study. Personal Assessment, Part 2

Continue this case study with the same subject from Part 1. You or your subject should keep an activity record using the form in Table 2 of Appendix B. Follow the directions to keep track of daily activity and calculate caloric expenditure. Perform the calculations required to determine the caloric value of 0.9–9 METs. If your subject has a pedometer, wear it when ambulating or exercising and try to keep track of steps taken during a given period of time.

- What kinds of difficulties did you or your subject have using the physical activity record?
- How might some of these difficulties be overcome with clients?
- What surprised you about the estimated daily caloric expenditure?
- If it was higher than you thought it should be, to what do you attribute this?
- If your subject wore a pedometer and attempted to determine activity intensity, how did it compare with estimated energy expenditure on the physical activity record?

REFERENCES

1. Wells, J. C., Fuller, N. J., Dewit, O., Fewtrell, M. S., Elia, M., & Cole, T. J. (1999). Four-component model of body composition in children: Density and hydration of fat-free mass and comparison with simpler models. *American Journal of Clinical Nutrition*, 69(5), 904–912.

2. Cameron, N. (1986). The methods of auxological anthropometry. In F. Faulkner, & J. M. Tanner (Eds.), *Human growth: A comprehensive treatise.* Vol. 3. Methodology, ecological, genetic, and nutritional effects on growth. As cited in Gibson, R.S. (1990). *Principles of nutritional assessment.* New York: Oxford University Press. (pp. 3–46). New York: Plenum Press.

3. Chumlea, W. C., Roche, A. F., & Mukherjee, D. (1984). *Nutritional assessment of the elderly through anthropometry.* Columbus, OH: Ross Laboratories.

4. National Heart, Lung, and Blood Institute. (1998). *Clinical guidelines on the identification, evaluation, and treatment for overweight and obesity in adults: The evidence report.* Bethesda, MD: National Institutes of Health.

5. Lara-Esqueda, A., Aguilar-Salinas, C. A., Velazquez-Monroy, O., Gomez-Perez, F. J., Rosas-Peralta, M., Mehta, R., et al. (2004). The body mass index is a less-sensitive tool for detecting cases with obesity-associated co-morbidities in short stature subjects.

International Journal of Obesity and Related Metabolic Disorders, 28(11), 1443–1450.

6. Willett, W. C., Stampfer, M., Manson, J., & VanItallie, T. (1991). New weight guidelines for Americans: Justified or injudicious? *American Journal of Clinical Nutrition, 53*(5), 1102–1103.

7. Bellisari, A., & Roche, A. F. (2005). Anthropometry and ultrasound. In S. Heymsfield, T. Lohman, Z. Wang & S. Going (Eds.), *Human body composition* (2nd ed., pp. 109–127). Champaign, IL: Human Kinetics.

8. Wadden, T. A., Stunkard, A. J., Johnston, F. E., Wang, J., Pierson, R. N., Van Itallie, T. B., et al. (1988). Body fat deposition in adult obese women. II. Changes in fat distribution accompanying weight reduction. *American Journal of Clinical Nutrition, 47*(2), 229–234.

9. American College of Sports Medicine (ACSM). (2006). *ACSM's guidelines for exercise testing and prescription* (7th ed.). Philadelphia: Lippincott Williams and Wilkins.

10. Callaway, C. W., Chumlea, W. C., Bouchard, C., Himes, J. H., Lohman, T. G., Martin, A. D., et al. (1988). Circumferences. In T. G. Lohman, A. F. Roche & R. Martorell (Eds.), *Anthropometric standardization reference manual* (pp. 39–54). Champaign, IL: Human Kinetics.

11. National Health and Nutrition Examination Survey. (2004). *Anthropometry procedures manual.* Atlanta, GA: Centers for Disease Control and Prevention.

12. Pischon, T., Boeing, H., Hoffmann, K., Bergmann, M., Schulze, M. B., Overvad, K., et al. (2008). General and abdominal adiposity and risk of death in Europe. *New England Journal of Medicine, 359*(20), 2105–2120.

13. Ardern, C. I., Janssen, I., Ross, R., & Katzmarzyk, P. T. (2004). Development of health-related waist circumference thresholds within BMI categories. *Obesity Research, 12*(7), 1094–1103.

14. NHLBI. (2000). *The practical guide: Identification, evaluation, and treatment of overweight and obesity in adults.* Bethesda, MD: National Institutes of Health.

15. Schneider, H. J., Glaesmer, H., Klotsche, J., Bohler, S., Lehnert, H., Zeiher, A. M., et al. (2007). Accuracy of anthropometric indicators of obesity to predict cardiovascular risk. *Journal of Clinical Endocrinology and Metabolism, 92*(2), 589–594.

16. Maher, V., O'Dowd, M., Carey, M., Markham, C., Byrne, A., Hand, E., et al. (2009). Association of central obesity with early carotid intima-media thickening is independent of that from other risk factors. *International Journal of Obesity, 33*(1), 136–143.

17. Lohman, T. G., Roche, A. F., & Martorell, R. (1988). *Anthropometric standardization reference manual.* Champaign, IL: Human Kinetics.

18. Jackson, A. S., & Pollock, M. L. (1978). Generalized equations for predicting body density of men. *British Journal of Nutrition, 40*(3), 497–504.

19. Jackson, A. S., Pollock, M. L., & Ward, A. (1980). Generalized equations for predicting body density of women. *Medicine and Science in Sports and Exercise, 12*(3), 175–181.

20. Williams, D. P., Going, S. B., Lohman, T. G., Harsha, D. W., Srinivasan, S. R., Webber, L. S., et al. (1992). Body fatness and risk for elevated blood pressure, total cholesterol, and serum lipoprotein ratios in children and adolescents. *American Journal of Public Health, 82*(3), 358–363.

21. Sun, S. S., Chumlea, W. C., Heymsfield, S. B., Lukaski, H. C., Schoeller, D., Friedl, K., et al. (2003). Development of bioelectrical impedance analysis prediction equations for body composition with the use of a multicomponent model for use in epidemiologic surveys. *American Journal of Clinical Nutrition, 77*(2), 331–340.

22. Varady, K. A., Santosa, S., & Jones, P. J. (2007). Validation of hand-held bioelectrical impedance analysis with magnetic resonance imaging for the assessment of body composition in overweight women. *American Journal of Human Biology, 19*(3), 429–433.

23. Heyward, V. H., & Stolarczyk, L. M. (1996). *Applied body composition assessment.* Champaign, IL: Human Kinetics.

24. Chumlea, W. C., & Sun, S. S. (2005). Bioelectrical impedance analysis. In S. Heymsfield, T. Lohman, Z. Wang & S. Going (Eds.), *Human body composition* (2nd ed., pp. 79–88). Champaign, IL: Human Kinetics.

25. Spalding, K. L., Arner, E., Westermark, P. O., Bernard, S., Buchholz, B. A., Bergmann, O., et al. (2008). Dynamics of fat cell turnover in humans. *Nature, 453* (7196), 783–787.

26. Wadden, T. A., & Foster, G. D. (1992). Behavioral assessment and treatment of markedly obese patients. In T. A. Wadden, & T. B. VanItallie (Eds.), *Treatment of the seriously obese patient* (pp. 290–330). New York: Guilford Press.

27. American Diabetes Association. (2006). Standards of medical care in diabetes—2006. *Diabetes Care, 29* (Suppl 1), S4–S42.

28. Rizzo, M., Kotur-Stevuljevic, J., Berneis, K., Spinas, G., Rini, G. B., Jelic-Ivanovic, Z., et al. (2009). Atherogenic dyslipidemia and oxidative stress: A new look. *Translational Research: The Journal of Laboratory and Clinical Medicine, 153*(5), 217–223.

29. Pennington, J. A., Stumbo, P. J., Murphy, S. P., McNutt, S. W., Eldridge, A. L., McCabe-Sellers, B. J., et al. (2007). Food composition data: The foun-

dation of dietetic practice and research. *Journal of the American Dietetic Association, 107*(12), 2105–2113.

30. Maurer, J., Taren, D. L., Teixeira, P. J., Thomson, C. A., Lohman, T. G., Going, S. B., et al. (2006). The psychosocial and behavioral characteristics related to energy misreporting. *Nutrition Reviews, 64*(2 Pt 1), 53–66.

31. Rebro, S. M., Patterson, R. E., Kristal, A. R., & Cheney, C. L. (1998). The effect of keeping food records on eating patterns. *Journal of the American Dietetic Association, 98*(10), 1163–1165.

32. Mahabir, S., Baer, D. J., Giffen, C., Subar, A., Campbell, W., Hartman, T. J., et al. (2006). Calorie intake misreporting by diet record and food frequency questionnaire compared to doubly labeled water among postmenopausal women. *European Journal of Clinical Nutrition, 60*(4), 561–565.

33. Goris, A. H., & Westerterp, K. R. (2000). Improved reporting of habitual food intake after confrontation with earlier results on food reporting. *British Journal of Nutrition, 83*(4), 363–369.

34. Goris, A. H., Westerterp-Plantenga, M. S., & Westerterp, K. R. (2000). Undereating and underrecording of habitual food intake in obese men: Selective underreporting of fat intake. *American Journal of Clinical Nutrition, 71*(1), 130–134.

35. Heitmann, B. L., & Lissner, L. (1995). Dietary underreporting by obese individuals—is it specific or nonspecific? *British Medical Journal, 311*(7011), 986–989.

36. Singh, R., Martin, B. R., Hickey, Y., Teegarden, D., Campbell, W. W., Craig, B. A., et al. (2009). Comparison of self-reported, measured, metabolizable energy intake with total energy expenditure in overweight teens. *American Journal of Clinical Nutrition, 89*(6), 1744–1750.

37. Olendzki, B. C., Ma, Y., Hebert, J. R., Pagoto, S. L., Merriam, P. A., Rosal, M. C., et al. (2008). Underreporting of energy intake and associated factors in a Latino population at risk of developing type 2 diabetes. *Journal of the American Dietetic Association, 108*(6), 1003–1008.

38. Lanctot, J. Q., Klesges, R. C., Stockton, M. B., & Klesges, L. M. (2008). Prevalence and characteristics of energy underreporting in African-American girls. *Obesity, 16*(6), 1407–1412.

39. Haines, P. S., Hama, M. Y., Guilkey, D. K., & Popkin, B. M. (2003). Weekend eating in the United States is linked with greater energy, fat, and alcohol intake. *Obesity Research, 11*(8), 945–949.

40. Black, A. E., Prentice, A. M., Goldberg, G. R., Jebb, S. A., Bingham, S. A., Livingstone, M. B., et al. (1993). Measurements of total energy expenditure provide insights into the validity of dietary measurements of energy intake. *Journal of the American Dietetic Association, 93*(5), 572–579.

41. Swinburn, B., & Ravussin, E. (1993). Energy balance or fat balance? *The American Journal of Clinical Nutrition, 57* (Suppl), 766S–771S.

42. Pereira, M. A., FitzerGerald, S. J., Gregg, E. W., Joswiak, M. L., Ryan, W. J., Suminski, R. R., et al. (1997). A collection of physical activity questionnaires for health-related research. *Medicine and Science in Sports and Exercise, 29*(6 Suppl), S1–205.

43. Westerterp, K. R. (2009). Assessment of physical activity: A critical appraisal. *European Journal of Applied Physiology, 105*(6), 823–828.

44. Jakicic, J. M., Polley, B. A., & Wing, R. R. (1998). Accuracy of self-reported exercise and the relationship with weight loss in overweight women. *Medicine and Science in Sports and Exercise, 30*(4), 634–638.

45. Welk, G. J., Differding, J. A., Thompson, R. W., Blair, S. N., Dziura, J., & Hart, P. (2000). The utility of the Digi-walker step counter to assess daily physical activity patterns. *Medicine and Science in Sports and Exercise, 32*(9 Suppl), S481–8.

46. Tudor-Locke, C., Williams, J. E., Reis, J. P., & Pluto, D. (2002). Utility of pedometers for assessing physical activity: Convergent validity. *Sports Medicine, 32*(12), 795–808.

47. Hendelman, D., Miller, K., Baggett, C., Debold, E., & Freedson, P. (2000). Validity of accelerometry for the assessment of moderate intensity physical activity in the field. *Medicine and Science in Sports and Exercise, 32*(9 Suppl), S442–9.

48. Holbrook, E. A., Barreira, T. V., & Kang, M. (2009). Validity and reliability of Omron pedometers for prescribed and self-paced walking. *Medicine and Science in Sports and Exercise, 41*(3), 670–674.

49. Tudor-Locke, C., McClain, J. J., Hart, T. L., Sisson, S. B., & Washington, T. L. (2009). Pedometry methods for assessing free-living youth. *Medicine and Science in Sports and Exercise, 80*(2), 175–185.

50. Melanson, E. L., Knoll, J. R., Bell, M. L., Donahoo, W. T., Hill, J. O., Nysse, L. J., et al. (2004). Commercially available pedometers: Considerations for accurate step counting. *Preventive Medicine, 39*(2), 361–368.

CHAPTER

3

Energy Metabolism

L iving beings continuously expend energy to sustain vital life functions and, when food is available, they periodically consume energy from fat, carbohydrate, and protein. The difference between energy intake and energy expenditure determines whether, over time, body weight increases, decreases, or remains constant. This is why understanding energy processes is very important in any study of weight management.

The relationship between energy in and energy out is referred to as the energy balance equation and is expressed as: Energy Intake = Energy Expenditure ± Energy Storage.

As illustrated by Figure 3-1, a person whose weight is stable successfully balances energy intake and expenditure. Weight loss occurs when expenditure exceeds intake or, put another way, when intake is less than expenditure. Weight gain occurs when intake exceeds expenditure or when expenditure is less than intake.

Your total daily energy expenditure (TEE) has three principal components: basal or resting metabolic rate (BMR or RMR), which is also called resting energy expenditure (REE); adaptive thermogenesis; and activity energy expenditure (AEE). Of these, and as shown in Figure 3-2, for the typical sedentary person, BMR accounts for an estimated 65–75% of total daily energy output, adaptive thermogenesis accounts for about 10–15%, and physical activity is responsible for the rest. In fact, activity energy expenditure accounts for most of the variability in TEE and has the greatest potential for increasing energy expenditure.

Energy intake occurs when we ingest foods that contain the energy nutrients (fat, carbohydrate, and protein). When intake is adequate, all of the energy nutrients can be stored. Fat, or lipid, is by far the most abundant source of stored energy. One pound of

stored body fat is equal to approximately 3,500 kcal of energy. Even a 140-lb person who is only 15% fat stores 21 lb of fat—the equivalent of more than 73,000 kcal.

The energy balance equation suggests that a decrease in food intake and/or an increase in any or all components of energy expenditure to equal 3,500 kcal will result in a loss of one pound. Conversely, an increase in food intake and/or a decrease in energy expenditure to equal 3,500 kcal will yield a gain of one pound. This conclusion is logical but overly simplistic. Energy metabolism is neither so predictable nor consistent from person to person. The billion-dollar diet book industry is based on the assumption that *energy in* is the main culprit in weight gain: If overweight people could only be convinced to reduce their intake or to eat certain foods, then excess pounds would disappear. In fact, the often-overlooked aspect of *energy out* bears much responsibility for weight gain and regain, too. This chapter examines both sides of the energy balance equation and the relationship between energy in and energy out.

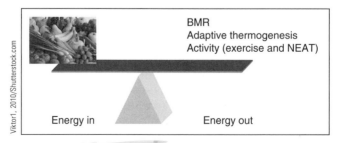

FIGURE 3-1 Energy Balance. Energy balance occurs when energy intake and output are the same.

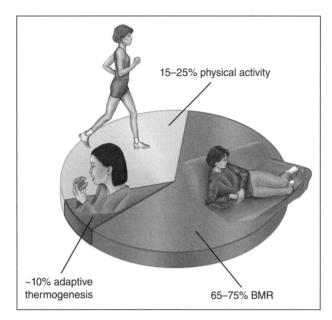

15–25% physical activity

~10% adaptive thermogenesis

65–75% BMR

FIGURE 3-2 The components of energy expenditure. An active person may expend a greater proportion of energy in activity and less in basal metabolism.

Source: From Sizer/Whitney, Nutrition, 10E. p. 317. © 2006 Brooks/Cole, a part of Cengage Learning Inc. Reproduced by permission. www.cengage.com/permissions.

energy nutrients
The nutrients (carbohydrate, fat, and protein) that contain calories. The energy nutrients are made up of carbon, hydrogen, oxygen, and (in protein) nitrogen.

kilocalories (kcal) or Calories (Cal)
Measure of the energy value of food; technically, the amount of heat required to raise the temperature of 1 kg of water 1° Celsius. A calorie is 1/1000 of a kcal.

bomb calorimeter *Device in which food samples can be burned to determine their caloric value.*

ENERGY IN: THE METABOLIC FATE OF INGESTED FOOD

Carbohydrates, lipids, and protein comprise the energy in component of energy balance. They are known as the **energy nutrients** because they contain calories; they are briefly characterized in Table 3-1. The caloric value of food is expressed in kilocalories (kcal). One **kilocalorie** is defined as the amount of heat required to raise the temperature of 1 kg of water 1° Celsius. One kilocalorie (or Calorie with a capital C) equals 1,000 cal. The terms calorie (with a small c) and kilocalorie are often used interchangeably.

You might wonder how the caloric value of food is determined. Food can be burned in a device known as a **bomb calorimeter**. A food sample is placed in the "bomb," which is a stainless-steel container surrounded by a water vessel, pictured in Figure 3-3. The heat generated as the food sample is ignited is transferred into a known quantity of water within the water vessel, and the rise in water temperature is measured to determine the energy value of the food. Fat, a very rich source of energy, provides 9 kcal/g, and protein and carbohydrate each provide 4 kcal/g. Today, when the proportion of carbohydrate, fat, and protein in a food are known, the caloric value of food is rarely determined using the bomb calorimeter, and government and food industry researchers simply calculate calories from recipes.

Clearly, the *quantity* of ingested kilocalories influences energy balance. Recent evidence indicates that the *composition* of the ingested kilocalories may also affect energy balance.

TABLE 3-1   The Energy Nutrients

Carbohydrates	**Energy value** 4 kcal/g
Types of carbohydrates	
Monosaccharides (glucose, fructose, galactose)	
Disaccharides (sucrose, lactose, maltose)	
Polysaccharides (glycogen, starch, fiber)	
Fats	**Energy value** 9 kcal/g
Types of fats	
Triglycerides comprise 95% of dietary fat	
1 glycerol + 3 fatty acids	
Fatty acids may be:	
Short-, medium-, or long-chain	
Saturated (no double bonds)	
Monounsaturated (one double bond)	
Polyunsaturated (> one double bond)	
Omega-3 polyunsaturated	
Omega-6 polyunsaturated	
Proteins	**Energy value** 4 kcal/g
Classification	
Amino acids (contain nitrogen)	
9 are essential (obtained through diet)	
11 are nonessential (may be synthesized)	

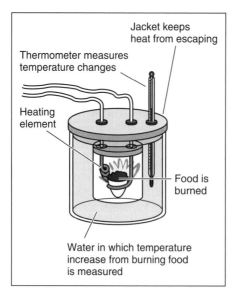

FIGURE 3-3   A bomb calorimeter is used to measure the caloric value of food.

TABLE 3-2 Aerobic and Anaerobic Metabolism

Aerobic metabolism	Anaerobic metabolism
Oxygen required	Oxygen not required
Theoretically, can continue forever	Cannot continue indefinitely
Carbohydrate, fat, and protein used	Only carbohydrate used
Low- to moderate-intensity activities	High-intensity activities

How Is Food Energy Transformed in the Body?

adenosine triphosphate (ATP) *High-energy compound made up of adenine (a carbon-nitrogen substance), ribose (a sugar), and three inorganic phosphates; ATP stores chemical energy from food in high-energy bonds between the phosphates.*

anaerobic *Not requiring or using oxygen.*

aerobic *Requiring or using oxygen.*

mitochondria *oval-shaped cell organelles often called the power-house of the cell because energy production, via the citric acid cycle, electron transport, and oxidative phosphorylation, occurs in them.*

The chemical energy in your breakfast cereal cannot be directly used by your body's cells. First, a compound called **adenosine triphosphate (ATP)** must be formed. ATP is a nucleotide composed of adenosine (made from adenine and ribose) and three phosphates (P_i):

Adenosine–Phosphate (P_i) ~ Phosphate (P_i) ~ Phosphate (P_i)

The symbol ~ indicates that the endmost phosphates are connected by high-energy bonds, which is where ATP holds the energy from nutrients. Later, ATP releases this energy to produce heat and to meet the needs of the body's cells—for digesting food, moving muscles, transmitting signals along nerve cells, as well as a thousand other biological activities. The sum of all the energy transformation processes in the human body is known as metabolism.

The energy from food is used continually to produce more ATP. This may occur anaerobically or aerobically. **Anaerobic** metabolic processes take place in the cytoplasm, or fluid portion, of each cell and do not require or use oxygen. **Aerobic** metabolism occurs in the cell **mitochondria**—specialized structures sometimes referred to as the powerhouse of the cell—and oxygen is required. You may be confused by the terms aerobic and anaerobic. Because humans require oxygen to sustain life, it would seem that all metabolic processes must be aerobic; however, some metabolic processes can occur quickly and without sufficient oxygen to sustain them aerobically. These processes use anaerobic metabolism. Carbohydrates are particularly versatile because they can be broken down either aerobically or anaerobically. The differences between aerobic and anaerobic metabolism are summarized in Table 3-2.

What Is the Role of Carbohydrates in Energy Metabolism?

Carbohydrates are vitally important in sustaining the body's metabolic machinery. Not only are they unique in their ability to generate ATP aerobically or anaerobically, but they also play a critical role in the oxidation of lipids. The expression "fats burn in a carbohydrate flame" refers to the essential contribution of carbohydrates in keeping aerobic energy systems running so that fats may be broken down completely. Incomplete breakdown of lipids results in accumulation of ketone bodies, which can have serious health complications.

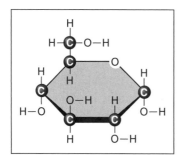

FIGURE 3-4   The chemical structure of glucose.

glycolysis

Metabolic process that may occur aerobically or anaerobically, in which glucose or glycogen is broken down to pyruvate (aerobic) or lactate (anaerobic) and ATP is produced.

citric acid cycle

Aerobic metabolic process that fully breaks down carbohydrate, fat, or protein to generate ATP, form carbon dioxide and water, and remove hydrogen atoms for the electron transport system; also known as the tricarboxylic or Krebs cycle.

Glycolysis

After dietary carbohydrates are digested and absorbed, they are reduced to glucose, the form in which carbohydrate circulates throughout the body and is used by the cells. The chemical formula for glucose is $C_6H_{12}O_6$, meaning that it is composed of six carbon atoms, twelve hydrogen atoms, and six oxygen atoms. The chemical structure of glucose is illustrated in Figure 3-4. Glucose may also form long chains to be stored in the muscles and liver as glycogen.

Glycolysis is the breakdown of glucose. It occurs very rapidly in the cytoplasm of the cell whether or not oxygen is present. As illustrated in Figure 3-5, glycolysis is a ten-step process (eleven steps when the process starts with glycogen, which is stored carbohydrate) that ends with the production of pyruvate. During anaerobic metabolism, when the body's energy demands are greater than the rate at which energy can be generated aerobically, lactic acid is formed from pyruvate.

Anaerobic glycolysis cannot continue indefinitely, in part because the production of lactic acid inhibits a key enzyme (phosphofructokinase) and ultimately slows metabolism. In addition, lactic acid accumulation causes pain or burning in the working muscles, which slows down all but the hardiest among us. Finally, anaerobic glycolysis releases only about 5% of the potential energy from glucose, so its ATP-generating capacity is limited. Fortunately, the energy that remains in glucose can be extracted via an aerobic process called the **citric acid cycle**.

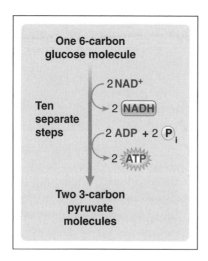

FIGURE 3-5   The original 6-carbon glucose molecule is split into two three-carbon sugars (pyruvate) during the 10-step process called glycolysis.

Source: From Sherwood. Human Physiology, 7E. p. 33. © 2010 Brooks/Cole, a part of Cengage Learning, Inc. Reproduced by permission. www.cengage.com/permissions.

Citric Acid Cycle

Carbohydrates, fats, and the amino acids that make up proteins synthesize ATP through the citric acid cycle. Carbohydrates enter the cycle as pyruvate, the end point of glycolysis. Accumulated lactic acid may be recycled to form more pyruvate as well as several other substances. The three-carbon pyruvate prepares to enter the citric acid cycle by losing a carbon atom and joining with a molecule of coenzyme A (CoA) to form the two-carbon acetyl CoA, which next combines with oxaloacetate, a 4-carbon compound, to form citrate. Citrate proceeds through a series of reactions in which hydrogen and carbon dioxide are released, ending with the formation of more oxaloacetate and continuation of the cycle. The citric acid cycle is pictured in Figure 3-6.

Carbohydrate, fat, and protein all contain carbon, hydrogen, and oxygen. As these nutrients progress through energy metabolism, they lose their hydrogen atoms. Hydrogen is an electron carrier and a potent source of energy for generating more ATP. Hydrogen atoms released in the citric acid cycle are transferred to oxygen, resulting in the formation of water and more ATP. This process is known as **electron transport**, or the respiratory chain.

electron transport *Series of reactions whereby hydrogens (electron carriers) are transferred by NADH and $FADH_2$ to a series of other substances on the inner membrane of the mitochondria and in the process generate energy to synthesize ATP; also known as the respiratory chain.*

oxidative phosphorylation *Final stage of aerobic energy metabolism, in which ATP is synthesized from the energy generated in the electron transport system.*

Electron Transport and Oxidative Phosphorylation

You have probably noticed the acronyms NAD, NADH, FAD, and $FADH_2$ in many of the figures presented so far. These elements are important for energy metabolism and are essential for electron transport and **oxidative phosphorylation**. The mitochondria in cells contain carriers that can remove the electrons from hydrogen atoms. These carriers are nicotinamide adenine dinucleotide (NAD) and flavin adenine dinucleotide (FAD). When NAD and FAD are bonded to hydrogen, the results are expressed as NADH + H, and $FADH_2$, respectively. In the process of electron transport or the respiratory chain, the electrons carried by NADH and $FADH_2$ are transferred to a series of other substances on the inner membrane of the mitochondria and generate energy to synthesize ATP. The ultimate synthesis of ATP is oxidative phosphorylation.

What Is the Role of Lipids in Energy Metabolism?

Lipids are a vast source of potential energy. Several hundred thousand calories of lipid may be stored in the adipose and the muscle tissue, significantly more than the few thousand calories of carbohydrate in the muscles and the liver. Although stored fat is lighter in weight than stored carbohydrate, it contains more than twice the calories per gram of weight.

Triglycerides, the primary fats in our diet, are the form in which fat is stored. A triglyceride molecule consists of one molecule of glycerol, a carbohydrate-like substance, in combination with three molecules of fatty acid. Glycerol proceeds through glycolysis like glucose, but fatty acids are the main sources of energy from lipids and must be broken down in the mitochondria. When energy is needed and aerobic conditions exist, fatty acids may enter the citric acid cycle, where they are metabolized just like carbohydrates. But, before fatty acids can enter the citric acid cycle, they are prepared in a process called beta oxidation.

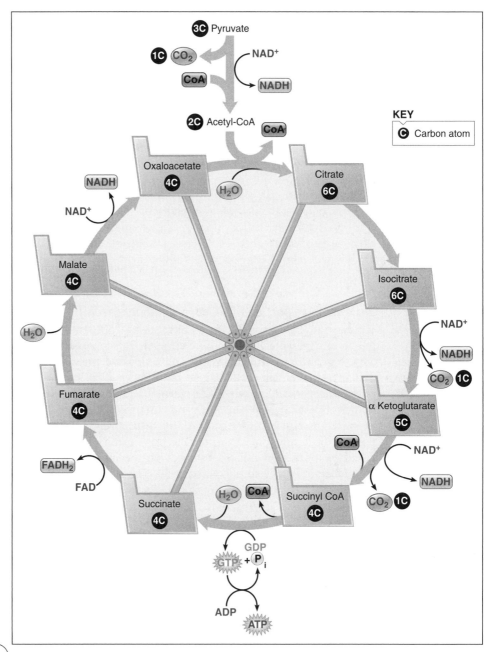

FIGURE 3-6 The citric acid cycle (also known as the Krebs cycle).

Source: From Sherwood. Human Physiology, 7E. p. 34. © 2010 Brooks/Cole, a part of Cengage Learning, Inc. Reproduced by permission. www.cengage.com/permissions.

Beta Oxidation

Unlike glucose, with its six carbons, the carbon chains of a fatty acid can be impressively long. The formula for stearic acid, a common dietary fatty acid, is $C_{57}H_{110}O_6$. Because lipids, like carbohydrates, enter the citric acid cycle as the two-carbon acetyl CoA, quite a transformation must occur to turn a 57-carbon compound (or even a 24-carbon compound) into several two-carbon compounds!

   Beta oxidation is an aerobic process that happens in the mitochondria. Carbon atoms are successively removed from the fatty acid carbon chain to produce acetyl, the two-carbon compound that joins with CoA to form acetyl CoA. Each fatty acid undergoes beta oxidation until all of its carbon atoms are removed in pairs. During the removal of carbons, hydrogen atoms are released and transported to the electron transport system, where energy is generated for oxidative phosphorylation. Beta oxidation's place in energy metabolism is pictured in Figure 3-7.

Fats Burn in a Carbohydrate Flame

Nutritionists and exercise physiologists frequently note that "fats burn in a carbohydrate flame." This observation refers to the crucial role of carbohydrate in keeping the citric acid cycle operating. When carbohydrates are insufficient (for example, in anorexia nervosa, fasting, and type 1 diabetes), oxaloacetate may be used to form glucose for cells

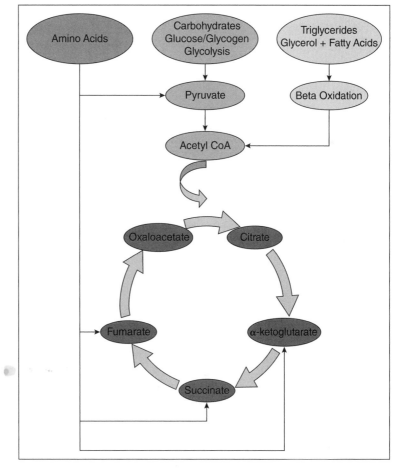

FIGURE 3-7 The interrelationships of energy nutrients in metabolism.

that require or prefer glucose as a fuel. The result is insufficient oxaloacetate to combine with acetyl CoA to form citrate. Under these conditions, acetyl CoA is instead converted to one of the ketones—acetoacetate, β-hydroxybutyrate, or acetone. Ketones are excreted in urine but may build up in the blood and tissues, resulting in the condition called **ketosis**. Acetone has a fruity smell that gives the breath of people in ketosis a characteristic odor. Uncontrolled build-up of ketones in body fluids may be fatal. So, although fats are a tremendous source of potential energy, they require carbohydrates to release their energy most efficiently.

ketosis *Build-up of ketone bodies in body tissues resulting from incomplete oxidation of fats.*

What Is the Role of Protein in Energy Metabolism?

The proteins that make up our bodies include muscle fibers, blood components, hormones, enzymes, and parts of cell structures. In the average adult, about 25% of total stored energy is in the form of protein. The body keeps its stored protein under fairly tight control and does not increase stores simply in response to an increased intake of dietary protein. Individuals who wish to increase their protein stores, such as bodybuilders trying to build muscle mass, require not only a good diet but also tissue-building drugs or intense physical training to do so.

Energy metabolism should never be the primary function of proteins. Amino acids, which are the building blocks of proteins, are better used for the growth and maintenance of body tissues. However, several amino acids can be used to support the body's energy needs. Amino acids can enter aerobic metabolic pathways as pyruvate, α-ketoglutarate, succinate, fumarate, and oxaloacetate.

Kzenon, 2010/Shutterstock.com

Physical training and adequate protein intake are needed to build muscle.

$$RQ = CO_2 \div O_2$$
RQ for carbohydrate $= 1.0$
RQ for fat $= 0.70$
RQ for protein $= 0.81$

FIGURE 3-8   The respiratory quotient reveals which nutrient—carbohydrate, fat, or protein—is being primarily burned to fuel metabolism.

How Do We Know Which Nutrient Is Being Used for Fuel?

respiratory quotient (RQ)
Indicator of the proportion of carbohydrate, fat, and protein being used to produce energy; calculated by dividing liters of carbon dioxide produced by liters of oxygen consumed (CO_2/O_2).

If you were to measure the volume of oxygen consumed and the volume of carbon dioxide produced as an individual went about daily activities, you could learn a great deal about his or her metabolic and physiologic functioning. One particularly interesting bit of information you could obtain is the relative contribution of carbohydrate, fat, and protein to energy expenditure. In other words, you could find out which nutrient is being primarily burned to fuel metabolism.

The **respiratory quotient (RQ)** provides this information. The RQ is calculated by dividing liters of carbon dioxide produced by liters of oxygen consumed, as illustrated in Figure 3-8. Scientists have determined that, at the cellular level, when only carbohydrates are being oxidized, the RQ is 1.0; when only fats are being oxidized, the RQ is 0.70; and, when only proteins are being oxidized, the RQ is 0.81.

Carbohydrate RQ

Calculating the RQ of carbohydrates is quite simple due to the "unity" of glucose—there are six carbon and six oxygen atoms and exactly twice as many (twelve) hydrogen atoms. The result is that the amount of oxygen added to oxidize the glucose molecule is exactly the same as the amount of carbon dioxide produced during metabolism:

$C_6H_{12}O_6$ + $6O_2$ → $6CO_2$ + $6H_2O$
glucose   oxygen consumed   carbon dioxide produced   water

$RQ = 6CO_2 \div 6O_2 = 1.0$

Fat RQ

The fatty acid molecule does not simply lack the unity of glucose; the amount of hydrogen relative to oxygen in the molecule is also disproportionately large. This means that a great deal more oxygen must be added to oxidize a fat than to oxidize a carbohydrate. Consequently, the fat RQ is lower, as in palmitic acid, for example:

$C_{16}H_{32}O_2$ + $23O_2$ → $16CO_2$ + $15H_2O$
palmitic acid   oxygen consumed   carbon dioxide produced   water

$RQ = 16CO_2 \div 23O_2 = 0.69$

Protein RQ

Because proteins contain nitrogen as well as carbon, hydrogen, and oxygen, calculating the protein RQ is not quite as straightforward as the RQs of carbohydrate and fat. Nitrogen, some oxygen, and some carbon in the protein molecule are used to form urea, which can ultimately be excreted in urine. But this means that not all of the components of protein are available for oxidation; albumin illustrates this fact:

$$C_{72}H_{112}N_2S + 77O_2 \rightarrow 63CO_2 + 38H_2O + SO_3 + 9CO(NH_2)_2$$

| albumin | oxygen | carbon dioxide produced | water | sulfur trioxide | urea |

$$RQ = 63CO_2 \div 77O_2 = 0.818$$

Caloric Value of Oxygen

People rarely use only one energy nutrient at a time to fuel their activities. Most people who consume a diet that contains carbohydrate, fat, and protein have an RQ of 0.82 as they go about their daily activities. Physiologists have determined that when the RQ = 0.82, each liter of oxygen consumed is equivalent to 4.825 kcal of energy expended. So, if you were to measure someone's oxygen consumption and carbon dioxide production while they slept, worked, or played, then you could calculate their caloric expenditure. For example, let's assume a recreational bowler whose RQ was 0.82 was using 0.85 l of oxygen per minute. If he bowled for 30 minutes, then he would use 25.5 l of oxygen.

$$(25.5 \text{ l } O_2)(4.825 \text{ kcal/l}) = 123 \text{ kcal expended in 30 minutes of bowling}$$

You may be surprised to learn that the caloric value of oxygen varies depending on whether carbohydrate or fat is the primary fuel being used. (We will assume that protein is not the primary fuel.) Table 3-3 shows how the caloric value of oxygen changes with different fuel combinations. For example, if the bowler had an RQ of 0.89, this would indicate that 62.6% of his fuel was being provided by carbohydrate, and 37.4% by fat. In this case, the caloric value of oxygen is 4.912 kcal/l. The result is an energy output of 147 kcal in 30 minutes. To calculate the caloric value of physical activity, an RQ of 0.82 is generally assumed. Still, this example demonstrates that the same activity may promote a slightly greater (or smaller) caloric expenditure in two similar people.

How Do Dietary Fats Act as Key Regulators of Energy Balance?

You are probably comfortable with the idea that caloric intake must be in balance with caloric output for body weight to remain stable. Now, consider an additional dimension to the energy balance equation: For body weight to remain stable, the nutrients that are burned to sustain metabolism and activity must be oxidized in proportion to their presence in the diet. In other words, the person who consumes a diet that is 45% fat must be oxidizing at least that much fat, or the excess will be stored. So, we can also state the

TABLE 3-3 Caloric Value of Oxygen at Different Respiratory Exchange Ratios

Nonprotein RQ	Proportion of calories from carbohydrate (%)	Proportion of calories from fat (%)	Caloric value of oxygen (kcal/liter)
0.70	0.0	100.0	4.686
0.71	1.4	98.6	4.690
0.72	4.8	95.2	4.702
0.73	8.2	91.8	4.714
0.74	11.6	88.4	4.727
0.75	15.0	85.0	4.739
0.76	18.4	81.6	4.752
0.77	21.8	78.2	4.764
0.78	25.2	74.8	4.776
0.79	28.6	71.4	4.789
0.80	32.0	68.0	4.801
0.81	35.4	64.6	4.813
0.82	38.8	61.2	4.825
0.83	42.2	57.8	4.838
0.84	45.6	54.4	4.850
0.85	49.0	51.0	4.863
0.86	52.4	47.6	4.875
0.87	55.8	44.2	4.887
0.88	59.2	40.8	4.900
0.89	62.6	37.4	4.912
0.90	66.0	34.0	4.924
0.91	69.4	30.6	4.936
0.92	72.8	27.2	4.948
0.93	76.2	23.8	4.960
0.94	79.6	20.4	4.973
0.95	83.0	17.0	4.985
0.96	86.4	13.6	4.997
0.97	89.8	10.2	5.010
0.98	93.2	6.8	5.022
0.99	96.6	3.4	5.034
1.00	100.0	0.0	5.047

Source: Adapted from Carpenter, T. M. (1921). *Tables, factors, and formulas for computing respiratory exchange and biological transformations of energy.* Washington, DC: Carnegie Institution of Washington, Publication 303, p. 104.

energy balance equation as: Energy intake = Energy expenditure ± (Fat intake – Fat oxidation).

Looking at it this way, obesity can develop from a failure to balance fat intake with fat oxidation.[1]

The factors that determine which nutrients are oxidized and in what proportion are dietary intake, energy needs, weight status, and probably heredity. An important point is that nutrient oxidation depends not only on which nutrients are consumed but also on which nutrients are burned, and this may be different for lean and obese individuals.

Dietary Protein and Carbohydrate Influence Nutrient Oxidation

Protein and carbohydrate are at the top of the heap when it comes to nutrients that are thermic (i.e., they generate heat), readily oxidized, and oxidized in proportion to their presence in the diet. The balance between protein and carbohydrate intake and oxidation is tightly regulated.

Let's look closely at carbohydrate because it should make up most of your caloric intake every day. After you eat, most dietary carbohydrate is either used immediately for cellular functions or converted to glycogen for storage in the liver and the muscles. On days when you eat a lot of carbohydrate, more carbohydrates are oxidized. A metabolic process called diet-induced thermogenesis (DIT), covered later in this chapter, may increase. The elevation in your blood glucose and the increase in your glycogen stores put the brakes on appetite (in most people) so you do not eat more than you can use. As a result, glycogen storage in the average person rarely reaches maximum capacity, which is about 1,500 carbohydrate calories.

Balance between carbohydrate intake and storage could also occur if excess carbohydrate were stored as fat. It is theoretically possible for this to occur. However, having enough excess carbohydrate to convert to fat would require an extremely high carbohydrate intake for several days. Some experts have estimated that an intake of 2,000–2,500 kcal of carbohydrate over several days, after carbohydrate stores were at maximum, would be required before carbohydrate would be converted to fat.[1,2] Because few people consistently eat this much carbohydrate, balancing the energy balance equation by storage of carbohydrate as fat is unlikely.

Dietary Fat Influences Nutrient Oxidation

Fat oxidation is not as tightly regulated as carbohydrate or protein oxidation. As you eat more fat, fat oxidation does not increase proportionately and appetite does not necessarily diminish. Fat stores are not limited in size. Therefore, dietary fat affects energy balance—particularly energy storage—to a far greater extent than does the intake of the other nutrients.

But what if you decreased carbohydrate intake? Would this force more fat to be burned? Many popular high-fat, low-carbohydrate diets would lead you to believe so. Unfortunately, replacing dietary carbohydrate with fat promotes obesity for several reasons:

1. If the diet is high in fat, then it is low in carbohydrate, which reduces the size of glycogen stores and stimulates hunger. Although the fat in your diet creates a feeling of stomach fullness, fats are overall less satiating than carbohydrates. Fats lack a

mechanism for gradually turning off appetite. As a result, people who regularly con-sume a high-fat diet tend to eat more.

2. Carbohydrates control the rate of fat oxidation ("fats burn in a carbohydrate flame"), and when carbohydrate stores run low, fat oxidation slows.

3. A diet rich in fat is also rich in energy. This increases the size of the fat stores unless energy expenditure is increased. And, remember, those fat stores are unlimited in their potential size.

4. Fat intake does not stimulate an equal and proportional increase in fat oxidation. When fat oxidation is less than fat intake, fat storage results.

Not all fats have the same metabolic effects. People who eat a diet rich in saturated fats oxidize less fat than those whose diet is based on polyunsaturated and monoun-saturated fats. Saturated fats are mainly found in animal products, like butter, cheese, and beef. Polyunsaturated fats are mainly found in plant products, like safflower oil and corn oil. Monounsaturated fats are also plant fats, found predominantly in nuts, olive oil, and canola oil. Fat oxidation is greatest when more polyunsaturated and monounsaturated fats are eaten than saturated fatty acids.[3-5]

Another type of fat that seems to vary the energy response to fat ingestion is the medium-chain triglyceride. Medium-chain triglycerides have a carbon chain length of six to ten carbons. Their thermic effect is 10–15% greater than the long-chain triglycerides that make up most of the diet.[6] Medium-chain triglycerides are much like carbohydrates in that they are rapidly oxidized and promote a strong satiety effect. In studies in which medium-chain triglycerides have been used as diet supplements (because they are not found in common dietary fats), energy intake is reduced and fat oxidation increased.[7-8] These effects on fat oxidation have not yet been shown to be adequate to cause signifi-cant weight loss.

Energy Need Influences Nutrient Oxidation

Although the amount of fat in the diet may not affect lipid oxidation, energy need certainly does. People who expend more energy in exercise and physical activity oxidize more fat. There are two reasons for this: (1) Fat stores are much larger than glycogen stores and can be called on to support increased energy needs for a considerable time; and (2) the adi-pocytes (fat cells) are sensitive to catecholamines, which are the hormones released when people exercise. Low-intensity activities like walking, moderate-intensity exercise like slow jogging, and even strength training are effective in increasing fat oxidation and lowering carbohydrate oxidation.[2,9]

There are, however, differences between people in the nutrients oxidized during exercise. Some people appear to be "high fat oxidizers," and others oxidize less fat and more carbohydrate. High fat oxidizers expend more energy during and after exercise and, because they burn more fat than carbohydrate, are less stimulated to eat after exercise.[10] Exercise training may help people become more fat-oxidizing. Moderate intensity exercise has been shown to be the most fat-oxidizing.[11] One pro-gram that combined moderate caloric intake with aerobic exercise and strength train-ing three times per week lowered the RQ of overweight women after 12 weeks.[12] Calorie reduction alone or calorie reduction plus only aerobic exercise did not change their RQs.

Corey Sundahl/iStockphoto

Physical activity can increase fat oxidation and aid in weight management.

Lean and Obese People Differ in Nutrient Oxidation

slow-twitch muscle fibers
Skeletal muscle fibers that are rich in mitochondria and preferentially use aerobic metabolism; also called type I muscle fibers.

fast-twitch muscle fibers
Skeletal muscle fibers that have a high capacity for anaerobic metabolism; also known as type II fibers.

Chapter 1 proposed that one explanation for increased rates of obesity in the American population is a higher intake of dietary fats. But intake of fat kilocalories is only part of the problem. If fats are oxidized, then they are not stored. Research indicates that some people, mainly those who remain lean throughout life, just seem better able to increase fat oxidation in response to a high fat intake.[13] The reasons for this are not understood, although there may be biological factors that account for differences in fat oxidation between lean and obese individuals. Some of these factors are under genetic control:

- Muscle fiber type may vary between lean and obese people. The skeletal muscles are made up of either **slow-twitch** (type I) muscle fibers, which tend to be more capable of aerobic metabolism and oxidize more fat, or **fast-twitch** (type II) muscle fibers, which are more anaerobic and oxidize more carbohydrate. The person with a lower ratio of slow-twitch to fast-twitch muscle fibers may have reduced fat oxidation. Some researchers have found fatter people to have more fast-twitch fibers.[14]
- Muscle lipoprotein lipase, an important enzyme that controls the use of fats as fuels in muscle, varies between individuals.[15]

- Obese individuals may produce less epinephrine, the catecholamine responsible for increasing the release of fat from adipose tissue. Or they may have less sensitivity to it.[16]

So, what does all of this information mean to the person trying to lose weight or prevent weight gain? Put simply, people who oxidize less fat relative to carbohydrate are more likely to gain weight. Carbohydrate oxidation increases in proportion to carbohydrate intake, but fat oxidation does not. Fat oxidation may be increased through exercise, although there is individual variation in this. The relationship between low fat oxidation and obesity has been called as strong as the relationship between obesity and coronary artery disease.[17]

The Metabolic Fate of Ingested Food: Summary

Carbohydrate, fat, and protein may be used to provide energy for aerobic metabolic processes. Only carbohydrate generates ATP anaerobically. In the laboratory, measurement of oxygen consumption and carbon dioxide production allows us to determine which nutrient combination is being used for energy and thereby to determine the caloric value of particular activities. While carbohydrate and protein are metabolized in close proportion to their intake, a high fat intake does not similarly result in increased oxidation of fat, and energy storage results.

ENERGY EXPENDITURE: METABOLIC RATE

18

basal metabolic rate (BMR) *Sum of the oxygen consumption of the active tissues of the body.*

Metabolic rate is defined as the rate at which the body expends energy to support its vital functions. **Basal metabolic rate**, or BMR, represents the activity of the body's active tissues when an individual is resting but is not asleep. An individual who has not eaten or exercised recently and who is reclining on a sofa, remote control in hand, would be just about at the basal level of activity. But, even though she is engaging in no obvious physical exertion, much cellular activity is under way—for example, heartbeat, production of ATP, and enzymatic reactions.

The BMR is measured under conditions of low cellular activity:

- Upon first waking (low physical activity)
- Twelve to fourteen hours after a moderate meal (no digestive activity)
- In a comfortable environment (no shivering or excessive perspiration)

The BMR is usually slightly lower than the RMR, which is measured at any time of day, even several hours after a meal or physical activity. The RMR is used more commonly than the BMR to determine energy expenditure because it is so much more practical to measure under less stringent conditions. The term resting energy expenditure (REE) is also used to refer to this component of energy out.

How Is Resting Energy Expenditure Determined?

When the cells do their work, heat is produced. Remember the definition of a kilocalorie: the amount of heat required to raise the temperature of 1 kg of water 1° Celsius. In addition to computing the caloric value of food with a bomb calorimeter, human caloric expenditure can be measured using a calorimetry chamber.

Direct calorimetry is the measurement of heat generated as a result of metabolism. It is the most accurate technique for measuring resting energy expenditure. The individual being studied remains in a small chamber called a **room calorimeter**. Water circulates within the walls of the chamber. For measurement of BMR, the person lies quietly on a cot for 30–60 minutes while heat production is determined by changes in the water temperature. When individuals stay in the chamber for 24 hours, caloric expenditure needed to support the typical activities of daily living—total daily energy expenditure—can also be determined. This accurate but very expensive method of assessing energy expenditure is reserved for the research laboratory.

Indirect calorimetry is a more practical method of estimating heat production. The quantity of oxygen inhaled and carbon dioxide produced is measured during short periods of time as the individual breathes through a face mask, a mouthpiece, or a hood system while resting quietly in the laboratory. Caloric expenditure can be calculated from the RQ or from the caloric value of oxygen, using the formula 1 l of O_2 = 4.825 kcal. Indirect calorimetry provides an acceptable estimate of metabolic rate because there is a known relationship between heat production and oxygen consumption. Portable spirometers can also be used to estimate energy expenditure during a variety of activities.

Special metabolic chambers, similar to room calorimeters, are also available at research facilities. Oxygen consumption and carbon dioxide production of a person living in the chamber for a short time are measured and used to estimate heat production. In addition, the doubly labeled water technique, discussed later in this chapter, is a form of indirect calorimetry that allows measurement of energy expenditure for almost a month.

The RMR can also be estimated with mathematical equations, several of which are presented in Table 3-4. These equations may over- or underestimate RMR by a considerable amount, and they should therefore be interpreted with caution. The most accurate equations are considered to be those from Mifflin and colleagues. Researchers found these to estimate RMR within ±10% in 82% of men and women studied; the equations were as likely to overestimate as underestimate RMR.[18] The Owen equations were more likely to underestimate RMR and were ±10% accurate in 73% of cases. Measuring RMR is always preferable when an accurate determination of caloric expenditure is needed.

direct calorimetry
Determination of caloric expenditure by measurement of heat generated as a result of metabolism.

room calorimeter *Small chamber in which heat production can be measured as an individual goes about normal activities of daily living.*

indirect calorimetry
Estimation of caloric expenditure by measurement of oxygen consumption alone or in combination with carbon dioxide production; also called spirometry.

What Factors Affect RMR?

Body size accounts for about 80% of our metabolic rates.[19] Our metabolic engine is primarily driven by the brain, liver, heart, and kidneys, four organs that make up about 6% of body weight but that account for about 60% of metabolic rate.[20] The variability in metabolic rate that exists between people is accounted for by the weight of these organs, in addition to several other things, which are summarized in Table 3-5. Keep these in mind when you interpret RMR derived from prediction equations.

Body Size and Composition Are Important Determinants of RMR

You may have noticed in Table 3-4 that prediction equations for RMR incorporate weight and sometimes height. This is because body size is usually considered to be the most

TABLE 3-4   Prediction Equations for Calculating RMR (kcal/day)

Males	Females
World Health Organization equations[1]	
Ages 10–18 years	Ages 10–18 years
(17.5 X weight in kg) + 651	(12.2 X weight in kg) + 746
Ages 18–30 years	Ages 18–30 years
(15.3 X weight in kg) + 679	(14.7 X weight in kg) + 496
Ages 30–60 years	Ages 30–60 years
(11.6 X weight in kg) + 879	(8.7 X weight in kg) + 829
Over 60 years	Over 60 years
(13.5 X weight in kg) + 487	(10.5 X weight in kg) + 596
Modified Harris-Benedict equations[2]	
65 + (13.40 X weight in kg) +	447 + (9.25 X weight in kg) +
(4.96 X height in cm) − (5.82 X age in years)	(3.10 X height in cm) − (4.33 X age in years)
Owen formulas	
Ages 18–82 years[3]	Ages 18–65 years[4]
879 + (10.2 X weight in kg)	Athletes: 50.4 + (21.1 X weight in kg)
	Nonathletes: 795 + (7.18 X weight in kg)
Mifflin et al.[5]	
(9.99 X weight in kg) + (6.25 X height in cm) − 4.92 X age + 5	(9.99 X weight in kg) + (6.25 X height in cm) − 4.92 X age − 161

[1]Source: Report of a Joint FAO/WHO/USU Expert Consultation. (1985). *Energy and protein requirements.* WHO Geneva, Switzerland.

[2]Source: Frankenfield, D. C., Muth, E. R., & Rowe, W. A. (1998). The Harris-Benedict studies of human basal metabolism: History and limitations. *Journal of the American Dietetic Association, 98*(4), 439–444.

[3]Source: Owen, O. E., Holup, J. L., D'Alessio, D. A., Craig, E. S., Polansky, M., Smalley, K. J., Kavle, E. C., Bushman, M. C., Owen, L. R., Mozzoli, M. A., Kendrick, Z. V., & Boden, G. H. (1987). A reappraisal of the caloric requirements of men. *American Journal of Clinical Nutrition, 46,* 875–885.

[4]Source: Owen, O. E., Kavle, E., Owen, R. S., Polansky, M., Caprio, S., Mozzoli, M. A., Kendrick, Z. V., Bushman, M. C., & Boden, G. (1986). A reappraisal of caloric requirements in healthy women. *American Journal of Clinical Nutrition, 44,* 1–19.

[5]Source: Mifflin, M. D., St. Jeor, S. T., Hill, L. A., Scott, B. J., Daugherty, S. A., & Koh, Y. O. (1990). A new predictive equation for resting energy expenditure in healthy individuals. *American Journal of Clinical Nutrition, 51,* 241–247.

TABLE 3-5   Factors that Affect Metabolic Rate

- Body size
- Thyroid hormone production
- Nutritional status
- Heredity
- Gender
- Age
- Race

important determinant of metabolic rate. The higher RMRs of large people, whether obese or extremely muscular, are explained by greater muscle mass, larger heart and liver, greater volume of blood, and, in obesity, enlarged fat mass. The body's lean body mass (LBM), made up of organs and muscles, is particularly metabolically active. Skeletal muscles account for about one-quarter of the RMR. This explains at least partly the higher RMRs of males and the reduced RMRs seen in aging. Conversely, a reduction in body mass decreases RMR. For example, when people lose weight, including the severe weight loss seen in anorexia nervosa, their RMRs decline.

Thyroid Hormone

Thyroid activity has a pronounced effect on the RMR. Using dietary iodine, the follicular cells of the thyroid gland produce thyroxine, also called tetraiodothyronine (T_4), and triiodothyronine (T_3), which are collectively known as thyroid hormone. The thyroid gland has both rich blood supply, which permits high levels of thyroid hormone to be circulated rapidly, and rich sympathetic innervation, which permits rapid change in the rate of hormone secretion. Thyroid hormone secretion affects virtually every tissue, organ, and physiological function, resulting in increased oxygen consumption and energy expenditure.

Excessive secretion of thyroid hormone causes a condition called hyperthyroidism. Usually hyperthyroidism results from Grave's disease, in which excessive production of thyroid-stimulating immunoglobulin causes thyroid hormone to be oversecreted. Individuals with hyperthyroidism have increased appetite but experience weight loss due to greatly accelerated metabolic rate. When too little thyroid hormone is secreted, the result is hypothyroidism, characterized by reduced RMR. Hypothyroidism may be accompanied by weight gain and obesity if activity levels are also depressed.

Nutritional Status

Undernutrition caused by starvation, fasting, or low-calorie diets will lower a person's resting energy expenditure. This probably occurs for two reasons: (1) loss of weight or muscle tissue resulting from undernutrition reduces body mass and, therefore, reduces the RMR; and (2) the brain responds to insufficient caloric intake by attempting to conserve energy. Exercise does not necessarily counteract the metabolic effects of food restriction. In contrast, overeating, particularly with carbohydrates, will elevate the metabolic rate even when weight gain does not occur.

Genetic Influences on RMR

Between one-fourth and one-half of an individual's metabolic rate can be explained by heredity.[21] This might suggest to you that in families where obesity is common, an obese parent with a low RMR could pass on this tendency to the children. Some researchers have found lower RMRs in people with one obese parent, but studies of Native Americans and other population groups in which obesity is common have not consistently shown a familial tendency toward reduced RMR. Lower RMR is probably one of many genetic characteristics that interact to promote obesity in certain families.

Other Factors

Gender, age, and race also influence metabolic rate. Men have slightly higher RMRs than women, even when their greater muscle mass is considered. Male caloric expenditure is greater than female by about 50 kcal/day.[21] Children have higher RMRs per kg of body weight than adults for reasons not completely understood. Older adults generally have lower RMRs than younger people, partly due to declining physical activity with age, which reduces muscle mass. There is also an aging effect, over and above any loss of muscle mass, which lowers RMR by approximately 1–3% per decade after age 20.[20]

There may also be racial/ethnic differences in metabolic rate. Some studies of children have identified RMRs that are 4–7% lower in African American than in Caucasian girls when factors like age, fat mass, and **fat-free mass** are comparable.[22] Recently researchers found that African American women have less trunk LBM and more limb LBM than white women.[23] Because LBM in the trunk includes those extremely metabolically active organs mentioned earlier in this chapter, this might explain racial differences in the prevalence of obesity and the ability to lose weight.

fat-free mass *Proportion of body weight made up of bone and other dense connective tissue, body water, and the fat-free portion of the body's cells.*

Does Exercise Elevate RMR?

Energy expenditure increases during exercise, which may tempt you to believe that this has a carry-over effect on metabolic rate. Certainly, energy expenditure remains elevated immediately after physical activity, and sometimes for hours. This is attributed to **excess postexercise oxygen consumption (EPOC)**. Sometimes called oxygen debt or recovery oxygen, EPOC is defined as oxygen used during recovery from exercise in excess of resting oxygen consumption. EPOC occurs for two reasons: (1) Body temperature and cardiorespiratory functions are heightened from the exercise, which increase oxygen consumption; and (2) additional energy is needed to return body systems to their pre-exercise condition—to restore ATP and remove lactic acid and hormones, for example. Strenuous anaerobic exercise, including high-intensity weight training, causes a greater rise in EPOC than moderate aerobic exercise.

excess postexercise oxygen consumption (EPOC) *Increased oxygen consumption (and, therefore, energy expenditure) for a period of minutes or hours after physical activity; sometimes called recovery oxygen or oxygen debt.*

Of great interest is whether REE eventually becomes higher as a result of regular participation in exercise. If elevated REE persisted for several days after exercise, then this might suggest that exercise could have long-term effect on metabolic rate. Some researchers have reported elevated RMRs up to 48 hours after strenuous (but not moderate) exercise, resulting in an increased caloric expenditure of 100–200 kcal/day.[24] Studies with sedentary older adults who engage in weight training or aerobic exercise also demonstrate increases in RMR of 7–10% after training.[25–27] Most impressive is that the increased RMR occurs without an equivalent increase in muscle mass, possibly due to higher protein turnover in the muscle and elevated secretion of norepinephrine.

But, in many cases, the RMR is the same 2–3 days after exercise as it was before. There are several explanations for the contradictions in research about the effects of exercise on RMR:

- If RMR is measured too soon after exercise, EPOC, not RMR, accounts for increased energy expenditure.
- If RMR is measured too long after exercise, then any possible effect will be missed.
- Variations in the type, intensity, and duration of exercise sessions, as well as the overall length of the training program, no doubt have different effects on RMR.

- Effects on RMR may be greater among older adults, who have reduced RMRs due to the age effect, and in previously sedentary people, and may be less noticeable among younger or fitter adults.
- The type and quantity of food ingested after exercise may affect RMR.

Until new research settles this debate, we can conclude that exercise training may exert an influence on REE in some people. Scientific evidence is not yet conclusive enough to prescribe a specific exercise program for a particular individual to promote a predictable rise in REE.

What Is the Relationship Between RMR and Obesity?

Overweight/obese individuals have higher absolute RMRs than people who are not overweight or obese. This is not surprising because we know that body size is one factor accounting for metabolism. But is the increased RMR proportional to body size? And is the RMR per kilogram of fat-free mass (the most metabolically active tissues) the same, less than, or greater than that of nonobese people?

Most obesity researchers have discovered that even when RMR is expressed per kilogram of fat-free mass, few differences exist between most average-weight and overweight individuals. Nevertheless, some obese and formerly obese people do have lower RMRs. A meta-analysis of data from over 100 formerly obese individuals found the formerly obese to have RMRs 3%–5% lower than the never-obese.[28] Which came first—obesity or lower RMR—is not clear.

Among Pima Indians, low RMR (relative to body weight) is considered to be a good predictor of weight gain, which suggests depressed metabolism might come first.[29] Low metabolic rate may also predict difficulty in losing weight. A classic study of obese women who were long-time dieters found that those who had the lowest BMRs were least likely to lose weight.[30] Among the women who agreed to have their adipose tissue biopsied, a relationship existed between fat cell number and BMR. The women with more fat cells had higher BMRs; those who had fewer fat cells but the same percentage of body fat had lower BMRs and more difficulty losing weight. This finding supports the idea that obesity is a heterogenous condition, and some obese individuals may have lower RMRs.

Could repeated dieting lower REE? Animals that experience repeated cycles of food restriction and weight loss followed by access to food and weight regain become more metabolically efficient. Once food-restricted animals are allowed access to food after their "diet," they require fewer kilocalories to regain lost weight and, when they are again put on a cycle of food restriction to lose weight, they lose much more slowly.

This effect may apply to humans, too. The term yo-yo dieting is used to describe repeated cycles of dieting—weight loss—refeeding—weight regain—dieting. Wrestlers who practice yo-yo dieting have lower RMRs than nondieting wrestlers.[31] Obese dieters may also become more metabolically efficient. Researchers at one weight-loss clinic reviewed data from obese inpatients and outpatients who had lost weight at least twice on prescribed very-low-calorie-diet cycles over a 9-year period. Patients were included in the study only if they had regained at least 20% of the weight lost in the first weight-loss cycle. Obese individuals in this study lost weight more slowly in the second cycle of weight loss than in the first. Researchers surmised that energy expenditure was reduced, probably as a result of dieting.[32]

Case Study. The Lawyers, Part I

Mavis and Monique are both 30-year-old lawyers working for the same firm in a large metropolitan area. Mavis weighs 150 lb and is 5′ 3″ tall. Monique weighs 130 lb and is 5′ 7″ tall. Monique has been a dedicated runner since college, running at least 50 miles a week in the early morning hours before going to work. She also enjoys going to a health club on weekends to do weight training. Mavis rarely exercises, although she occasionally walks in her neighborhood on weekends. She goes on a calorie-restricted diet several times a year, when she is feeling particularly fat. Mavis and Monique occasionally have lunch together. Mavis frequently comments on Monique's ability to eat a hearty lunch and never gain weight. "If I ate that lasagna, it would go straight to my hips," Mavis remarked on Thursday.

1. Calculate Mavis's and Monique's BMIs. What BMI classification is each woman?
2. Estimate Mavis's and Monique's BMRs using the formulas in Table 3-3. Note that to use these formulas, convert weight to kg and height to cm.
3. What was the highest estimate of RMR? The lowest?
4. Are there factors not reflected in the formulas that might influence their metabolic rates?

Other studies of overweight, obese, and post-obese dieters show that caloric restriction lowers RMR and total daily energy expenditure. These reductions in energy expenditure are larger than would be expected from loss in weight.[33] Resting energy expenditure may still be reduced six years after caloric restriction.[34]

Metabolic Rate: Summary

adaptive thermogenesis
Heat production that follows exposure to the cold or food intake.

Basal or resting metabolic rate, also called resting energy expenditure, is the caloric expenditure needed to support the activities of the body's cells. Body size, thyroid hormone activity, nutritional status, gender, age, race, and heredity are among the factors known to influence the RMR. In addition, regularly engaging in high-intensity physical activity may permanently increase RMR. Some obese individuals may have lower RMRs, perhaps due to heredity or, in a few cases, to repeated dieting. Such a reduction in energy expenditure would make weight maintenance or weight loss difficult.

ENERGY EXPENDITURE: ADAPTIVE THERMOGENESIS

diet-induced thermogenesis (DIT) *Component of adaptive thermogenesis in which heat is produced by body cells, particularly those in brown adipose tissue, in response to eating.*

Adaptive thermogenesis is the production of heat that occurs either after exposure to cold or after eating, which contributes to total daily energy expenditure. Impaired adaptive thermogenesis has been identified as a factor in the development of obesity.[35]

Diet-induced thermogenesis (DIT) is the component of adaptive thermogenesis that occurs after eating. DIT is sometimes called the specific dynamic action of food or the thermic effect of food. Here is a brief summary of what is known about energy expenditure in response to eating, overfeeding, and underfeeding:

- When people or animals eat, both sympathetic nervous system activity and total daily energy expenditure increase.
- When people overeat, they do not necessarily gain weight in direct proportion to excess calories consumed.

brown adipose tissue (BAT)
Specialized form of adipose tissue that has the capacity to produce heat, which can result in dissipation of the energy from food as heat, rather than storage of energy.

positron emission tomography (PET)/computed tomography (CT) *An imaging technique in which an injected radioactive tracer will emit positively charged particles (positrons), permitting organs or body areas to be scanned to create a visual representation; when PET scans are superimposed on CT scans (or, with new machines, done at the same time), a detailed image is produced.*

uncoupling protein (UCP) *Proton transporter found in the mitochondria of several tissues; UCP "uncouples" the normal process of oxidative phosphorylation so that less ATP is formed and more heat is produced. UCP-1 is found in BAT, and UCP-2 and UCP-3 have been identified in storage fat as well as brain and muscle tissue.*

- When people fail to eat enough to support metabolism, their energy expenditure sometimes decreases more than would be expected, and they do not lose predicted amounts of weight.

Scientists have known for decades that small rodents exposed to the cold would increase their energy expenditure by a significant amount, sometimes two- to four-times normal levels. Humans also increase energy expenditure in response to the cold. Recently a study of obese men found that their energy expenditure increased—although it increased less than that of lean men—after spending an hour at 15°C (59°F) with only their face, hands, and ankles exposed.[36] This has led to speculation that our comfortably heated homes may be contributing to the increase in obesity rates!

Adaptive thermogenesis probably accounts for not more than 15% of daily energy expenditure. This amount is small but, if thermogenesis were faulty, a slow but steady increase in fat stores might occur over the course of many years.

What Is the Mechanism for Adaptive Thermogenesis?

Not everyone who overeats becomes obese. In some people, weight gain is staved off by engaging in regular physical activity. But adaptive thermogenesis might also prevent weight gain, due to a kind of fat tissue called **brown adipose tissue (BAT)**.

BAT is a Key Player in Adaptive Thermogenesis

BAT is a specialized form of fat tissue that is abundant in some animals, including hibernating bears and rodents that live in cold environments. Human infants have long been believed to have some BAT to help them maintain body temperature. A human autopsy study done in the early 1970s confirmed the presence of brown fat in children and adults, although it was found in small quantities.[37] Whether the BAT remains active or remains at all in adult humans has been a topic of disagreement for years. **Positron emission tomography/computed tomography (PET/CT)** scanning seems to have resolved the issue. PET/CT scans have clearly identified BAT in humans,[38] and BAT deposits are largest when people are exposed to cold.[39]

Brown fat is uniquely different from white adipose tissue, more commonly called storage fat. We are born with brown fat fully formed and ready to function, whereas storage fat develops after birth in response to energy surplus. Brown fat cells are smaller than storage fat cells, contain less fat, and are very rich in mitochondria. The mitochondria contain an **uncoupling protein (UCP-1)** that "uncouples" hydrogen atoms from the respiratory chain, producing heat but no ATP for work. The mechanism works like this:

- The animal needs heat to maintain core body temperature.
- When the animal is exposed to the cold, its sympathetic nervous system is activated and norepinephrine is released from nerve endings in the BAT.
- UCP-1 in brown fat is activated and uncouples oxidative phosphorylation, so that ATP is not generated for work but energy is released as heat.

(2lb) Food consumption can activate UCP-1 in a way that is similar to cold exposure. We all know people who appear to be able to eat anything and still maintain a trim physique. Thermic activity in brown fat could partly explain their failure to gain weight.

In humans, most BAT is located between the shoulder blades. When brown fat is stimulated, it produces heat and eventually may grow larger by increasing the number of cells (**hyperplasia**). This enlargement of BAT is what PET/CT scans have revealed in some people. Hyperplasia is also seen in animals, where it can be documented by dissection. Rats that live in a cold environment or that eat a palatable, varied diet (sometimes called a cafeteria diet) develop hyperplasia of brown fat. Conversely, rats kept in a thermoneutral environment or on calorie-restricted diets experience shrinkage of BAT. Hyperplasia occurs as already present precursor cells (preadipocytes) become mature cells.

A protein found in BAT, PRDM16, seems to be involved in preventing the formation of white adipose tissue from preadipocytes and enhancing the formation of brown fat cells.[40] PRDM16 promotes the expression of the genes that give brown fat its unique characteristics. If PRDM16 production could somehow be enhanced, scientists are intrigued to think that this might increase production of BAT, enhance thermogenesis, and treat or prevent obesity.

hyperplasia
Excess number of something—in this case, fat cells.

Diet-Induced Thermogenesis

When you eat, two components of DIT may contribute to increased energy expenditure: (1) obligatory thermogenesis, which involves increased energy expenditure immediately after eating and for a period of several hours to support the ingestion, digestion, absorption, and transport of food; and (2) facultative thermogenesis in which heat is produced in BAT. The magnitude of DIT may depend on the number of calories ingested. In a study of nonobese young men who consumed their daily caloric requirement in two meals, oxygen consumption was measured at an additional 10.21 l of oxygen per meal, for a total of 20.42 l of oxygen per day.[41] When their daily caloric intake was spread across four smaller meals, oxygen consumption per meal was smaller (8.31 l/meal), but overall oxygen consumption was greater, 33.4 l/day (8.31 × 4 meals). Translated into caloric expenditure, these men expended 99 kcal after eating two large meals and 161 kcal after eating four smaller meals (assuming the caloric value of 1 l of oxygen equals 4.825 kcal).

The type of nutrient ingested also affects DIT, with carbohydrates thought to have a greater thermic effect than fats.[42] This factor may be related to the metabolic energy required to convert glucose into its storage form—glycogen. Proteins have the greatest thermic effect of all the energy nutrients, probably due to synthesis of body proteins that occurs after protein ingestion.

Specific foods may also have a thermic effect. Caffeine, a stimulant in some soft drinks, chocolate, coffee, tea, and several over-the-counter remedies, can elevate metabolism by 3–4%, even when as little as 100 mg is consumed. This amount of caffeine is present in a cup of coffee or approximately two sodas. The chemical that gives hot red peppers their punch—capsaicin—is also known to have a pronounced thermic effect.[43] Both caffeine and capsaicin exert their metabolic effect by stimulating the sympathetic nervous system; however, it is unlikely that a diet of cola and hot peppers might effectively treat or prevent obesity. At least with caffeine, the thermic response differs markedly among obese people.[44]

Is Thermogenesis Defective in Obesity?

When some rats are overfed, they gain less weight than is expected. Since the early 1980s, scientists have recognized that this finding can be attributed to increased sympathetic nervous system activity in BAT, which causes accelerated metabolism. The amount of brown fat also increases in these overfed rats.[45] Conversely, some other rodents inherit a tendency to become obese at least in part because they have inactive BAT. Studies with humans have been less definitive than studies with animals.

Relationship Between Diet-Induced Thermogenesis and Weight Gain

If weight gain were a function of caloric intake alone, then we could feed people an excess quantity of kilocalories, based on their body size, and sit back and wait for a predictable increase in weight. When researchers did this at the Vermont State Prison, they were surprised by the results.

A prison population can be monitored more easily than a free-living population, allowing for better control of some of the variables that may affect studies of the thermic effect of food. In the early 1970s, a group of nonobese male volunteers from the Vermont State Prison were given varied diets to induce experimental obesity.[42] Nineteen of the men gained 21% of their initial weight. As the men gained weight, it became more and more difficult for them to maintain the gain without continually increasing caloric intake. Two men lost weight even while consuming 2,700 kcal/m^2. For a 5′6″ person, this would have required almost 7,000 kcals.

There was great overall variability in each individual's ability to gain weight, even under conditions of similar physical activity. Although DIT was not measured, the researchers concluded that a thermic mechanism must explain this variability. Like your friend who seems to eat constantly and never gains an ounce, some of these men had a metabolic protection against weight gain. The researchers speculated that such a mechanism might have developed to allow our ancestors to overeat foods that contained small amounts of essential nutrients without becoming overfat.

DIT in Obesity

About half of the studies of DIT have found a lower thermic effect of food in obese individuals than lean, and half have found no differences.[46] The discrepancy in results may mean that there *are* thermogenic defects in the obese or that there are *not* thermogenic defects in the obese; or it may reflect the difficulties in studying people, even in a laboratory setting. The thermic effect of food can be assessed quite simply in a laboratory equipped to measure RMR. First, a standard assessment of RMR is made, usually by measuring oxygen consumption. Then the individual is given a meal or a nutritional supplement, and the RMR is measured again every 30 minutes for a predetermined time period. DIT is assumed to be the difference between post-meal REE and fasting REE.

However, there is no standard research methodology for determining the thermic effect of food. This and several other factors make the task of comparing findings of many studies extremely challenging. For example:

- Was obesity defined based on BMI or measured body composition?
- How overfat were the research subjects? (Many studies document greater reductions in DIT in the very obese.)[47]

- What was the size and composition of the test meal?
- Was oxygen consumption measured continuously (with a comfortable ventilated canopy) or intermittently (with a less comfortable facemask or mouthpiece)?
- How long was RMR measured after the test meal?
- Did all the individuals being studied have the same nutritional status or smoking and exercise habits?
- Were the research subjects insulin resistant, which is known to lower DIT?[47]

In addition, we would assume that an individual demonstrates substantial day-to-day variations in DIT, even when the diet immediately preceding the measurement of thermogenesis is the same. In studies with a small number of people, this variation could create error and lead to incorrect conclusions.

This gives you several points to ponder. There appear to be metabolic mechanisms that protect certain overeaters against weight gain, but numerous problems complicate the investigation of DIT in humans, and therefore conclusions are limited. Although some studies suggest that thermogenesis may be impaired in obesity, it is still impossible to determine whether such a metabolic flaw causes weight gain or results from weight gain.[48] Table 3-6 summarizes some of the factors thought to affect adaptive thermogenesis.

Adaptive Thermogenesis: Summary

Adaptive thermogenesis is the production of heat that occurs either after exposure to cold or after eating, which contributes to total daily energy expenditure. Adaptive thermogenesis has been identified as a factor in the development of obesity. Heat production occurs in brown adipose tissue (BAT) in response to sympathetic nervous system stimulation after eating or exposure to cold. Thus, adaptive thermogenesis may serve as a kind of energy buffer, preventing weight gain during periods of overfeeding or living in a cold environment. Evidence for impaired adaptive thermogenesis in humans is thought provoking. People apparently do respond differently to the ingestion of excess calories or exposure to cold, and a failure of energy-regulating mechanisms may account for weight gain in some.

TABLE 3-6 Factors that Affect Adaptive Thermogenesis

Quantity of brown adipose tissue
Thermic activity of brown adipose tissue

- Cold exposure
- Caloric intake
- Meal frequency
- Type of nutrients consumed
- Obesity
- Insulin resistance

Case Study. The Lawyers, Part 2	Mavis and Monique have radically different eating habits. Since her days as a competitive swimmer in college, Monique has eaten a diet rich in carbohydrates—pasta, legumes, vegetables, and breads. While she is not a vegetarian and consumes some eggs and cheese, she eats meat sparingly and prefers chicken and fish to other animal products. Because her work days are long and busy, she tends to eat cold cereal or oatmeal for a quick breakfast, has yogurt and fruit at mid-morning, brings a salad for lunch, and has a half sandwich or some nuts in the late afternoon. This keeps her from being ravenous at dinner, which is typically a meal of fish or chicken, rice or pasta, and a cooked vegetable. Monique is weight stable.

Mavis does not like to cook and relies heavily on prepared foods from the supermarket or take-out. She rarely has time for breakfast and relies on coffee to get through the morning, occasionally supplemented with a doughnut if someone brings them to the morning meeting. Lunch is a take-out sandwich or fast-food meal. Dinner is a pizza, fried chicken meal, or Mexican food item from the drive-through on her way home. Periodically, Mavis goes on a strict diet where she drastically reduces her caloric intake by skipping both breakfast and lunch and having a salad or soup for dinner. She typically loses about 10 pounds and then resumes her regular routine, regaining lost weight.

- You can probably summarize several factors in Mavis's and Monique's eating behavior that affect energy *IN*. But what factors might affect energy *OUT*?

ACTIVITY ENERGY EXPENDITURE: FIDGETING, EXERCISE, AND OTHER PHYSICAL ACTIVITIES

We have already seen how REE and adaptive thermogenesis contribute to energy expenditure, accounting for approximately 70% and 10%–15% of calories expended, respectively. Metabolism above the resting level accounts for the remainder of energy expenditure, and it is the most variable of the three components of energy out. Some people barely manage to expend 400 kcal above resting metabolism a day, while others zoom along at high speed, expending in excess of 1,500 additional kilocalories each day.

activity energy expenditure

Energy expended in all of the activities that occupy our time when we are not at the basal level of energy expenditure.

Activity energy expenditure includes all of the activities that occupy our time when we are not "at rest." Absent-mindedly twisting your hair while watching television (fidgeting), walking to the mailbox (physical activity), and running on a treadmill (exercise) are examples of nonresting activities. The cost of this activity varies according to:

- Body size: Bigger people expend more kilocalories doing just about everything. This is an advantage for overweight people.
- Type of activity: Working at a computer costs fewer kilocalories than doing housework, and lifting weights is more calorically costly than walking.
- Volume of activity: The combination of an activity's duration and intensity determine caloric cost.

Activity energy expenditure may help to unbalance the energy balance equation. If RMR and DIT are indeed suppressed in some obese individuals, then increased physical activity may provide the only opportunity to increase caloric expenditure so that weight can be lost or maintained.

How Is Activity Energy Expenditure Measured?

We could estimate someone's daily energy expenditure by having them wear portable equipment to measure oxygen consumption and carbon dioxide production. By calculating the RQ, we could determine caloric expenditure. However, because wearing portable respiratory gas collection equipment is cumbersome and not practical during all daily activities, less intrusive methods of estimating energy expenditure have been developed. The most unobjectionable and scientific methods are the **doubly labeled water method** and heart-rate monitoring. These techniques are primarily used in research, but they are the basis for many of the tables that estimate caloric cost of various physical activities. Accelerometers, which were described in Chapter 2, may someday be valid and reliable measures of activity energy expenditure. They are particularly useful for measuring the cost of walking and running, but determination of the caloric cost of other activities with accelerometers is not as accurate as the methods described below.[49]

> **doubly labeled water method**
>
> *Method of estimating total daily energy expenditure from the difference in excretion of two isotopes that serve as markers of carbon dioxide production.*

- Doubly labeled water method: This procedure estimates carbon dioxide production instead of measuring it continuously in a chamber or with a spirometer. Research subjects receive a dose of two stable isotopes and provide a daily urine sample to track the rate at which the isotopes are eliminated from the body. The difference in rate of elimination of the two isotopes represents carbon dioxide production, from which caloric expenditure can be mathematically calculated. This method can be used to assess total daily energy expenditure for up to 4 weeks. When BMR is subtracted from TEE, the remainder is assumed to be activity energy expenditure (minus about 10% for DIT).

- Heart-rate monitoring: The relationship between heart rate and oxygen consumption is approximately linear. This means that as oxygen consumption rises, so does heart rate, in a proportional manner. So, we do not have to measure someone's oxygen consumption to estimate how much air they are inhaling; we can use their heart rate as a guide. Because oxygen consumption is a good indicator of energy expenditure, heart rate can also help us to estimate how many kilocalories someone is expending.

What Is the Relationship Between Activity Energy Expenditure and Weight?

National surveys have documented the declining level of physical activity among Americans while average adult weight is increasing. This certainly suggests a relationship between low levels of activity energy expenditure and obesity. What exactly is the relationship between activity and weight?

Activity Energy Expenditure and Body Weight

Energy expended in daily activity can help to balance the energy balance equation (by making energy out equal to energy in), or it can unbalance the equation (by allowing energy out to exceed energy in). Not only is activity energy expenditure the most unpredictable part of total daily energy expenditure, it has the greatest capacity to raise daily energy expenditure. Given that BMR is fairly similar between two people of the same body size and age and that adaptive thermogenesis accounts for about the same loss of heat in most people, activity energy expenditure is the component of energy out that makes the most difference in calories out between people.

Exercise is an obvious way to increase energy expenditure. Less obvious is nonexercise physical activity, like activities of daily living, fidgeting, and random muscle movements. Many experts have begun to refer to this component of activity energy expenditure as NEAT—**non-exercise activity thermogenesis**.

NEAT

non-exercise activity thermogenesist (NEAT) *Energy expended in all activities except formal exercise.*

NEAT is energy expended for everything but formal exercise. Standing, walking, climbing stairs, dancing around the kitchen, and even chewing gum increase energy expenditure above resting levels. While exercise has obvious health benefits and the effects of bed rest on health have been well documented, NEAT—standing and moving around—has only recently been appreciated for its value in weight management.

What factors influence NEAT?

- **Occupation**. What we do every day in our jobs is probably the most important factor in potentially raising our energy expenditure. People in primarily sitting jobs may be functioning at 20 to 40% above the basal level, those with standing jobs are up to 80% above basal level, and those with the most strenuous jobs are expending twice or more of their BMR in work-related NEAT.
- **Leisure activities.** Like work, leisure activities that involve primarily sitting will result in lower NEAT than those that involve movement. People who have a great deal of leisure time and choose to spend it seated in a recliner either reading or passively watching television will expend few calories, just like a seated office worker during working hours. Those who choose to use their leisure in more active pursuits, like gardening, will expend considerably more energy.
- **Labor-saving devices**. Electronic communication, drive-through restaurants, and washing machines promote more efficient use of time but also reduce NEAT. One estimate says that we expend 111 fewer kilocalories each day using labor-saving devices than we would performing the same tasks manually.[19] In addition, communities built without sidewalks and inaccessible or unsafe recreational facilities keep people in their homes and cars and limit energy expenditure.

Several other factors affect non-exercise activity thermogenesis: age (NEAT tends to decline with age), gender (in some cultures women are expected to spend more time in the home and therefore have less opportunity for activity), and heredity (genetic factors may incline some people to a lower level of spontaneous activity).

Some obesity researchers argue that we would probably expend more energy each day by increasing NEAT than by periodic bursts of exercise.[19,50] One researcher determined that obese individuals sit about 2-½ hours longer each day than sedentary lean individuals.[19] Note that the lean individuals being used for comparison were sedentary. Yet, they stood or moved about 150 minutes more than their obese counterparts.

Because walking is the largest component of NEAT, walking more could have a significant impact on energy expenditure. Unfortunately, as weight increases, movement and ambulation decline, and obese adults have been found to walk about 2 hours less than lean adults over the course of a day.[51] In fact, when lean people were overfed for two months in one study, their activity levels declined, too.[51]

Undernutrition also affects NEAT. People who are on calorie-restricted diets not only experience decreased BMR, but their physical activity levels decline, too.[33] When people reduce their body mass as a result of weight loss, they will naturally expend fewer kilocalories doing the same activity than they engaged in when they were heavier. However, caloric restriction causes down-regulation of both metabolism and energy-expending behaviors, which can persist for up to 6 years—and the reduction in energy expenditure is more than you would expect simply due to weight loss.[34]

Balancing the Energy Balance Equation with Physical Activity

For weight maintenance, it is necessary to balance calories in with calories out. By now you should realize that this involves more than simply reducing caloric intake and hoping for a proportional loss of weight. The "energy out" side of the equation is just as important as "energy in." In addition, scientists know that people who expend some energy every day in physical activities enjoy better health. People who expend about 2,000 kcal/week (approximately 286 kcal/day) have a reduced risk of cardiovascular disease and may even live longer.[52] So we need to be encouraging people to be active.

How much physical activity—exercise and NEAT—do people need if they want to avoid gaining or regaining weight? The ratio of total energy expenditure (TEE) to RMR—sometimes called the **physical activity index (PAI)**—can help answer this question. Someone who is completely sedentary has a PAI of about 1.2. For example, Mr. Smith is found to have a TEE of 1,800 kcal/day, and his RMR is measured as 1,500 kcal/d. His PAI = 1,800 ÷ 1,500, or 1.2. In other words, he is expending just 20% more than his RMR every day in activity. Compare Mr. Smith to a competitive athlete, who probably has a PAI above 2.5 (if the athlete had a RMR of 1,500 kcal/day and TEE of 3,750 kcal/day, 3,750 ÷ 1500 = 2.5). The average American has a PAI of about 1.5.[53] Most children have PAIs of around 1.7.[49]

A study of nonobese, formerly obese, and obese women found that weight gain over the course of a year occurred when the PAI was less than 1.75.[54] Table 3-7 presents information about the weights, RMRs, and other components of energy expenditure for the formerly obese women as well as for nonobese and obese women.

As you study Table 3-7, you will notice the following:

- The first column indicates that the data in the table are based on measurements of 10 nonobese, 33 previously obese, and 23 obese women.

physical activity index (PAI) *Ratio of total daily energy expenditure to resting metabolic rate. Lower PAIs signify more time spent in sedentary pursuits.*

TABLE 3-7 PAIs (TEE ÷ RMR) Protective Against Weight Gain

Subjects	Total Daily Energy Expenditure (TEE)					
	Weight (kg)	RMR (kcal)	DIT [a] (kcal)	AEE[b] (kcal)	Total (kcal)	PAI (TEE ÷ RMR[c])
Nonobese (10)	56 ± 4.4	1,270	155	914	2,299	1.8
Previously obese (33)	67 ± 9.3	1,334	111	768	2,213	1.66
Obese (23)	92 ± 9.8	1,522	133	1,002	2,657	1.75

[a]DIT was calculated based on the assumption that it accounts for 5% of TEE.
[b]AEE = activity energy expenditure
[c]PAI calculated by dividing TEE by RMR.

Source: Adapted from Schoeller, D. A. (1998). Balancing energy expenditure and body weight. *American Journal of Clinical Nutrition, 68*(Suppl), 956S–961S.

- The nonobese women weighed on average 56 kg, with a range of ± 4.4 kg; the previously obese weighed on average 67 kg, with a wider weight range of ± 9.3 kg; and the obese women averaged 92 kg, with a range of ± 9.8 kg.
- The nonobese women had an average daily TEE of 2,299 kcal, of which 115 kcal came from DIT, 914 kcal from activity, and 1,270 kcal from RMR. Dividing 2,299 kcal by 1,270 kcal gives a PAI of 1.8.
- The previously obese women had an average daily TEE of 2,213 kcal, of which 111 kcal came from DIT, 768 kcal from activity, and 1,334 kcal from RMR. Dividing 2,213 kcal by 1,334 kcal gives a PAI of 1.66
- The obese women had an average daily TEE of 2,657 kcal, of which 133 kcal came from DIT, 1,002 kcal from activity, and 1,522 kcal from RMR. Dividing 2,657 kcal by 1,522 kcal gives a PAI of 1.75.

We may need PAIs of over 1.8 to prevent obesity, and the previously obese women in this study appear to be at risk of future weight gain due to their lower energy expenditure. However, consider the obese person who decides to increase NEAT. As described in the previous section, a person who intentionally focuses on reducing sitting time and increasing standing and ambulation time by 150 minutes/day could potentially increase caloric expenditure by 300 kcal/day. An obese woman who increases TEE by this much could easily have a PAI of 1.8 or higher.

Formerly obese people may need to expend more energy to avoid weight regain. In another study, formerly obese women required a physical activity caloric expenditure of 11 kcal/kg body weight/day to prevent weight gain.[55] For a 64-kg (140-lb) woman, this equals about 700 kcal/day, substantially more activity than most nonathletes voluntarily engage in each day. So to prevent and treat obesity, interventions need to be aimed at increasing both exercise and NEAT. Factors that affect activity energy expenditure are summarized in Table 3-8.

Figuring Out How Much to Eat Every Day

Athletes may expend 25–50% more energy than nonathletes due to their active lifestyles. Cyclists and swimmers in hard training, for example, may expend 2,500–6,000 kcal/day in physical training alone. To obtain sufficient calories during periods of heavy training or competition, some athletes will have to increase fat consumption to the maximum recommended (35% of calories) or more. Failure to get enough calories may result in muscle wasting, loss of bone mineral density, cessation of menstruation, and chronic illness.

TABLE 3-8 Factors that Affect Activity Energy Expenditure

Non-exercise activity thermogenesis (NEAT), affected by
• Occupational and leisure energy expenditure
• Use of labor-saving devices
• Age
• Gender
• Heredity
• Obesity
• Caloric restriction
• Exercise

Case Study. The Lawyers, Part 3	Both Mavis and Monique recently participated in a study at a local university where they had their TEE measured using doubly labeled water. Mavis had an average TEE of 1,820 kcal/d and Monique an average TEE of 2,340/day.

1. Calculate the PAI for each woman, using the RMR value from Mifflin et al. that you previously determined.
2. Considering the study outlined in Table 3-5, what conclusions can you reach about their PAIs?
3. Mavis typically comes home from work around 8:00 p.m., having grabbed something to eat at a drive-through near home, and sits in her recliner and watches television until midnight. Refer back to the PAI for a sedentary activity, and estimate her caloric expenditure for those four hours.

Active nonathletes interested in weight loss or weight maintenance have different concerns about calories—making sure that caloric intake is either close to or less than caloric expenditure. Because two people of the same age, gender, height, weight, and lean body mass may have very different caloric requirements for weight loss or maintenance, recommending a standard caloric intake for everyone is impossible. The diet and activity assessments described in Chapter 2 can be used to estimate present caloric intake and expenditure, which will help you to establish a realistic target for caloric intake.

As a general rule, to minimize loss of lean body mass, any combination of reduced caloric intake and increased energy expenditure through physical activity should never exceed 1,000 kcal/day.[56] So, if your client appears to be consuming and expending about 2,500 kcal/day, and if you are recommending a walking program that will cost about 200 kcal/day, you should not lower caloric intake by more than 800 kcal (200 + 800 = 1,000). In addition, no one should consume fewer than 1,200 kcal/day without medical supervision. People need nutrients to keep normal physiological processes running, and it is almost impossible to obtain sufficient nutrients in 1,200 kcal.

Activity Energy Expenditure: Summary

Activity energy expenditure is the most variable of the three components of energy expenditure and provides an effective way to increase energy out. The amount of activity needed to promote health or reduce the risk of cardiovascular disease is probably considerably less than what is needed to reduce weight or prevent weight regain after weight loss. Metabolic tendencies toward low nonresting energy expenditure may make it very difficult for some people to elevate their activity to a level needed for permanent weight loss.

CONCLUSIONS ABOUT ENERGY METABOLISM

The most fundamental physiological principle of weight management is that manipulation of the energy balance equation is necessary for weight change. People who maintain stable weights over a long period of time are successful in balancing food intake and energy expenditure. Those who gain weight do so as a result of either small imbalances in the elements of the energy balance equation accumulated over a long period of time or large imbalances accrued over a short time.

Traditionally, imbalances in the energy balance equation resulting in weight gain have been attributed to excess caloric intake. No one would argue that a taste for rich food could tip the scale upward.

However, given that many nonobese people overeat without weight change, there must be factors in energy metabolism that account for weight gain. Four that are particularly important, which have been discussed in this chapter, are impaired fat oxidation, low RMR, impaired adaptive thermogenesis, and insufficient physical activity. Perhaps the issue is not that obese people eat too much but that they eat too much based on their energy requirements.

Those who have successfully lost weight face additional challenges. Permanent calorie reduction—below that of same-weight, never-obese individuals—may be required to maintain their reduced body weight. Not only is RMR lower, but fat oxidation may be suppressed and, at the same time, hormones and neurochemicals, discussed in the next chapter, encourage the reduced-weight person to eat. Sustaining weight loss can be extraordinarily difficult for some formerly obese people—not because they are lazy gluttons but because their own physiology drives them to regain weight.

This look at energy metabolism may have suggested several things to you:

- People must eat to maintain normal metabolism. Dietary restriction that reduces caloric intake below the RMR is counterproductive.
- Exercise promotes fat oxidation, may have a long-term effect of raising RMR, and burns calories.
- Non-exercise physical activity can contribute significantly to energy out and should be encouraged.
- There may be differences in energy metabolism between lean and obese people. In addition, two people of similar weights may have quite different energy requirements.

The information in this chapter is linked to three concepts that will be explored further in upcoming chapters:

1. To prevent excessive food intake and fat storage, fat intake must be controlled (Chapter 4).
2. Physical activity is the key to increased fat oxidation (Chapter 8).
3. Maintenance of weight loss is difficult, and people who need or want to permanently lose weight require support (Chapter 9).

Because permanent weight loss is a physiological challenge for some people, weight loss is clearly not the answer for everyone. Overweight individuals may be better off focusing on health habits that can be changed, not body weight, which perhaps cannot.

REFERENCES

1. Swinburn, B., & Ravussin, E. (1993). Energy balance or fat balance? *American Journal of Clinical Nutrition, 57*(5 Suppl), 766S–770S; discussion 770S–771S.
2. Flatt, J. P. (1995). Use and storage of carbohydrate and fat. *American Journal of Clinical Nutrition, 61*(Suppl), 952S–959S.
3. Jones, P. J. H., & Schoeller, D. A. (1998). Polyunsaturated: saturated ratio of diet fat influences energy substrate utilization in the human. *Metabolism, 37,* 145–151.
4. Clandinin, M. T., Wang, L. C., Rajotte, R. V., French, M. A., Goh, Y. K., & Kielo, E. S. (1995). Increasing the dietary polyunsaturated fat content alters whole-body utilization of 16:0 and 10:0. *American Journal of Clinical Nutrition, 61*(5), 1052–1057.

5. Piers, L. S., Walker, K. Z., Stoney, R. M., Soares, M. J., & O'Dea, K. (2002). The influence of the type of dietary fat on postprandial fat oxidation rates: Monounsaturated (olive oil) vs saturated fat (cream). *International Journal of Obesity and Related Metabolic Disorders, 26*(6), 814–821.

6. Stubbs, R. J., & Harbron, C. G. (1996). Covert manipulation of the ratio of medium- to long-chain triglycerides in isoenergetically dense diets: Effect on food intake in ad libitum feeding men. *International Journal of Obesity and Related Metabolic Disorders, 20*(5), 435–444.

7. St-Onge, M. P., & Bosarge, A. (2008). Weight-loss diet that includes consumption of medium-chain triacylglycerol oil leads to a greater rate of weight and fat mass loss than does olive oil. *American Journal of Clinical Nutrition, 87*(3), 621–626.

8. St-Onge, M. P., Ross, R., Parsons, W. D., & Jones, P. J. (2003). Medium-chain triglycerides increase energy expenditure and decrease adiposity in overweight men. *Obesity Research, 11*(3), 395–402.

9. Hawley, J. A. (1998). Fat burning during exercise: Can ergogenics change the balance? *Physician and Sportsmedicine, 26*(9), 56–63.

10. Almeras, N., Lavallee, N., Despres, J. P., Bouchard, C., & Tremblay, A. (1995). Exercise and energy intake: Effect of substrate oxidation. *Physiology & Behavior, 57*(5), 995–1000.

11. Hansen, K., Shriver, T., & Schoeller, D. (2005). The effects of exercise on the storage and oxidation of dietary fat. *Sports Medicine, 35*(5), 363–373.

12. Kraemer, W. J., Volek, J. S., Clark, K. L., Gordon, S. E., Incledon, T., Puhl, S. M., et al. (1997). Physiological adaptations to a weight-loss dietary regimen and exercise programs in women. *Journal of Applied Physiology, 83*(1), 270–279.

13. Westerterp, K. R., Smeets, A., Lejeune, M. P., Wouters-Adriaens, M. P., & Westerterp-Plantenga, M. S. (2008). Dietary fat oxidation as a function of body fat. *American Journal of Clinical Nutrition, 87*(1), 132–135.

14. Wade, A. J., Marbut, M. M., & Round, J. M. (1990). Muscle fiber type and aetiology of obesity. *Lancet, 335*, 805–808.

15. Ferraro, R. T., Eckel, R. H., Larson, D. E., Fontvieille, A. M., Rising, R., Jensen, D. R., et al. (1993). Relationship between skeletal muscle lipoprotein lipase activity and 24-hour macronutrient oxidation. *Journal of Clinical Investigation, 92*(1), 441–445.

16. Astrup, A., Buemann, B., Christensen, N. J., & Madsen, J. (1992). 24-hour energy expenditure and sympathetic activity in postobese women consuming a high-carbohydrate diet. *American Journal of Physiology, 262*(3 Pt 1), E282–E288.

17. Zurlo, F., Lillioja, S., Esposito-Del Puente, A., Nyomba, B. L., Raz, I., Saad, M. F., et al. (1990). Low ratio of fat to carbohydrate oxidation as predictor of weight gain: Study of 24-h RQ. *American Journal of Physiology, 259*(5 Pt 1), E650–E657.

18. Frankenfield, D., Roth-Yousey, L., & Compher, C. (2005). Comparison of predictive equations for resting metabolic rate in healthy nonobese and obese adults: A systematic review. *Journal of the American Dietetic Association, 105*(5), 775–789.

19. Levine, J. A., Vander Weg, M. W., Hill, J. O., & Klesges, R. C. (2006). Non-exercise activity thermogenesis: The crouching tiger hidden dragon of societal weight gain. *Arteriosclerosis, Thrombosis, and Vascular Biology, 26*(4), 729–736.

20. Gallagher, D., & Elia, M. (2005). Body composition, organ mass, and resting energy expenditure. In S. Heymsfield, T. Lohman, Z. Wang & S. Going (Eds.), *Human body composition* (2nd ed., pp. 219–239). Champaign, IL: Human Kinetics.

21. Goran, M. I. (1997). Genetic influences on human energy expenditure and substrate utilization. *Behavior Genetics, 27*(4), 389–399.

22. Sun, M., Gower, B. A., Bartolucci, A. A., Hunter, G. R., Figueroa-Colon, R., & Goran, M. I. (2001). A longitudinal study of resting energy expenditure relative to body composition during puberty in African American and white children. *American Journal of Clinical Nutrition, 73*(2), 308–315.

23. Byrne, N. M., Weinsier, R. L., Hunter, G. R., Desmond, R., Patterson, M. A., Darnell, B. E., et al. (2003). Influence of distribution of lean body mass on resting metabolic rate after weight loss and weight regain: Comparison of responses in white and black women. *American Journal of Clinical Nutrition, 77*(6), 1368–1373.

24. Hunter, G. R., Weinsier, R. L., Bamman, M. M., & Larson, D. E. (1998). A role for high intensity exercise on energy balance and weight control. *International Journal of Obesity and Related Metabolic Disorders, 22*(6), 489–493.

25. Treuth, M. S., Hunter, G. R., Weinsier, R. L., & Kell, S. H. (1995). Energy expenditure and substrate utilization in older women after strength training: 24-h calorimeter results. *Journal of Applied Physiology, 78*(6), 2140–2146.

26. Pratley, R., Nicklas, B., Rubin, M., Miller, J., Smith, A., Smith, M., et al. (1994). Strength training increases resting metabolic rate and norepinephrine levels in

healthy 50- to 65-yr-old men. *Journal of Applied Physiology*, 76(1), 133–137.

27. Poehlman, E. T., & Danforth, E., Jr. (1991). Endurance training increases metabolic rate and norepinephrine appearance rate in older individuals. *American Journal of Physiology*, 261(2 Pt 1), E233–9.

28. Astrup, A., Gotzsche, P. C., van de Werken, K., Ranneries, C., Toubro, S., Raben, A., et al. (1999). Meta-analysis of resting metabolic rate in formerly obese subjects. *American Journal of Clinical Nutrition*, 69(6), 1117–1122.

29. Ravussin, E. (1995). Metabolic differences and the development of obesity. *Metabolism: Clinical and Experimental*, 44(9 Suppl 3), 12–14.

30. Miller, D. S., & Parsonage, S. (1975). Resistance to slimming: Adaptation or illusion? *Lancet*, 1(7910), 773–775.

31. Steen, S. N., Opplinger, R. A., & Brownell, K. D. (1987). Weight cycling and basal metabolic rate in wrestlers. *Medicine and Science in Sports and Exercise*, 19(2), S69.

32. Blackburn, G. L., Wilson, G. T., Kanders, B. S., Stein, L. J., Lavin, P. T., Adler, J., et al. (1989). Weight cycling: The experience of human dieters. *American Journal of Clinical Nutrition*, 49(5 Suppl), 1105–1109.

33. Redman, L. M., Heilbronn, L. K., Martin, C. K., de Jonge, L., Williamson, D. A., Delany, J. P., et al. (2009). Metabolic and behavioral compensations in response to caloric restriction: Implications for the maintenance of weight loss. *PLoS One*, 4(2), e4377.

34. Rosenbaum, M., Hirsch, J., Gallagher, D. A., & Leibel, R. L. (2008). Long-term persistence of adaptive thermogenesis in subjects who have maintained a reduced body weight. *American Journal of Clinical Nutrition*, 88(4), 906–912.

35. Wijers, S. L., Saris, W. H., & van Marken Lichtenbelt, W. D. (2009). Recent advances in adaptive thermogenesis: Potential implications for the treatment of obesity. *Obesity Reviews*, 10(2), 218–226.

36. Claessens-van Ooijen, A. M., Westerterp, K. R., Wouters, L., Schoffelen, P. F., van Steenhoven, A. A., & van Marken Lichtenbelt, W. D. (2006). Heat production and body temperature during cooling and rewarming in overweight and lean men. *Obesity*, 14 (11), 1914–1920.

37. Heaton, J. M. (1972). The distribution of brown adipose tissue in the human. *Journal of Anatomy*, 112 (Pt 1), 35–39.

38. Cypess, A. M., Lehman, S., Williams, G., Tal, I., Rodman, D., Goldfine, A. B., et al. (2009). Identification and importance of brown adipose tissue in adult humans. *New England Journal of Medicine*, 360(15), 1509–1517.

39. Nedergaard, J., Bengtsson, T., & Cannon, B. (2007). Unexpected evidence for active brown adipose tissue in adult humans. *American Journal of Physiology*, 293 (2), E444–E452.

40. Seale, P., Kajimura, S., Yang, W., Chin, S., Rohas, L. M., Uldry, M., et al. (2007). Transcriptional control of brown fat determination by PRDM16. *Cell Metabolism*, 6(1), 38–54.

41. Belko, A. Z., Barbieri, T. F., & Wong, E. C. (1986). Effect of energy and protein intake and exercise intensity on the thermic effect of food. *American Journal of Clinical Nutrition*, 43(6), 863–869.

42. Sims, E. A. H., Danforth, E., Jr., Horton, E. S., Bray, G. A., Glennon, J. A., & Salans, L. B. (1973). Endocrine and metabolic effects of experimental obesity in man. In R. O. Greep (Ed.), *Recent progress in hormone research*, 29, (pp. 457–496).

43. Lim, K., Yoshioka, M., Kikuzato, S., Kiyonaga, A., Tanaka, H., Shindo, M., et al. (1997). Dietary red pepper ingestion increases carbohydrate oxidation at rest and during exercise in runners. *Medicine and Science in Sports and Exercise*, 29(3), 355–361.

44. Yoshida, T., Sakane, N., Umekawa, T., & Kondo, M. (1994). Relationship between basal metabolic rate, thermogenic response to caffeine, and body weight loss following combined low calorie and exercise treatment in obese women. *International Journal of Obesity and Related Metabolic Disorders*, 18(5), 345–350.

45. Himms-Hagen, J. (1984). Thermogenesis in brown adipose tissue as an energy buffer—Implications for obesity. *New England Journal of Medicine*, 311(24), 1549–1558.

46. Granata, G. P., & Brandon, L. J. (2002). The thermic effect of food and obesity: Discrepant results and methodological variations. *Nutrition Reviews*, 60(8), 223–233.

47. de Jonge, L., & Bray, G. A. (2002). The thermic effect of food is reduced in obesity (letter to the editor). *Nutrition Reviews*, 60(9), 295–297.

48. Tentolouris, N., Pavlatos, S., Kokkinos, A., Perrea, D., Pagoni, S., & Katsilambros, N. (2008). Diet-induced thermogenesis and substrate oxidation are not different between lean and obese women after two different isocaloric meals, one rich in protein and one rich in fat. *Metabolism: Clinical and Experimental*, 57(3), 313–320.

49. Westerterp, K. R., & Plasqui, G. (2004). Physical activity and human energy expenditure. *Current Opinion in Clinical Nutrition and Metabolic Care*, 7(6), 607–613.

50. Hamilton, M. T., Hamilton, D. G., & Zderic, T. W. (2007). Role of low energy expenditure and sitting in

obesity, metabolic syndrome, type 2 diabetes, and cardiovascular disease. *Diabetes, 56*(11), 2655–2667.

51. Levine, J. A., McCrady, S. K., Lanningham-Foster, L. M., Kane, P. H., Foster, R. C., & Manohar, C. U. (2008). The role of free-living daily walking in human weight gain and obesity. *Diabetes, 57*(3), 548–554.

52. Lee, C. D., Blair, S. N., & Jackson, A. S. (1999). Cardiorespiratory fitness, body composition, and all-cause and cardiovascular disease mortality in men. *American Journal of Clinical Nutrition, 69*(3), 373–380.

53. Roberts, S. B., Krupa Das, S., & Saltzman, E. (2004). Energy expenditure in obesity. *American Journal of Clinical Nutrition, 79*(2), 181–182.

54. Schoeller, D. A. (1998). Balancing energy expenditure and body weight. *American Journal of Clinical Nutrition, 68*(4), 956S–961S.

55. Schoeller, D. A., Shay, K., & Kushner, R. F. (1997). How much physical activity is needed to minimize weight gain in previously obese women? *American Journal of Clinical Nutrition, 66*(3), 551–556.

56. Yang, M. U., & Van Itallie, T. B. (1992). Effect of energy restriction on body composition and nitrogen balance in obese individuals. In T. A. Wadden, & T. B. Van Itallie (Eds.), *Treatment of the seriously obese patient* (pp. 83–106). New York: Guilford Press.

Regulation of Eating Behavior and Body Weight

CHAPTER 4

W hy do we eat? Why do we stop eating? What makes us choose particular foods? You have just explored in Chapter 3 the notion that metabolic differences may explain some of the variation between lean and obese individuals. This chapter looks at regulatory mechanisms that provide additional clues to the causes of obesity and eating disorders, which may help in the development of more effective prevention and treatment programs.

HOMEOSTASIS AND REGULATION OF WEIGHT

set point *Level of body weight or fatness maintained by a complex combination of body systems; after weight loss, a person whose body weight keeps creeping back up to the previously higher level may have a high set point.*

homeostasis *Stability of the body's internal environment.*

adipose tissue *Storage fat.*

Regulation of body weight is a remarkably complex process governed by numerous interconnected body systems, metabolites, hormones, and neurochemicals and influenced by environmental and genetic factors. Weight is maintained by achieving a balance between energy intake (consumption of food) and energy expenditure (metabolism and physical activity). The component of weight of greatest interest in weight management is, of course, fat. Some authors compare the regulation of stored body fat to the regulation of heat, where a thermostat-like adipostat attempts to keep the adipose (fat) mass constant. Such an adipostat would have to detect physiological conditions as they exist at a given time, compare this information with a **set point** (reference value) for the condition, and stimulate either energy intake or energy expenditure to return physiological conditions to the set point.

Scientists have argued about the existence of an adipostat/set point for over 40 years. Research available today indicates that there is not a single adipostat but rather a combination of systems that may create a set point for body fat. The complexity of these systems and the set point they create could ultimately defeat efforts to lose weight by driving the overweight or weight-reduced person to eat more or expend less energy.

The concept of **homeostasis**, the body's ability to maintain balance or stability of the internal environment and thereby ensure survival, may help you to understand how body weight is regulated. Figure 4-1 illustrates the concept of homeostasis, which occurs as follows:

- Receptors detect a stimulus and send signals to an integrator. For body weight regulation, receptors are nutrient sensors located primarily in the brain, gastrointestinal tract, and **adipose tissue**.

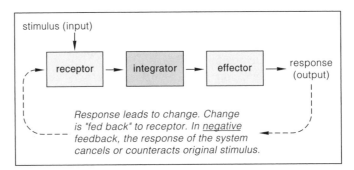

FIGURE 4-1 How homeostasis is maintained. Just as mechanisms ensure homeostasis in many metabolic functions, there are probably mechanisms for homeostasis of body fat.

- The integrator collects information from receptors and coordinates a response. The response is based not just on what is happening now but also on what should be happening (in other words, based on the set point). Several areas of the brain serve to integrate information about hunger, satiety, fat stores, and energy needs.
- Effectors carry out the response, which sends new information called feedback to the integrator. In body weight regulation, effectors may be endocrine (hormones that increase or decrease activity of body tissues or cause increased or decreased food intake), autonomic (sympathetic nervous stimulation that increases or decreases brown adipose tissue activity and/or metabolic rate), or behavioral (increased or decreased physical activity or intake of specific types of foods). The result: changes in energy in and energy out.

This chapter looks in detail at the physiological factors that regulate body weight and body fat. The chapter starts with the brain, which is a key integrator of information about energy balance. The digestive system and both white (storage) and brown adipose tissue also play roles in energy homeostasis. Finally, the chapter considers genetic and environmental influences on food intake, energy expenditure, and fat storage. After reading this chapter, you may agree with many clinicians and scientists who believe that obesity is not simply the result of gluttony and sloth but rather the outcome of complex interactions between biologically controlled mechanisms for energy regulation. In an environment characterized by staggering quantities of readily available food and more disincentives than incentives to be active, you might find yourself wondering why we aren't all obese.

THE BRAIN: CENTRAL REGULATOR OF WEIGHT

hunger *Physiological drive to eat.*

appetite
Psychological desire to eat, whether there is hunger or not; includes selection of particular nutrients, a taste for specific foods, and food cravings in response to the sight, smell, or thought of food.

satiation
Termination of eating after hunger has

satiety *A relatively long-term feeling of fullness or satisfaction that remains after eating, which inhibits further consumption of food.*

Because our ancestors were more likely to experience famine than plenty, most internal mechanisms that regulate body weight turn *on* eating, rather than turning it *off*. The biological logic of this is clear—it is far better for the species to overeat than to undereat, especially when our ancestors rarely lived longer than 36 years and were unlikely to develop obesity, diabetes, or cardiovascular disease. These on-off mechanisms also control energy expenditure.

Food intake is prompted by hunger and appetite and is terminated when satiation occurs. **Hunger** has been described as "a subjective sensation associated with the drive to obtain food."[1] **Appetite** represents a psychological desire to eat, whether or not there is hunger, and includes selecting particular nutrients, having a taste for specific foods, and craving specific foods in response to the sight, smell, or thought of them. A person who has not eaten in several hours may experience the discomfort associated with a physiological need to eat, which we know as hunger. Fulfilling that hunger with particular desired foods, or simply eating in response to a food craving, reflects appetite. **Satiation** is the termination of eating after hunger has been satisfied. Satiation is then followed by a relatively longer period of feeling full and satisfied, which is called **satiety**.[2]

The brain is the primary integrator of the body systems that initiate and terminate feeding. Studies on rats have shown the brain's influence to be so powerful that it can drive animals to continue feeding even when peripheral signals (like stomach distention) indicate fullness.

What Is the Role of the Brain in Regulating Weight?

If there is a drive to eat, then the brain is the chauffeur. Several areas of the brain play roles in energy homeostasis. The most important of these is the **hypothalamus**, a small vital structure in the center of the brain that integrates information from many organs and systems. In addition to weight management, the hypothalamus is also integrally involved in emotional response, energy expenditure, temperature regulation, water equilibrium, and hormone secretion. Regarding the hypothalamus, it has been observed, "There are few structures in the mammalian brain so small in size and yet involved in such a wide variety of functions."[3] The importance of this structure cannot be overstated.

The hypothalamus receives input from sensors inside and outside of the brain, it integrates this information, and it sends appropriate output to target organs and tissues outside the brain via two pathways: the autonomic nervous system and the endocrine system. You might think of the hypothalamus as a link between these two important systems:

- The **autonomic nervous system** is the branch of the nervous system that regulates physiological functions outside of voluntary control, like sweating, circulation, and digestion. Autonomic nerves carry information to and from smooth muscles (those found lining the walls of organs), cardiac muscles, and the endocrine glands. The relationship of the autonomic nervous system to the central nervous system is pictured in Figure 4-2.
- The **endocrine system** includes the glands as well as the hormones they secrete to deliver chemical messages to target cells throughout the body.

Five areas of the hypothalamus, studied mainly in animals but increasingly in humans with the aid of magnetic resonance imaging (MRI), positron emission tomography (PET), and computer tomography (CT), have been found to play a role in eating behaviors and energy metabolism. Neurons in these areas *receive* information via receptors sensitive to hormones (like insulin and leptin), metabolites (like glucose), and various neurotransmitters. These neurons also *produce* neurotransmitters and deliver them via projections into other areas of the brain.

- Arcuate nucleus (ARC): This critical hub contains neurons that produce appetite-stimulating neurotransmitters and neurons that produce appetite-dampening neurotransmitters. This area also contains many insulin and leptin receptors.

hypothalamus
Area of the brain that serves as a link between the endocrine and the autonomic nervous systems; it promotes homeostasis by regulating temperature, water balance, food intake, hormonal balance, and emotional responses.

autonomic nervous system
Branch of the nervous system that regulates involuntary physiological functions, such as sweating, circulation, and digestion.

endocrine system *System of glands and the hormones they secrete, which affect target cells throughout the body.*

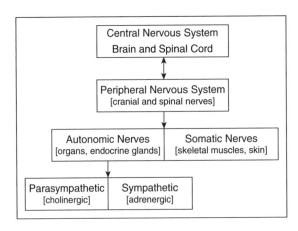

FIGURE 4-2   The nervous system.

- Paraventricular nucleus (PVN): Appetite-stimulating neurons from the ARC project into this area and stimulate feeding.
- Ventromedial hypothalamic nucleus (VMH): The VMH is sensitive to leptin, promotes satiety and, if destroyed, results in uncontrolled eating, reduction of sympathetic nervous system activity, and extreme weight gain.
- Dorsomedial hypothalamic nucleus (DMN): Although this hypothalamic area is less well understood than the other areas, it probably integrates information from other areas.
- Lateral hypothalamic area (LHA): The lateral hypothalamic area is a feeding center and very sensitive to low circulating glucose. Neurons in this area produce neurotransmitters called orexins, which stimulate appetite. Destruction of the lateral area results in failure to eat and extreme weight loss.

Which Neurotransmitters Are Involved in Regulating Weight?

The hypothalamic areas just described could not do their jobs without being able to communicate with each other and other brain regions. Neurotransmitters can be thought of as the messengers on the brain's superhighway. More than 50 neurotransmitters and their receptors are found throughout the regions of the hypothalamus and serve a critical role as regulators of food intake, food choice, hormone secretion, and, ultimately, body weight.[4] In addition, hormones produced in adipose tissue and the gastrointestinal system make their way to the brain to provide further information. These are discussed in later sections of this chapter.

Neurotransmitters that Turn on Feeding

peptide *A protein made up of two or more amino acids. Peptides secreted in the hypothalamus are neuropeptides and include neuropeptide Y, melanin-concentrating hormone, agouti gene-related protein, the orexins, galanin, and dynorphin. Cholecystokinin, insulin, and glucagon are also peptides.*

Neuropeptides are neurotransmitters found in the central nervous system (a **peptide** is a compound made from two or more amino acids). Several key neuropeptides secreted in the hypothalamus that are known to increase food intake and, in some cases, decrease energy expenditure are: neuropeptide Y, melanin-concentrating hormone, orexins, galanin, agouti gene-related protein, and dynorphin. Several of these are presented in more detail here. In addition, the neurotransmitters dopamine and norepinephrine, which have actions that reach far outside the brain, are involved in food intake, as is ghrelin, a peptide produced in the gut. The peptides involved in promoting food intake are classified as orexigenic and are summarized in Table 4-1.

- Neuropeptide Y (NPY) and agouti gene-related protein (AgRP). NPY, the most abundant neuropeptide in humans, seems to drive consumption of high carbohydrate foods, whereas AgRP is associated with increased consumption of high fat foods.[4] NPY is mainly produced by neurons in the arcuate nucleus (ARC) and not only stimulates intake of food but also reduces energy expenditure by inhibiting sympathetic nervous system activity in the brown adipose tissue. As food is consumed, both NPY and AgRP levels fall, and the desire to eat diminishes. Animal studies demonstrate that NPY levels in the hypothalamus rise in obesity, diabetes, weight loss, food deprivation, and food restriction. The power of NPY is so great that when it is injected into the rat hypothalamus, a sated animal will eat and

TABLE 4-1   Peptides that Regulate Food Intake

Turn On Feeding	Turn Off Feeding
Orexigenic	Anorexigenic
Neuropeptide Y (NPY)	α-melanocyte-stimulating hormone
Melanin-concentrating hormone (MCH)	Pro-opiomelanocortin
Orexins	Cocaine-and-amphetamine-regulated
Galanin	transcript
Agouti gene-related protein (AgRP)	Serotonin
Dynorphin	Cholecystokinin*
Dopamine	Peptide YY*
Norepinephrine	Amylin*
Ghrelin*	Bombesin*
Insulin* (when blood glucose is low)	Insulin* (when blood glucose is high)

*These are digestive system peptides.

gain weight. So NPY promotes fat storage in two ways: It stimulates eating and it reduces thermogenesis.

- Melanin-concentrating hormone (MCH). MCH is produced in the arcuate nucleus (ARC) and lateral hypothalamic area (LHA). Mice that lack MCH become very lean, whereas mice that produce too much MCH overeat, particularly high fat foods, and become obese.[5] MCH binds to a receptor identified as melanocortin receptor subtype 4 (MC4R). Genetic defects in this receptor, discussed later in this chapter, cause up to 6% of morbid obesities.
- The orexins are also produced in the LHA, and while they stimulate food intake, they may have a more important role in regulating sleep.
- Galanin. In animals, galanin synthesis increases at puberty, which corresponds with an increase in fat consumption and body weight. Like NPY, galanin secretion increases during times of food restriction.
- The opioid peptide, dynorphin, promotes eating and has pain-reducing properties. This morphine-like substance acts on the feeding centers of the hypothalamus to stimulate the consumption of highly palatable foods, such as fats. It may also play a role in stress-induced eating, although this concept is less well understood because not everyone under stress eats more.
- Norepinephrine acts on the paraventricular nucleus (PVN) to stimulate eating, particularly by increasing the size of a meal and the intake of carbohydrates. In addition to norepinephrine's activity in the brain, it is released by sympathetic nervous system fibers throughout the body. In animal and some human studies, overeating increases the release of norepinephrine in the brown adipose tissue, which causes diet-induced thermogenesis. The opposite occurs during food restriction, resulting in reduced energy expenditure and, possibly, weight gain.

catecholamines
Compounds produced from the amino acid tyrosine that include dopamine (a neurotransmitter), epinephrine (a hormone secreted by the adrenal medulla), and norepinephrine (secreted by the adrenal medulla and released from sympathetic nerve endings). Norepinephrine and dopamine are involved in eating behavior.

Neurotransmitters that Turn Off Feeding

The hypothalamus also produces neuropeptides and **catecholamines** that turn off eating and, in some cases, increase energy expenditure. Such actions promote fat loss. These are known as anorexigenic peptides and neurotransmitters and include: pro-opiomelanocortin

(POMC), α-melanocyte-stimulating hormone, cocaine- and amphetamine-regulated transcript, serotonin, and three peptides that originate in the gut but find their way to the brain—cholecystokinin, peptide YY, and amylin, which are discussed in the next section of this chapter.

A variety of neuropeptides derived from pro-opiomelanocortin (POMC) precursor peptide are produced in the arcuate nucleus, work in opposition to melanin-concentrating hormone, and inhibit appetite. One of these is α-melanocyte-stimulating hormone. When people do not ingest enough calories on a given day, or diet for a long period of time, production of POMC drops, and melanin-concentrating hormone and other hunger-inducing neuropeptides drive people to eat.

Serotonin is a catecholamine that works in opposition to norepinephrine. When brain serotonin levels are high, the rate of eating, the quantity of food consumed, and the duration of eating all are reduced. Some drugs that will be covered in Chapter 10 increase the secretion of serotonin and are effective appetite suppressants.

How Else Does the Brain Regulate Energy Balance?

Recently scientists have become intrigued by what some call non-homeostatic mechanisms that also influence energy balance. The largely "unconscious" actions of the hypothalamus are influenced by inputs from brain circuits that are less attuned to physiologic need and far more wired into the expectations and pleasures associated with food. These brain areas do not need hunger signals to stimulate food intake, and they can override hunger signals when, by choice, food is ignored. They give us food "memories" and help us interpret what an appropriate quantity of food is, what foods are healthy, and may even help people make decisions related to achieving weight goals. Some research suggests that there are important differences between lean and obese individuals in these brain areas.

Several non-homeostatic brain areas that play a role in food intake and energy expenditure are currently being studied. The largest of these is the prefrontal cortex, which makes up about one-third of the human brain. Damage to this area is connected to an eating disorder known as gourmand syndrome, discussed later in Chapter 11. The prefrontal cortex integrates a variety of information from sensory and visceral organs and helps perform complex behaviors and high-level thinking. Connections with the hypothalamus permit this area to potentially override hunger. However, a meal may make less of an impression on the prefrontal cortex of an obese individual than of a lean individual, resulting in more food consumption before satiation occurs.[6]

circumventricular organs *organs within the brain that surround the ventricular system, outside the blood brain barrier.*

A group of brain structures collectively known as **circumventricular organs** are also candidates for integrating sensory information and transmitting that information to other parts of the brain, including the hypothalamus. The circumventricular organs are, as the name implies, located around the ventricles of the brain, specifically the third and fourth ventricles, and close to the hypothalamus (see Figure 4-3 for an illustration). They have receptors for many of the peptides included in Table 4-1. Because they are outside the blood-brain barrier, changes in circulating levels of hormones, peptides, and even glucose can elicit an immediate response, which is then transmitted to other brain areas. The role of the circumventricular organs in diet-induced thermogenesis, while still speculative, provides another potential link to obesity.[7]

limbic system *a ring of interconnected structures that include parts of the hypothalamus, thalamus, and cerebral cortex, which play a role in motivation, emotions, and learning.*

The **limbic system**, which includes the striatum, amygdale, and some areas of the hippocampus and hypothalamus, has been studied for many years in relation to drug

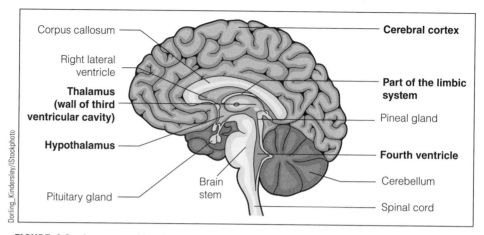

FIGURE 4-3 Location of key brain areas involved in hunger, appetite, and satiety.

Source: From Sherwood. Human Physiology, 7E. p. 155. © 2010 Brooks/Cole, a part of Cengage Learning, Inc. Reproduced by permission. www.cengage.com/permissions.

and tobacco addiction. Recently obesity researchers have become interested in the limbic system's role in the rewards associated with eating. When people abuse drugs or smoke, dopamine is released from neurons and provides a pleasurable sensation. This also occurs after people eat. So food can do more than end metabolic hunger; it can also make one feel good. High fat food is especially rewarding, from the standpoint of the brain. But dopamine only provokes this response when there are dopamine receptors available—and the limbic system includes such receptors. Some studies have found that obese individuals, especially those with BMIs above 40 kg/m^2 have fewer dopamine receptors.[8] **Insulin** is suspected to decrease the pleasurable response to food.[9] As a result, in obesity there might be a drive to ingest more food so that more dopamine is released to activate fewer dopamine receptors, and high insulin levels would make the drive even more compelling.

Figure 4-4 summarizes homeostatic mechanisms that affect food intake and energy expenditure in response to changes in nutritional need or size of the fat stores. Scientists believe that drugs which can either block or enhance the action of neurotransmitters in the brain will ultimately be developed. Such pharmacological aids could help treat obesity and eating disorders.

insulin *Hormone secreted by the pancreas in response to an increase in the blood glucose level; insulin permits glucose to be taken up by the cells and is responsible for fat storage.*

The Brain: Summary

The brain and its neurochemicals play important roles in regulating body weight. Within the hypothalamus there are areas that stimulate food intake and those that promote satiety. These hypothalamic centers are stimulated by various hormones and neurochemicals, including some that promote eating (neuropeptide Y, agouti gene-related protein, galanin, norepinephrine, dopamine) and some that put the brakes on eating (α-melanocyte-stimulating hormone, cocaine-and-amphetamine-regulated transcript, and serotonin). In addition, other areas of the brain that interact with the hypothalamus add a different dimension to food-seeking and avoidance—pleasure, reward, and memory. Overactive feeding centers or increased production of certain neuropeptides could lead to excess consumption of food, binge eating, food obsession, and reduced energy expenditure. Some day these neurochemicals or their relatives may prove to be effective in treating eating disorders and obesity.

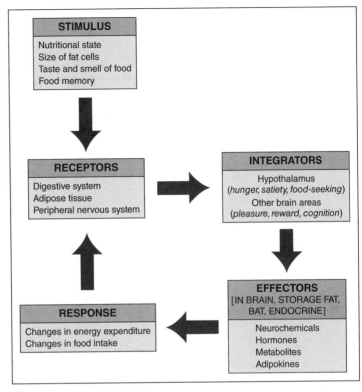

FIGURE 4-4 Energy Homeostasis. Changes in nutritional need or size of fat stores will affect food intake and energy expenditure.

THE DIGESTIVE SYSTEM: RECEPTOR AND EFFECTOR IN REGULATION OF WEIGHT

The digestive system provides the brain with important information about hunger, appetite and, in particular, satiety. Digestive system structures, the senses of taste and smell, and peptides in the digestive tract are the receptors that provide cues to the central nervous system. The digestive system also carries out the eating responses directed by the brain. Figure 4-5 provides an overview of the structures of the digestive system.

How Do the Taste and the Smell of Food Contribute to Weight Regulation?

gustatory system *Structures and pathways involved in the sense of taste.*

olfactory system *Structures and pathways involved in the sense of smell.*

The **gustatory system** is responsible for the sense of taste and includes the tongue and several areas of the brain, most notably the cortex and the hypothalamus. Gustatory sensors in the mouth serve as a sort of "gatekeeper," potentially rejecting substances with a sour or bitter taste and accepting foods with a more pleasurable flavor, even preparing the stomach and intestines for what is to come. Taste perception is strongly influenced by smell, which is the responsibility of the **olfactory system**. Millions of olfactory receptors in the nose send information to the brain, allowing smell discrimination and

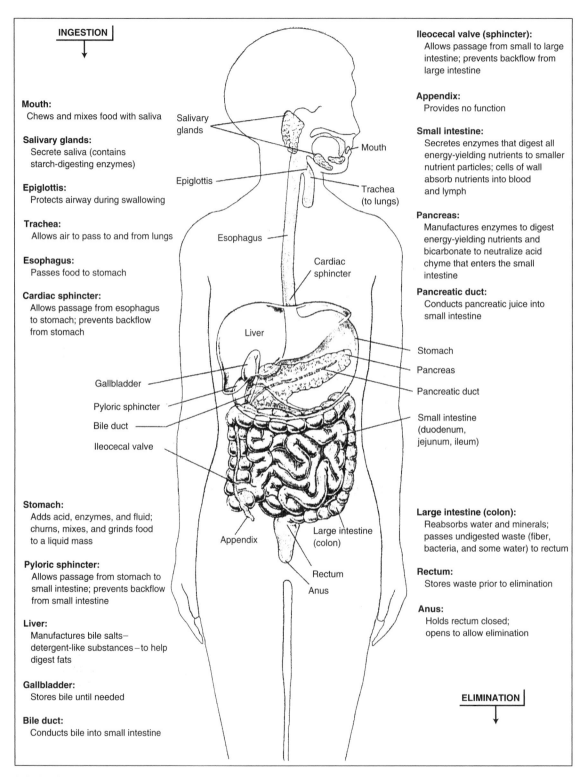

INGESTION

Mouth:
Chews and mixes food with saliva

Salivary glands:
Secrete saliva (contains starch-digesting enzymes)

Epiglottis:
Protects airway during swallowing

Trachea:
Allows air to pass to and from lungs

Esophagus:
Passes food to stomach

Cardiac sphincter:
Allows passage from esophagus to stomach; prevents backflow from stomach

Salivary glands

Mouth

Epiglottis

Trachea (to lungs)

Esophagus

Cardiac sphincter

Liver

Gallbladder

Pyloric sphincter

Bile duct

Ileocecal valve

Stomach:
Adds acid, enzymes, and fluid; churns, mixes, and grinds food to a liquid mass

Pyloric sphincter:
Allows passage from stomach to small intestine; prevents backflow from small intestine

Liver:
Manufactures bile salts— detergent-like substances—to help digest fats

Gallbladder:
Stores bile until needed

Bile duct:
Conducts bile into small intestine

Ileocecal valve (sphincter):
Allows passage from small to large intestine; prevents backflow from large intestine

Appendix:
Provides no function

Small intestine:
Secretes enzymes that digest all energy-yielding nutrients to smaller nutrient particles; cells of wall absorb nutrients into blood and lymph

Pancreas:
Manufactures enzymes to digest energy-yielding nutrients and bicarbonate to neutralize acid chyme that enters the small intestine

Pancreatic duct:
Conducts pancreatic juice into small intestine

Stomach

Pancreas

Pancreatic duct

Small intestine (duodenum, jejunum, ileum)

Appendix

Large intestine (colon)

Rectum

Anus

Large intestine (colon):
Reabsorbs water and minerals; passes undigested waste (fiber, bacteria, and some water) to rectum

Rectum:
Stores waste prior to elimination

Anus:
Holds rectum closed; opens to allow elimination

ELIMINATION

FIGURE 4-5 The Digestive System.

appropriate behavioral responses. So, the gustatory and olfactory systems bring you to the table.

The sight, smell, and taste of food are important contributors to appetite. As you know from your own experiences, the presence of tasty food in the mouth tends to promote the consumption of more food. This phenomenon has a physiological basis. In rats, ingestion of foods containing sugar or fat elevates dopamine levels in the hypothalamus, stimulating further intake. There is, unquestionably, a psychological basis for appetite, too. The appearance of a dessert cart after a filling restaurant meal has prompted many a sated human to consume additional quantities of palatable food.

The gustatory and olfactory systems are also involved in satiation. Studies with animals and humans have documented that preference for a food diminishes as more of that food is consumed in the same meal. Food odors considered to be pleasant at the beginning of a meal become less pleasant as the meal progresses. This process, known as sensory-specific satiety, may ensure that we consume a varied diet. But, when a variety of foods are available (that dessert cart, for example, or an all-you-can-eat buffet), the quest for variety may promote overeating.[10]

How Is the Digestive System Involved in Regulating Intake?

The digestive system has a fairly obvious function in food intake—it includes structures responsible for getting foods into the body and processing them. In addition, the entire gastrointestinal tract contains both mechanical and chemical sensors that provide information to the brain. The contractions of an empty stomach remind you that you haven't eaten for some time (although, as the previous section on the brain may have suggested, you don't actually need a stomach to experience hunger). The stomach and the intestines also play an important role in satiation and satiety if one is willing to acknowledge the signals they provide.

An empty stomach has a volume of approximately 50 ml. As food is consumed, the stomach's smooth muscle walls are able to stretch and relax simultaneously to accommodate an enormous increase in volume—up to about 1,000 ml. Stomach distention has two effects:

- It promotes stomach emptying into the duodenum (the upper part of the small intestine).
- It signals fullness. Activation of stretch receptors from a distended stomach sends signals to the brain that slow the rate of eating. A high-fiber diet promotes this function by absorbing and holding more water in the stomach.

Do obese people have larger stomachs and, therefore, a greater capacity to ingest food before the stretch receptors are activated? The main determinant of stomach size is thought to be meal size, with chronic overeaters and people who routinely consume large meals having greater stomach capacities.[11] Average weight bulimics who engage in binge eating have larger stomachs than obese individuals who do not routinely overeat. If the stomach gets *larger* in overeaters, could this mean that the opposite occurs and the stomach actually *shrinks* after calorie restriction? One group of researchers found the stomach had a reduced capacity after dieting. In theory, this could result in faster activation of the stretch receptors that signal satiety and would actually help someone to avoid overeating. But, as you can imagine, this measurement is difficult to obtain in humans, and little evidence supports the notion of the stomach actually shrinking.[12]

Digestive tract differences between lean and obese people have been observed.[12] For example, the stomach may empty more rapidly in obese individuals. One researcher determined that the small intestine was significantly longer in obesity.[13] This could allow the ingestion of more food and the absorption of more nutrients before the brain says, "Stop!"

Hunger and the Digestive System

Only one hormone in the gastrointestinal tract stimulates food intake—ghrelin, secreted mainly in the stomach and reaching peak levels just before eating. As soon as the full stomach begins to empty into the duodenum, ghrelin levels drop. Ghrelin activates the limbic (pleasure and reward) system of the brain, including areas involved in memory, which has led some researchers to speculate that ghrelin may be involved in helping people find food.[14] For our ancient ancestors, secretion of ghrelin in an empty stomach might have helped them sniff out new food sources in their environment. Today, this more likely leads us to forage in the pantry for long-forgotten cookies.

Ghrelin levels increase after fasting or caloric restriction, a very inconvenient physiological situation for those trying to lose weight by dieting. However, ghrelin levels decrease following gastric bypass surgery, which may be one of the factors that suppress appetite post-surgery and allow substantial weight loss.[5]

Satiety and the Digestive System

All of the other signals that arise in the gastrointestinal system are associated with satiety and should put the brakes on eating. Cholecystokinin (CCK) is a peptide secreted by the duodenum. When fat is present in the digestive tract, CCK performs a digestive function by stimulating the secretion of enzymes needed to digest fats. It also acts on the brain by inhibiting neuropeptide Y secretion in the hypothalamus to slow eating. Some recent evidence suggests that chronic high-fat diets can reduce the strength of this feedback system to the brain.[15] This means that NPY secretion would not diminish after a meal and an important satiety factor would be lost. In addition, obese rats and some humans may have a reduced number of brain CCK receptors, which could limit satiety signals and lead to ingestion of more food during a high-fat meal and, ultimately, weight gain.

Peptide YY is secreted farther along the intestine after eating and inhibits NPY secretion similarly to CCK. Other hormones and peptides released throughout the gastrointestinal tract may serve as additional influences on satiety. Among these are bombesin, secreted by the stomach, and glucagon and amylin, secreted by the pancreas; these produce a short-term feeling of fullness in humans (satiation). The usefulness of these chemical satiety signals cannot be overstated. Because digestion and absorption are time consuming and complex, the nutrients ingested in a meal do not appear in the bloodstream for hours. If the presence of these nutrients in the bloodstream were the primary signal to stop eating, then everyone would be extraordinarily overfat from continued overeating, particularly those with large stomach capacity. Peptides can send a satiety signal to the brain within minutes.

What Is the Function of Insulin in Hunger and Satiety?

The pancreas secretes several hormones: The pancreatic α-cells secrete glucagon, the delta- (Δ-) cells secrete somatostatin, PP-cells secrete pancreatic polypeptide, and β-cells secrete insulin. Insulin acts on absorbed nutrients to convert them to stored carbohydrate, protein, and fat.

Insulin has been called the "master metabolic switch" that moves us from "I'm full" to "I'm hungry."[5] Insulin secretion is at its lowest when someone hasn't eaten in many hours and rises after eating. The most well-known stimulus for insulin secretion is an increase in blood glucose level, such as would occur after a meal. After insulin helps glucose get into the cells that need it, insulin levels gradually drop. But blood glucose levels drop, too, which once again stimulates hunger.

Recall from Chapter 1 that hyperinsulinemia is often associated with obesity. This occurs both as a response to high carbohydrate consumption and as a consequence of insulin resistance resulting from enlarged fat cells. When insulin resistance occurs, hyperinsulinemia worsens.

There are insulin receptors in the hypothalamus, and insulin can cross the blood–brain barrier. In the laboratory, high glucose and insulin levels in the ventromedial hypothalamic nucleus cause animals to gradually slow, and finally stop, eating. Increased food intake can occur for two insulin-related reasons: (1) high insulin levels coupled with low blood glucose, which might occur in hyperinsulinemia or type 2 diabetes, stimulates eating; and (2) low insulin levels (and low blood glucose) have been shown to increase synthesis of neuropeptide Y in the hypothalamus, resulting in increased intake.[16] Table 4-1 summarizes these effects.

Insulin may also play a role in restoring the set point after people lose weight through calorie restriction. When enlarged fat cells shrink in response to dieting, plasma insulin levels become more normal. This may lower insulin binding in the hypothalamus, which will promote the production of neuropeptide Y. The result: increased food intake, which may rapidly restore the pre-diet fat cell size.

The Digestive System: Summary

The digestive system helps regulate body weight through its role as receptor of information about hunger, appetite, and satiety, and its role as effector of appropriate responses. The taste and the smell of food bring you to the table and keep you there until satiety signals from elsewhere in the digestive system slow and finally halt eating.

Physical characteristics of the stomach and the small intestine could promote weight gain or weight maintenance in susceptible people. Some obese individuals and binge-eaters may be able to ingest more food than nonobese or normal eaters. In addition, a lack of functioning CCK receptors in the brain could prevent this duodenal peptide from doing its job—inducing satiety.

Insulin has a broader and more complex role. Insulin secretion increases after a meal or in response to signals from the brain. High levels of insulin cross the blood–brain barrier and promote satiety but, under certain conditions, may induce hunger, such as when insulin levels are high and blood glucose is low (type 2 diabetes) or when both insulin and glucose levels are low (food deprivation). The resultant consumption of excess calories could promote weight gain.

Case Study. The Hungry Family, Part 1	Mr. and Mrs. Rhys are both obese. Mrs. Rhys has a BMI of 35 and recently developed type 2 diabetes. Mr. Rhys has a BMI of 33 and takes medication for high blood lipids and hypertension. They are worried about their 7-year-old son, Joey, who is in the 97th BMI percentile.

Everyone in the family loves to eat. On a typical busy evening, one parent stops at a pizza restaurant on the way home and brings in two large pizzas for dinner. No one stops eating until both pizzas are gone. And by 8:30 p.m. at least one family member is foraging for cookies in the kitchen.

- Consider the potential roles of the brain and digestive system in this behavior.
- Are there strategies that the Rhys family could take to reduce the impact of physiological factors on eating behavior?

STORAGE FAT: AN ACTIVE PARTICIPANT IN WEIGHT REGULATION

cortisol *Hormone secreted by the adrenal cortex that is important in the body's response to stress. Cortisol secretion results in increased blood glucose levels and release of fat and protein from body stores. Cortisol has been associated with increased storage of abdominal fat.*

Storage fat is known as white adipose tissue, or simply adipose tissue. It is made up of fat cells (called adipocytes), blood vessels, nerves, and connective tissue. Since the 1950s, scientists have suspected that fat tissue is not simply a passive recipient of excess lipid but is an active participant in maintenance of its size. Much recent research supports such an interrelationship among storage fat, brain, and body weight.

Storage fat acts as a receptor and an effector in the regulation of body weight. As a receptor, white adipose tissue detects its own size and sends information to the brain via several substances secreted by adipose tissue itself. As an effector, storage fat responds to hormonal or neural signals that stimulate its growth or cause it to release stored triglycerides. Only recently has the role of adipose tissue as an endocrine organ and as a component of the body's inflammation system been appreciated.

How Do the Body Fat Stores Develop?

Adipocytes arise from pluripotential stem cells, which are cells that can give rise to a variety of types of daughter cells. Once a stem cell has committed to become an adipocyte, it forms an adipoblast, which then becomes a preadipocyte—a fat cell that contains no lipid stores. Upon accumulation of lipids, an immature and then a mature adipocyte is formed. Both the glucocorticoid hormone **cortisol** and insulin are required for the differentiation of precursor cells into adipocytes. Adipose tissue is pictured in the photo given below.

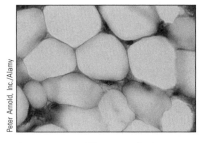
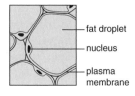

fat droplet

nucleus

plasma membrane

Adipose tissue. The cells of adipose tissue are specialized for fat storage.

By the second trimester of pregnancy, fetal development of white adipose tissue is well under way. After birth, the proliferation of preadipocytes from adipoblasts continues. Some of those preadipocytes will be incapable of ever becoming adipocytes.[17] Others, however, are poised to become full-fledged fat cells. When these preadipocytes accumulate detectable lipid, they can be counted as adipocytes. Fat cells serve an important purpose. When food is abundant, they remove lipids from the bloodstream for storage in almost limitless quantities. When energy is needed, such as during food shortages or to sustain physical activity, they release fatty acids to be used as fuel.

The average nonobese adult has between 30 and 50×10^9 (30–50 billion) adipocytes.[18] Although there is individual variation, a single adipocyte typically stores between 0.4 and 0.5 µg fat. In obesity, fat cell sizes of 0.6–1.2 µg are common. The enlargement of fat cells is called **hypertrophy** and is reversible. Most adult-onset obesity is due to hypertrophy. A critical mass of approximately 0.8 µg may stimulate the filling of additional preadipocytes in susceptible individuals.[17] This probably occurs because fat cells that reach a particular critical size release substances that stimulate adipocyte proliferation.[19] The resultant increase in the number of fat cells is known as **hyperplasia** and is permanent. Figure 4-6 illustrates the development and proliferation of fat cells.

Recall from Chapter 2 that most adipose tissue is found under the skin (subcutaneous). The remainder of stored fat is located within the abdomen and is known as internal visceral or intraabdominal fat.

hypertrophy
Enlargement; in this chapter, of fat cells.

hyperplasia
Excess number; in this chapter, of fat cells.

What Determines Fat Cell Number and Size?

What stimulates the development of preadipocytes? Why do some preadipocytes become fat cells and others remain unfilled? What is the signal generated by very large adipocytes

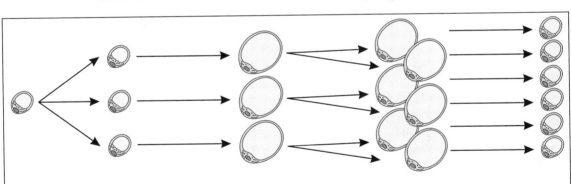

During growth, fat cells increase in number.

When energy intake exceeds expenditure, fat cells increase in size.

When fat cells have reached their maximum size and energy intake continues to exceed energy expenditure, additional preadipocytes will be filled, increasing fat cell number again.

With fat loss, the size of the fat cells shrinks, but the number does not.

FIGURE 4-6 Fat cell development.

Peter Arnold, Inc./Alamy

that stimulates hyperplasia? These are among the unanswered research questions that scientists continue to investigate. Research about the proliferation of fat cells is conducted with both rodents and cloned cells in culture. Some findings from these studies have also been documented in human cells.

One important observation about the replication of fat cells that applies to humans and other animals is that, if conditions are right, the *size* of white adipose cells can grow throughout life, and the *number* of white adipose cells may grow at least into early adulthood. For many years, it was thought that hyperplasia could occur only during three critical periods—the prenatal period, early childhood, and adolescence. More sophisticated methods to identify and count adipocytes and preadipocytes are now available, and scientists believe that fat cells probably can increase in number for a longer time period. One group of researchers theorizes that proliferation of fat cells may end at about age 20, but even this group acknowledges that they just do not know.[20]

It is also likely that in humans, as in rodents, the tendency to increase fat cell number varies anatomically. Certain fat deposits may simply be more sensitive to hormonal, neural, or nutritional conditions that stimulate the proliferation of more preadipocytes such that adipocyte number increases.

Lipogenesis, the accumulation or production of fat, occurs by the uptake of fatty acids into adipocytes. It is regulated by lipoprotein lipase and acylation-stimulating protein; insulin also promotes uptake of fat into the adipocytes. The breakdown of fat, known as **lipolysis**, results in the release of fatty acids into the bloodstream after they have undergone hydrolysis. Lipolysis is regulated by hormone-sensitive lipase. The balance between lipogenesis and lipolysis determines fat cell size. Adipose cells have both lipolytic and antilipolytic adrenergic receptors that play a key role in fat cell size, number, and function.

Adrenergic Receptors

As seen in Figure 4-2, the autonomic nervous system has two divisions: sympathetic and parasympathetic. Parasympathetic nerves release acetylcholine and are called **cholinergic**. Sympathetic nerves release norepinephrine (noradrenaline) and are called **adrenergic**. Adrenergic receptors are found in many tissues, including fat cells.

Adrenergic receptors on fat cells include five subtypes: three β-adrenergic receptors (β_1-, β_2-, and β_3-) and two α-adrenergic receptors (α_1- and α_2-). The β-receptors stimulate lipolysis and are most abundant in intraabdominal fat deposits and less abundant in gluteofemoral fat. The role of the β_3-adrenergic receptors in humans is controversial, but evidence increasingly suggests that these receptors have thermic (heat-generating) and anti-diabetic effects. Mutations in the β_3-adrenergic receptor gene are associated with obesity, insulin resistance, and type 2 diabetes. Low sympathetic nervous system activity may predispose someone to gain weight and prevent an obese individual from losing weight. Several experimental drugs that stimulate the β_3-receptors are being tested; these drugs will be discussed in Chapter 10.

The α-adrenergic receptors outnumber the β-receptors in fat cells. The α-receptors are antilipolytic and, therefore, promote fat storage. They also stimulate proliferation of preadipocytes, contributing to increased fat cell number. These receptors are more abundant in gluteofemoral than in intraabdominal fat. Since excess fat cells never go away, this explains in part why fat in the hips is more difficult to reduce than fat around the waist.

lipogenesis
Accumulation of fat in the adipocytes.

lipolysis
Breakdown of fat; a lipolytic process increases fat breakdown, and an antilipolytic process stops fat breakdown.

cholinergic receptors
Receptors on cells that are sensitive to parasympathetic activation; they respond to acetylcholine; fat cells do not contain cholinergic receptors.

adrenergic receptors
Receptors on fat cells that are sensitive to sympathetic activation; they respond to norepinephrine (α-adrenergic receptors) or epinephrine (β-adrenergic receptors) to regulate fat cell size and function.

Lipoprotein Lipase

lipoprotein lipase *Enzyme found in various body tissues, including adipose tissue; its purpose is to split apart fatty acids and glycerol in a process known as hydrolysis and promote fat storage.*

Lipoprotein lipase is an enzyme produced in many tissues to help remove fatty acids from the bloodstream. This enzyme is particularly active in white adipose tissue. It hydrolyzes the triglycerides being transported through the bloodstream and then stores fatty acids. It has been referred to as the gatekeeper of the fat cells because of its role in fat storage.

As you would expect, high levels of lipoprotein lipase promote the growth in size of fat cells. Lipoprotein lipase levels in the fat cells vary under different conditions:

- Levels increase after eating, peaking at about 2–3 hours after a meal.
- Levels increase in the presence of insulin.
- At least initially, levels decrease in fasting or dieting. After several weeks or months of low-calorie diet and weight loss, lipoprotein lipase levels in fat cells throughout the body actually increase in some people. Fat storage is then enhanced when eating returns to normal.
- Levels increase with exercise in some people, perhaps as a regulatory mechanism to prevent shrinkage of fat cell size below a certain volume.
- Levels decrease in insulin resistance (associated with type 2 diabetes) or insulin deficiency (type 1 diabetes). This reduces the clearance of triglycerides from the bloodstream, which is one explanation for high blood triglyceride levels seen in diabetics.

Lipoprotein lipase activity also varies among the different fat deposits:

- Higher activity in the gluteofemoral deposits, especially in women, which promotes greater fat cell size in the hips and the thighs (and which may ultimately lead to hyperplasia in this region).
- Lower gluteofemoral activity in men and higher activity in the abdomen, promoting more central fat storage, probably due to the influence of reproductive hormones.

Acylation-Stimulating Protein

acylation-stimulating protein *Small protein found in the plasma that acts as a strong stimulator of triglyceride synthesis in the fat cells.*

Recently, researchers discovered another substance—**acylation-stimulating protein**—that may enhance the uptake of fat in white adipose tissue. After a meal that contains fats, acylation-stimulating protein levels rise as triglycerides appear in the bloodstream and fall as triglyceride levels drop. Acylation-stimulating protein levels rise proportionately with the body mass index (BMI) and are therefore higher in obesity.[21] Acylation-stimulating protein levels are also higher in individuals with insulin resistance and cardiovascular disease, but levels can be reduced with weight loss and exercise. The effects on fat cells of lipoprotein lipase, acylation-stimulating protein, and adrenergic receptors are summarized in Table 4-2.

What Are the Other Functions of Adipose Tissue?

Discoveries within the past 10 years have broadened our knowledge of adipose tissue. It is no longer regarded as simply a fat storage depot but plays a larger role in eating, energy expenditure, weight management, and disease.

Adipose Tissue as Endocrine Organ

The endocrine system is comprised of glands that secrete hormones, which act like messengers. However, unlike the neurotransmitter messengers of the brain and nervous

TABLE 4-2 How Fat Cell Size Changes

These increase fat cell size:	These decrease fat cell size:
Lipoprotein lipase	**Adrenergic stimulation**
• Promotes triglyceride storage in fat cells (gatekeeper function) • Increases in the presence of insulin • Decreases in insulin resistance and insulin deficiency • Becomes more active in gluteofemoral fat deposits • May increase after prolonged dieting or in response to exercise	β-adrenergic receptors • Are lipolytic and predominate in internal visceral abdominal fat β_3 adrenergic receptors • May have thermogenic and anti-diabetic actions
Acylation-stimulating protein	
• Promotes triglyceride storage in fat cells • Increases in obese individuals • May be maintained at higher levels after weight loss	
Adrenergic stimulation	
α-adrenergic receptors • Are anti-lipolytic and predominate in gluteofemoral fat	

leptin *Protein produced by the fat cells that provides a satiety signal to the brain. The name "leptin" is derived from the Greek "leptos," which means "thin" because people who secrete normal amounts of leptin and whose hypothalamus is responsive to leptin are frequently lean.*

system, which are relatively fast-acting, hormonal messengers tend to have a long-term effect. Both the nervous and endocrine systems help to maintain homeostasis.

Adipose tissue secretes the hormone **leptin**, which has generated great interest among both researchers and the general public. For more than 20 years, researchers theorized about a satiety factor that seemed to be lacking in genetically obese (ob/ob) mice. In 1994, experiments confirmed that the ob/ob mouse has a defective or missing ob gene needed for the fat cells to produce leptin. Within about a month after genetically obese mice were injected with leptin, they lost 40% of their weight, almost all of which was fat. They accomplished this remarkable weight loss by eating less food and expending more energy. Suddenly, the mice that could not be sated had lost their profound drive to eat. Since that initial research, scientists have determined that mouse and human leptin have similar amino acid sequences.

Leptin is now understood to be a messenger between fat cells and brain that provides vital information about the size of the body's energy stores. Leptin receptors are found on neurons in several areas of the hypothalamus as well as in the circumventricular organs. Leptin levels rise when we eat. This kicks off a series of events that prevents overeating:

- production of pro-opiomelanocortin (POMC) is stimulated, which then increases production of α-melanocyte-stimulating hormone (α-MSH).
- simultaneously the melanocortin MC4R receptors are activated.[22]
- activity of neuropeptide Y (NPY) and agouti gene-related protein (AgRP) are reduced, and sensitivity to cholecystokinin (CCK) and bombesin are increased.[5,7]

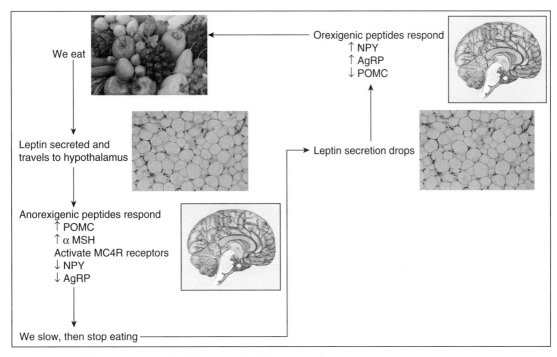

We eat

Leptin secreted and travels to hypothalamus

Anorexigenic peptides respond
↑ POMC
↑ α MSH
Activate MC4R receptors
↓ NPY
↓ AgRP

We slow, then stop eating

Orexigenic peptides respond
↑ NPY
↑ AgRP
↓ POMC

Leptin secretion drops

FIGURE 4-7   The negative feedback loop in which leptin works.

Thus, leptin increases activity of the anorexigenic effectors. Leptin levels rise more—and stay elevated for up to 24 hours—following low-fat, high-carbohydrate meals than high-fat, low-carbohydrate meals.[23] When leptin levels are low, none of these events can occur, the brain interprets this as starvation, and orexigenic effectors are activated. The drive to eat increases and energy expenditure decreases to conserve fat stores. The negative feedback loop of leptin is illustrated in Figure 4-7.

Mice producing too little leptin or an inactive form of leptin are at risk of weight gain. However, only a few extremely obese people are truly leptin deficient. Most obese humans produce amounts of leptin in expected amounts, based on their BMI and level of body fat, particularly their subcutaneous fat.[24] In humans, either leptin transport across the blood–brain barrier is defective, or leptin receptors that should be present in the brain are absent.

Or, perhaps hypothalamic leptin receptors are resistant to leptin, and this may have something to do with insulin. In laboratory animals, leptin deficiency is associated with high levels of circulating insulin. When these animals are given leptin injections, insulin levels fall, and body fat and weight are lost. However, diabetic, genetically obese mice (db/db) are not responsive to leptin. When (db/db) mice are given leptin injections, no reduction in food intake or body weight occurs, possibly because of nonresponsive hypothalamic leptin receptors. Obese humans with type 2 diabetes may be similarly leptin resistant.

Leptin plays a part in increased eating after dieting. When people lose weight or fast, plasma leptin levels fall. If you understand the negative feedback loop, then you may correctly conclude that reduced leptin drives people to eat. Leptin plays an additional role in

TABLE 4-3   Relationship Between Leptin and Weight Gain in Humans

Too little leptin is produced (only a small number of cases).
Expected amounts of leptin are produced:
• Leptin transport across the blood-brain barrier is impaired.
• Leptin receptors in the brain are absent or severely reduced.
• Hypothalamic receptors are insensitive to leptin.
Low leptin levels stimulate food-seeking behaviors.

animals. Leptin- or leptin-receptor-deficient mice have enormously increased capacity to smell and locate buried food, compared to wild-type mice.[25] Whether this accounts for some formerly obese or obese humans' ability to root out potato chips hidden in the pantry remains to be seen. The relationship between leptin and weight gain in humans is summarized in Table 4-3.

Recent studies indicate that leptin may also play a role in bone formation and the synthesis of a peptide produced by bone cells, called osteocalcin.[26] Leptin-deficient ob/ob mice have increased bone mass. Recall from Chapter 1 that overweight individuals seem to maintain bone mass better than the underweight, which has been attributed to the mechanical stress of carrying excess weight. This research suggests that leptin is also a factor.

Adipose Tissue and Inflammation

In Chapter 1, inflammation was highlighted as a significant factor in promoting cardiovascular disease. When plaque begins to form in an artery, markers of inflammation—including small proteins known as cytokines—are released into the bloodstream. Some of these cytokines are produced in adipose tissue and have been referred to as adipokines. Those that are associated with increased inflammation include tumor necrosis factor-α, interleukin-6, resistin, and leptin, discussed in the previous section. Levels of these are increased in obesity. Adiponectin is a cytokine associated with reduced inflammation, and its production is reduced in obesity.

Obesity has been characterized as a "chronic low-grade inflammatory state," like heart disease, type 2 diabetes, and metabolic syndrome.[27] Adipokines could be responsible for this. Other than leptin, which is found in greater amounts in subcutaneous fat cells, the pro-inflammatory adipokines are seen mainly in enlarged internal visceral abdominal adipose tissue. Their presence or, in the case of adiponectin, absence may be involved in insulin resistance.[5,7] Chapter 1 pointed out that, in addition to type 2 diabetes, central obesity is associated with cardiovascular disease and the metabolic syndrome. Shrinking hypertrophic fat cells reverses these comorbidities. Table 4-4 summarizes the actions of some adipokines.

What Determines Where Body Fat Is Deposited?

As noted earlier in this section of the chapter, fat may be stored centrally—in the internal visceral abdominal and subcutaneous abdominal fat deposits—or peripherally in the subcutaneous deposits of the hips, the thighs, and the arms. Much evidence points to visceral abdominal fat as a culprit in many disorders associated with obesity.

TABLE 4-4   Some Adipokines and Their Effects

Adipokine	Effect	Levels in Obesity
Leptin	Satiety ↑ Energy expenditure ↑ Insulin ↑ Inflammation	Increases with fat cell size, but receptors in brain may be defective
Adiponectin	Anti-diabetic ↓ Inflammation	Reduced
Resistin	↑ Insulin sensitivity	Reduced
Tumor necrosis factor-α	Insulin resistance	Increases

To a great extent, gender determines where fat is preferentially stored—peripherally in the hips and the thighs for most women, and centrally in the abdomen for most men. Gender differences in fat deposition are seen as early as 4 years of age.[28] These gender-based fat storage patterns undoubtedly evolved to sustain survival. Because lower body fat is broken down much more slowly than internal visceral abdominal fat and therefore, is a more permanent source of stored energy, this pattern of storage would support the female role of childbearing and lactation. Visceral abdominal fat, in contrast, is more readily mobilized for energy and therefore would better support the male role of hunting.[17]

Of course, not all women tend toward the classic pear-shaped pattern of lower-body obesity; nor do all men show the apple-shaped pattern of upper-body obesity. The number of fat cells in a particular area and the size that these cells eventually attain ultimately determine fat pattern. These outcomes are genetically and hormonally regulated.

Upper and Lower Body Fat Deposits Are Different

The adipocyte deposits in different anatomical regions vary in many respects. This is the good news and the bad news of fat storage: The lower-body subcutaneous fat pattern (once called "gynoid"), seen in more women than men, is far less associated with health problems than the upper-body visceral abdominal fat pattern. However, gluteofemoral fat is much more difficult to reduce. The upper-body fat pattern (once called "android"), seen in more men than women, is associated with numerous health complications, but it is easier to reduce fat stored intraabdominally. Some people with more fat around the middle actually have greater amounts of subcutaneous—not visceral—abdominal fat. Those people have been termed "metabolically benign," because even with a large waist circumference they are not insulin resistant.[29] Table 4-5 summarizes the differences between upper- and lower-body fat.

The most obvious difference between the fat deposits is the size and the number of fat cells. Lower-body deposits are likely to be composed of a greater number of smaller adipocytes. Smaller adipocytes are more insulin sensitive, but the fact that there are more of them makes reductions of overall levels of fat difficult. And in women, both the size and the number of gluteofemoral cells are larger than in men. The internal visceral abdominal adipocytes are likely to be composed of cells that are larger in size but not in number.[28] Their larger size makes them less insulin sensitive and more likely to pose health risks. Hyperplasia of abdominal adipocytes is not impossible, however. Several

TABLE 4-5 Characteristics of Upper- and Lower-Body Fat Distribution Patterns

	Lower-body fat pattern	Upper-body fat pattern
Also known as	Gluteofemoral; pear shape; gynoid; peripheral fat	Internal abdominal; visceral; central; intraabdominal; apple shape; android
Primarily seen in	Women	Men
Fat cell characteristics	Hyperplasia	Hypertrophy
Innervation	Less sympathetic innervation; more α_2- and fewer β-adrenergic receptors	Greater sympathetic innervation; fewer α_2- and more β-adrenergic receptors
Hormonal sensitivity	Estrogens enhance fat storage	Androgens enhance fat storage; sensitive to cortisol
Metabolic activity	Higher lipoprotein lipase (LPL) activity promotes fat storage	Higher rate of glucose metabolism and lipolysis
Effects of weight loss	Less change evident in weight loss	Greater change seen after weight loss; mobilized during exercise
Health implications	Not associated with health complications until large amounts are accumulated	Associated with hyperinsulinemia, elevated blood lipids, and hypertension

experts have pointed out that more dormant preadipocytes may exist in the central fat deposits than in peripheral deposits.[17,30]

Lower-body subcutaneous adipocytes are richer in α-adrenergic than β-adrenergic receptors, making these areas more likely to store fat than to break it down. Central nervous system stimulation of α_2-adrenergic receptors causes proliferation of pre-adipocytes, which explains why adipocyte number is larger in the gluteofemoral region. Intraabdominal adipocytes, on the other hand, have more β-adrenergic receptors, which are far more susceptible to circulating catecholamines. Catecholamine levels are elevated during exercise—one reason that exercise is so effective in shrinking internal abdominal fat.

Glucocorticoid receptors also differ between regional fat deposits. People who have a disease known as **Cushing's syndrome** secrete higher than normal levels of cortisol, a glucocorticoid, and exhibit a pronounced abdominal pattern of fat deposition. The internal visceral abdominal adipocytes are very sensitive to cortisol, which in turn causes increased lipoprotein lipase activity in the abdominal fat cells. People who are under stress or depressed frequently have elevated cortisol levels, which may promote greater abdominal fat storage.

Sex steroid levels differ between those with different patterns of fat storage. The principal sex steroid in males is testosterone (an androgen); in females, it is estrogen. The internal visceral abdominal fat cells have more androgen receptors, and subcutaneous fat cells have more estrogen receptors. At puberty, increased testosterone production in boys is accompanied by a more evident pattern of abdominal fat deposition and lower levels of a protein that binds to testosterone—sex hormone–binding globulin. Increases in estrogen and progesterone in girls produce a more evident lower-body pattern of fat distribution. Women who deposit fat intraabdominally have higher circulating levels of free testosterone and lower levels of sex hormone–binding globulin. Men who deposit more fat in the gluteofemoral region have higher levels of estradiol, a form of estrogen secreted

Cushing's syndrome
Metabolic disorder in which oversecretion of adrenocortical hormones results in increased fat storage in the upper body, high blood glucose levels, diabetes, impaired immunity, and osteoporosis.

in small quantities by the testes. The hormones received by men and women undergoing sex change affect fat patterning similarly.[17] In other words, women receiving testosterone injections develop a more central pattern of fat deposition, and men receiving estrogen begin to deposit more fat in the lower body.

Can Fat Pattern Be Changed?

Can professionals help people change their shape to improve their health? Yes and no. Because subcutaneous gluteofemoral fat is more stable and less likely to be mobilized even during exercise, individuals with the lower-body fat pattern have a shape more resistant to change. Internal abdominal fat, in contrast, mobilizes more readily and may be more likely to change.

Storage Fat: Summary

Far from being a passive recipient of excess lipid, storage fat (or white adipose tissue) is an active participant in the maintenance of its size and location. Under the right neural, hormonal, or nutritional conditions, the number of fat cells can increase throughout life. Fat cells get larger when levels of lipoprotein lipase or acylation-stimulating protein increase, and they shrink under β-adrenergic stimulation. Because both exercise and caloric restriction may elevate lipoprotein lipase, attempts to maintain weight loss through diet and exercise may be sabotaged in some individuals. Adipose tissue also has endocrine and immune functions. It secretes the hormone leptin, which plays a significant role in providing information to the brain about size of fat stores. Adipose tissue also secretes adipokines that promote inflammation and increase the risk of cardiovascular disease, type 2 diabetes, and metabolic syndrome.

The sex hormones regulate where fat is deposited. Males as well as females who produce more testosterone and less sex hormone–binding globulin store fat primarily in the visceral abdominal deposits; females and males who produce more estradiol store fat primarily in the hips and the thighs. A reduction in body fat may be especially difficult for those with gluteofemoral fat, an excess number of fat cells, leptin resistance, or reduced response to adrenergic stimulation.

BAT: EFFECTOR OF ENERGY EXPENDITURE

As discussed in Chapter 3, brown adipose tissue (BAT) is a specialized form of fat tissue that contributes to energy expenditure through its capacity for thermogenesis, or the production of heat. In humans most BAT is located in the neck and shoulder area. Scientists believe that brown fat develops prenatally and is fully functional at birth. If all of the brown fat in a typical adult's body were combined, it would weigh less than 1 kg.

The genes that give brown fat its unique characteristics are regulated by a protein in BAT, called PRDM16. When activated, PRDM16 initiates the sequence of events that form brown adipocytes from preadipocytes. In rodents, this can actually happen in white adipose tissue, so a cold-exposed mouse experiences a proliferation of brown fat cells within storage fat. In humans this same effect is not observed, and PRDM16 only increases the size of BAT within brown fat depots. But the fact that scientists now know that BAT can increase in size in humans is important news.

How Does Thermogenesis Occur?

BAT consists of densely packed mitochondria, substantial sympathetic innervation in the form of many β-adrenergic receptors, and—unique to BAT—a specialized protein called uncoupling protein. When uncoupling protein is stimulated, it short-circuits normal energy metabolism, bypassing oxidative phosphorylation. The result is heat. BAT also contains a rich blood supply, which allows it to dissipate the heat that it produces.

Because of its numerous β-adrenergic receptors, brown fat is very sensitive to norepinephrine. When BAT is stimulated by cold stress or food intake, thermogenesis occurs. Most humans who have central heating and proper clothing are not routinely subjected to cold stress and mainly rely on food to activate BAT.

The hypothalamus is involved in regulating both body temperature and food intake, so it should be no surprise to learn that the hypothalamus is involved in regulating brown fat activity. Several of the peptides listed in Table 4-1 increase or decrease thermogenesis. In addition, the heat produced by BAT may be another mechanism that signals satiety.[31]

Is BAT Defective in Obesity?

Mice with the ob/ob gene cannot survive at temperatures below 48° Celsius. Even though norepinephrine is secreted during cold stress, the brown fat is unresponsive. In addition, the brown fat of some genetically obese animals is not responsive to adrenergic stimulation, and obese and average-weight rats that are fasting have either a loss or a deactivation of uncoupling protein. Instead of producing heat from ingested nutrients, food energy is conserved and stored as fat.

You learned in Chapter 3, that positron emission tomography (PET) combined with computer tomography (CT) only recently has permitted the study of BAT in humans. Like animals, humans may also have defects in BAT that result in energy conservation and weight gain. A study of 18- to 32-year-old lean and overweight or obese men found significantly reduced response to cold in the overweight or obese men, compared to the lean.[32] Others have also found lower BAT response as body fat increases.[33] Whether the defect is within the brown fat itself, within the hypothalamus, or a combination of both is not yet clear. Also not known is how prevalent inactive brown fat is or if there are gender and age differences.

Although diet-induced thermogenesis accounts for only a small proportion of daily energy use, if this process were faulty, a slow but steady increase in fat stores might occur over the course of many years. Unfortunately, BAT activation is not voluntarily controlled. Exercise, a high-carbohydrate diet, and frequent, small meals have been shown to increase thermogenesis in rodents, but experts are not in agreement that these approaches work similarly in humans.

BAT: Summary

Unlike the tissues and systems discussed thus far, BAT is involved in weight regulation by promoting caloric expenditure. Brown fat provides a handy way to burn excess calories through the production of heat. Defective BAT, known to occur in some obese rodents and believed to occur in some obese humans, could contribute to weight gain in humans, too.

GENETIC FACTORS AND BODY COMPOSITION

An explosion of evidence over the past decade substantiates that there are genetic influences on hunger and satiety that affect body weight, fat storage, fat patterning, energy expenditure, food preferences, eating behavior, and possibly even response to different diet programs. Studies of genetically obese animals, human twins, adoptees, and population groups, such as the Arizona Pima Indians and South Pacific islanders, have contributed to the knowledge about hereditary aspects of weight gain.

Some human obesities are known to result from defects in specific gene locations on various chromosomes. These include Prader-Willi, Bardet-Biedl, Alstrom, and Cohen syndromes. In **Prader-Willi syndrome**, for example, which is the most common of the genetically linked human obesities, the defect lies in a specific segment of chromosome 15. The 2005 update of the Human Obesity Gene Map identifies 11 different single-gene mutations that have led to a small number of reported cases of severe obesity.[34] In most cases, however, obesity does not result from a single genetic cause, but rather from a multiplicity of factors, probably involving interactions among several genes and most certainly having environmental influences.

Prader-Willi syndrome
Chromosomal disorder characterized by voracious appetite, obesity, and developmental delays or mental retardation.

How Are Traits Inherited?

Obesity is generally considered to be a polygenic trait, which means that it comes from the expression of more than one gene. An observable trait, such as weight, height, or fatness, is called a **phenotype**. The genetic combination that causes the observable trait is known as a **genotype**. The term heritability is used to describe the degree to which a phenotype is related to a genotype, and it is expressed as a percentage, from 0% to 100%. Height, body weight, and fatness all show a range of values.

Genes are the carriers of information about an individual. Genes are made up of deoxyribonucleic acid (DNA) and are located at various points along the **chromosomes** in places called gene loci. There are 46 chromosomes arranged in 23 pairs in the nucleus of every human cell, except the reproductive cells (sperm and ova), each of which contains just one set of 23 chromosomes. Chromosome pairs are called homologues or homologous chromosomes. Within a given chromosome pair, each chromosome looks about the same and carries genes for the same traits, although on one chromosome the eye-color gene may stipulate "blue" and on the other chromosome the eye color gene may stipulate "brown." Different genes for the same trait that are found on corresponding loci of homologous chromosomes are known as alleles. Your set of alleles represents your genotype. Figure 4-8 illustrates some of this terminology on a pair of homologous chromosomes.

A genome is an organism's complete set of DNA. In April 2003 the Human Genome Project completed mapping the human genome sequence, and now the International HapMap Project is moving forward to identify genes that are associated with health and disease. A group of scientists at the Pennington Biomedical Research Center has created a Human Obesity Gene Map and believes that 20 to 30 genes identified so far are involved in human obesity (www.pbrc.edu).[34] Mutations or alleles of any of these genes could result in genotypes for different patterns of fat storage, energy expenditure, peptide secretion, and response to neurochemicals as well as different responses to overfeeding, underfeeding, and physical activity. The outcome: multiple obesity phenotypes.

phenotype
Observable trait, such as eye color or pattern of fat distribution.

genotype
Combination of genes that causes an observable trait.

chromosomes
Single elongated DNA molecules. There are 46 chromosomes arranged in 23 pairs in the nucleus of every human cell except the reproductive cells. The chromosomes in a pair are known as homologous chromosomes. DNA segments known as genes are located at various points along each chromosome.

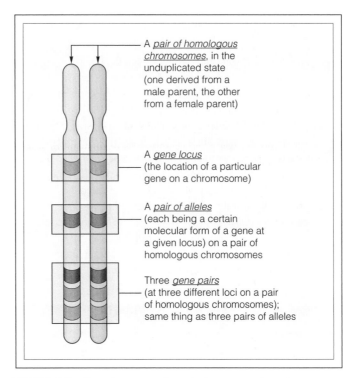

A *pair of homologous chromosomes*, in the unduplicated state (one derived from a male parent, the other from a female parent)

A *gene locus* (the location of a particular gene on a chromosome)

A *pair of alleles* (each being a certain molecular form of a gene at a given locus) on a pair of homologous chromosomes

Three *gene pairs* (at three different loci on a pair of homologous chromosomes); same thing as three pairs of alleles

FIGURE 4-8 Genetic terminology. In humans, pairs of genes are found on pairs of homologous chromosomes. One chromosome is inherited from each parent.

What Genes Might Be Obesity-Promoting?

Logic suggests that heredity would favor weight gain to ensure survival of the species. In human history, those who survived famine (and passed on their genes to their offspring) were able to store enough energy to get through hard times. Genetically obese mice, for example, are energy-thrifty and live 25% longer during a fast than lean mice.[35] People who inherit such a "thrifty" gene would have a survival advantage when food is scarce. However, when food is plentiful over a long period of time, such a genetic advantage would quickly turn into a disadvantage, with the development of obesity and related comorbidities.

epidemiological studies *Studies of the frequency of disease and factors associated with health or changes in health.*

Epidemiological studies of twins, adopted children, and large families have found heritability values for both BMI and body composition to range from about 40% to 70%.[36] Monozygous ("identical") twins raised apart have BMIs that are as similar to each other as the BMIs of twins raised together. In fact, adopted twins have BMIs much more like their co-twins raised in other households than their adoptive families.[37] Studies of families in which there are obese young children have determined that when both parents are obese, the child has a 10-fold greater probability of becoming obese by age 7 years.[38] While environment surely plays a role, these studies provide compelling evidence for the existence of an underlying genetic predisposition for body weight.

All of the obesity-promoting genes identified to date act on the brain and its anorexigenic or orexigenic neuropeptides—those substances that are linked to hunger and satiety and can protect us from overeating. The phenotypes influenced by these genes include:

- Anthropometric phenotypes: BMI, percent of weight made up of fat, fat distribution, sum of skinfolds, and waist and other circumferences.
- Metabolic phenotypes: Blood glucose level, plasma insulin level, resting metabolic rate, lipoprotein lipase production, leptin, and neuropeptide receptors.
- Behavioral phenotypes: Food preferences and physical activity level.

Even ten years ago scientists were unaware of single-gene mutations that might cause human obesity, but today genes responsible for leptin, leptin receptors, POMC, melanocortin receptor subtype 4R (MC4R), and a protein known as FTO are implicated in obesity. The MC4R receptor gene probably accounts for most cases—up to 6 percent—of severe obesity in children and adults.[39] This mutation increases hunger, drives people to overeat and binge, and is linked to severe hyperinsulinemia. However, over 94% of severe obesities have other causes.

Congenital leptin-receptor deficiency has been reported in 12 humans.[40] All developed severe obesity in early childhood. Genome scans of individuals in the Québec Family Study and the Samoan Family Study of Overweight and Diabetes have also found linkages between morbid obesity and leptin.[41]

Products of Adipocytes Are Inherited

Several additional lines of inquiry are looking at excess production of substances that enhance fat storage. Lipoprotein lipase is needed to store fat, and high levels of the enzyme are associated with larger fat cells. In addition, exercise may increase lipoprotein lipase levels, making it more difficult for fat cells to shrink. Some evidence indicates that lipoprotein lipase production in adipose cells is inherited and that the response of lipoprotein lipase to exercise is also genetically influenced.[42] Identical twins respond very similarly in lipoprotein lipase levels following overfeeding or exercise.[17]

tumor necrosis factor-a (TNF-a)
A substance produced in all cells, including fat cells, that is known to play a role in immunity and is believed to have a role in insulin resistance.

Another fat cell component associated with increased adipocyte size is the adipokine **tumor necrosis factor-**α (TNF-α), a substance produced in all human cells and known to have a role in immunity and fighting infection. As learned earlier in this chapter, adipose tissue TNF-α is elevated in obesity and promotes inflammation, as well as hyperinsulinemia and insulin resistance. Researchers studying TNF-α believe that enlarged fat cells may cause the gene responsible for TNF-α to be overproduced in obesity.[43]

Fat cell receptors that regulate the differentiation of preadipocytes to adipocytes are also under genetic control. Overexpression of a key gene involved in adipocyte development, the peroxisome-proliferator-activated receptor gene, accelerates the proliferation of fat cells.[44]

What Other Genetic Factors Might Promote Weight Gain?

A low–energy-expenditure phenotype may make some people more susceptible to obesity. The studies of twins and families have led researchers to conclude that at least a portion of what determines our metabolic rate is inherited.[45] Because more than half of our daily caloric expenditure occurs via the metabolic rate, a person with low metabolic rate expends fewer calories and, therefore, is at risk of weight gain.

This tendency to conserve calories may begin very early in life. One study of infants discovered that half of babies born to overweight mothers became overweight during their first year of life.[46] No babies born to lean mothers during the same time period became overweight. The doubly labeled water method was used to determine the infants' 24-hour energy expenditure at 3 months of age. Surprisingly, the infants who became overweight by the time they were one year old expended less energy at 3 months. Because they would probably be too young to imitate the behavior of physically inactive family members, this finding suggests that a tendency toward lowered activity might be inherited. Recall that low leptin levels, associated with obesity, also reduce energy expenditure, so perhaps that was a factor in low activity levels among these babies.

Development of the Human Fitness Gene Map began in 2001 and continues to report new genes that influence physical activity or exercise-related traits. Genetic studies have identified genotype-phenotype associations for endurance performance, muscular strength, heart rate and blood pressure responses to exercise, and changes in bone density, fatness, and body weight following physical activity. Even heart transplant patients who have particular gene variants respond differently to exercise.[47] For a variety of genetic reasons, people may be more or less resistant to exercise as an obesity treatment or preventive strategy. In addition, there is evidence of genotypes that influence how active—or sedentary—people are. Physical activity/inactivity is estimated to be 46% to 57% heritable.[47]

There may also be genetic influences on the effectiveness of various diets in promoting weight loss. Stanford University researchers analyzed DNA from 101 overweight women for genes needed to metabolize fats and carbohydrates and classified women as having a low-carbohydrate diet responsive genotype, a low-fat diet responsive genotype, or a balanced diet responsive genotype. These women were previously in a 2005 study that examined the effectiveness of several low-fat or low-carbohydrate diets in reducing weight over one year. In that study women were randomly assigned to the Atkins, Zone, Weight Watchers, or Ornish diets. While adherence to the diets was low, those who stuck with the program for a year lost small amounts of weight, and there were no noticeable differences between diets. After the researchers reanalyzed 2005 study findings with information about genotype, they reported at an American Heart Association conference in March 2010 that women who had been in the diet group which matched their genotype lost 2–3 times more weight than women who followed a diet that did not match their genotype.[48]

Genetic Influences on Food Choice

Most food choices are based on learned preferences for the flavor, texture, temperature, and appearance of various foods. If there is any hereditary influence on food choice, then it is more likely related to nutrient intake than to eating specific foods. In other words, people may inherit a preference for carbohydrate but probably do not inherit a preference for specific carbohydrate-containing foods (like spaghetti instead of potatoes).

This chapter has already discussed several neuropeptides that may regulate nutrient intake. Neuropeptide Y, for example, seems to entice people to eat more carbohydrate foods, whereas agouti gene-related protein, melanin-concentrating protein, and galanin

Case Study. The Hungry Family, Part 2

The Rhyses recently read about a family obesity study in which all family members would undergo genetic testing and counseling. The study is looking at several "obesity genes," including the MC4R receptor. The family attended an information session at the university that was conducting the research with funding from a federal agency. They learned that they would all undergo several medical tests, as well as genetic testing. After being informed in more detail about the risks and benefits of the study, Mr. and Mrs. Rhys signed informed consents. Joey also received information about the study, and both he and his parents signed his consent form. All procedures were consistent with university and federal requirements for research involving human subjects.

Do some reading about genetic testing guidelines (both the American Academy of Pediatrics and American Medical Association have guidelines) and specifically about the MC4R gene.

- How is this type of genetic testing done? Is it painful or invasive?
- Consider some of the issues that arise when parents decide for a child that he/she should undergo genetic testing.
- Do the benefits of learning about one's MC4R gene status outweigh the risks of having such information?
- Should Joey be informed about the results of his genetic testing? Should others in the Rhys family be informed?
- How do you think the outcome of genetic testing might affect Joey's psychological status; self esteem; future health risks?

are associated with overconsumption of high fat foods. So any genetic factors that influence the neuropeptides might explain genetic influences on food choice:

How Do Heredity and Environment Interact?

Because of genetic factors, people whose family history indicates a predisposition toward overweight may have to be particularly diligent to prevent excess weight gain. Nevertheless, even when a genetic susceptibility exists, for most people there must be a favorable environment for genes to be expressed. This makes control of nongenetic influences extremely important in weight management.

Fat Promoting Dietary Factors

Dietary fat promotes fat gain. Fats contain more than twice as much energy as carbohydrate or protein. So, when the same weight of food is eaten, foods rich in fat are twice as caloric as foods rich in carbohydrate. Some people may be genetically inclined to consume higher fat foods. Fats also tend to be metabolized more efficiently than carbohydrates. In other words, the body expends fewer calories processing fats than processing carbohydrates or proteins and, therefore, more energy is left over to be stored. When rats are fed an equal number of calories, with one group getting a high-fat diet and the other a low-fat diet, the high-fat group gains more weight.[49]

Most studies with humans have shown a weak relationship, but a relationship nevertheless, between fat intake and obesity. Dietary fat is particularly obesity promoting when it is part of a high-calorie diet. Americans' caloric intake has increased steadily since the early 1980s. Not only is food widely available in all types of fast-food restaurants, convenience stores, grocery stores, and even gas stations, but today much of our food is purchased ready-to-eat. Little energy has to be expended to locate or prepare food. Our intake of tasty, low-fat foods and fat substitutes increased dramatically in the 1990s, leading people to assume that caloric intake has declined. This is not necessarily the case.

Food intake alone, however, cannot account for obesity. Leading obesity researcher Jules Hirsch once noted that, over a lifetime, mildly obese people probably consume about the same quantity of food as nonobese people.[50] Lack of sufficient physical activity must be acknowledged as a second environmental factor promoting weight gain.

Inactivity and Weight

People in traditional cultures who adopt Western diets and low activity levels increase their susceptibility to weight gain. The Mexican Pima Indians have both lower BMIs and prevalence of type 2 diabetes than the Arizona Pima. In addition to consuming a diet lower in fats, the Mexican Pima are engaged in heavy labor more than 40 hours per week, whereas the Arizona Pima consume more fat and are less active.[51] Nobody is suggesting that the average person should sustain more than 5 hours of vigorous physical activity each day, but a level of activity between almost total sedentism and vigorous exercise is indicated for obesity prevention and may be important in treatment.

Neurotransmitters which influence energy expenditure, as well as inclination toward physical activity itself, are under some degree of genetic control. Patterns of inactivity that develop at a young age and are predictive of weight gain are influenced by additional factors. One of the most significant promoters of physical inactivity among children is television, and childhood weight gain has been linked to time spent watching television.

Exercise and physical activity don't simply burn calories; they also have a number of other positive effects in the prevention and treatment of obesity, including improved health and the fitness needed to keep being active throughout life. The role of physical activity in obesity prevention and treatment is discussed fully in Chapters 7 and 8.

Family Factors Influencing Weight

Inherited factors may promote obesity, but the family influences eating and activity habits of its members in other ways, too. Cultural values affect dietary behaviors, such as how particular foods are valued, the way foods are prepared, and the importance of eating at social, cultural, or religious events; how activity is valued; and an acceptable level of weight. As covered in Chapter 1, the socioeconomic status of a family may also influence the body weight of its members.

Environment and Genetic Susceptibility Collide: Obesity Among American Indians and Pacific Islanders

Obesity and its comorbidity, type 2 diabetes mellitus, have been increasing in frequency since the 1960s among several native populations. Because of the high prevalence of type 2 diabetes among the Arizona Pima Indians, they have been studied extensively since 1965. Thousands of miles away, the Nauru Islanders of the western Pacific, who have similar problems, have been studied almost as closely. Findings from these investigations offer insights into the relationships among obesity, disease, environment, and heredity.

Arizona Pima Indians   The Pima Indians have lived since 300 B.C. in what is now southern Arizona and northwestern Mexico. The Pima survived in their arid environment by hunting, gathering indigenous plants and seeds, and developing elaborate irrigation systems to support the growth of crops, including maize, pumpkins, beans, and gourds. The traditional Pima diet was low in fats and rich in carbohydrates and plant sources of protein, with the occasional addition of birds, rabbit, or deer. The Pima lived in this manner, coping with an occasional drought or flood, for almost 2,000 years.

In 1854, the United States acquired southern Arizona from Mexico and, by 1900, was diverting the area's water supply to support white settlers. With completion of the Coolidge Dam on the Gila River in 1929, some crops were reestablished, but today little agriculture remains and most crops are sold. As fast-food restaurants and convenience stores have proliferated near Indian lands, the modern Pima diet is very much like the typical American diet—low in fiber, high in fat (especially animal fat), and high in calories.

Today the Arizona Pima have very high rates of obesity, and half of them have type 2 diabetes. This is in striking contrast to the Mexican Pima, who live in a mountainous, rural area of southeastern Sonora. The Mexican Pima are leaner, extremely active in their daily livelihood, and have retained a traditional diet of beans, corn tortillas, and potatoes—a diet high in fiber and about 23% fat, primarily from plant sources.

Pacific Islanders   The Pacific is home to three major ethnic groups—Melanesians, Polynesians, and Micronesians. Epidemiologists note that the prevalence of type 2 diabetes is low among most Melanesians and among Polynesians and Micronesians who adhere to a traditional lifestyle. However, in urban areas, the rate of type 2 diabetes and obesity has soared.[52] The Nauruans are of Micronesian ancestry. Their formerly agricultural society was transformed in the 1970s as a result of the discovery of high-grade phosphate ore. The mining industry moved the country from predominately rural into an urban nation that at one time gave its inhabitants one of the highest per-capita incomes in the world. Nauru now imports most of its foods. Studies of Polynesians living on the island of Western Samoa, Tonga, Niue, the Cook Islands, and French Polynesia have also shown higher diabetes rates among city dwellers, compared to rural.

What Is the Shared Susceptibility of These Populations?   People who live in unindustrialized societies, where physical activity is necessary and consumption of excess fat and calories is rare, do not exhibit the same prevalence of obesity and overweight seen in more developed societies. Heredity cannot be discounted as one explanation for the high rates of type 2 diabetes and obesity seen among Pima Indians and Pacific islanders. However, if obesity-promoting genes were solely responsible for fat gain, then it is unlikely that these genes would be seen only in industrialized, affluent populations. Therefore, other factors must explain obesity and diabetes:

- Diets high in animal fat and calories and low in fiber.
- Reductions in physical activity due to more sedentary recreational and occupational pursuits.
- Prenatal environment: Babies born to Pima Indian women who have diabetes during pregnancy have higher birth weights and are more likely to be obese and to remain obese through adulthood than babies born to nondiabetic women. By 29 years old, 83% of these higher-birth-weight babies are diabetic.[53]

These factors present a strong case for the role of environment in enhancing the expression of obesity and type 2 diabetes genotypes. The result of this interaction between environment and heredity is a cycle of obesity and diabetes that repeats itself through each successive generation.

Genetic Factors and Body Composition: Summary

Sufficient evidence exists to conclude that a small number of obesities are clearly linked to genetic factors and that the susceptibility to becoming obese is heritable. All of the obesity-promoting genes identified to date act on the brain and the neuropeptides that regulate hunger, satiety, and energy balance. These include genes responsible for leptin, leptin receptors, POMC, melanocortin receptor subtype 4R (MC4R), and the FTO protein. The MC4R receptor gene accounts for up to 6 percent of severe obesity in children and adults. Additional genes are undoubtedly responsible for other factors that promote fat gain, including those that regulate energy expenditure, exercise behavior and response to exercise, size and proliferation of fat cells, macronutrient metabolism, and preference for particular nutrients. A single gene does not explain obesity in the great majority of cases. But individuals with genetic susceptibility to obesity may gain weight easily in the presence of a favorable environment, and treatments may be ineffective through no fault of the individual.

CONCLUSION

This chapter has examined a number of factors known to influence food intake, energy expenditure, body weight, and fat deposition. You should now be able to begin to address the questions posed at the beginning of the chapter: Why do we eat? Why do we stop eating? What makes us choose particular foods? Why do some people seem to be able to eat all they want and never gain weight while others have to watch every calorie?

Part of the answer to these questions is that we follow our noses, so to speak, and our own free will. After all, no one really needs to buy a recliner with a built-in cooler. But we cannot ignore the growing body of evidence that there are biological forces capable of overriding our so-called will power. Differences between lean and obese individuals exist in receptors and effectors—brain and digestive tract, neurotransmitters, adipose tissue, brown fat, leptin, sympathetic nervous system activity—as well as the genes that regulate these things, which may confound best efforts at weight management. The modern lifestyle is enough to overwhelm homeostatic regulation in the genetically susceptible person.

As health professionals, we need to acknowledge that some of the factors that promote obesity are simply beyond the individual's control. Knowledge of this can be empowering to the person who has believed for so long that he or she is just weak-willed. From a public health standpoint, we should support initiatives that are obesity-preventing, such as walking paths, sidewalks, parent education, food and menu labeling, and healthy school foods. From an individual health standpoint, our efforts must be directed at helping people develop the knowledge and skills to attain healthy diet and activity patterns in an obesity-promoting environment.

Case Study. The Hungry Family, Part 3	Genetic testing indicated that Joey has one defective MC4R gene, inherited from one of his parents. The family obesity study researchers referred the Rhys family to a registered dietitian so they could get some assistance in dealing with their obesity-related health problems and could help Joey. Mr. Rhys continues to believe that his and Joey's continued weight gain is inevitable and that changing their diet will have no effect. Mrs. Rhys thinks that the family's dietary choices and inactivity might play a larger role than heredity.

- What environmental factors might be contributing to this family's weight and health problems?
- Do you think that obesity is inevitable for the Rhys family?

REFERENCES

1. King, N. A., Tremblay, A., & Blundell, J. E. (1997). Effects of exercise on appetite control: Implications for energy balance. *Medicine and Science in Sports and Exercise, 29*(8), 1076–1089.

2. Tataranni, P. A., Gautier, J. F., Chen, K., Uecker, A., Bandy, D., Salbe, A. D., et al. (1999). Neuroanatomical correlates of hunger and satiation in humans using positron emission tomography. *Proceedings of the National Academy of Sciences of the United States of America, 96*(8), 4569–4574.

3. Luiten, P. G., ter Horst, G. J., & Steffens, A. B. (1987). *Progress in Neurobiology, 28*(1), 1–54.

4. Williams, G., Harrold, J. A., & Cutler, D. J. (2000). The hypothalamus and the regulation of energy homeostasis: Lifting the lid on a black box. *Proceedings of the Nutrition Society, 59*(3), 385–396.

5. Flier, J. S. (2004). Obesity wars: Molecular progress confronts an expanding epidemic. *Cell, 116*(2), 337–350.

6. Le, D. S. N., Pannacciulli, N., Chen, K., Salbe, A. D., Del Parigi, A., Hill, J. O., et al. (2007). Less activation in the left dorsolateral prefrontal cortex in the reanalysis of the response to a meal in obese than in lean women and its association with successful weight loss. *American Journal of Clinical Nutrition, 86*(3), 573–579.

7. Fry, M., Hoyda, T. D., & Ferguson, A. V. (2007). Making sense of it: Roles of the sensory circumventricular organs in feeding and regulation of energy homeostasis. *Experimental Biology and Medicine, 232*(1), 14–26.

8. Epstein, L. H., Wright, S. M., Paluch, R. A., Leddy, J. J., Hawk, L.W.Jr , Jaroni, J. L., et al. (2004). Relation between food reinforcement and dopamine genotypes and its effect on food intake in smokers. *American Journal of Clinical Nutrition, 80*(1), 82–88.

9. Isganaitis, E., & Lustig, R. H. (2005). Fast food, central nervous system insulin resistance, and obesity. *Arteriosclerosis, Thrombosis, and Vascular Biology, 25*(12), 2451–2462.

10. Rolls, B. J., Rolls, E. T., Rowe, E. A., & Sweeney, K. (1981). Sensory specific satiety in man. *Physiology & Behavior, 27*(1), 137–142.

11. Geliebter, A., Schachter, S., Lohmann-Walter, C., Feldman, H., & Hashim, S. A. (1996). Reduced stomach capacity in obese subjects after dieting. *American Journal of Clinical Nutrition, 63*(2), 170–173.

12. Read, N., French, S., & Cunningham, K. (1994). The role of the gut in regulating food intake in man. *Nutrition Reviews, 52*(1), 1–10.

13. Backman, L. & Hallberg, D. (1974). Small-intestinal length. *Acta Chirugica Scandinavica, 140,* 57.

14. Berthoud, H. R. (2008). Vagal and hormonal gut-brain communication: From satiation to satisfaction. *Neurogastroenterology and Motility, 20 Suppl 1,* 64–72.

15. Little, T. J., Horowitz, M., & Feinle-Bisset, C. (2007). Modulation by high-fat diets of gastrointestinal function and hormones associated with the regulation of energy intake: Implications for the pathophysiology of obesity. *American Journal of Clinical Nutrition, 86*(3), 531–541.

16. Schwartz, M. W., Figlewicz, D. P., Baskin, D. G., Woods, S. C., & Porte, D., Jr. (1992). Insulin in the brain: A hormonal regulator of energy balance. *Endocrine Reviews, 13*(3), 387–414.

17. Kissebah, A. H., & Krakower, G. R. (1994). Regional adiposity and morbidity. *Physiological Reviews, 74*(4), 761–811.

18. Leibel, R. L., Berry, E. M., & Hirsch, J. (1983). Biochemistry and development of adipose tissue in man.

In H. L. Conn, E. A. DeFelice & P. Kuo (Eds.), *Health and obesity*. New York: Raven Press.

19. Hausman, D. B., DiGirolamo, M., Bartness, T. J., Hausman, G. J., & Martin, R. J. (2001). The biology of white adipocyte proliferation. *Obesity Reviews, 2(4)*, 239–254.

20. Spalding, K. L., Arner, E., Westermark, P. O., Bernard, S., Buchholz, B. A., Bergmann, O., et al. (2008). Dynamics of fat cell turnover in humans. *Nature, 453(7196)*, 783–787.

21. Cianflone, K., Paglialunga, S., & Roy, C. (2008). Intestinally derived lipids: Metabolic regulation and consequences—an overview. *Atherosclerosis. Supplements, 9(2)*, 63–68.

22. List, J. F., & Habener, J. F. (2003). Defective melanocortin 4 receptors in hyperphagia and morbid obesity. *New England Journal of Medicine, 348(12)*, 1160–1163.

23. Klok, M. D., Jakobsdottir, S., & Drent, M. L. (2007). The role of leptin and ghrelin in the regulation of food intake and body weight in humans: A review. *Obesity Reviews, 8(1)*, 21–34.

24. Ruhl, C. E., & Everhart, J. E. (2001). Leptin concentrations in the United States: Relations with demographic and anthropometric measures. *American Journal of Clinical Nutrition, 74(3)*, 295–301.

25. Shin, A. C., Zheng, H., & Berthoud, H. R. (2009). An expanded view of energy homeostasis: Neural integration of metabolic, cognitive, and emotional drives to eat. *Physiology & Behavior, 97(5)*, 572–580.

26. Wolf, G. (2008). Energy regulation by the skeleton. *Nutrition Reviews, 66(4)*, 229–233.

27. Cancello, R., & Clement, K. (2006). Is obesity an inflammatory illness? Role of low-grade inflammation and macrophage infiltration in human white adipose tissue. *BJOG : An International Journal of Obstetrics and Gynaecology, 113(10)*, 1141–1147.

28. Leibel, R. L., Edens, N. K., & Fried, S. K. (1989). Physiologic basis for the control of body fat distribution in humans. *Annual Review of Nutrition, 9*, 417–443.

29. Stefan, N., Kantartzis, K., Machann, J., Schick, F., Thamer, C., Rittig, K., et al. (2008). Identification and characterization of metabolically benign obesity in humans. *Archives of Internal Medicine, 168(15)*, 1609–1616.

30. Hauner, H., & Entenmann, G. (1991). Regional variation of adipose differentiation in cultured stromal-vascular cells from the abdominal and femoral adipose tissue of obese women. *International Journal of Obesity, 15(2)*, 121–126.

31. Himms-Hagen, J. (1984). Thermogenesis in brown adipose tissue as an energy buffer—Implications for obesity. *New England Journal of Medicine, 311(24)*, 1549–1558.

32. van Marken Lichtenbelt, W. D., Vanhommerig, J. W., Smulders, N. M., Drossaerts, J. M. A. F. L., Kemerink, G. J., Bouvy, N. D., et al. (2009). Cold-activated brown adipose tissue in healthy men. *New England Journal of Medicine, 360(15)*, 1500–1508.

33. Saito, M., Okamatsu-Ogura, Y., Matsushita, M., Watanabe, K., Yoneshiro, T., Nio-Kobayashi, J., et al. (2009). High incidence of metabolically active brown adipose tissue in healthy adult humans: Effects of cold exposure and adiposity. *Diabetes, 58(7)*, 1526–1531.

34. Rankinen, T., Zuberi, A., Chagnon, Y. C., Weisnagel, S. J., Argyropoulos, G., Walts, B., et al. (2006). The human obesity gene map: The 2005 update. *Obesity, 14*, 529–644.

35. Albu, J., Allison, D., Boozer, C. N., Heymsfield, S., Kissileff, H., Kretser, A., et al. (1997). Obesity solutions: Report of a meeting. *Nutrition Reviews, 55(5)*, 150–156.

36. Katzmarzyk, P., & Bouchard, C. (2005). Genetic influences on human body composition. In S. Heymsfield, T. Lohman, Z. Wang & S. Going (Eds.), *Human body composition* (2nd ed., pp. 243–258). Champaign, IL: Human Kinetics.

37. Stunkard, A. J., Sorensen, T. I., Hanis, C., Teasdale, T. W., Chakraborty, R., Schull, W. J., et al. (1986). An adoption study of human obesity. *New England Journal of Medicine, 314(4)*, 193–198.

38. Reilly, J. J., Armstrong, J., Dorosty, A. R., Emmett, P. M., Ness, A., Rogers, I., et al. (2005). Early life risk factors for obesity in childhood: Cohort study. *BMJ, 330(7504)*, 1357.

39. Farooqi, I. S., Keogh, J. M., Yeo, G. S., Lank, E. J., Cheetham, T., & O'Rahilly, S. (2003). Clinical spectrum of obesity and mutations in the melanocortin 4 receptor gene. *New England Journal of Medicine, 348(12)*, 1085–1095.

40. Farooqi, I. S., Wangensteen, T., Collins, S., Kimber, W., Matarese, G., Keogh, J. M., et al. (2007). Clinical and molecular genetic spectrum of congenital deficiency of the leptin receptor. *New England Journal of Medicine, 356(3)*, 237–247.

41. Dai, F., Keighley, E. D., Sun, G., Indugula, S. R., Roberts, S. T., Åberg, K., et al. (2007). Genome-wide scan for adiposity-related phenotypes in adults from American Samoa. *International Journal of Obesity, 31*, 1832–1842.

42. Bouchard, C., Despres, J. P., & Mauriege, P. (1993). Genetic and nongenetic determinants of regional fat distribution. *Endocrine Reviews, 14(1)*, 72–93.

43. Hotamisligil, G. S., & Spiegelman, B. M. (1994). Tumor necrosis factor alpha: A key component of the obesity–diabetes link. *Diabetes, 43(11)*, 1271–1278.

44. Ristow, M., Müller-Weiland, D., Pfeiffer, A., Krone, W., & Kahn, R. (1998). Obesity associated with a mutation in a genetic regulator of adipocyte differentiation. *New England Journal of Medicine, 339,* 953–959.

45. Ravussin, E., & Bogardus, C. (1989). Relationship of genetics, age, and physical fitness to daily energy expenditure and fuel utilization. *American Journal of Clinical Nutrition, 49(5 Suppl)*, 968–975.

46. Roberts, S. B., Savage, J., Coward, W. A., Chew, B., & Lucas, A. (1988). Energy expenditure and intake in infants born to lean and overweight mothers. *New England Journal of Medicine, 318(8)*, 461–466.

47. Bray, M. S., Hagberg, J. M., Perusse, L., Rankinen, T., Roth, S. M., Wolfarth, B., et al. (2009). The human gene map for performance and health-related fitness phenotypes: The 2006–2007 update. *Medicine and Science in Sports and Exercise, 41(1)*, 35–73.

48. Nelson, M. D., Prabhakar, P., Kondragunta, V., Kornman, K. S., & Gardner, C. (2010). *Genetic phenotypes predict weight loss success: The right diet does matter.* Abstract from the American Heart Association Joint Conference: Nutrition, Physical Activity and Metabolism and 50th Cardiovascular Disease Epidemiology and Prevention. Accessed December 16, 2010 from http://americanheart.org/downloadable/heart/1267139896704EPI_NPAM_FinalProg_ABSTRACTS.pdf.

49. Boozer, C. N., Schoenbach, G., & Atkinson, R. L. (1995). Dietary fat and adiposity: A dose–response relationship in adult male rats fed isocalorically. *American Journal of Physiology, 268(4 Pt 1)*, E546–50.

50. Hirsch, J. (1994). Herman award lecture, 1994: Establishing a biologic basis for human obesity. *American Journal of Clinical Nutrition, 60(4)*, 615–618.

51. Ravussin, E., Valencia, M. E., Esparza, J., Bennett, P. H., & Schulz, L. O. (1994). Effects of a traditional lifestyle on obesity in Pima Indians. *Diabetes Care, 17(9)*, 1067–1074.

52. Zimmet, P., Faaiuso, S., Ainuu, J., Whitehouse, S., Milne, B., & DeBoer, W. (1981). The prevalence of diabetes in the rural and urban polynesian population of Western Samoa. *Diabetes, 30(1)*, 45–51.

53. Knowler, W. C., Pettitt, D. J., Saad, M. F., & Bennett, P. H. (1990). Diabetes mellitus in the Pima Indians: Incidence, risk factors and pathogenesis. *Diabetes/Metabolism Reviews, 6(1)*, 1–27.

The Energy Nutrients and Weight Management

CHAPTER OUTLINE

R. Gino Santa Maria/Shutterstock.com

cappi thompson/Shutterstock.com

lorga Studio/Shutterstock.com

For most people, eating—and perhaps also selecting, preparing, and thinking about food—is one of life's most enjoyable activities. Food meets our physiological needs and has social and cultural meanings. Even people with little discretionary income mark special occasions, the arrival of friends, and religious holidays with food.

Most people regard recommendations to alter their diets with dread, and they see the need for dietary change as a temporary state and can't wait to return to their "normal" diets. Food preferences are difficult to alter, and access to healthy food varies by income and place of residence. In addition, for people who have not been given time to prepare for change, who are not helped to develop skills for diet change, or who are given diets that have little relevance to their preferences, food access, or time, permanent dietary change is unlikely.

This chapter provides an overview of the energy nutrients—carbohydrate, fat, and protein—with particular attention to the role of each in weight management. The role of sugar and fat replacers in weight management, as well as the potential health effects of overreliance on these products, is included. Finally, dietary modifications that enhance weight management and health are presented.

COMPONENTS OF A HEALTHY DIET

energy nutrients
The nutrients (carbohydrate, fat, and protein) that contain calories. The energy nutrients are made up of carbon, hydrogen, oxygen, and (in protein) nitrogen.

The **energy nutrients**—carbohydrate, fat, and protein—not only provide kcals but also serve several physiological functions and are sources of essential micronutrients needed to support health. Dietary guidelines from the federal government, the Food and Nutrition Board of the Institute of Medicine, and national health organizations like the American Heart Association, the American Cancer Society, and the American Diabetes Association generally recommend an adult diet that provides—

- 45–65% of calories from carbohydrate
- 20–35% of calories from total fat and no more than 7% from saturated fat
- No more than 300 mg of cholesterol each day
- 10–35% of calories from protein
- 25–38 g of dietary fiber daily

Most Americans do not follow this recommended diet.

Where Are the Problems in the American Diet?

The U.S. Department of Agriculture (USDA) publishes The Healthy Eating Index, a "report card" on the American diet. The Healthy Eating Index-2005 (HEI-2005) rates twelve dietary components: total fruit; whole fruit (forms other than juice); total vegetables; dark green and orange vegetables and legumes; total grains; whole grains; milk (including soy beverages); meat and beans (which also includes fish, non-beverage soybean products, nuts, and seeds); oils (nonhydrogenated vegetable oils and the oils in fish, nuts, and seeds); saturated fat; sodium; and calories from solid fats, alcoholic beverages, and added sugars. A perfect score is 100.

The most recent total HEI-2005 score for Americans was 58.2, unimproved from 1994-96.[1] While scores for total grains were high (10 out of a possible 10 points), scores for whole grains were low (1 out of 5 points). Scores for meat and beans were also at the maximum, but scores for whole fruit and total vegetables actually declined since 1994–96. Scores for dark green and orange vegetables and legumes were low (1.4 out of 5 possible points). Despite thirty years of government recommendations through the Dietary Guidelines for Americans, people still do not consume enough fruits, vegetables, whole grains, and fat-free or low-fat milk.

Americans do, however, eat a lot more than they used to. The Dietary Guidelines for Americans recommend balancing food and beverage calories with activity calories. For example, sedentary 31- to 50-year-old women are advised to consume no more than 1,800 calories/day, and sedentary men of the same ages no more than 2,200 calories/day.[2] Between 1971–2000 average daily caloric intake in the United States rose from 2,450 to 2,618 kcal in men and from 1,542 to 1,877 kcal in women.[3] Most of that increase was due to greater carbohydrate intake. Given results of HEI-2005, it is probably safe to surmise that products like snack foods and sugars accounted for most of those carbohydrates.

Components of a Healthy Diet: Summary

A diet that is 45–65% carbohydrate, 10–35% protein, and less than 35% fat and that provides 25–38 g of fiber, not more than 300 mg of cholesterol, and limited saturated fatty acids is widely recommended for health. Moderate caloric intake also promotes weight management. Most Americans do not meet these recommended guidelines.

CARBOHYDRATES: PRIMARY CONSTITUENTS OF THE DIET

monosaccharide
Simple carbohydrate made up of a single unit of sugar (glucose, fructose, and galactose).

disaccharide
Carbohydrate made up of two monosaccharides (sucrose, lactose, and maltose).

Most of the energy in each day's diet should come from carbohydrates, which are energy nutrients classified as **monosaccharides**, **disaccharides**, and **polysaccharides**. The monosaccharides (glucose, fructose, and galactose) and disaccharides (sucrose, lactose, and maltose) are collectively known as sugars, while the polysaccharides include starch, glycogen, and fiber. Figure 5-1 gives an overview of the sugars and polysaccharides. All carbohydrates except fiber provide 4 kcal/g.

Why Do We Need Dietary Carbohydrates?

Disaccharides and polysaccharides make up most of the carbohydrate in our diets. Free monosaccharides are not abundant in foods, other than in honey and some fructose-based

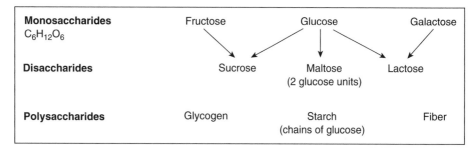

FIGURE 5-1 The Carbohydrates.

polysaccharide
Complex carbohydrate composed of repeating glucose units (starch, fiber, and glycogen).

sweeteners. The monosaccharide glucose is, however, the most nutritionally significant carbohydrate because it is a component of all of the disaccharides and polysaccharides, it is the form in which carbohydrate circulates in the bloodstream, and virtually every cell needs it for energy.

Carbohydrates as an Energy Source

After we eat foods that contain carbohydrates, our blood glucose level gradually rises. The hormone insulin is then released from the β cells of the pancreas, and insulin receptors on the body's cells respond by allowing glucose to enter the cells. Without insulin, or with insufficient insulin receptors, clearance of glucose from the bloodstream and use of glucose as a fuel are severely impaired. Before the development of injectable insulin, type 1 diabetics (who do not secrete insulin) died at early ages.

glycogen *Storage form of carbohydrate, consisting of one unit of glucose and three units of water.*

Glucose not needed right away can be stored as **glycogen** in the liver and muscle cells. Approximately 100 g of glycogen can be stored in the liver, and the skeletal muscles can hold 300–400 g. Each gram of glycogen is stored with approximately 3 g of water. As the body puts glucose into storage as glycogen, about 5% of its energy content is lost.[4] Because carbohydrate yields 4 kcal/g, 500 g of glucose is worth about 1,900 kcal of stored glycogen. During moderate- to high-intensity exercise, fat cannot be oxidized quickly enough to meet energy needs, so stored carbohydrate becomes very important. Theoretically, excess dietary carbohydrate could be converted to and stored as fat, but carbohydrate is rarely consumed in a high enough quantity for this to occur.

dietary fiber
Nondigestible carbohydrate and lignin that makes up plant cell walls.

Several hours after a meal, as blood glucose levels drop, the hormone glucagon is released from α cells of the pancreas. Glucagon stimulates release of stored glycogen to maintain normal blood glucose levels. Falling blood glucose levels are also a stimulus to eat, because few of us store more than a day's worth of glycogen.

Fiber: Much Value for Few Calories

functional fiber
Nondigestible carbohydrate extracted from plant or animal sources (like pectin, gums, and chitosan) or commercially synthesized.

Total fiber includes both dietary and functional fiber. **Dietary fiber** is the nondigestible carbohydrate and lignin that makes up plant cell walls. It is found in conjunction with the digestible carbohydrate in fruits, vegetables, and whole grains. **Functional fiber** is nondigestible carbohydrate that is either extracted from plant or animal sources (such as pectin, gums, and chitosan) or commercially synthesized.[5] Because fiber is not digestible, it contributes bulk but not calories to the diet.

soluble fiber *Type of dietary fiber abundant in oat bran and legumes and associated with a decrease in blood cholesterol.*

The terms soluble and insoluble are used to refer to fiber. The Food and Nutrition Board of the Institute of Medicine, which establishes recommended dietary allowances for Americans, no longer uses these terms. However, they are briefly described here because food labels and Web sites still refer to them.

insoluble fiber
Type of dietary fiber found in wheat bran, fruits, and vegetables and associated with a reduced risk of diverticulosis, hemorrhoids, and colon cancer.

- **Soluble fiber**, abundant in oat bran, barley bran, psyllium, and legumes, dissolves in water and may form a gel in the digestive tract. It is often linked to reductions in blood cholesterol.
- **Insoluble fiber**, found in wheat bran, fruits, vegetables, and legumes, is associated with speeding the passage of waste materials through the colon.

In fact, there is not definitive evidence supporting the different effects of soluble and insoluble fiber. Both are at least partly fermented in the large intestine by the bacteria that live there. Both play a role in preventing cardiovascular disease, gastrointestinal

TABLE 5-1   Health Benefits of Fiber: Summary of the Evidence

For prevention of...	Fiber offers these benefits...
Cardiovascular disease	↓ total cholesterol ↓ LDL cholesterol ↓ blood pressure
Type 2 diabetes	Improved glucose control ↓ insulin secretion after meals
Cancer	↓ risk of colorectal cancer Effect on risk of other cancers inconclusive
Gastrointestinal conditions	↓ risk of diverticulosis May help treat diverticulosis and irritable bowel syndrome
Obesity	May promote weight loss by decreasing hunger and creating sensation of fullness

disease, diabetes, and cancer. The evidence for fiber's health impact is summarized in Table 5-1.

A daily fiber intake of 14 g/1,000 kcal, or 25 g for adult women and 38 g for adult men, is recommended, quite a bit more than the average American consumes.[5] Children over the age of 2 years should eat an amount of fiber that corresponds with their age in years plus 5 g/day.[6]

In addition to its health benefits, fiber is an important adjunct to weight management. Because high-fiber foods add bulk but not calories, they create a sensation of fullness without caloric cost. High-fiber foods require more chewing and are therefore consumed at a slower rate than highly processed foods, and fiber empties slowly from the gastric tract, which promotes satiety and delays the onset of hunger. Most studies find that hunger decreases and satiety increases with increased fiber consumption, resulting in a 10% reduction in caloric intake and weight losses of about 4½ lb over four months.[7] These effects can occur from consumption of either whole food fiber or viscous fibers added to drinks.

Do Sugars Harm Health and Cause Obesity?

Humans love the sweet taste of sugar, a fact duly noted by the food industry. In 1970 the U.S. food supply provided 119 lb/person of added sugars and caloric sweeteners, which increased to 142 lb/person in 2005.[8] While this is not a direct measure of sugar consumption, the U.S. Department of Agriculture considers it a valid indicator of sugar intake in America. Soft drink intake alone has increased almost seven-fold in the past 50 years. In 1942 per person consumption of soft drinks was approximately 90 eight-ounce servings/year (equivalent to just under 2 servings/week), while in 2000 per person consumption increased to 600 servings/year, or about 2 servings/day.[9]

Both the monosaccharides and disaccharides are classified as "sugars" and exist side-by-side with other nutrients in many of the foods we eat, such as fruit and milk. "Added sugars" are those added as foods are processed and tend to be found in foods with lower nutritional value. Most added sugars today are either from sucrose (table sugar) or from high-fructose corn syrup. The average American ingests 22 tsp (equivalent to about 355 kcal) of added sugars each day, with 14- to 18-year-old consuming an average of 34 tsp/day (about 549 kcals).[10] Common sources of added sugars are provided in Table 5-2.

TABLE 5-2 Common Sources of Added Sugars

Foods	Added sugars in a typical serving (in grams)
Table sugar, 1 tsp	4
Bagel, cinnamon-raisin	3
Bread, mixed-grain, slice	2
Cake, chocolate, from pudding-type dry mix, slice	43
Cake, snack-type, cream-filled, 1	19
Brownie, 1	8
Animal crackers, 12	4
Doughnut, cake-type, chocolate, sugared, or glazed	15
Muffin, blueberry	10
Pie, pecan, slice	34
Pie, pumpkin, slice	16
Toaster pastry, fruit	7
Cola drink, 12 oz	38
Ginger ale, 12 oz	32
Chocolate syrup, 2 T	19
Citrus fruit drink, from frozen concentrate, 8 oz	44
Cranberry juice cocktail, bottled, ½ c	10
Corn flakes, 1 c	2
Yogurt, fruit, non-fat, 1 c	28
Fast-food French toast sticks	21
Salad dressing, Caesar, low-calorie, 2 T	5
Applesauce, canned, sweetened, ½ c	8
Fruit cocktail, canned in heavy syrup, ½ c	10
Peanut butter, smooth, low-fat, 1 T	1
Breakfast bar with oats, sugar, raisins	8
Granola bar with oats, fruit, and nuts	11
Candy, hard, 1 piece	1
Chocolate pudding, ready-to-eat, ½ c	19

Added sugars include sugar, fructose, lactose, maltose, glucose/dextrose, syrups, honey, molasses, and fruit juice concentrates

Source: Added sugars in a typical serving were calculated by the author using the USDA database for the added sugars content of selected foods. Release 1. (2006). Beltsville, MD: Nutrient Data Laboratory, Beltsville Human Nutrition Research Center, U.S. Department of Agriculture. Available from: http://www.ars.usda.gov/SP2UserFiles/Place/12354500/Data/SR22/nutrlist/sr22w269.pdf. Accessed December 19, 2010.

Sucrose

Sucrose, or table sugar, is a popular sweetener frequently blamed for obesity and the development of diabetes. Speculation about a sucrose-obesity link arises from observations that obesity and sugar consumption often increase at the same time in a population. This is probably more coincidental than causal. When people overeat, whether they overeat lasagna, chicken, or jelly beans, obesity may result. Because sweets are often small, portable, and combined with fat in foods, overeating sugar-containing foods—and getting lots of extra calories—is easy.

The association between sucrose and diabetes relates to the effect of increased blood glucose on insulin secretion. The theorized scenario is that excess sugar consumption could cause the pancreatic β-cells to overproduce insulin and, eventually, to burn out. Sugar alone is unlikely to cause this to happen. All carbohydrates, except perhaps those with high fiber content, have the potential to increase blood glucose and stimulate insulin secretion. The amount of carbohydrate, rather than the type, determines insulin response and blood glucose control in a diabetic or prediabetic. The person whose diabetes is under control does not need to avoid sugar but should use it in moderation.

Fructose

high-fructose corn syrup (HFCS) *A sucrose-like product manufactured from processed corn and corn syrup.*

Fructose is a component of the widely used **high-fructose corn syrup (HFCS)**, which has recently been touted by the Corn Refiners Association as a product that is as natural as the fructose found in fruits and honey. While HFCS is made from a natural substance—corn—it is highly processed to increase its fructose content and then blended with pure corn syrup to yield a product that costs about 46% less than refined sugar.[8] HFCS first came on the scene as an added sugar in 1967. Domestic production of HFCS increased dramatically between 1980–2000, with a concurrent 25% rise in per capita HFCS consumption in the same period.[11] Sugar-sweetened beverages use HFCS almost exclusively.

HFCS is similar to sucrose in its chemical composition, but it has been linked to several serious health problems—hypertension, elevated blood lipids, and type 2 diabetes mellitus. In animals, fructose increases fat storage.[12] In humans, weight gain and obesity are more likely in both children and adults who consume the most sugar-sweetened soft drinks.[13] Conversely, overweight individuals who reduce intake of sugar-sweetened beverages may lose weight—in one study, cutting out one serving/day for 18 months resulted in weight losses of almost 1.5 lb more than weight lost by those who reduced intake of solid calories by an equivalent amount.[14]

While obesity rates in the United States have increased at the same time as HFCS consumption, a direct link with human obesity is unproven. What is clear is that HFCS is in many products, some obvious (like soft drinks and fruit drinks) and some not so obvious (bread and cereal). Unless you are a diligent label reader, you may consume a significant amount of added sugars every day from HFCS.

Why does high-fructose corn sweetener cause health problems? One theory is that fructose, alone among the sugars, can elevate uric acid levels, which increases the risk of cardiovascular disease.[11] Fructose also promotes both high blood lipids and increased fat storage. Recall from Chapter 3 that glucose breakdown is regulated by an enzyme, phosphofructokinase (PFK). The breakdown of glucose yields three-carbon units, which could be metabolized for energy or, if a great deal of carbohydrate is consumed, could be used to synthesize fat in the form of triglycerides. But PFK, a rate-limiting enzyme, prevents that from happening. Fructose is metabolized in the same metabolic pathway as glucose, but fructose enters that pathway "downstream" from PFK, so its breakdown is not closely regulated, and it is metabolized quickly. As a result, excess fructose can—and does—promote triglyceride synthesis.[12] Fructose also has a lower thermic effect that other carbohydrates, so less heat (and therefore energy) is released when it is metabolized.[15]

Recommended Sugar Intake

Sugars provide 4 kcal of energy per gram of weight, just like starches. If all you want is energy, it does not matter if you eat whole wheat bread or jelly beans, because they have the same caloric value per gram. However, added sugars deliver little nutritional value per calorie. The 2005 Dietary Guidelines for Americans recommends that we "choose and prepare foods and beverages with little added sugars and caloric sweeteners."[2] The 2010 Dietary Guidelines for Americans, currently under revision, is likely to echo this recommendation. Other health organizations provide more specific guidelines for sugar consumption, although you will notice that recommendations are not consistent.

- The Institute of Medicine suggests a maximum of 25% of calories from added sugars.[5]
- The World Health Organization recommends less than 10% of calories from added sugars and sugars naturally present in foods and dairy.[16]
- The American Heart Association recommends no more than 5% of total calories from added sugars (fewer than 100 kcal for women and 150 kcal for men).[17]

A person consuming 1,800 kcals/day who is trying to limit added sugars to 10% of caloric intake would have a "budget" of about 180 sugar kcals (45 g) each day. Look again at Table 5-2 and see how difficult this might be!

The information about carbohydrates presented in this section is summarized in Table 5-3.

TABLE 5-3 Overview of Carbohydrates

Classification

    Monosaccharides (glucose, fructose, galactose)
    Disaccharides (sucrose, lactose, maltose)
    Polysaccharides (glycogen, starch, fiber)

Energy value 4 kcal/g

Functions

    Provide energy
    Maintain normal blood glucose level
    Aid in normal functioning of gastrointestinal tract (fiber)
    Reduce blood lipids (fiber)

Food sources

    Breads, cereals, and grains
    Vegetables and fruits
    Milk and dairy
    Legumes and nuts
    Sweets and sweeteners

Food sources of fiber

Wheat bran, oats, oat bran, barley, psyllium, fruits and vegetables, legumes, seeds

Recommended intake

Lower limit: 45% of caloric intake
Upper limit: 65% of caloric intake
Added Sugars: recommendations vary, between 5–25% of caloric intake
Fiber: 25–38 g/day (adults)

How Does Carbohydrate Modification Promote Health and Weight Management?

Because carbohydrates are readily burned for energy, they are unlikely to be incorporated into storage fat (HFCS may be an exception to this). Some people are under the impression that carbohydrate calories are "free" and may be eaten in unlimited quantities, but this belief is incorrect.

High-Carbohydrate Diets and Weight Management

Not so many years ago, people who wanted to lose weight were advised to stay away from bread, potatoes, and other starches, which were regarded as fattening. Today, we know that carbohydrates possess many excellent qualities for weight management:

- A low-fat diet rich in plant foods has been shown to promote weight loss.
- Carbohydrates account for much of the thermic effect of food, and metabolism accelerates slightly more after a high-carbohydrate meal than a high-fat meal, as you learned in Chapter 3.
- With the exception of fructose, excess carbohydrate is immediately oxidized or stored as glycogen and not converted to fat for storage.
- Carbohydrate makes people feel full and, in people who "listen" to their bodies, may help prevent overeating.

Carbohydrates are not, however, blameless in weight gain. When large amounts of carbohydrates are eaten, the body does not need to burn dietary fat or release stored fat for energy. So, an excess of carbohydrate calories can be fat maintaining or, indirectly, increase fat stores. The key is to avoid over consuming carbohydrates.

High-Carbohydrate Diets and Prevention of Chronic Disease

phytochemicals
Compounds produced by plants that have antioxidant or hormone-like actions; include isoflavones, phytoestrogens, carotenoids, flavonoids, terpenes, steroles, indoles, and phenols.

In addition to the usefulness of carbohydrates in weight management, they may also reduce the risk of chronic disease. Whole grains (from wheat, oats, barley, millet, rye, and bulgur) are particularly helpful in preventing cardiovascular disease. Unlike highly processed sugars and starches, whole grains are rich in both fiber and vitamin E and are excellent sources of vitamins and minerals that are lost in the refinement process but not added back through enrichment (such as zinc, magnesium, vitamin B_6, and chromium). In addition, whole grains contain **phytochemicals**, fiber, and other substances that favorably affect plasma glucose, insulin, and lipids.

High-carbohydrate diets (more than 55% of calories) will raise plasma triglyceride levels in some people, increasing the risk of cardiovascular disease.[18] People who already have high triglyceride levels may be advised to limit carbohydrate intake, particularly fructose, to avoid further risk of cardiovascular disease.

The "Diabetic Diet"

Manipulating dietary carbohydrates has been the cornerstone of the so-called diabetic diet for many years. The reason for this is that certain dietary carbohydrates seem to increase blood glucose level more rapidly than others do, and they promote a stronger insulin response. The **glycemic index** describes the extent to which specific foods raise blood glucose as compared to white bread, which has a glycemic index of 100. Refined starchy foods are generally high on the glycemic index, whereas oats, fruits, nonstarchy vegetables, and legumes are low. Some unrefined starches also have high glycemic index values, like baked potatoes (85) and raw carrots (71); while foods that you might think of as sweets have low values because of their protein and fat content, such as ice cream (51). You can learn more about the glycemic index value of foods by consulting the Revised International Table of Glycemic Index and Glycemic Load Values, available as an online appendix (see reference list at the end of this chapter for the link).[19]

The glycemic effect of a food also depends on how it is cooked, its protein and fat content, digestibility, and whether a food is eaten alone or as part of a meal. For example, when carrots are cooked, their glycemic index score drops to 39. In addition, the same food may affect a person's blood glucose level differently on different days.

Type 2 diabetics who follow a low–glycemic index diet often have lower blood lipids and better glucose control than those who do not.[20] Should people with diabetes eliminate high–glycemic index foods from their diets? Not necessarily. The amount of carbohydrate probably has a greater effect on glycemic control than the type of carbohydrate, but the American Diabetes Association acknowledges that there may be an added benefit from considering the glycemic index as well. And we can certainly all benefit from a diet lower in processed foods and higher in whole grains, vegetables, fruits, and legumes.

There is no "diabetic diet" today, and most type 2 diabetics maintain good glycemic control without following a special diet. The proportion of calories from fat, carbohydrate, and protein recommended for diabetics is generally within the same ranges advocated for nondiabetics. Although carbohydrates, especially sugars, were once restricted in diabetic diets, today's recommendations generally permit the enjoyment of sugars, starches, and sweeteners as long as total allowable carbohydrates are not exceeded. Dietary guidelines for type 2 diabetics are summarized in Table 5-4.

glycemic index
System of classifying carbohydrates based on their ability to raise the blood glucose level; bread has a glycemic index of 100.

TABLE 5-4 Dietary Guidelines for People with Type 2 Diabetes Mellitus

- Monitor carbohydrate intake by counting carbohydrate grams or using carbohydrate exchanges
- Consume recommended amounts of fiber (25–38 g/day for adults)
- Limit intake of saturated fats to less than 7% of daily caloric intake
- Limit intake of cholesterol to 200 mg/day
- Have 2 servings of fish per week
- Consume recommended amounts of protein (10–35% of calories), with an emphasis on plant proteins
- Use alcohol in moderation (≤1 drink/day for women; ≤2 drinks/day for men)

The American Diabetes Association recommends that the emphasis for people with type 2 diabetes be on normalizing blood glucose, blood lipids, and blood pressure, and that weight management is a key component of managing diabetes and pre-diabetes.[20] Meal spacing, choosing healthy foods, and reducing saturated fat are important nutritional goals, in addition to increasing physical activity. Small weight losses (5–10% of body weight) that result from these dietary modifications can significantly improve hemoglobin A_{1c} levels.

The Impact of Sugar Replacers on Health and Weight Management

Sugar replacers—also known as nonnutritive sweeteners and sugar substitutes—provide either no calories or a very small number of calories. Because of this, they might be useful to people trying to satisfy cravings for sweets while limiting intake of caloric sugars.

What Sugar Replacers Are in Common Use?

The Food and Drug Administration (FDA) has approved five sugar replacers: saccharine, sucralose, aspartame, acesulfame-K, and neotame; and a family of substances known as sugar alcohols, or polyols.

- Sugar alcohols include sorbitol, mannitol, xylitol, and maltitol. They are poorly absorbed and therefore have a caloric value below 4 kcal/g, providing between 1.5 and 3 kcal/g. They are not as sweet as table sugar and are found in sugar-free candies, chewing gum, ice cream, and baked goods.
- Saccharine (common name: Sweet and Low) has 200 to 700 times the sweetness of sugar but at no caloric cost. It is the most commonly used sweetener in the world.[22] You may have read that saccharine consumption is associated with cancer in animals, and until recently a warning label was required on saccharine-containing products. That requirement was lifted by the FDA in 2001.
- Sucralose, commonly known as Splenda, has 600 times the sweetness of sugar. It is not absorbed by the digestive tract. Sucralose is heat stable so can be used in baked goods as well as sweetened beverages.
- Acesulfame-K (common name: Sunett) is 200 times sweeter than sugar and, like sucralose, can be heated.
- Aspartame (Nutrasweet; Equal) has 160 to 220 times the sweetness of table sugar. While aspartame was approved by the FDA in 1996 as a general purpose sweetener, 70% of its use in the United States continues to be in soft drinks.[21] Aspartame is derived from the amino acid phenylalanine, which cannot be metabolized by individuals who have a condition called phenylketonuria (PKU). For this reason the FDA requires that products containing aspartame indicate this on the label. When aspartame was initially marketed in the United States. there were many reports of adverse health effects (headaches, dizziness, mood changes, gastric distress), so considerable post-marketing research was undertaken by the Centers for Disease Control and

Prevention and the FDA. This research found no adverse health effects and affirmed the safety of aspartame.[22]

- Neotame is the newest sweetener to be approved by the FDA. It is 7,000 to 13,000 times sweeter than sucrose so it can be used in even smaller amounts than other sugar replacers to sweeten foods and beverages. Neotame is derived from two amino acids, phenylalanine and aspartic acid. Unlike aspartame, it does not release enough phenylalanine to pose a risk for individuals who have PKU, so there is no labeling requirement.

Do Sugar Replacers Provide Any Health Benefits (or Detriments)?

Only the sugar alcohols offer any *direct* health benefits. They help reduce the risk of dental decay and have a lower glycemic response than sugar, so are useful for diabetics. However, when consumed in excess (more than 50 g/day of sorbitol or 20 g/day of mannitol) they may cause diarrhea, because they are not absorbed well in the digestive tract.

Sugar replacers may have a role in weight management. When used in place of sugar, some people have reduced their energy intake by 5–15% per day.[23] Nutritionists sometimes speculate that use of sugar replacers could both increase appetite and increase an individual's preference for sweet foods and beverages, which could lead to greater caloric intake. There is insufficient evidence to support these claims.[23]

Unlimited consumption of sugar replacers is not recommended, and several government agencies have suggested an acceptable daily intake (ADI) for these products. Table 5-5 summarizes these recommendations.

TABLE 5-5 Acceptable Daily Intakes (ADI) of Sugar Replacers, Based upon Recommendations of National and International Groups

Sugar replacer	JECFA[1] ADI	EFSA[2] ADI	FDA[3] ADI
	All values are mg/kg body weight/day		
Saccharine	5	5	5
Sucralose	15	15	5
Acesulfame-K	15	9	15
Aspartame	40	40	50
Neotame	0–2	1	18

[1]Joint Commission of Experts on Food Additives of the World Health Organization and the Food and Agriculture Organization (JECFA)
[2]European Food Safety Agency (EFSA)
[3]Food and Drug Administration (FDA)

Source: Mattes, R. D., & Popkin, B. M. (2009). Nonnutritive sweetener consumption in humans: Effects on appetite and food intake and their putative mechanisms. *American Journal of Clinical Nutrition, 89*(1), 1–14; American Dietetic Association. (2004). Position of the American Dietetic Association: Use of nutritive and nonnutritive sweeteners. *Journal of the American Dietetic Association, 104*(2), 262–265.

Carbohydrates: Summary

triglyceride
Principal constituent of dietary fat, consisting of one molecule of glycerol bound to three fatty acids.

Carbohydrates are needed by all cells for energy and can be stored in small quantities as glycogen. Starches, whole grains, fruits, and vegetables are excellent sources of dietary carbohydrate. A carbohydrate-rich diet that does not provide excess calories or added sugars can promote weight management and reduce the risk of chronic disease. Sugar replacers may play a role in weight management by reducing caloric intake.

LIPIDS: KEY PLAYERS IN HEALTH AND WEIGHT MANAGEMENT

Lipids fulfill several important functions in addition to providing energy. They help keep us warm, maintain cell membranes, produce many necessary body constituents, and provide padding and protection.

fatty acids *Any of a number of organic acids made of carbon chains of varying lengths and degrees of saturation with hydrogen.*

The family of lipids includes triglycerides, phospholipids, and cholesterol. **Triglycerides** are the fats (solid at room temperature) and oils (liquid at room temperature) that make up 95% of the lipids in our diet. A triglyceride consists of one molecule of a carbohydrate-like alcohol called glycerol that is bound to three **fatty acids**. These fatty acids account for the differences between the various fats and oils in our diet.

What Are Fatty Acids?

saturated fatty acid *Fatty acid in which all carbons have the maximum number of hydrogens; this type predominates in animal products.*

Fatty acids may be classified in several different ways according to their chemical composition. These classifications have important health ramifications:

- By chain length: Short-chain fatty acids have fewer than six carbon atoms, medium-chain fatty acids have six to twelve carbons, and long-chain fatty acids have more than twelve carbons. Most dietary triglyceride is made up of eighteen-carbon (long-chain) fatty acids.

monounsaturated fatty acid

Fatty acid having one double bond on the carbon chain where hydrogen is missing; the main constituent of olive and canola oils.

- By degree of saturation with hydrogen atoms: Each carbon atom has four bonding places where other atoms must attach. When every carbon atom has two attached hydrogens, as shown in Figure 5-2a, the fatty acid is said to be **saturated**. Saturated fatty acids have been linked to obesity, cardiovascular disease, and some cancers. When hydrogens are missing, adjacent carbon atoms must form a double bond, resulting in a **monounsaturated fatty acid** (one double bond) or a **polyunsaturated fatty acid** (two or more double bonds) (see Figure 5-2b and c). Unsaturated fatty acids offer many health benefits, which are discussed later in this chapter.

polyunsaturated fatty acid *Fatty acid in which there are two or more double bonds on the carbon chain where hydrogen is missing; found in most vegetable oils and plant fats.*

- By location of double bonds: When the first double bond in a polyunsaturated fatty acid is located next to the sixth carbon (counting from the CH_3 end of the chain), the fatty acid is known as an **omega-6 fatty acid**. Linoleic acid is an omega-6 fatty acid that is essential (it can be obtained through the diet only). When the first double bond is located next to the third carbon, the fatty acid is an **omega-3 fatty acid**. Linolenic acid is an essential omega-3 fatty acid. These two groups of fatty acids are involved in regulating blood lipids, clotting, and pressure.

The fats and oils that comprise our diet contain varying proportions of saturated and unsaturated fatty acids. For example, olive oil is about 14% saturated fat, 78% monounsaturated, and 8% polyunsaturated. Figure 5-3 illustrates the fatty acid composition of other fats in the diet.

omega-6 fatty acid
Polyunsaturated fatty acid having its first double bond next to the sixth carbon; linoleic, arachidonic, and docosatetraenoic acids are omega-6 fatty acids.

omega-3 fatty acid *Polyunsaturated fatty acid having its first double bond next to the third carbon; linolenic, eicosapentaenoic, and docosahexaenoic acids are omega-3 fatty acids.*

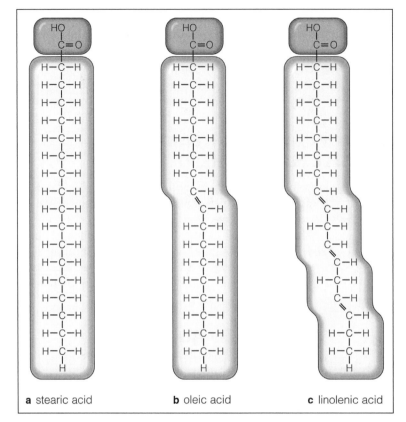

FIGURE 5-2 Three fatty acids: (a) stearic acid is fully saturated with hydrogens; (b) oleic acid, with its one double bond, is monounsaturated; (c) linolenic acid is polyunsaturated.

Source: From Starr/Mcmillan. Human Biology, 8E. p. 30. © 2010 Brooks/Cole, a part of Cengage Learning, Inc. Reproduced by permission. www.cengage.com/permissions.

Do We Need Cholesterol?

Cholesterol belongs to the sterol family, which is a type of lipid characterized by a multiple-ring structure. Cholesterol is found only in animal products, with meat, egg yolk, and dairy foods being the richest dietary sources. Significant amounts of cholesterol (as much as 1,000 mg/day) can also be produced by most body cells. When people consume a lot of cholesterol, the body's cholesterol production slows down, although not enough to prevent accumulation of excessive amounts of cholesterol. The result is often cardiovascular disease.

This possibility does not mean that cholesterol is bad. On the contrary, it is needed for many important physiological functions. Cholesterol is a component of cell walls and is a constituent of **bile**, sex hormones, adrenal hormones, and vitamin D. To limit the risk of cardiovascular disease, current recommendations call for a dietary cholesterol intake of no more than 300 mg/day. People who already have CVD or type 2 diabetes or who are at high risk for CVD or type 2 diabetes are recommended to consume less than 200 mg of cholesterol daily.[24]

Because animal products are the only source for cholesterol, a diet that contains a lot of plant foods can help control cholesterol intake. The fiber in plant foods lowers blood

bile *Liver secretion stored in the gallbladder; bile secretion rises when fat is consumed and acts to emulsify dietary fats so they can be absorbed.*

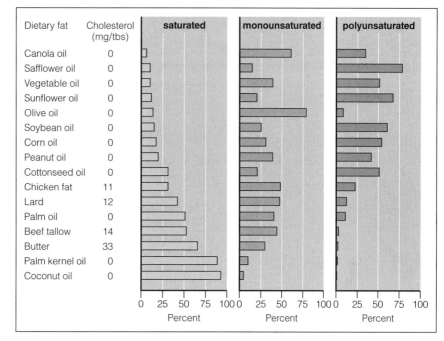

FIGURE 5-3   Fatty acid composition of common food fats.

chylomicron
Lipoprotein that transports dietary triglycerides and cholesterol from the intestines to the liver and body cells.

very-low-density lipoprotein (VLDL)
Lipoprotein that transports cholesterol and triglycerides from the liver to the cells; VLDL particles become LDLs when they lose most of their triglycerides.

low-density lipoprotein (LDL) *A fat and cholesterol carrier that transports cholesterol out of the liver to the tissues; high levels are associated with increased risk of cardiovascular disease.*

high-density lipoprotein (HDL) *A fat and cholesterol carrier involved in transporting cholesterol out of the tissues and back to the liver, where it can be disposed of.*

cholesterol by forming a gel in the lower intestines that binds to and eliminates bile, which contains cholesterol.

Lipoproteins

Picture what happens when you combine oil and water to make salad dressing. Lipids are not water soluble, so the oil in your salad dressing needs help to mix with the water. Body tissues have a high water content, so dietary triglycerides and cholesterol must be transported in water-soluble vehicles. Lipoproteins, with their protein coating and lipid core, are ideal for moving lipids through the bloodstream. Lipoproteins important in health are briefly described here and are illustrated in Figure 5-4:

- **Chylomicrons**: These large lipoproteins transport incoming dietary triglycerides and cholesterol from the intestines to the liver and other tissues.
- **Very-low-density lipoproteins (VLDLs)**: VLDLs transport to the cells the cholesterol and triglycerides that are synthesized in the liver. As VLDLs gradually give up their triglycerides, they become cholesterol-rich, low-density lipoproteins.
- **Low-density lipoproteins (LDLs)**: As covered in Chapter 1, LDLs are sometimes called bad cholesterol because these particles are small enough to be taken up into the lining of damaged arteries and, if oxidized, can form atherosclerotic plaques. LDLs escape the bloodstream by binding to special liver receptors.
- **High-density lipoproteins (HDLs)**: HDLs may counteract LDLs by removing excess cholesterol from the cells and transporting it back to the liver for disposal as bile. For their role in reverse-cholesterol transport, HDLs are sometimes referred to as good cholesterol.

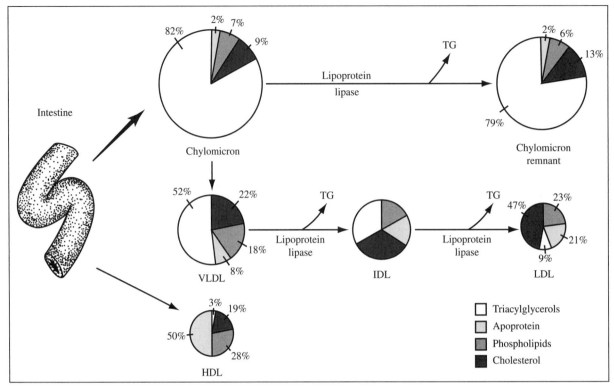

FIGURE 5-4 Lipoproteins.

More important than cholesterol alone in predicting cardiovascular disease risk is the amount of saturated fat in the diet. Saturated fatty acids increase blood lipids, especially triglycerides and LDLs, and decrease production of LDL receptors on liver cells, which slows clearance of LDLs from the bloodstream. For this reason, saturated fats should make up no more than 7% of total calories.

The information about lipids presented in this section is summarized in Table 5-6.

TABLE 5-6 Overview of Lipids

Classification
Triglycerides comprise 95% of dietary fat
I glycerol + 3 fatty acids
Fatty acids may be:
Short-, medium-, or long-chain
Saturated (no double bonds)
Monounsaturated (one double bond)
Polyunsaturated (more than one double bond)
Omega-3 polyunsaturated
Omega-6 polyunsaturated
Cholesterol
Phospholipids

Energy value 9 kcal/g

Functions

Provide energy

Serve as energy reserve

Assist in transport and absorption of fat-soluble vitamins

Serve as source of essential fatty acids

Insulate and protect organs

Cholesterol maintains cell walls and plays role in making vitamin D, bile, and hormones

Food sources

Oils, sauces, spreads, butter

Meat, poultry, fish, eggs, nuts

Milk, cheese, dairy

Breads, cereals, and grains with added fat

Recommended intake

Less than or equal to 35% of calories from fat

Less than or equal to 7% of calories from saturated fat

Less than or equal to 1% of calories from *trans* fat

No more than 300 mg cholesterol/day (less than 200 mg/day for people with CVD or at high risk of CVD or type 2 diabetes)

How Do *Trans* Fatty Acids Affect Health?

trans fatty acids
Fatty acids that are created when unsaturated fatty acids go through a manufacturing process called hydrogenation, which extends their shelf life by stabilizing double bonds; these fatty acids increase risk of cardiovascular disease.

The double bonds that exist in unsaturated fatty acids may have either a *cis* or a *trans* configuration. Most fatty acids that occur naturally have the *cis* configuration, which means that both hydrogens bound to carbons adjacent to the double bond are on the same side of the carbon chain (see Figure 5-5). This then allows the *cis* fatty acid to be folded at its double bond into a U-shape. ***Trans* fatty acids** have hydrogens on opposite sides of the carbon chain at the double bond, which makes them more linear and rigid, like saturated fats. The *trans* bond typically occurs between the fourth and tenth carbon.

Although some *trans* fatty acids are found naturally in animal products, those of greatest health concern are produced in a manufacturing process called hydrogenation. During hydrogenation, some *cis* unsaturated fatty acids are converted to *trans* fatty acids to make the fat more solid and, therefore, more stable. Many vegetable oils are partially hydrogenated, such as those used to produce margarine, shortening, crackers, snack foods, and baked goods, giving them a longer shelf life. *Trans* fatty acids increase the risk of heart disease, and in 2006 the FDA required that food labels include information about *trans* fatty acids in food. To prevent cardiovascular disease, it is prudent not only to lower cholesterol and saturated fats but also to limit the use of foods rich in *trans* fatty acids.

Does Dietary Fat Make People Obese?

Recall from Chapter 3 that humans tend to oxidize as much carbohydrate as they consume each day because carbohydrate stores are limited. Fat oxidation does not have to be as well regulated because we have an almost unlimited capacity for fat storage. For this reason, the dietary manipulation that first comes to mind when discussing the prevention or treatment of obesity is to lower dietary fat intake. When fat intake is reduced (and particularly when energy expenditure is increased), logic suggests that the body cells are more likely to burn the fat consumed without having any excess to store.

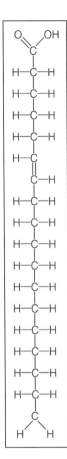

FIGURE 5-5 Trans fatty acids are unhealthy because of the arrangement of the hydrogens at the double bonds on the carbon chain, which makes them more like saturated than unsaturated fats.

Source: From Starr/Mcmillan. Human Biology, 8E. p. 31. © 2010 Brooks/Cole, a part of Cengage Learning, Inc. Reproduced by permission. www.cengage.com/permissions.

High-Fat Diets Promote Fat Storage

Biological mechanisms promote fat storage in response to a high-fat diet. In animals, weight gain almost always occurs when the fat content of the diet is increased. Many studies of humans have noted a similar relationship between the diet's fat content and the prevalence of obesity in the population. Why?

- It's easy to overeat fat (and calories) because fats are such a dense source of energy. Fat is the most calorically dense of all the nutrients, providing 9 kcal/g. Eating an amount of high-fat food equal in weight to a high-carbohydrate food would provide more than twice the calories.

- Fats taste good. People who routinely consume a high-fat, high-energy diet have a tendency to overeat. Scientists call this passive overconsumption, or **hyperphagia**.

- Fats are transferred from the digestive tract slowly, so satiety signals take a relatively long time to reach the brain.

- Fat oxidation does not increase proportionately with fat ingestion. Dietary fat that is not used for energy or other cellular functions heads directly to the fat cells, which are almost totally composed of triglycerides. The number and/or size of adipocytes increases when more fat is consumed than is burned to meet energy needs.

- **Lipoprotein lipase** (a fat-storing enzyme) is activated by increased fat intake.

hyperphagia
Tendency to overeat fat-rich foods, also called passive overconsumption.

lipoprotein lipase *Enzyme found in adipose tissue that is needed for fat storage.*

- Conversion of dietary carbohydrate to stored fat results in a considerable loss of energy, but dietary fat is easily converted to storage fat with minimal loss of energy.[25]

Almost every extra fat calorie eaten becomes a stored fat calorie.

Low-Fat Diets and Weight Loss

When the fat content of the diet is reduced to 20–25% of calories, people tend to lose weight, regardless of whether they decrease their caloric intake, too.[26–27] Sometimes, replacing dietary fat with carbohydrates even allows people to eat more calories and still lose weight. These observations correctly suggest that reducing dietary fat is a useful strategy for weight management.

Still, when calories are reduced, people don't necessarily have to be on a low-fat diet to lose weight. The research literature includes many studies of individuals consuming low-calorie diets who lost weight, whether fat made up 10%, 20%, 30%, or 40% of calories.[28–30] The problem with a low-calorie diet in which fat makes up a substantial proportion of calories is the likelihood of long-term compliance, not whether the diet will promote weight loss. People who wish to reduce calorie intake but keep fat intake moderately high can eat only a very small quantity of food.

People tend to prefer low-fat diets over low-calorie ones. Individuals on reduced-fat diets allowing as little as 20 g of fat per day have reported that their diets are more acceptable than those low in calories.[31] Very-low-fat diets have demonstrated significant weight losses and reasonable adherence.[32–33] One comparison of numerous studies of low-fat and low-calorie diets found the low-fat diets to be more effective after six months, mainly due to better compliance with and acceptance of low-fat programs.[34]

Low-Fat Diets and Prevention of Chronic Disease

Most experts agree that a diet containing no more than 35% of calories as fat, less than 7% of calories as saturated fats, less than 1% of calories as *trans* fats, and less than 300 mg of cholesterol per day offers protection against cardiovascular disease. When low-fat diets are combined with physical activity, they are particularly effective in reducing LDL cholesterol. Limiting dietary fat can also decrease blood pressure and may prevent cancers of the colon and breast, although you will see in the next section that the type of fat may be more significant than the quantity of fat in disease prevention.

A potentially negative effect of low-fat diets is that they sometimes decrease HDL cholesterol. Remember that higher is better with HDL cholesterol, and HDL levels above 35 mg/dL protect against cardiovascular disease. The client on a low-fat diet might rightly ask, "Why should I reduce my fat intake if my HDL cholesterol drops, too?" This question may be addressed with three points:

- Reduced HDL cholesterol in people on low-fat diets may simply be a physiological adaptation that reflects less need to transport fat.[35]
- Increased HDL cholesterol is only one of many factors offering protection against cardiovascular disease. In rural populations where people are lean, physically active, and consume a low-fat diet, rates of cardiovascular disease are low despite decreased HDL.[35]
- Decreased HDL cholesterol in a person with a low intake of fat, cholesterol, and animal protein may not be as unhealthy as decreased HDL cholesterol in a person consuming a more fat-rich diet.[36]

In other words, given the health benefits of a diet low in fat, the fact that HDL levels drop may not be enough of a negative factor to rule out low-fat diets. Participating in daily physical activity may offset some of the drop in HDL, which is discussed further in Chapter 7.

Type of Dietary Fat and Risk of Chronic Disease

The type of fat may be even more important than the amount of fat when evaluating the impact of fat on health. Greeks and Greenland Eskimos eat a diet that provides a great deal more fat than the typical American diet. Yet, the prevalence of cardiovascular disease in these populations is low. The explanation for this apparent contradiction lies in the type of fats that predominate in their diets.

Trans fatty acids are the most blood cholesterol–raising types of fats, followed by saturated fatty acids. Both *trans* fats and saturated fats increase LDLs and decrease HDLs; *trans* fatty acids also increase triglycerides, decrease the size of LDL particles, promote inflammation, and have even been implicated in sudden cardiac death.[37] Denmark has banned *trans* fats in food processing, and Canada is considering such a ban. In the United States, manufacturers are gradually reducing or removing *trans* fats from processed foods. Fortunately, U.S. food labels (and some restaurant menus) now allow consumers to evaluate the *trans* fat content of foods, as well as the saturated fat content, when choosing what to purchase.

Polyunsaturated fatty acids are less likely to promote cardiovascular disease than saturated fats. Omega-3 polyunsaturated fatty acids, which are found in fish, reduce blood lipid levels and prevent blood platelets from aggregating and forming clots. Greenland Eskimos, who eat enough seafood to provide 30–40 g of oil each day, have low rates of heart disease and stroke. Even one fish meal each week offers protection against cardiovascular disease. Scientists are finding that not all fish have the same health effects. For example, fatty fish—salmon, mackerel, sardines, and herring—also provide greater protection against breast and prostate cancer than leaner fish, due to their high concentrations of eicosapentaenoic acid and docosahexaenoic acid.[38] Some evidence suggests that fish intake reduces the risk of colorectal cancer, while foods containing animal fats increase it.[39]

Monounsaturated fatty acids raise HDL levels even more than polyunsaturated fatty acids and also lower LDLs.[40] In addition, they prevent the oxidation of LDLs, which reduces the likelihood that LDLs will become part of atherosclerotic plaque. Olive and canola oil are two common monounsaturated fatty acids. In Greece, fat intake is approximately 40% of total calories, and half of dietary fat comes from olive oil. This so-called Mediterranean diet, discussed later in this chapter, is also rich in low glycemic index foods and low in animal protein and saturated fats.

Health benefits of the Mediterranean diet include lower prevalence of cardiovascular disease and some cancers than in the United States, despite sometimes inferior health care and higher smoking rates in countries where this diet is typical.[41–43] In fact, the Adult Treatment Panel III (ATP III) of the National Cholesterol Education Program (NCEP) recommends that while polyunsaturated fats comprise up to 10% of total calories, monounsaturated fats can comprise up to 20%.[44] Sources of monounsaturated fats are given in Table 5-7.

TABLE 5-7   Common Food Sources of Monounsaturated Fats

Olive oil
Canola oil
Peanut oil
Monounsaturated varieties of safflower and sunflower oil
Margarine made from canola oil
Avocados
Pecans
Almonds
Macadamia nuts
Cashews
Pistachio nuts
Hazelnuts
Animal products that provide some monounsaturated fat include pork, duck, chicken, veal, salmon, kippers, herring, and ocean perch

Studies of the Mediterranean diet have documented that, not only can adherents lose weight, they can lower glucose and insulin levels, reduce markers of inflammation associated with cardiovascular disease, and even lower mortality rates from cardiovascular disease, cancer, and other causes.[42–43,45–46] Adherence rates to this type of diet are higher than to a traditional low-fat diet.[41,45]

Type of Dietary Fat and Weight Loss

Both the length of the carbon chain and the degree of saturation of fatty acids have generated interest in the treatment of obesity. Studies comparing monounsaturated-fat-rich diets (olive oil) to saturated-fat-rich diets (animal fats) have found that both body weight and fat mass are reduced in those consuming more olive oil.[47]

Long-chain triglycerides, which have a carbon chain length greater than twelve and make up most dietary lipid and storage fat, may be oxidized at a lower rate when people are obese.[48] This means that less dietary fat is used to meet energy needs and more remains in storage.

Triglycerides with six to ten carbons (the longer short-chain triglycerides and the medium-chain triglycerides) appear to have a higher thermic effect than other fats and are less likely to be deposited in the fat cells.[49] In a recent study, overweight and obese men and women received dietary counseling, a reduced-calorie diet (1,500 kcal/day for women; 1,800 kcal/day for men), and 12% of calories from either olive oil or medium-chain triglyceride (MCT) oil. Those using the MCT oil reduced fat mass and lost an average of about 4 lb more than the ones using olive oil.[50]

Does this mean that if people switch from saturated fats to unsaturated fats or from long-chain triglycerides to medium-chain triglycerides they will lose weight? Current evidence is not conclusive, but is suggestive that not all fats are similarly obesity promoting. Some may even be obesity fighting. Saturated fats may tend to be stored, whereas unsaturated fats are oxidized. Long-chain fatty acids may be less thermic, while MCT increase energy expenditure.

Case Study. Personal Diet Analysis, Part I	As with Chapter 2, you will be the subject of this case study. (If you prefer, and with permission of your instructor, you may select someone else to be your subject. If you select someone else, remember that all information collected for the case study is to be treated with confidentiality).

- Provide a brief description of the subject of this case (you already did this in Chapter 2, unless you are using a different subject, and you may repeat it here).
- Report your subject's BMI and waist circumference. What was the health risk associated with this waist circumference and BMI?
- Using results from the Chapter 2 Physical Activity Assessment, report your subject's estimated daily caloric expenditure.
- Use the Interactive DRI for Healthcare Professionals (http://fnic.nal.usda.gov/interactiveDRI/) to estimate your subject's recommended daily caloric and macronutrient needs. You will need height, weight, age, and estimated activity level for this.
- Assuming that your subject should limit intake of added sugars to 15% of calories, calculate grams of added sugars that your subject could consume daily. Use Table 5-2 to list some foods that would provide this much added sugar.
- What nutritional concerns does your subject have? Is he/she interested in losing, gaining, or maintaining weight?

What Is the Impact of Low-Fat Foods and Fat Substitutes on Health and Weight Management?

Fat has several characteristics that make it a valuable ingredient in foods. The richness and creamy texture of foods result from their fat content. (Think of how the consistency of frosting would change if you completely eliminated butter or shortening and only used confectioner's sugar and water.) Fats can be heated at high temperatures, which make them ideal for baking and frying. Fat also gives consistency and stability to foods, acts as an emulsifier and separating agent, and controls viscosity.

Fat substitutes, also called fat replacers, have been developed to mimic the chemical properties of fat. Because fat replacers have less than half the calories of fat, they also offer the potential to reduce fat—and calorie—intake. Thousands of fat-modified foods and beverages have been introduced into the U.S. marketplace since the late-1980s, and the Calorie Control Council (caloriecontrol.org) reports that over 80% of American consumers use low- or reduced-fat products. What effect will all of these fat substitutes have on our health?

The Most Widely Used Fat Replacers

So far, no single fat substitute has been developed that performs all of the functions of fat. Consequently a variety of fat replacers are available for use in different types of foods. Fat replacers are developed from carbohydrate, protein, or fat. The food products in which they are typically used are listed in Table 5-8.

TABLE 5-8 Fat Replacers in Foods

Fat replacers	Food products in which they are primarily used
Carbohydrate-based	Dairy products, frozen desserts, salad dressing, puddings,
Carbohydrate polymers	gelatins, spreads, nutrition bars, chewing gum, dry cake and cookie mixes, frosting, baked goods, processed meat products
Cellulose and fiber-based products	Mayonnaise, salad dressing, frozen desserts, processed cheese, processed meats, various dairy products, soups, cookies, muffins, fruit bars, fried foods
Protein-based	Salad dressing, butter, margarine, spreads, sour cream,
Microparticulated protein product	mayonnaise, cheese, yogurt, soups, sauces, coffee creamer, baked goods
Protein blends	Frozen desserts, baked goods
Fat-based	
Reduced-calorie triglycerides	Baked goods, chocolate chips, candy
Olestra	Chips, crackers, snack foods

Carbohydrate- and Protein-Based Fat Replacers

The first fat substitutes were carbohydrate based, and most fat replacers in use today are in this category. They were originally developed to improve food quality, but carbohydrate-based fat replacers are now also used to lower calories. Replacing some of the fat in a product with carbohydrate results in reduction in over half of the calories because carbohydrates have only 4 kcal/g. Most carbohydrate-based fat replacers are completely digested. Protein-based fat replacers provide 1–3 kcal/g.

- Carbohydrate polymers (modified food starches, dextrins, maltodextrins, and polydextrose): Remember how glucose combines with a great deal of water to form glycogen? Fat replacers made from carbohydrate polymers also hold water and, because of this, they create texture—thickness and creaminess—and hold moisture in foods. They cannot be fried, but many can be heated.
- Cellulose and fiber-based products (Z-Trim, gums, carrageenan, various forms of cellulose, gel, pectin): These add not only thickness and texture but also structure to foods. They have no calories. Applesauce and puréed prunes, which can be used as fat substitutes in recipes, are in this category.
- Microparticulated protein product: Whey protein or egg white and milk protein are heated and processed to form microscopic particles that create the sensation of fat (known as "mouthfeel") when eaten. Simplesse, which is made from eggs and cow's milk, provides 1–2 kcal/g. Dairy-Lo is produced from whey protein and has a caloric value of 4 kcal/g.

Fat-Based Fat Replacers

The concept of fat-based fat substitutes is a relatively new area of development for the fat-replacement industry but one with the potential to have tremendous impact on the food supply and health.

- Reduced-calorie triglycerides: Caprenin and Salatrim (BENEFAT) are examples of compounds whose triglycerides have been manipulated to prevent complete absorption, giving them a caloric value of 5 kcal/g. Neither can be used for frying.
- Olestra: Olestra is marketed by Procter & Gamble under the brand name Olean. It was approved as a food additive for snack foods in January 1996. Olestra is unique in that it has a sucrose core bound to six to eight fatty acids. The fatty acids may come from soybean, corn, palm, coconut, and cottonseed oils. The olestra molecule is too large and its fatty acid bonding sites too inaccessible to be broken down during digestion. It has all of the properties of a dietary triglyceride but none of the calories. Unlike other fat replacers, olestra can be subjected to very high heat, even frying, making it ideal for the production of chips and snack foods.

Health Benefits (and Detriments) of Fat Replacers

One of the most obvious benefits of fat replacers is that they allow people to maintain variety and palatability in the diet while reducing fat intake. This may contribute to weight management. For example, individuals in one study who consumed a diet for two weeks in which olestra had been covertly substituted for conventional dietary fat had a lower intake of both calories (8%) and fat (11%).[51] In another study moderately overweight men consuming a 25% fat diet in which olestra had been substituted for conventional fat lost significantly more weight and body fat over nine months than men following a 25% or 33% fat diet.[52]

Use of fat replacers may also help people reduce the risk of cardiovascular disease, while allowing them to occasionally enjoy foods like sour cream, chocolate chip cookies, and potato chips. The Olestra Post-Marketing Surveillance Study reported that, even when caloric intake was not reduced, people consuming the most olestra (more than 2 g/day) reduced energy intake from fat by almost 3% and intake from saturated fats by 1.1%.[53] Olestra is also reported to have cholesterol-lowering effects.[54]

carotenoids Red, orange, and yellow pigments found in plant foods; nutritionally significant carotenoids include beta-carotene, alpha-carotene, lycopene, lutein, beta-cryptoxanthin, and zeaxanthin.

Olestra stands alone among the fat replacers in the controversy it has generated, primarily related to concerns about gastrointestinal side effects and depletion of the fat-soluble vitamins A, D, E, and K. Heavy users of olestra do have lower circulating levels of **carotenoids**, which are plant sources of vitamin A.[55] This is not considered to be significant enough to impair health. In addition, neither Procter & Gamble's consumer telephone line nor independently conducted research has yielded evidence of gastrointestinal effects that are more common than would be expected in a given population.[22] Initially, the FDA imposed a requirement that the labels of olestra-containing foods contain this statement: "This product contains olestra. Olestra may cause abdominal cramping and loose stools. Olestra inhibits the absorption of some vitamins and other nutrients. Vitamins A, D, E, and K have been added." While the FDA continues to mandate that products containing olestra be supplemented with vitamins A, D, E, and K, the labeling requirement was withdrawn in 2003.

No adverse health effects are associated with the carbohydrate-based fat replacers or with the reduced-calorie triglycerides, caprenin, and salatrim. The only issue with protein-based fat replacers is the potential for food allergy, and simplesse should be avoided by people with allergies to eggs and milk.

We can conclude that fat replacers may safely play a role in weight management and promotion of cardiovascular health. However, use of fat replacers alone is probably insufficient for significant weight loss, and individuals need to consider additional prudent dietary and activity strategies.

Lipids: Summary

Lipids are needed for energy and other vital functions but, if they are consumed in excess, can promote obesity, cardiovascular disease, and some cancers. Most experts recommend a diet containing no more than 35% of calories as fat, less than 7% of calories as saturated fats, few *trans* fatty acids, and less than 300 mg of cholesterol. Lower-fat diets (20–25% fat) and diets rich in monounsaturated fats may offer even greater protection against cardiovascular disease and hinder weight gain. Medium-chain triglycerides may become more widely available to help prevent weight gain and aid weight loss. There is some evidence that fat replacers like simplesse and olestra aid in achieving cardiovascular health and weight management.

PROTEIN: BUILDER AND MAINTAINER OF BODY TISSUES

amino acid
Building block of protein; each amino acid has an amino group, an acid group, and a unique side chain.

essential amino acids *Amino acids that are needed for vital body functions and that can be obtained only through the diet; also called indispensable.*

nonessential amino acids
Amino acids needed for physiological functions that can be manufactured from available nitrogen, carbohydrate, and fat fragments; also called dispensable.

Proteins are made of amino acids. An **amino acid** consists of a central carbon atom with attached hydrogen, amino group (NH_2), acid group (COOH), and a side chain that gives each amino acid its unique identity (see Figure 5-6). There are twenty common amino acids, which are classified as essential or nonessential. The nine **essential amino acids** (sometimes called indispensable) cannot be made by the body in sufficient quantities to support its needs and must be obtained through the diet. The eleven **nonessential amino acids** (sometimes called dispensable) are either supplied by the diet or manufactured by the body from available nitrogen, carbohydrate, and fat fragments. A list of the amino acids can be found in Table 5-9.

What Do Proteins Do?

Although proteins contain 4 kcal/g, just like carbohydrates, their primary function is to support growth of body tissues in childhood, pregnancy, lactation, wound healing, and bodybuilding, and maintenance of body tissues throughout life. Amino acids are also needed for the synthesis of hormones, enzymes, and neurochemicals. If sufficient calories are not available from dietary carbohydrate and fat, then protein will be used for fuel, resulting in muscle wasting and weight loss. Adults need about 0.8 g of protein per kilogram of body weight every day to support health, whereas infants, children, and adolescents need quite a bit more (1–2 g/kg).

Excess protein presents two health problems: First, protein not needed for tissue maintenance or other physiological functions must be disassembled and processed. The

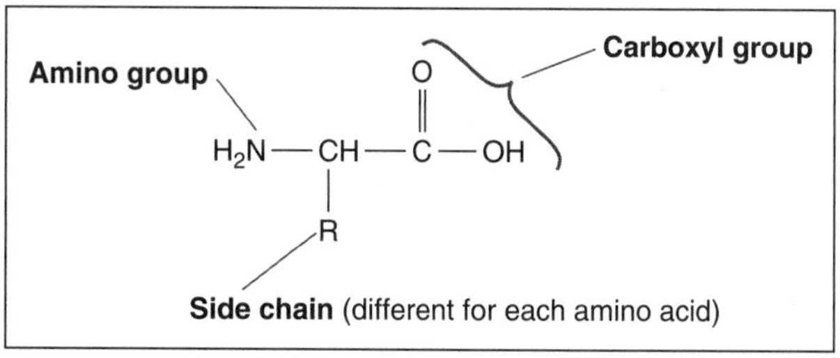

FIGURE 5-6 Amino acid structure.
Source: From Sherwood. Human Physiology, 7E. p. A-14. © 2010 Brooks/Cole, a part of Cengage Learning, Inc. Reproduced by permission. www.cengage.com/permissions.

TABLE 5-9 Overview of Proteins

Classification

Proteins are made up of amino acids, which are classified as:

Essential amino acids (indispensible)	Nonessential amino acids (dispensible)
Histidine	Alanine
Isoleucine	Arginine*
Leucine	Aspargine
Lysine	Aspartic acid
Methionine	Cysteine*
Phenylalanine	Glutamine acid
Threonine	Glutamine*
Tryptophan	Glycine*
Valine	Proline*
	Serine
	Tyrosine*

These amino acids are sometimes called "conditionally indispensible," meaning that food sources may be needed when body synthesis cannot keep up with need.

Energy value 4 kcal/g

Functions

Growth and maintenance of tissues
Formation of enzymes, hormones, antibodies, and other essential components
Maintenance of fluid and electrolyte balance
Maintenance of acid-base balance
Energy

Food sources

Meat, poultry, fish, eggs
Milk, cheese, dairy
Legumes, nuts
Breads, cereals, grains
Vegetables

Recommended intake

Lower limit: 10% of caloric intake
Upper limit: 35% of caloric intake
0.8 g/kg body weight/day (adults)
1.2–1.7 g/kg body weight/day for athletes
1.3–1.8 g/kg body weight/day for vegetarian athletes

kidneys are responsible for eliminating the nitrogen removed from protein and may eventually be damaged from processing too much nitrogen. Second, excess animal protein is associated with excess animal fats (and calories), which contributes to cardiovascular and other chronic diseases. For these reasons, protein consumption should generally not exceed 35% of total calories or 1.6 g/kg body weight.

Protein is found in both animal and plant foods. Animal proteins have the advantage of being excellent sources of all of the essential amino acids and are called

complete protein *Protein that contains all of the essential amino acids in necessary proportions to support health; all animal proteins are complete.*

incomplete protein *Protein missing one or more of the essential amino acids; all plant proteins are incomplete.*

complete proteins. Plant proteins are called **incomplete proteins** because they are missing one or more of the essential amino acids. People who eat a variety of plant proteins will get all of the essential amino acids, just as if they had consumed a complete protein.

How Does Protein Modification Promote Health and Weight Management?

A high-protein diet does not increase the rate of weight loss or the likelihood of maintaining weight loss. However, for people on calorie-restricted diets, keeping protein intake at least at 15% of calories can prevent or minimize the drop in energy expenditure typically seen when calories are reduced, because protein is thermic (it generates heat).[56] More important, the type of dietary protein consumed can influence health.

Plant-Based Diets and Risk of Chronic Disease

Adopting a vegetarian or modified vegetarian diet can have beneficial effects on both weight and health. Plant-based diets increase fiber, decrease saturated fat and cholesterol, and help people obtain recommended servings of fruits, vegetables, and whole grains. Fruits, vegetables, and grains contain more than 100 vitamins, minerals, and other nutrients with cancer-fighting properties and other health advantages. In 1995, the Dietary Guidelines for Americans acknowledged for the first time that vegetarian diets could meet nutritional recommendations.

Because plant proteins have lower fat content, they may be lower in calories. This combination of a low-fat, high-fiber diet can help in weight management. Animal proteins, in contrast, contain more saturated fat and cholesterol, which are known to increase the risk of cardiovascular disease and certain cancers. Overall mortality, as well as morbidity from cardiovascular disease, diabetes, hypertension, and possibly colorectal cancer, is lower in vegetarians.[57–58] Obesity rates are also lower among vegetarians.[58]

Recommendations for Plant-Based Diets

vegans *Vegetarians who exclude all animal products from their diets.*

lacto-ovo vegetarians *Vegetarians who exclude meat, fish, and poultry from their diets but who eat dairy and eggs.*

A plant-based diet is not necessarily a vegetarian diet. Many people on plant-based diets continue to use meat products and/or fish but in smaller quantities. **Vegans** exclude all animal products from their diets, and **lacto-ovo vegetarians** eat eggs and dairy products but not meat, poultry, or fish. Clients who are interested in pursuing a plant-based diet might want to start by cutting back on meat, fish, and poultry and gradually adopt a lacto-ovo-vegetarian eating plan. Unless a person is very motivated by health or ethical concerns, beginning with a vegan diet is challenging.

Vegetarians need to be careful to obtain enough protein, iron, calcium, vitamin D, and vitamin B_{12}. Table 5-10 summarizes good food sources for these nutrients.

Legumes and Soy

Perhaps the best part of a vegetarian diet (as well as the Mediterranean diet) is its reliance on legumes for protein and other nutrients. Legumes grow in plant pods and include dried beans, peas, pinto beans, lentils, and soybeans. They are a meat-alternative source of protein but do not contain all of the essential amino acids. Although they lack vitamins C, A, and B_{12}, they are excellent sources of starch, fiber, calcium, iron, and folate. Legumes contain negligible fat.

TABLE 5-10   Good Sources of Nutrients for Vegetarians

Nutrient	Plant source	Animal source*
Protein	Tofu, legumes, nuts, seeds, texturized vegetable protein	Eggs, milk, yogurt, cheese
Iron	Legumes; bulgur and some other grains; leafy vegetables like chard and spinach (absorption is enhanced by consuming these with vitamin C-rich foods or by occasionally cooking in an iron skillet)	Occasionally cook in an iron skillet
Calcium	Kale, collard greens, broccoli, tofu	Dairy products
Vitamin D	Fortified grains; exposure to sunlight promotes formation of vitamin D	Fortified dairy products, eggs
Vitamin B_{12}	Fortified soy "milk" and meat replacements	Dairy products, eggs

*For lacto-ovo vegetarians

A half cup serving of most legumes provides about 110 kcal, 8 g protein, 20 g carbohydrate, 7 g fiber, 20–60 mg calcium, 1.8–2 mg iron, and 130 mg folate. Legumes are inexpensive and can be purchased dry (and then soaked and cooked before being used in recipes) or, as a time saver, already cooked in cans. (Canned legumes are usually high in sodium.) Legumes are easily substituted for meat in chili, enchiladas, and many casseroles.

Soybeans can be used to prepare many different kinds of foods. Tofu, texturized soy protein, and other soy-derived products can be incorporated into traditional recipes with a little guidance. Soy products reduce blood cholesterol and lower the risk of heart disease and have also been linked to lower cancer rates, particularly among Asians, who consume 20 to 80 times more soy than Americans. The health benefits of soy foods are conferred by several phytochemicals, including an important group of compounds known as **isoflavones**. Many researchers believe that the combination of soy protein and isoflavones is what provides protection against cancer and cardiovascular disease.[59] Soy isoflavones have also been implicated in the prevention and treatment of osteoporosis. Soy seems to both prevent the resorption of bone, so less bone is lost, and to promote bone formation, so bone density is maintained as we age.[59] Health concerns about soy have focused on the estrogen-like effects of isoflavones—could isoflavones promote estrogen-related cancers? Fortunately, an intake of 40–80 mg/day of isoflavones—the recommended dose—does not increase the risk of cancer.[60]

isoflavones *Plant-based compounds with estrogen-like effects; they are in a category of substances called the phytoestrogens.*

What About Fish?

Fish (both finfish and shellfish) are an excellent source of high-quality protein. Three ounces of most fish provides between 14 and 24 g of protein. Earlier in this chapter you learned that two omega-3 polyunsaturated fatty acids found in fish—eicosapentaenoic acid (EPA) and docosahexaenoic acid (DHA)—have beneficial effects on blood lipids. DHA is also associated with positive neurodevelopmental outcomes in children, such as improved cognition and language.[61] However, fish contain mercury in varying amounts, which raises health concerns.

Mercury is a heavy metal that occurs naturally in the environment but that is also a byproduct of industrial pollution, released into the atmosphere where it finds its way into lakes, streams, and the ocean. Bacteria in water convert mercury to methylmercury, which is then absorbed by fish. Larger fish that live longer accumulate the most. Human tissues absorb methylmercury from the fish that we eat. Health concerns focus on mercury's impact on neurological development in children, cognitive function in adults, and cardiovascular disease risk.

Most research finds the benefits of DHA from fish to outweigh the risks of methylmercury.[61] Still, prudent consumption of fish, particularly by children and women considering becoming pregnant or who are pregnant, is recommended by the FDA and the Environmental Protection Agency (EPA). The EPA and FDA advise pregnant women to avoid fish high in mercury and consume no more than one serving per week (about 6 ounces) of fish with intermediate levels of mercury. Approximately 12 ounces per week of fish with lower levels of mercury is considered safe for pregnant women and others.[62] Table 5-11 lists fish with varying levels of mercury and highlights those fish that are also rich in omega-3 polyunsaturated fatty acids.

TABLE 5-11 Methylmercury in Fish (Finfish and Shellfish)

Fish with lower levels of mercury 12 ounces/week can be safely consumed	Fish with intermediate levels of mercury 6 ounces/week can be safely consumed	Fish having high levels of mercury Avoid these
Anchovies* (.043)	Grouper (.465)	Mackerel, king (.730)
Catfish (.049)	Halibut (.252)	Shark (.988)
Cod (.095)	Lobster, northern (.310)	Swordfish (.976)
Crab (.060)	Monkfish (.180)	Tilefish (1.450)
Haddock, Atlantic (.031)	Orange roughy (.554)	
Herring* (.044)	Skate (.137)	
Lobster, spiny (.09)	Snapper* (.189)	
Mackerel, Atlantic* (.050)	Tuna, canned, albacore* (.353)	
Mackerel chub, Pacific (.088)	Tuna (.383)	
Oyster* (.013)		
Pollock (.041)		
Salmon* (.014)		
Sardines (.016)		
Scallops (.050)		
Squid (.070)		
Tilapia (.010)		
Trout, freshwater* (.072)		
Tuna, canned, light (.118)		
Whitefish (.069)		
Fast food fish and fish sticks are usually low in mercury		

*These fish are rich in omega-3 polyunsaturated fatty acids
Values given in parentheses are average for each fish, in parts per million (PPM)
Source: FDA. (2009). Mercury levels in commercial fish and shellfish. Available from: http://www.fda.gov/Food/FoodSafety/Product-SpecificInformation/Seafood/FoodbornePathogensContaminants/Methylmercury/ucm115644.htm. Accessed July 13, 2010.

Protein: Summary

A diet that is 15–35% protein, primarily from plant sources, can contribute to health and weight management. Plant sources of protein are lower in saturated fat and higher in fiber than animal proteins. Fish, an excellent source of protein and healthy omega-3 polyunsaturated fatty acids, should be consumed in moderate amounts to avoid risks associated with methylmercury.

DIETARY APPROACHES TO WEIGHT MANAGEMENT

You now have some information about the role the energy nutrients play in weight management. This section of the chapter describes two diet planning aids–food guides/pyramids and the Exchange System—and two dietary approaches—the DASH and the Mediterranean diets—that may help you put together a diet that promotes health and weight management.

How Are Food Guides Used for Diet Planning?

Two diet planning aids—food guide pyramids and the Exchange System—can be helpful in pulling dietary recommendations together into a plan for healthy eating. The U.S. Food Guide Pyramid gives a general outline of a healthy diet based on the Dietary Guidelines for Americans and the Dietary Reference Intakes from the National Academy of Sciences. The Exchange System organizes foods in a way that makes planning for a specific nutrient or caloric intake easier. Both are valuable tools to assist clients in viewing their diets holistically.

Diet Planning with Food Pyramids and Food Guides

The U.S. Food Guide Pyramid has been the official food guide for the United States since 1992. It is a tool for helping people over the age of 2 years to meet the U.S. Dietary Guidelines. The U.S. Food Guide Pyramid is now a Web-based interactive system called MyPyramid (www.mypyramid.gov). An overview of MyPyramid can be found in Figure 5-7. (Note that this will probably be revised when the 2010 Dietary Guidelines for Americans are released). MyPyramid recommends a range of servings from each food group. For a 2,000-calorie diet, this is: 6 ounces of grains (at least 3 oz should be whole grains), 2½ cups of vegetables, 2 cups of fruit, 3 cups of milk products, and 5½ ounces of meat and beans. Users can use the Web-based pyramid to calculate appropriate amounts of foods from each group according to their own caloric needs.

Although the pyramid provides a readily understandable framework for diet planning, it does require that the user have access to the Internet. And without the Internet, the pyramid itself provides no information, as you can see from Figure 5-7.

Alternatives to MyPyramid

The Healthy Eating Pyramid was developed by faculty at Harvard University's School of Public Health, as an alternative to MyPyramid. Scientists at Harvard, led by Dr. Walter Willett, were concerned that MyPyramid was based on outdated evidence and overly influenced by food business interests, particularly the beef and dairy industries. The Healthy Eating Pyramid is illustrated in Figure 5-8. Note that exercise is the foundation and that

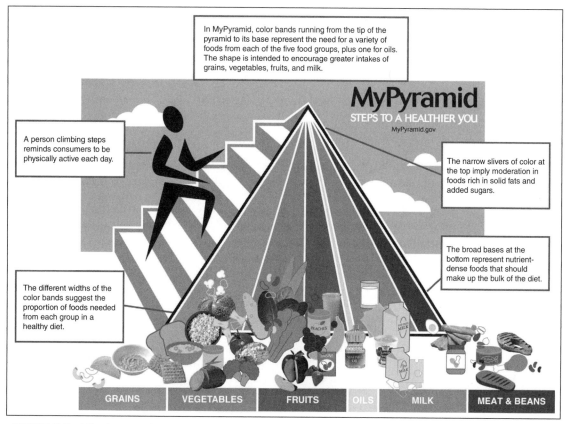

FIGURE 5-7 The Internet-based MyPyramid provides a visual representation of the role of physical activity and diet in promoting health.

Source: From Sizer/Whitney, Nutrition, 10E. p. 41. © 2006 Brooks/Cole, a part of Cengage Learning Inc. Reproduced by permission. www.cengage.com/permissions.

this pyramid focuses on a plant-based diet. Critics of the Healthy Eating Pyramid point to several concerns: it includes alcohol (which can increase caloric intake); plant oils appear to occupy about the same dietary prominence as whole grains (and even good fats are caloric); dairy is de-emphasized (which could compromise calcium and vitamin D levels); and all refined grains are at the pyramid's tip (although these do provide nutrients).

The U.S. Food Exchange System

The U.S. Food Exchange System was developed by the American Diabetes Association and the American Dietetic Association to help diabetics with meal planning. The system is also very helpful to nondiabetics who are trying to devise a diet that is varied and balanced and controls calories.

The Exchange System classifies foods into seven lists according to their carbohydrate, fat, and protein content. A serving size for each food on the list is determined so that one serving of any foods on the same list will have approximately the same number of kilocalories. There is also a list of free foods, which contain fewer than 20 kcal/serving,

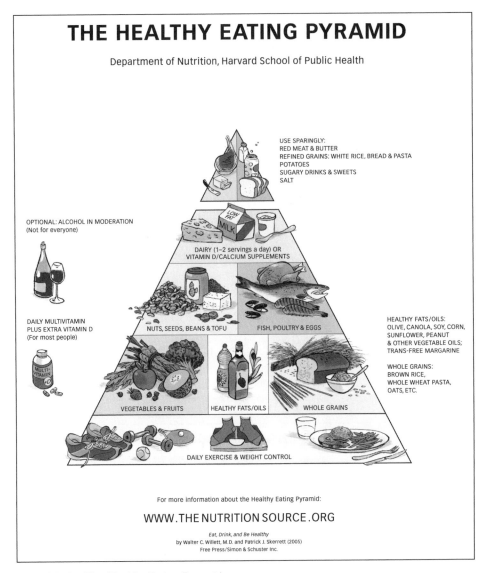

FIGURE 5-8 The Healthy Eating Pyramid.

Source: Copyright © 2008 Harvard University. For more information about The Healthy Eating Pyramid, please see The Nutrition Source, Department of Nutrition, Harvard School of Public Health, http://www.thenutrition-source.org, and *Eat, Drink, and Be Healthy*, by Walter C. Willett, M.D. and Patrick J. Skerrett (2005), Free Press/Simon & Schuster Inc.

and a list of combination foods. The exchange lists and their carbohydrate, fat, protein, and energy content are presented in Table 5-12. Foods and serving sizes in each exchange list are detailed in Appendix C.

Table 5-13 illustrates how the exchange lists can be used to plan a calorie-controlled diet. A person who wants to limit caloric intake to 1,800 kcal/day is allowed eight starch, four fruit, two milk, five vegetable, six lean meat, and six fat exchanges per day. Dozens

TABLE 5-12   The Exchange Lists

List	Carbohydrate (g)	Protein (g)	Fat (g)	Energy (kcal/serving)
Carbohydrates				
Starch: breads, cereals, and grains; starchy vegetables; crackers and snacks; and beans, peas, and lentils	15	0–3	0–1	80
Fruit	15	–	–	60
Milk				
Fat-free, lowfat, 1%	12	8	0–3	100
Reduced-fat, 2%	12	8	5	120
Whole	12	8	8	150
Sweets, desserts, and other carbohydrates	15	Varies	Varies	Varies
Nonstarchy vegetables	5	2	–	25
Meat and meat substitutes				
Lean	–	7	0–3	45
Medium-fat	–	7	4–7	75
High-fat	–	7	8+	100
Plant-based proteins	–	7	Varies	Varies
Fats	–	–	5	45
Alcohol	Varies	–	–	100

Source: The Exchange Lists are the basis of a meal planning system designed by a committee of the American Diabetes Association and The American Dietetic Association. While designed primarily for people with diabetes and others who must follow special diets, the Exchange Lists are based on principles of good nutrition that apply to everyone. © 2008 by the American Diabetes Association and The American Dietetic Association.

of different diets can be constructed within these guidelines. Although the Exchange System is much more complicated than MyPyramid, its great value is in letting people select foods that they like within a framework of calorie control.

You might wonder where alcohol fits in with the Exchange System. Because it inhibits fat oxidation, alcohol is generally counted like fat in diet planning. In the Exchange System, one alcoholic beverage is equivalent to slightly more than two fat exchanges.

Problems with the U.S. Food Exchange System

One complicating factor in the Exchange System is that people may confuse "one exchange" with "one serving." For example, the food labels of some ready-to-eat cereals indicate that a serving size is anywhere from 2/3 cup to 1 cup. In the Exchange System, 1/2 cup of sugar-sweetened cereal constitutes one starch exchange. The consumer who measures 1 cup of cereal would need to know to count this quantity as two starch exchanges, even though the food label records this as one serving. Otherwise, the client might over consume calories. Clients who are using Exchange System will have to keep exchange lists with

TABLE 5-13 Diet Planning with the Exchange System

Exchange	Caloric value of the diet					
	1,200	1,500	1,800	2,000	2,200	2,600
Starch*	6	7	8	9	11	13
Fruit	2	3	4	4	4	5
Milk (nonfat)†	2	2	2	3	3	3
Vegetable	3	4	5	5	5	6
Meat (lean)‡	4	5	6	6	6	7
Fat	3	5	6	7	8	10

*If foods containing added fats are chosen, then their fat content must be subtracted from allowable servings from the "Fat" list. Foods with extra fat are counted with 1 fat exchange each.

†Be aware of the added fat content from reduced-fat and whole milk. This could add up to 60 calories per serving.

‡If medium-fat and high-fat meats and meat substitutes are used, be aware that a serving could add 1–5 g (or more) of additional fat, for up to 55 added calories.

Source: Adapted from Sizer, F., & Whitney, E. (1997). *Nutrition: Concepts and controversies* (7th ed.). Belmont, CA: West/Wadsworth.

them until they are familiar with what constitutes an exchange portion size of the foods they eat.

What Are Dietary Approaches that Address Health Concerns in Overweight/Obesity?

The food guides described in the previous section give you a good overview of what constitutes a healthy diet. For individuals trying to manage their weight and address health problems, including hypertension, type 2 diabetes, and CVD, two diet plans may be particularly attractive—the DASH diet and the Mediterranean diet. A preliminary report of the 2010 Dietary Guidelines for Americans notes that these lower-sodium, plant-based diets have an important role in reducing CVD risk.[24]

Dietary Approaches to Stop Hypertension (DASH)

The Dietary Approaches to Stop Hypertension (DASH) research trial found that men and women of various ethnicities lowered their blood pressure when they ate a diet with limited meat, poultry, and eggs; rich in fruits, vegetables, and low-fat dairy products; and with reduced sodium and total and saturated fats.[63] The diet provided approximately 27% of kcals from fat (6% saturated, 13% monounsaturated, 8% polyunsaturated), 55% of kcals from carbohydrates, and 31 g of fiber. Individuals were permitted to consume 3,000 mg of sodium and two alcoholic drinks per day. Eight to ten servings of fruits and vegetables were eaten daily.

Perhaps most of us don't get ten servings of fruits and vegetables each day, but the advice to increase fruits, vegetables, and whole grains and to limit intake of sodium and highly refined starches and sugars is sound. Table 5-14 gives an overview of the DASH

TABLE 5-14 The DASH Diet Plan

	Number of servings for 1,600–3,100 kcal diets	Number of servings for 2,000 kcal diets
Grains and grain products (include at least 3 whole grain foods every day	6–12	7–8
Fruits	4–6	4–5
Vegetables	4–6	4–5
Lowfat or nonfat dairy foods	2–4	2–3
Lean meats, fish, poultry	1.5–2.5	2 or less
Nuts, seeds, and legumes	3–6/week	4–5/week
Fats and sweets	2–4	limited

Source: The DASH Diet Eating Plan. Accessed at http://www.dashdiet.org/default.asp. Accessed March 3, 2011.

diet. You will notice that it includes (in addition to lots of fruits and vegetables) at least three whole grain foods each day, several servings of nuts, seeds, and legumes every week, and limited fats and sweets. A book is available that provides 28 days of meal plans and a lower-calorie diet plan (visit the DASH Web site for ordering information: www.dashdiet.org/default.asp)

The Mediterranean Diet

The Mediterranean diet is low in red meat, rich in vegetables, and incorporates olive oil and nuts as the main added fats. It is characterized by:

- A high ratio of monounsaturated to saturated fats (unlimited use of olive oil, as long as energy balance is maintained)
- Moderate consumption of wine at meals
- High intake of legumes, breads, cereals, and grains
- High intake of fruits and vegetables
- Little use of meat and eggs (once a week as a main meal)
- Moderate use of dairy products (mainly yogurt)

The Mediterranean Diet Pyramid, pictured in Figure 5-9, has several similarities to the Healthy Eating Pyramid mentioned earlier in this chapter. In the Mediterranean Diet Pyramid, updated in 2008, all plant foods are listed in a single group at the pyramid's base (this group also includes herbs and spices); consumption of fish and shellfish is recommended at least twice per week; and meats and sweets occupy a small space at the top of the pyramid. Researchers are not yet clear about whether the diet offers health benefits only when it is followed in totality, or whether individual elements of the diet (like olive oil) can convey benefits independent of the others.

Putting the Elements of a Healthy Diet Together

While both the Mediterranean diet and DASH are well-researched diets for reducing blood pressure, blood lipids, and even weight, they require a considerable adjustment in

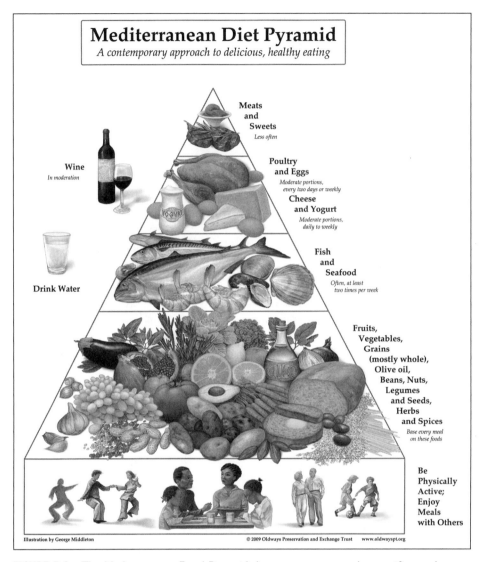

Mediterranean Diet Pyramid
A contemporary approach to delicious, healthy eating

Meats
and
Sweets
Less often

Poultry
and Eggs
*Moderate portions,
every two days or weekly*

Cheese
and Yogurt
*Moderate portions,
daily to weekly*

Wine
In moderation

Fish
and
Seafood
*Often, at least
two times per week*

Drink Water

Fruits,
Vegetables,
Grains
(mostly whole),
Olive oil,
Beans, Nuts,
Legumes
and Seeds,
Herbs
and Spices
*Base every meal
on these foods*

Be
Physically
Active;
Enjoy
Meals
with Others

Illustration by George Middleton   © 2009 Oldways Preservation and Exchange Trust   www.oldwayspt.org

FIGURE 5-9 The Mediterranean Food Pyramid does not recommend a specific number of servings from each food group, but suggests foods to be eaten at every meal, daily, and weekly.

Source: http://www.oldwayspt.org/mediterranean-diet-pyramid

eating for most people. Chapter 9 will cover behavioral approaches to weight management that might be helpful in guiding clients to gradually change their diets. Table 5-15 provides a sample of a lower-sodium, calorie-controlled, plant-based diet that incorporates elements of the Exchange system, Mediterranean diet, and DASH. This can be used as an example for people trying to find ways to incorporate more fruits and vegetables into a satisfying eating plan.

TABLE 5-15 Using the Exchange System to Create a DASH- and Mediterranean-like Diet

Exchange lists	Breakfast	Lunch	Dinner	Snacks	Approx. cals.
Starch (7)	1 slice whole wheat bread	2 slices whole wheat bread ½ c chickpeas (for hummus)	⅓ c cooked orzo ½ c peas, cooked	3 c popcorn, no salt, no added fat	560
Nonstarchy Vegetables (4)	1 c raw spinach (for omelet)	1 c raw tomato	½ c mushrooms, cooked ½ c asparagus, cooked		100
Fruit (4)	½ c orange juice or a small orange	Small apple	1¼ c strawberries (to make yogurt parfait)	Nectarine	240
Meat (5)	2 egg whites (for omelet) 1 oz lowfat cheese		3 oz salmon (to cook and add to orzo and vegetables)		225
Dairy (2)		½ c skim milk	6 oz plain Greek yogurt, lowfat (for yogurt parfait)		200
Fat (5)	1 tsp olive oil for cooking	2 tsp olive oil for hummus	1 tsp olive oil for orzo, 1 T reduced fat salad dressing		225
Free Foods		Several pieces of lettuce for hummus sandwich	1 c salad greens ½ c diced cucumber		
Alcohol				5½ oz wine, if desired	100
					1,550–1,650 kcals

Dietary Approaches to Weight Management: Summary

Food guides, such as MyPyramid, the Healthy Eating Pyramid, and the U.S. Food Exchange System, are tools that can be used to create a framework for a healthy, varied diet that promotes weight management. The Mediterranean and DASH diets are excellent diet plans that not only promote weight control but weight loss.

Case Study. Personal Diet Analysis, Part 2	Your subject should use the food record form in Appendix B to monitor his/her intake of all foods and beverages for three typical days (two weekdays and one weekend day are recommended). • Using either nutrition software (preferred) or paper and pencil, calculate your subject's average intake of carbohydrate, fat, and protein. Nutrition software will also provide information on micronutrient intake. Save this for the Chapter 6 case studies. • What difficulties did your subject have using the food record? How might these difficulties be overcome with clients? • How close did your subject come to the dietary recommendations presented in this chapter and to the recommendations of the Interactive DRI?

REFERENCES

1. Guenther, P. M., Juan, W. Y., Reedy, J., Britten, P., Lino, M., Carlson, A., et al. (2008 August). Diet quality of Americans in 1994–96 and 2001–02 as measured by the Healthy Eating Index–2005. *Nutrition Insight 37.* Alexandria, VA: CNPP. Available online from http://www.cnpp.usda.gov/Publications/NutritionInsights/Insights37.pdf. Accessed July 13, 2010.

2. U.S. Department of Health and Human Services & U.S. Department of Agriculture. (2005). *Dietary guidelines for Americans, 2005* (6th ed.). Washington, D.C.: U.S. Government Printing Office.

3. Wright, J. D., Kennedy-Stephenson, J., Wang, C. Y., McDowell, M. A., & Johnson, C. L. (2004). Trends in intake of energy and micronutrients – United States, 1971–2000. *Morbidity and Mortality Weekly Reports,* 53(4), 80–82.

4. Flatt, J. P. (1995). Use and storage of carbohydrate and fat. *American Journal of Clinical Nutrition, 61* (Suppl), 952S–959S.

5. Otten, J. J., Hellwig, J. P., & Meyers, L. D. (Eds.). (2006). *Dietary reference intakes: The essential guide to nutrient requirements.* Washington, D.C.: The National Academies Press.

6. Slavin, J. L. (2008). Position of the American Dietetic Association: Health implications of dietary fiber. *Journal of the American Dietetic Association, 108*(10), 1716–1731.

7. Howarth, N. C., Saltzman, E., & Roberts, S. B. (2001). Dietary fiber and weight regulation. *Nutrition Reviews,* 59(5), 129–139.

8. Wells, H. F., & Buzby, J. C. (2008). High-fructose corn syrup usage may be leveling off. *Amber Waves,* 6(1), 4–5.

9. Vartanian, L. R., Schwartz, M. B., & Brownell, K. D. (2007). Effects of soft drink consumption on nutrition and health: A systematic review and meta-analysis. *American Journal of Public Health,* 97(4), 667–675.

10. *Usual Intake of Added Sugars.* Risk Factor Monitoring and Methods Branch Web site. Applied Research Program. National Cancer Institute. Available from http://riskfactor.cancer.gov/diet/usualintakes/pop/t35.html. Accessed July 13, 2010.

11. Johnson, R. J., Segal, M. S., Sautin, Y., Nakagawa, T., Feig, D. I., Kang, D., et al. (2007). Potential role of sugar (fructose) in the epidemic of hypertension, obesity and the metabolic syndrome, diabetes, kidney disease, and cardiovascular disease. *American Journal of Clinical Nutrition,* 86(4), 899–906.

12. Rutledge, A. C., & Adeli, K. (2007). Fructose and the metabolic syndrome: Pathophysiology and molecular mechanisms. *Nutrition Reviews,* 65(6 Pt 2), S13–23.

13. Malik, V. S., Schulze, M. B., & Hu, F. B. (2006). Intake of sugar-sweetened beverages and weight gain: A systematic review. *American Journal of Clinical Nutrition,* 84(2), 274–288.

14. Chen, L., Appel, L. J., Loria, C., Lin, P., Champagne, C. M., Elmer, P. J., et al. (2009). Reduction in consumption of sugar-sweetened beverages is associated with weight loss: The PREMIER trial. *American Journal of Clinical Nutrition,* 89(5), 1299–1306.

15. Saris, W. H. (2003). Glycemic carbohydrate and body weight regulation. *Nutrition Reviews,* 61(5 Pt 2), S10–6.

16. Nishida, C., Uauy, R., Kumanyika, S., & Shetty, P. (2004). The joint WHO/FAO expert consultation on diet, nutrition and the prevention of chronic diseases: Process, product and policy implications. *Public Health Nutrition,* 7(1A), 245–250.

17. Johnson, R. K., Appel, L. J., Brands, M., Howard, B. V., Lefevre, M., Lustig, R. H., et al. (2009). Dietary sugars intake and cardiovascular health: A scientific statement from the American Heart Association. *Circulation, 120*(11), 1011–1020.

18. Garg, A., Bantle, J. P., Henry, R. R., Coulston, A. M., Griver, K. A., Raatz, S. K., et al. (1994). Effects of varying carbohydrate content of diet in patients with non-insulin-dependent diabetes mellitus. *JAMA, 271*(18), 1421–1428.

19. Atkinson, F. S., Foster-Powell, K., & Brand-Miller, J. C. (2008). International tables of glycemic index and glycemic load values: 2008. *Diabetes Care, 31*(12), 2281–2283. (The online appendix can be found at http://care.diabetesjournals.org/content/31/12/2281/suppl/DC1)

20. American Diabetes Association. (2008). Nutrition recommendations and interventions for diabetes. *Diabetes Care, 31*(Suppl 1), S61–S78.

21. American Dietetic Association. (2004). Position of the American Dietetic Association: Use of nutritive and nonnutritive sweeteners. *Journal of the American Dietetic Association, 104*(2), 255–275.

22. Hepburn, P., Howlett, J., Boeing, H., Cockburn, A., Constable, A., Davi, A., et al. (2008). The application of post-market monitoring to novel foods. *Food and Chemical Toxicology, 46*(1), 9–33.

23. Mattes, R. D., & Popkin, B. M. (2009). Nonnutritive sweetener consumption in humans: Effects on appetite

and food intake and their putative mechanisms. *American Journal of Clinical Nutrition, 89*(1), 1–14.

24. *Report of the Dietary Guidelines Advisory Committee on the Dietary Guidelines for Americans, 2010.* (2010). Available from http://www.cnpp.usda.gov/DGAs2010-DGACReport.htm. Accessed July 12, 2010.

25. Acheson, K. J., Schutz, Y., Bessard, T., Ravussin, E., Jequier, E., & Flatt, J. P. (1984). Nutritional influences on lipogenesis and thermogenesis after a carbohydrate meal. *American Journal of Physiology, 246*(1 Pt 1), E62–70.

26. Prewitt, T. E., Schmeisser, D., Bowen, P. E., Aye, P., Dolecek, T. A., Langenberg, P., et al. (1991). Changes in body weight, body composition, and energy intake in women fed high- and low-fat diets. *American Journal of Clinical Nutrition, 54*(2), 304–310.

27. Lyon, X. H., Di Vetta, V., Milon, H., Jequier, E., & Schutz, Y. (1995). Compliance to dietary advice directed towards increasing the carbohydrate to fat ratio of the everyday diet. *International Journal of Obesity and Related Metabolic Disorders, 19*(4), 260–269.

28. Powell, J. J., Tucker, L., Fisher, A. G., & Wilcox, K. (1994). The effects of different percentages of dietary fat intake, exercise, and calorie restriction on body composition and body weight in obese females. *American Journal of Health Promotion, 8*(6), 442–448.

29. Gardner, C. D., Kiazand, A., Alhassan, S., Kim, S., Stafford, R. S., Balise, R. R., et al. (2007). Comparison of the Atkins, Zone, Ornish, and LEARN diets for change in weight and related risk factors among overweight premenopausal women: The A TO Z weight loss study: A randomized trial. *JAMA, 297*(9), 969–977.

30. Dansinger, M. L., Gleason, J. A., Griffith, J. L., Selker, H. P., & Schaefer, E. J. (2005). Comparison of the Atkins, Ornish, Weight Watchers, and Zone diets for weight loss and heart disease risk reduction: A randomized trial. *JAMA, 293*(1), 43–53.

31. Jeffery, R. W., Hellerstedt, W. L., French, S. A., & Baxter, J. E. (1995). A randomized trial of counseling for fat restriction versus calorie restriction in the treatment of obesity. *International Journal of Obesity and Related Metabolic Disorders, 19*(2), 132–137.

32. Franklin, T. L., Kolasa, K. M., Griffin, K., Mayo, C., & Badenhop, D. T. (1995). Adherence to very-low-fat diet by a group of cardiac rehabilitation patients in the rural southeastern United States. *Archives of Family Medicine, 4*(6), 551–554.

33. Astrup, A. (2005). The role of dietary fat in obesity. *Seminars in Vascular Medicine, 5*(1), 40–47.

34. Bray, G. A., & Popkin, B. M. (1998). Dietary fat intake does affect obesity! *American Journal of Clinical Nutrition, 68*(6), 1157–1173.

35. Sacks, F. M., & Willett, W. W. (1991). More on chewing the fat. the good fat and the good cholesterol. *New England Journal of Medicine, 325*(24), 1740–1742.

36. Ornish, D. (1998). Low-fat diets. *New England Journal of Medicine, 338*(2), 127; author reply 128–9.

37. Mozaffarian, D., Katan, M. B., Ascherio, A., Stampfer, M. J., & Willett, W. C. (2006). *Trans* fatty acids and cardiovascular disease. *New England Journal of Medicine, 354*(15), 1601–1613.

38. Terry, P. D., Rohan, T. E., & Wolk, A. (2003). Intakes of fish and marine fatty acids and the risks of cancers of the breast and prostate and of other hormone-related cancers: A review of the epidemiologic evidence. *American Journal of Clinical Nutrition, 77*(3), 532–543.

39. World Cancer Research Fund, & American Institute for Cancer Research. (2007). *Food, nutrition, physical activity, and the prevention of cancer: A global perspective.* Washington, D.C.: American Institute for Cancer Research.

40. Dengel, J. L., Katzel, L. I., & Goldberg, A. P. (1995). Effect of an American Heart Association diet, with or without weight loss, on lipids in obese middle-aged and older men. *American Journal of Clinical Nutrition, 62*(4), 715–721.

41. Trichopoulou, A., Costacou, T., Bamia, C., & Trichopoulos, D. (2003). Adherence to a Mediterranean diet and survival in a Greek population. *New England Journal of Medicine, 348*(26), 2599–2608.

42. Knoops, K. T., de Groot, L. C., Kromhout, D., Perrin, A. E., Moreiras-Varela, O., Menotti, A., et al. (2004). Mediterranean diet, lifestyle factors, and 10-year mortality in elderly European men and women: The HALE project. *JAMA, 292*(12), 1433–1439.

43. Esposito, K., Marfella, R., Ciotola, M., Di Palo, C., Giugliano, F., Giugliano, G., et al. (2004). Effect of a Mediterranean-style diet on endothelial dysfunction and markers of vascular inflammation in the metabolic syndrome: A randomized trial. *JAMA, 292*(12), 1440–1446.

44. Grundy, S. M., Cleeman, J. I., Merz, C. N., Brewer, H. B., Jr, Clark, L. T., Hunninghake, D. B., et al. (2004). Implications of recent clinical trials for the National Cholesterol Education Program Adult Treatment Panel III guidelines. *Circulation, 110*(2), 227–239.

45. McManus, K., Antinoro, L., & Sacks, F. (2001). A randomized controlled trial of a moderate-fat, low-energy diet compared with a low fat, low-energy diet for weight loss in overweight adults. *International Journal of Obesity and Related Metabolic Disorders*, 25(10), 1503–1511.

46. Shai, I., Schwarzfuchs, D., Henkin, Y., Shahar, D. R., Witkow, S., Greenberg, I., et al. (2008). Weight loss with a low-carbohydrate, Mediterranean, or low-fat diet. *New England Journal of Medicine*, 359(3), 229–241.

47. Piers, L. S., Walker, K. Z., Stoney, R. M., Soares, M. J., & O'Dea, K. (2003). Substitution of saturated with mono-unsaturated fat in a 4-week diet affects body weight and composition of overweight and obese men. *British Journal of Nutrition*, 90(3), 717–727.

48. Binnert, C., Pachiaudi, C., Beylot, M., Hans, D., Vandermander, J., Chantre, P., et al. (1998). Influence of human obesity on the metabolic fate of dietary long- and medium-chain triacylglycerols. *American Journal of Clinical Nutrition*, 67(4), 595–601.

49. St-Onge, M., & Jones, P. J. H. (2002). Physiological effects of medium-chain triglycerides: Potential agents in the prevention of obesity. *Journal of Nutrition*, 132(3), 329–332.

50. St-Onge, M. P., & Bosarge, A. (2008). Weight-loss diet that includes consumption of medium-chain tri-acylglycerol oil leads to a greater rate of weight and fat mass loss than does olive oil. *American Journal of Clinical Nutrition*, 87(3), 621–626.

51. Hill, J. O., Seagle, H. M., Johnson, S. L., Smith, S., Reed, G. W., Tran, Z. V., et al. (1998). Effects of 14 d of covert substitution of olestra for conventional fat on spontaneous food intake. *American Journal of Clinical Nutrition*, 67(6), 1178–1185.

52. Bray, G. A., Lovejoy, J. C., Most-Windhauser, M., Smith, S. R., Volaufova, J., Denkins, Y., et al. (2002). A 9-mo randomized clinical trial comparing fat-substituted and fat-reduced diets in healthy obese men: The Ole study. *American Journal of Clinical Nutrition*, 76(5), 928–934.

53. Patterson, R. E., Kristal, A. R., Peters, J. C., Neuhouser, M. L., Rock, C. L., Cheskin, L. J., et al. (2000). Changes in diet, weight, and serum lipid levels associated with olestra consumption. *Archives of Internal Medicine*, 160(17), 2600–2604.

54. Senanayake, S. P. J., & Shahidi, F. (2005). Dietary fat substitutes. In F. Shahidi (Ed.), *Bailey's industrial oil and fat products* (6th ed., pp. 503–534). John Wiley & Sons, Inc.

55. Neuhouser, M. L., Rock, C. L., Kristal, A. R., Patterson, R. E., Neumark-Sztainer, D., Cheskin, L. J., et al. (2006). Olestra is associated with slight reductions in serum carotenoids but does not markedly influence serum fat-soluble vitamin concentrations. *American Journal of Clinical Nutrition*, 83(3), 624–631.

56. Whitehead, J. M., McNeill, G., & Smith, J. S. (1996). The effect of protein intake on 24-h energy expenditure during energy restriction. *International Journal of Obesity and Related Metabolic Disorders*, 20(8), 727–732.

57. Appleby, P. N., Thorogood, M., Mann, J. I., & Key, T. J. (1999). The Oxford vegetarian study: An overview. *American Journal of Clinical Nutrition*, 70(3 Suppl), 525S–531S.

58. Fraser, G. E. (2009). Vegetarian diets: What do we know of their effects on common chronic diseases? *American Journal of Clinical Nutrition*, 89(5), 1607S–1612.

59. Omoni, A. O., & Aluko, R. E. (2005). Soybean foods and their benefits: Potential mechanisms of action. *Nutrition Reviews*, 63(8), 272–283.

60. Mahady, G. B. (2005). Do soy isoflavones cause endometrial hyperplasia? *Nutrition Reviews*, 63(11), 392–397.

61. Mozaffarian, D., & Rimm, E. B. (2006). Fish intake, contaminants, and human health: Evaluating the risks and the benefits. *JAMA*, 296(15), 1885–1899.

62. FDA/EPA. (2004) *What you need to know about mercury in fish and shellfish*. Available from: http://www.epa.gov/waterscience/fish/advice/. Accessed July 13, 2010.

63. Obarzanek, E., Sacks, F. M., Vollmer, W. M., Bray, G. A., Miller, E. R., 3rd, Lin, P. H., et al. (2001). Effects on blood lipids of a blood pressure-lowering diet: The dietary approaches to stop hypertension (DASH) trial. *American Journal of Clinical Nutrition*, 74(1), 80–89.

Vitamins, Minerals, and Water

R. Gino Santa Maria/Shutterstock.com

cappi thompson/Shutterstock.com

Iorga Studio/Shutterstock.com

Vitamins and minerals are collectively known as micronutrients because they are needed in only small amounts to ensure optimal body functioning. With a few exceptions, outright micronutrient deficiencies are uncommon in the United States. Still, some people—the elderly, pregnant women, people following low-calorie diets, smokers, strict vegetarians, and those who consume excessive amounts of alcohol—may not be obtaining or absorbing vitamins and minerals in the amounts needed for optimal health. Also, new evidence indicates that micronutrients not only prevent deficiency diseases but also play a significant role in preventing chronic diseases associated with overweight/obesity, such as cardiovascular disease and cancer.

DETERMINING OPTIMAL MICRONUTRIENT INTAKE

recommended dietary allowances (RDAs) *Average amounts of each nutrient needed to prevent inadequacy without causing toxicity in practically all healthy people. RDAs are one set of reference values that make up the dietary reference intakes (DRIs).*

Dietary Reference Intakes (DRIs) *Set of reference values for nutrient intake for healthy people that is adequate but not excessive; includes estimated average requirements (EARs), recommended dietary allowances (RDAs), adequate intakes (AIs), and tolerable upper intake levels (ULs).*

Since 1941, the **recommended dietary allowances** (RDAs) have been the standard against which the adequacy of energy and nutrient intake are measured. The RDAs are average amounts of each nutrient needed to prevent inadequacy without causing toxicity in practically all healthy people. Until recently, dietary "inadequacy" was indicated mainly by symptoms of deficiency disease.

The RDAs are periodically reviewed and updated by an expert committee of the Food and Nutrition Board of the National Academy of Sciences' Institute of Medicine, which is funded by the U.S. government. In 1993, the Food and Nutrition Board held a symposium titled "Should the Recommended Dietary Allowances Be Revised?" Following commentary by the general public and nutrition professionals, the board decided to reassess the RDAs and create a broader set of reference standards, the **dietary reference intakes (DRIs)**. Since 1997, the Food and Nutrition Board has issued thirteen DRI reports and one summary volume, covering macronutrients, micronutrients, energy, fiber, antioxidants, nutrition labeling, dietary planning, and dietary assessment. You can access all these reports at the Web site of the USDA's Food and Nutrition Information Center (http://fnic.nal.usda.gov/nal_display/index.php?info_center=4&tax_level=3&tax_subject=256&topic_id=1342&level3_id=5141).

What Are DRIs?

The DRIs encompass four sets of reference values:

- Estimated average requirements (EARs): The amount of each nutrient that meets the requirements of half of healthy people in particular life-stage and gender groups.
- Recommended dietary allowances (RDAs): The amount of each nutrient that meets the nutrient needs of about 97% of healthy people in particular life-stage and gender groups. The RDAs are calculated from the EARs; if no EAR can be determined, then no RDA will be established. For diet planning, healthy individuals should aim to meet the RDA for each nutrient.
- Adequate intakes (AIs): The average intake of each nutrient needed to sustain health, based on studies of people in particular life-stage and gender groups. AIs are used when scientific information is insufficient to determine a nutrient's EAR and RDA. For diet planning, healthy individuals should aim to meet the AI if no RDA is available.

- Tolerable upper intake levels (ULs): The upper limit of safe, chronic, daily intake for a nutrient. Although regularly consuming the UL for a nutrient should not cause adverse health effects, the UL is not a recommended level of intake. Individuals should avoid exceeding the UL to avoid possible toxic effects.

DRIs and Tolerable upper intake levels (ULs) for vitamins and minerals are found in tables in Appendix D.

How Should DRIs Be Used?

The DRIs remind us that, in diet planning, people should aim to consume the RDA or AI for each nutrient. For nutrients with an RDA, the requirement is set sufficiently high to meet or exceed needs of most of the healthy population. There is less certainty about the AI as an indicator of adequate intake for those nutrients that do not have an RDA, but AIs have been set based on best evidence to date. EARs, RDAs, and AIs were set at levels thought to prevent both nutrient deficiency diseases and chronic diseases.

The DRIs also provide an estimate of the risk for developing an adverse effect from taking too much of a particular nutrient from any source (naturally occurring in food or from fortified foods and beverages or supplements). ULs are available for eight vitamins (vitamins A, C, D, E, and B_6, choline, folate, and niacin) and fifteen minerals (boron, calcium chloride, copper, fluoride, iodine, iron, magnesium, manganese, molybdenum, nickel, phosphorus, selenium, sodium, and zinc). Nutrients for which there are no ULs may still be hazardous at high doses, but scientific evidence of this danger is not available (and may never be available, pending consideration of the hazards presented by giving research subjects high levels of these nutrients).

bioavailability

Extent to which a micronutrient is absorbed from the digestive tract and available to the cells and the tissues.

Bioavailability of micronutrients complicates determination of DRIs. Nutrient bioavailability refers to how efficiently a nutrient is absorbed from the gastrointestinal tract. This can be enhanced or impeded by a variety of factors: the chemical form of the nutrient, how absorbable the nutrient is in varying concentrations of gastric juice, presence of naturally occurring substances in the food, how food is prepared. At present the DRIs do not reflect bioavailability for most nutrients. Folate is an exception. Folate naturally occurring in food is not absorbed as readily as synthetic folic acid in supplements and fortified foods. In other words, the bioavailability of naturally occurring folate is lower than the bioavailability of folic acid from a supplement. For this reason, the RDA for folate is reported in dietary folate equivalents (DFEs), where 1 mg of food folate is equivalent to 0.6 mg of synthetic folate.

Determining Optimal Micronutrient Intake: Summary

The new DRIs—RDAs, EARs, AIs, and ULs—provide scientists' best estimates of vitamin and mineral intake that supports good health and lowers the risk of disease.

FAT-SOLUBLE VITAMINS

Fat-soluble vitamins—A, D, E, and K—are absorbed, transported, and stored along with lipids. Like dietary lipids, fat-soluble vitamins are incorporated into chylomicrons following digestion. After entering the bloodstream, the fat-soluble vitamins are gradually given

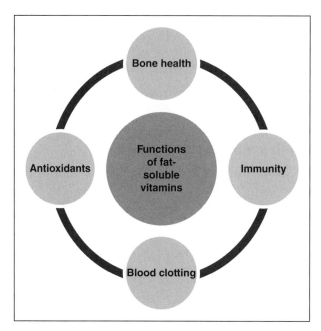

FIGURE 6-1   Some functions of the fat-soluble vitamins.

retinol *Active form of vitamin A.*

carotenoids *Red, orange, and yellow pigments found in plant foods; beta-carotene, alpha-carotene, and beta-cryptoxanthin are vitamin A precursors, and other nutrition-ally significant caro-tenoids include lycopene, lutein, and zeaxanthin.*

antioxidant *Substance in foods that protects cell membranes from destruction resulting from exposure to chemicals, radiation, ozone, cigarette smoke, and smog; may play a role in the prevention of cancer and cardiovascular disease.*

up to the cells and the tissues that need them. Figure 6-1 summarizes the functions of fat-soluble vitamins.

Vitamin A and Carotenoids

The two dietary forms of vitamin A are **retinol**, found in animal products, and more than 600 different **carotenoids**, found in plants, about 50 of which are typically com-ponents of the diet.[1] Only a few carotenoids are actually absorbed from food. In humans these are most likely to be beta-carotene (β-carotene), alpha-carotene (α-carotene), beta-cryptoxanthin (β-cryptoxanthin), lycopene, lutein, and zeaxanthin. Retinol is an active form of vitamin A (sometimes called preformed), meaning that it does not have to be transformed to function as a vitamin. Unlike retinol, the carotenoids have to be converted to vitamin A in the body to function as the vitamin. They are called vitamin precursors, and of those common in our diets only β-carotene, α-carotene, and β-cryptoxanthin can form retinol.

Functions of Vitamin A and the Carotenoids

Vitamin A stimulates the growth of cells, so its functions extend to many tissues and sys-tems and include immunity, reproduction, bone development and maintenance, and pro-motion of vision. Vitamin A deficiency is the leading cause of childhood blindness throughout the world.

Only one function of the carotenoids has been consistently documented—to produce vitamin A. Other potential carotenoid functions stem from their strong **antioxidant** effects, which means that they protect cell membranes from destruction caused by expo-sure to chemicals, radiation, ozone, cigarette smoke, and smog. This might suggest that

TABLE 6-1   Components of Fruits and Vegetables that Reduce Disease Risk

Food component	Function
Carotenoids	Antioxidants that reduce the risk of cancer and cardiovascular disease
Fiber	Carbohydrate that speeds transit time of wastes through the colon, reducing the risk of colorectal cancer, and lowers blood cholesterol by binding to cholesterol in the digestive tract
Vitamins A, C, and E	Antioxidants that lower risk of cancer, cataracts, and cardiovascular disease
Folic acid	B vitamin that lowers plasma homocysteine, which is associated with heart and blood vessel disease
Polyphenols	Large family of antioxidant compounds occurring in plants that are useful in preventing cardiovascular disease; also give plants their bitterness and astringency
Flavonoids	Most common and widely studied group of polyphenols; useful in the treatment of hypertension and hypercholesterolemia; possible anticancer properties.
Tannins	Polyphenols that have a desirable effect on blood cholesterol

macular degeneration
Disorder in which the maculae of the retina progressively deteriorate, leading to blindness.

cataracts *Disorder in which the lens of the eye(s) lose their transparency; if untreated, blindness results.*

polyphenols
Group of compounds abundant in fruits, leaves, nuts, seeds, cereals, wine, cider, beer, tea, and cocoa that may prevent cardiovascular disease and cancer by acting as antioxidants; flavonoids and tannins are polyphenols.

they could play a role in the prevention of cancer, cardiovascular disease, lung disease, and eye diseases like **macular degeneration** and **cataracts**. However, research has produced inconsistent results, and one study even found the risk of cancer was higher in smokers who took a β-carotene supplement.[2]

To exert their beneficial effects, carotenoids may have to be consumed in foods as part of a diet that includes an abundance of fruits and vegetables. Dietary supplements might not mimic these effects. For example, a group of compounds called the **polyphenols**, available in some of the same foods as carotenoids (fruits, leaves, nuts, seeds, cereals, wine, cider, beer, tea, and cocoa), may prevent cardiovascular disease and cancer by acting as antioxidants. For weight management, an additional benefit of fruits and vegetables is their low caloric cost. Substituting half a grapefruit for one of the eggs you planned to have for breakfast or 1 cup of broccoli for 2 ounces of meat at dinner reduces the energy value of the meal while adding antioxidant nutrients and fiber. Table 6-1 lists several components of fruits and vegetables that might offer protection against cancer and heart disease.

Food Sources of Vitamin A and Carotenoids

Retinol is primarily found in animal products, such as liver, tuna, sardines, herring, fish liver oils, butter, whole milk, cheese, and fortified low-fat milk. Some cereal grains are also fortified with vitamin A. Preformed vitamin A is well-absorbed from foods.

Carotenoids are found in plants that produce red, orange, and yellow pigments as well as some green plants whose chlorophyll masks the pigment. Cheese, eggs, butter, and corn meal are additional sources of carotenoids. Table 6-2 lists common food sources of β-carotene, α-carotene, β-cryptoxanthin, lycopene, lutein, and zeaxanthin. Notice how the distribution of carotenoids varies considerably in foods. Carrots and pumpkin are particularly rich sources of α-carotene and β-carotene; orange juice and tangerines provide a lot of β-cryptoxanthin; collards, spinach, and corn are excellent sources of lutein and zeaxanthin; and watermelon, red grapefruit, and tomato products provide most of our dietary lycopene. People who eat a varied diet are most likely to

TABLE 6-2 Food Sources of Carotenoids

Food	Alpha-carotene (µg)	Beta-carotene (µg)	Beta-cryptoxanthin (µg)	Lutein and zeaxanthin (µg)	Lycopene (µg)
Fruits					
Apricots, raw, 3	0	2,554	0	0	50
Blueberries, raw, 1 c	0	50	0	0	*
Cantaloupe, raw, ¼	37	2,200	0.5	55	0
Grapefruit, pink and red, ½	6	742	15	16	1,798
Mango, raw, ½	8	222	5	0	*
Orange juice from concentrate, ½ c	1	30	123	171	0
Peach, raw, 1	1	97	24	57	0
Tangerine, raw, 1	12	60	407	204	0
Watermelon, raw, 1 c	0	448	157	26	7,445
Vegetables					
Beans, baked, canned, ½ c	187	518	0	0	*
Beans, snap green, cooked, ½ c	58	348	0	441	0
Broccoli, frozen, cooked, ½ c	0	920	0	764	0
Brussels sprouts, cooked, ½ c	0	363	0	1,006	0
Cabbage, cooked, ½ c	0	66	0	0	*
Cabbage, raw, 1 c	0	46	0	217	0
Carrot, raw, 1 large	3,347	6,362	0	0	0
Catsup, 1 T	0	110	0	0	2,551
Collards, cooked, ½ c	86	4,197	19	7,687	0
Corn, sweet yellow, cooked, ½ c	0	0	0	1,476	0
Lettuce, romaine, 1 c	0	712	0	1,476	0
Lettuce, iceberg, 1 c	0	106	0	194	0
Peas, green, canned, ½ c	0	272	0	1,148	0
Pepper, sweet, green, 1	16	147	0	0	*
Pepper, sweet, red, ½ c	44	1,784	1,654	0	*
Pumpkin, canned, ½ c	5,898	8,536	0	0	0
Spinach, cooked, ½ c	0	4,718	0	6,339	0
Squash, butternut, baked, ½ c	1,130	4,570	0	0	*
Sweet potato, baked in skin, 1	0	10,816	0	0	0

Tomato juice, canned, ½ c	0	522	0	73	11,275
Tomato, red, raw, 1	138	483	0	160	3,721
Vegetable juice cocktail, canned, ½ c	254	1,004	0	97	1,169
Other Foods					
Sauce, marinara, 1 c	0	1,096	0	398	3,981
Soup, minestrone, canned, 1 c	506	2,217	0	362	3,567
Beef stew w/vegetables, 1 c	1,715	4,361	0	147	740
Cheese, low fat cheddar or Colby, 2 oz	0	172	0	0	0
Cornmeal, degermed, enriched, yellow, 1 c	87	134	0	1,870	0

*Lutein composition of these foods is not yet available.

Source: Calculated by the author from data in Holden, J. M., Eldridge, A. L., Beecher, G. R., Buzzard, I. M., Bhagwat, S., Davis, C. S., et al. (1999). Carotenoid content of U.S. foods: An update of the database. *Journal of Food Composition and Analysis, 12*, 169–196. Available from http://www.nal.usda.gov/fnic/foodcomp/Data/Other/jfca12_169-196.pdf. Accessed March 3, 2010 (Table 3, pp. 175–186; and Table 4, pp. 187–188).

obtain all of the carotenoids. Carotenoid bioavailability varies considerably. Often, carotenoids are better absorbed when foods are heated and when fats are eaten with them.[1]

Recommended Intakes of Vitamin A

The RDA for vitamin A is stated in retinol activity equivalents or RAEs (it was previously stated in retinol equivalents or REs). This reflects the higher efficiency of absorption of vitamin A from retinol rather than carotenoids. One RAE is equivalent to 1 μg retinol, 12 μg β-carotene, and 24 μg α-carotene or β-cryptoxanthin. As food composition databases are revised, the quantity of vitamin A in various foods will be calculated to reflect this.

There is no DRI for carotenoids, due to insufficient information. For sufficient intake of these, the Food and Nutrition Board recommends consuming 5 or more servings of fruits and vegetables daily.

Retinol may cause harmful effects if consumed in excess. Like the other fat-soluble vitamins, vitamin A is stored with lipids, and excesses are not readily eliminated from body tissues. Ten times the RDA for vitamin A (7,000–9,000 μg RAE/day for several months) can cause hair loss, dry skin, headache, bone and muscle pain, liver damage, birth defects, bone abnormalities, and even death. One study of Northern Europeans discovered a connection between osteoporosis and high retinol intake from dietary supplements or a diet very rich in liver, fish oils, and fortified dairy products.[3] A UL of 3,000 μg RAE of preformed vitamin A for adults 19 years of age and older was set by the Food and Nutrition Board. There is no UL for carotenoids, and β-carotene is generally recognized as nontoxic.

Vitamin A and Carotenoid Supplements

Vitamin A deficiencies are rare in the United States. Only individuals with fat absorption disorders or gallbladder or liver disease might need to take a vitamin A supplement to prevent outright deficiency. Severe acne is treated with the drug Accutane, which is derived from vitamin A, but its effects cannot be mimicked by retinol supplements.

Carotenoid supplements are not advisable for several reasons. Most important, research has not found supplements to produce benefits equal or superior to the benefits of consuming carotenoids through food. Research has also not identified the upper level at which carotenoids become harmful.

Vitamin E

alpha-tocopherol *Most biologically available form of vitamin E.*

Vitamin E is a family of compounds known as tocopherols. The most common and biologically available tocopherol is **alpha-tocopherol** (α-tocopherol), which is often used synonymously with the term vitamin E.

Functions of Vitamin E

Vitamin E is an important antioxidant. It prevents the destruction of unsaturated fatty acids in cell membranes. In this role, it may reduce the risk of cardiovascular disease, prevent cataracts, and even prevent some cancers. Vitamin E may also play a role in immunity, wound healing, and the treatment of diabetes and neurological disorders.[4]

free radicals (reactive oxygen species) *Oxygen atoms having unpaired electrons that attack lipids, proteins, nucleic acids, and other cell components, causing oxidative damage, aging, and perhaps cancer, cataracts, and heart disease.*

Dietary antioxidants are substances present in foods that reduce the adverse effects of oxygen. You may be surprised to learn that oxygen, although needed to sustain life, can be toxic to human tissues. When exposed to smog, cigarette smoke, excessive sunlight and other forms of radiation, ozone, drugs, and various chemicals, oxygen atoms may end up with unpaired electrons and form **free radicals** (also called reactive oxygen species). Free radicals attack lipids, proteins, nucleic acids, and other cell components, causing oxidative damage. Oxidative damage has been linked to several chronic diseases—cancer, cataracts, heart disease—and even to the aging process. For example, recall from Chapter 1 that only low-density lipoproteins (LDLs) that have been oxidized are implicated in atherosclerosis. Preventing the oxidation of LDLs could stop or slow the development of atherosclerosis. Smokers have lower plasma levels of certain antioxidants than nonsmokers, which may explain their higher rates of stroke, cardiovascular disease, and cancer.[5] Even exercise can increase oxidant damage to cells. Vitamin E is able to donate an electron to free radicals that cause oxidant damage, helping to stabilize them.

In addition to vitamin E and the carotenoids, vitamin C and the mineral selenium are also antioxidants. Fruits, vegetables, and grains are excellent dietary sources of these antioxidants.

Food Sources and Recommended Intakes of Vitamin E

The best sources of vitamin E are plant oils and spreadable oils, followed by nuts, seeds, wheat germ, and wheat bran. Small amounts of vitamin E are also found in the fatty tissues of fish and animals. Individuals who eat a low-fat diet or who have fat absorption problems may not be able to obtain enough vitamin E through dietary sources.

The RDA for vitamin E for adult men and non-lactating women is 15 mg/day of α-tocopherol. The UL for vitamin E is 1,000 mg/day for adults over age 18 years. This value was established to prevent the risk of hemorrhage. To avoid over-consuming vitamin E taken in supplemental form, users may have to do some math. Sometimes the vitamin E content of a supplement is expressed in international units (IU). To convert

IUs to mg of α-tocopherol, use the following: (1) if the supplement is labeled as "natural" or *RRR*-α-tocopherol (sometimes incorrectly designated *d*-α-tocopherol), multiply IU times 0.67; (2) if the supplement is labeled *all rac*-α-tocopherol (sometimes designated *dl*-α-tocopherol), multiply IU times 0.45.

Vitamin E Supplements

Vitamin E supplementation is common in the United States, and supplements may be needed by individuals with fat malabsorption diseases. Studies conducted 15 years ago suggested that vitamin E supplements could reduce the risk of cardiovascular disease. More recent investigations do not support this claim, and both the American College of Cardiology and American Heart Association have taken positions opposing the use of vitamin E supplements for this purpose.[6]

Supplemental forms of vitamin E (and other antioxidants) may not have the same effects as the antioxidant naturally occurring in foods. As with the carotenoids, there are health-conveying properties of fruits, vegetables, and grains unrelated to their antioxidant content, and vitamins may need to be obtained in whole foods to get their maximum benefit.

Vitamin D

Vitamin D (sometimes called calciferol) is unique among vitamins in that it can be produced in the skin following sun exposure. It has received a great deal of attention recently, being linked to diseases from cancer to multiple sclerosis, and its recommended intake is currently under scrutiny.

Functions of Vitamin D

rickets *Softening of the bones that results from vitamin D deficiency in childhood; in addition to bowing of the legs, the skull and breastbone may be deformed.*

osteomalacia *Condition characterized by loss of bone mineralization and bone softening due to inadequate vitamin and mineral intake; sometimes called adult rickets.*

Vitamin D's primary function is to enhance the ability of the digestive tract to absorb calcium and phosphorus, which are needed for bone mineralization. Without sufficient vitamin D, children develop **rickets** and adults develop **osteomalacia**. Greater vitamin D intake also reduces the risk of hip fracture in the elderly.[7] Recent evidence links adequate vitamin D nutrition with reduced mortality, reduced risk of falls, and prevention of type 1 diabetes mellitus, multiple sclerosis, cardiovascular disease, stroke, and cancers of the breast, pancreas, esophagus, and colon.[8] Additional study is needed, but clearly vitamin D's role goes far beyond strong bones.

Food Sources of Vitamin D

Vitamin D is found in animal products, such as meat, eggs, dairy products, butter, and fish (particularly tuna and salmon). In the United States, milk and some cereals are fortified with vitamin D as a public health measure to prevent rickets in children. A form of vitamin D can be synthesized in the skin during exposure to sunlight. A fair-skinned adult with light clothing and no sunscreen can produce 250–500 µg of vitamin D in a day.[9] The ability to synthesize vitamin D is reduced with age, among individuals with dark skin, by application of sunscreens, and in the shorter days of winter. The skin cells cannot produce too much vitamin D, even in people who spend hours in the sun.

Vitamin D is very stable in cooking, storage, and processing. It may be destroyed by exposure to light.

Recommended Intakes of Vitamin D

While vitamin D formerly had an AI, the Food and Nutrition Board recently established an EAR and calculated an RDA for people over the age of 1 year. The RDA is 600 IU (15 µg/day). This recommendation assumes that no vitamin D will be formed from sun exposure. A UL was set at 2500 IU for 1–3 year olds, 3000 IU for ages 4–8, and 4000 IU for individuals over age 8 years.[9]

A blood test for serum 25-hydroxyvitamin D (25-OH-D) is considered the best indicator of vitamin D status. In adults, a level of at least 75 nmol/L is considered adequate.[8] Over half of adolescents and adults and up to 90% of elderly have 25-OH-D levels below this recommendation, which prompted reconsideration of the recommended intake of vitamin D.

Vitamin D and Sunshine

The fact that vitamin D can be synthesized in the skin following exposure to the sun creates a bit of a dilemma. On the one hand, recommending that people spend time in the sun could help assure adequate vitamin D nutrition. On the other hand, spending too much time in the sun increases the risk of skin cancer.

In Australia, which has the highest rates of skin cancer in the world, the Cancer Council of New South Wales recommends the following,[10] which has been endorsed by some dermatologists in the United States:

- In winter months, as temperature permits, expose the face, hands, and arms without coverage for about 30 minutes
- In spring and fall, reduce exposure to 15–20 minutes
- In summer months, limit exposure to about 10 minutes.

Vitamin D Supplements

Obtaining sufficient vitamin D through the diet can be a challenge. Vitamin D supplements may be needed by:

- People who spend little time outside, particularly older adults living in northern regions, unless dietary intake is excellent
- Individuals prescribed glucocorticoids, a category of anti-inflammatory drugs used for asthma, severe allergies, and adrenal gland dysfunction, which can impair vitamin D absorption
- People taking the fat-reducing drug orlistat, which affects the absorption of dietary fat; or people with malabsorption disorders like sprue, Crohn's disease, and severe liver disease.

In addition, people of any age who live farther from the equator, spend little time outdoors, and keep their skin covered with clothing and/or sunscreen will not synthesize much vitamin D. Supplementation may be essential for these individuals.

Vitamin D from supplements is often labeled in IUs. One µg of vitamin D is equivalent to 40 IUs. Excessive intake of vitamin D can have toxic effects, including nausea, vomiting, elevated blood pressure, and kidney failure, Nevertheless, one researcher believes that the UL could be five times higher (250 µg/day, or 10,000 IU) without ill effects.[11] In a rather extreme example of this, a study of fourteen men found no ill effects when 1,250 µg of supplemental vitamin D was consumed every day for eight weeks.[12]

Vitamin K

Phylloquinone (K_1) is the main form of dietary vitamin K. Like vitamin D, vitamin K can be synthesized by the body. In this case, a form of vitamin K called menaquinone (K_2) is produced by bacteria in the large intestine, although never in a large enough quantity to meet the daily need.

Functions of Vitamin K

Vitamin K works with vitamin D to promote bone uptake of calcium. Vitamin K is also required for blood clotting. Some people with cardiovascular disease or a risk of stroke are prescribed blood thinners to prevent blood clotting. These individuals generally have to limit their intake of vitamin K–rich foods. Recent studies have suggested a role for phylloquinone in promoting insulin sensitivity and improved glycemic control, although this could simply reflect the presence of vitamin K in foods that contain insulin-friendly nutrients, like fiber and potassium.[13]

Food Sources of Vitamin K

Most dietary vitamin K comes from plant foods, including broccoli, kale and other greens, turnip, cabbage, Brussels sprouts, and legumes, as well as canola and soybean oils. To preserve the vitamin K content of vegetable oils, they should be stored inside cabinets. Cheese is another source of vitamin K, in the form of menaquinone (100 g of cheese contains between 40–80 µg of vitamin K).

Common plant food sources of vitamin K are presented in Table 6-3. Notice that not all types of the same food are equivalent in vitamin K content. Butterhead lettuce

TABLE 6-3   Vitamin K Content of Foods

Food	Amount of vitamin K (µg)
Kale, cooked, ½ c	531
Spinach, cooked, ½ c	444
Spinach, raw, 1 c	145
Collards, cooked, ½ c	418
Brussels sprouts, frozen cooked, ½ c	150
Onions, spring or scallion, raw, 1 c	207
Broccoli, frozen, cooked, ½ c	92
Lettuce, butterhead (includes Boston and bibb types), 1 head	167
Lettuce, iceberg, 1 head	130
Cucumber with peel, raw, 1 large	49
Cucumber, peeled, raw, 1 large	20
Vegetable oil, canola, 1 T	17

Source: Vitamin K (phylloquinone) (µg) content of selected foods per common measure. *USDA National Nutrient Database for Standard Reference, Release 17.* Available from: http://www.nal.usda.gov/fnic/foodcomp/Data/SR17/wtrank/sr17w430.pdf. Accessed February 6, 2010

contains slightly more vitamin K than iceberg lettuce, and over half the vitamin K in a cucumber is in the peel.

Recommended Intake of Vitamin K

A recommended intake for vitamin K was first published in 1989. The AI is 90 μg for adult women and 120 μg for adult men. Bioavailability of phylloquinone from vegetables is probably not more than 20%, although absorption is improved when dietary fat is consumed at the same time.[14] Bioavailability from supplements is much greater than from foods.

Vitamin K Supplements

Adverse effects from overconsumption of vitamin K have not been reported. People who eat vegetables will get about 300–500 μg of vitamin K each day and do not need supplements. Under some circumstances, vitamin K supplementation is beneficial: (1) following long-term antibiotic therapy, which can destroy vitamin K-producing bacteria in the large intestine; and (2) when an individual has a fat absorption disorder or has had recent gastric bypass surgery. In addition, individuals using anticoagulant medications like warfarin may develop vitamin K deficiency and require supplements. Because of vitamin K's role in blood coagulation, these should only be taken under the direction and supervision of a health care provider.

Fat-Soluble Vitamins: Summary

Vitamins A, D, E, and K are absorbed, transported, and stored with lipids. Other than vitamin D, the fat soluble vitamins are easily obtained through diet, and supplements are not generally needed. However, people with fat absorption or digestive tract diseases may have difficulty absorbing adequate amounts of these vitamins and require supplements. Scientists continue to investigate whether supplements might lower the risk of several chronic diseases.

WATER-SOLUBLE VITAMINS

Water-soluble vitamins include the eight B vitamins, and vitamin C. Because they are water soluble, these vitamins are absorbed from the digestive tract directly into the bloodstream. With the exception of vitamin B_{12}, they are stored in small amounts in body tissues. When more water-soluble vitamins are consumed than needed, the excess is eliminated in the urine.

The B vitamins, which are discussed first in this section of the chapter, perform a variety of functions. They can be broadly categorized as being involved in energy metabolism or the production of blood cells. Some B vitamins do both. The classification of the water-soluble vitamins is pictured in Figure 6-2.

Thiamin, Riboflavin, and Niacin

When the B vitamins were first discovered, they were numbered, not named. Thiamin, the first B vitamin identified, is also known as vitamin B_1, riboflavin as vitamin B_2, and niacin as vitamin B_3. Although each of these vitamins has complex, individual functions, they are interrelated as important players in energy metabolism.

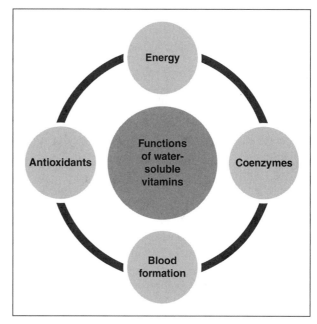

FIGURE 6-2   Some functions of the water-soluble vitamins.

Functions of Thiamin, Riboflavin, and Niacin

coenzyme

Nonprotein substance, such as the B vitamins, that helps to activate enzymes.

Each of these vitamins plays a role as a **coenzyme** in energy metabolism to help enzymes carry out their important functions. Without thiamin, riboflavin, and niacin, a person would feel unenergetic because the body's energy processes (described in Chapter 3) would slow down and operate less efficiently.

- Thiamin is a coenzyme in carbohydrate and amino acid metabolism and aids in the transmission of impulses through the nervous system. Inadequate thiamin leads to confusion, short-term memory loss, irritability, apathy, weakness, and cardiovascular complications.
- Riboflavin assists with electron transport, fatty acid oxidation, and the metabolism of folate and vitamin B_6. Signs of riboflavin deficiency include sores and swelling in the mouth and lips; dermatitis; and, in older adults, cloudiness of the lens of the eyes.
- Niacin is a key component of nicotinamide adenine dinucleotide (NAD) and other coenzymes involved in energy metabolism. It is also involved in the replication and repair of many kinds of body cells. The signs of niacin deficiency are rash, nausea, vomiting, diarrhea, depression, fatigue, and memory loss.

Overt deficiencies of thiamin, riboflavin, and niacin are rare in industrialized nations, such as the United States. Only chronic alcoholics, some older adults, individuals on severe calorie-restricted diets, and people with diseases that affect nutrient absorption exhibit outright deficiency symptoms.

enrichment

Adding nutrients lost in processing back into a food.

Food Sources of Thiamin, Riboflavin, and Niacin

When whole grains are processed to make white rice, white bread, cereal, and other products, several key nutrients are lost, including the B vitamins. Flour **enrichment** was introduced in the United States during World War II to require that manufacturers add back most, but not all, of the nutrients lost in processing. Today, both whole and enriched breads and grains remain an important source of thiamin, riboflavin, and niacin. Other good dietary sources of these nutrients include:

- Thiamin: meat (especially pork products), legumes, seeds, and peas
- Riboflavin: dairy products, eggs, meat, and legumes
- Niacin: meat, poultry, fish, seeds, and legumes

As long as protein intake is adequate, some niacin can be derived from the amino acid tryptophan. Tryptophan converted to niacin is expressed in niacin equivalents (NEs):

60 mg of tryptophan = 1 mg of niacin = 1 mg NE

Both niacin and riboflavin are thought to be highly bioavailable. Data on thiamine bio-availability are limited.[14]

Recommended Intakes of Thiamin, Riboflavin, and Niacin

RDAs have been set for thiamin, riboflavin, and niacin and can be found in the table Appendix D. Recommended intakes for all three vitamins are highest in pregnancy and lactation. Only niacin has a UL—35 mg/day in adult men and women. When taken in excess through foods in which it naturally occurs, niacin has no known adverse effects. However, supplemental or fortified niacin in excess can be hazardous.

Thiamin, Riboflavin, and Niacin Supplements

Because thiamin and riboflavin are poorly absorbed when ingested in large doses, neither is known to have any adverse effects even when taken in excess through foods or supplements. Still, individuals should be cautious when taking large doses of any vitamin because the long-term effects of supplements are unknown and there is no known benefit to excess intake for otherwise healthy people.

High doses of supplemental niacin, or excess niacin accidentally added to fortified foods, do have an adverse effect. Dosages of 2,000 mg/day, especially sustained-release niacin, cause flushing of the skin, headache, fatigue, nausea, vomiting, and in some cases liver dysfunction. Flushing can occur at doses as low as 50 mg/day.

Pharmacologic doses of nicotinic acid and niacin are effective treatments for high blood cholesterol. Daily doses of 1,000–3,000 mg of niacin preparations lower total cholesterol, LDL cholesterol, and triglycerides while increasing high-density lipoprotein (HDL) cholesterol. Both time-release and unmodified (immediate-release) preparations are available for this purpose by prescription and over the counter. Time-release niacin causes less flushing but has a greater association with liver toxicity.[15] Clients should be discouraged from using high-dose niacin preparations to self-treat high blood lipids due to potentially toxic effects.

Vitamin B₆, Pantothenic Acid, and Biotin

Vitamin B$_6$, pantothenic acid, and biotin—like thiamin, riboflavin, and niacin—are involved in energy metabolism. Vitamin B$_6$ is sometimes called pyridoxine, which is one of the forms in which the vitamin exists in food.

Functions of Vitamin B₆, Pantothenic Acid, and Biotin

The specific functions of these three vitamins are:

- Vitamin B_6: a coenzyme in glycogen breakdown and amino acid metabolism, including the synthesis of heme (the iron-containing component of hemoglobin) and the conversion of tryptophan to niacin
- Pantothenic acid: a component of coenzyme A (CoA) as well as a partner in a variety of protein metabolic activities and fatty acid synthesis
- Biotin: a coenzyme in many enzymatic reactions in energy metabolism

Outright deficiencies of these vitamins are rare. Biotin deficiency may occur in people who eat a lot of raw egg whites (because raw, but not cooked, egg inhibits biotin absorption). Alcoholics or people with malabsorption diseases who don't get supplemental forms of these vitamins might exhibit deficiency symptoms. Some researchers have reported poorer vitamin B_6 status among women taking oral contraceptives, but experts do not believe that true B_6 deficiency results from the use of birth-control pills.[16]

Food Sources and Recommended Intakes of Vitamin B₆, Pantothenic Acid, and Biotin

Both pantothenic acid and biotin are thought to be widely found in foods, but little research has been conducted to actually measure this. Neither is included in tables of food composition. The bioavailability of vitamin B_6 is good, but the bioavailability of pantothenic acid and biotin is not known. Common food sources of these vitamins are:

- Vitamin B_6: beef liver and plant foods, especially bananas, navy beans, starchy vegetables, walnuts, and fortified cereals;
- Pantothenic acid: meat, poultry, organ meats, egg yolk, tomato products, whole grains, broccoli, and potatoes. Processing foods lowers pantothenic acid content.
- Biotin: liver, soybeans, egg yolk, cereals, legumes, and nuts

An RDA has been set for vitamin B_6. Adult males and females need 1.3 mg/day until 50 years of age, when intake should be increased to 1.5 mg/day. Vitamin B_6 has an adult UL of 100 mg/day. When amounts in excess of this are consumed from food, there are no reported adverse effects.

Because of insufficient information to set an RDA, pantothenic acid and biotin have AIs. The adult AI for biotin is 30 mg/day. Although some biotin can be synthesized in the intestines, it is not enough to meet daily needs. The adult AI for pantothenic acid is 5 mg/day. Neither pantothenic acid nor biotin has a UL.

Vitamin B₆, Pantothenic Acid, and Biotin Supplements

Pyridoxine is the form of vitamin B_6 found in supplements. Limited evidence suggests that supplemental vitamin B_6 may relieve the symptoms of premenstrual syndrome, chronic fatigue syndrome, and carpal tunnel syndrome.[17] Although high intake of vitamin B_6 from food is not associated with any adverse effects, the same cannot be said for supplemental pyridoxine. Doses in excess of 2,000 mg/day have resulted in dermatological lesions and sensory neuropathy, in some cases leading to irreversible nerve damage. Because of this, a UL of 100 mg/day has been set for adults.

There are no reported adverse effects of pantothenic acid or biotin. Biotin is probably safe up to 10 mg/day, which is 300 times the AI.[16]

Folate and Vitamin B_{12}

Folate and vitamin B_{12} are involved in maintaining healthy blood and a healthy nervous system. Folate exists in several forms, including folic acid and folacin. What we call vitamin B_{12} is really a family of cobalt-containing compounds known as the cobalamins.

Functions of Folate and Vitamin B_{12}

Folate and vitamin B_{12} are linked by a common deficiency symptom—**megaloblastic anemia**, a blood disorder characterized by the formation of large, immature red blood cells that lack the ability to transport oxygen efficiently. Fatigue, irritability, depression, headache, heart palpitations, and general weakness are signs of this form of anemia. Folate deficiency can progress fairly rapidly to megaloblastic anemia. However, several years' worth of vitamin B_{12} is stored in the liver, so an insufficient vitamin B_{12} intake might go undiscovered for some time. Folic acid supplements will treat megaloblastic anemia but leave the vitamin B_{12} deficiency undetected. The result is irreversible damage to the nervous system—neuropathy, numbness in the legs and feet, difficulties walking, memory loss, and visual disturbances.

Sometimes vitamin B_{12} deficiency results from an inability to absorb the vitamin from the digestive tract. Normally, the stomach secretes intrinsic factor, a substance needed to absorb vitamin B_{12}. When there is insufficient **intrinsic factor** (usually due to heredity or aging), the megaloblastic anemia that develops is called **pernicious anemia**.

Vitamin B_{12} and folate are also involved in formation of the neural tube—the embryonic tissue from which the brain and spinal cord develop. Each year in the United States, about 1 in 1,000 pregnancies result in a baby born with a neural tube defect. **Anencephaly** and **spinal bifida** are the most common. In 1998, the federal government mandated **fortification** of enriched flour, rice, pasta, cornmeal, and other cereal grains with 140 µg of folate or folic acid per 100 g of grains. The purpose was to assure that women eating six servings of grains per day would receive up to 440 µg of folate, an amount thought sufficient to reduce the risk of neural tube defects. This has been an effective strategy. Between 1995 and 2000, the rate of spina bifida dropped 25%, and between 2000 and 2005, an additional drop of 13% occurred.[18]

A further function of folate and vitamin B_{12} is to promote normal **homocysteine** metabolism, which plays a role in the prevention of cardiovascular disease. Homocysteine is related to the amino acid cysteine. High homocysteine levels increase the likelihood of vascular disease by promoting blood clotting, inflammation, and the oxidation of LDLs. If either folate or vitamin B_{12} is unavailable, then homocysteine cannot be converted to methianone and its levels rise.[19]

Folate and vitamin B_{12} may also help to prevent cancer, especially colorectal cancer and perhaps breast cancer, although the relationship is a complex one.[20,21] Large amounts of dietary folate or moderate amounts of supplemental folic acid seem to decrease the risk of colorectal cancer. However, high levels of supplemental folic acid may actually promote tumor growth in people with polyps or the early stages of cancer. Additional factors may affect the relationship between folate and cancer. A study in

homocysteine *An amino acid linked to cardiovascular disease; homocysteine differs by one chemical group from the amino acid cysteine.*

Sweden found women with BMIs above 25 had a lower incidence of breast cancer when higher amounts of folate were consumed, either through diet or supplements. Folate offered no significantly reduced risk of breast cancer in women whose BMIs were below 25.[21] The Report of the Dietary Guidelines Advisory Committee on the 2010 Dietary Guidelines for Americans noted that there are some indications that, since folic acid fortification was mandated, the death rate from stroke has decreased. However, the rate of colorectal cancer has increased, perhaps due to higher folic acid intake.[22]

Food Sources of Folate and Vitamin B_{12}

Folate is found in "foliage" foods—green vegetables, mushrooms, broccoli, Brussels sprouts, and asparagus—as well as legumes, fortified grain products, and liver. It is easily lost during cooking. Folate naturally occurring in food is about 50% bioavailable; the folate from fortified foods is absorbed even better.

The best food sources of vitamin B_{12} are all animal products—shellfish, organ meats, game, fish, and dairy. Plant sources of vitamin B_{12} include algae and fortified cereals and grains. The bioavailability of vitamin B_{12} is assumed to be about 50%.

Recommended Intakes of Folate and Vitamin B_{12}

Of all the micronutrient recommendations published by the Food and Nutrition Board, the recommended intake for folate has changed the most. The RDA for folate is 400 µg/day for people over the age of 14 years, with more needed by pregnant and lactating women. This recommendation is substantially higher than the 1989 RDA of 180 µg for women and 200 µg for men and was increased mainly in response to scientific data about the prevention of neural tube defects in pregnancy.

The use of the term dietary folate equivalent (DFE) reflects varying bioavailability of food folate versus supplemental or fortified folic acid.

1 DFE = 1 µg food folate
       = 0.6 µg folic acid (from fortified food or supplements taken with a meal)
       = 0.5 µg supplemental folic acid taken on an empty stomach.

Stated another way, 1 µg of synthetic folic acid or fortified folate in food = 1.7 µg DFE. Taking a supplement or eating a fortified grain product that provides 400 µg folic acid is equivalent to consuming about 680 µg DFE of naturally occurring dietary folate. This equation reflects the greater bioavailability of supplemental folate than food folate. By the way, food labels do not use DFEs.

A UL for synthetic folate (the folate/folic acid obtained from supplements and fortified foods) of 1,000 µg/day has been recommended for adults (lower for children and adolescents). It is unlikely that your intake will exceed this UL. There are no reports of adverse effects from folate naturally occurring in food; however, excess supplemental or fortified folic acid will effectively treat pernicious anemia while masking any underlying vitamin B_{12} deficiency. High doses of folate are not advisable. In addition, many medications are known to interfere with folate metabolism, and the safety of taking folic acid supplements while using these medications is unstudied.

The RDA for vitamin B_{12} is 2.4 µg/day for adults and more for pregnant and lactating women. There is no evidence that even high doses of vitamin B_{12} cause adverse effects, so no UL for vitamin B_{12} has been set.

Folate and Vitamin B_{12} Supplements

Folic acid is the form of synthetic folate used in most supplements and fortified foods. Whether taken as a supplement or in enriched and fortified foods, folic acid is about equally well absorbed, and both forms are slightly better absorbed than food folate. Most supplements offer about 400 µg DFE, so caution is advised to avoid exceeding the UL.

Some people may need folic acid supplements. For example, alcoholics absorb folate poorly, and individuals with malabsorption disorders or chronic disease may not be able to obtain sufficient folate through diet alone.

Various reports have suggested that some people are hypersensitive to folic acid and should avoid supplements exceeding 1,000 µg/day. Epileptics taking anticonvulsants may experience poor seizure control when taking folic acid supplements.[23] In addition, people taking high daily doses of folate (15,000 µg) may experience irritability, sleeplessness, and hyperactivity, and depressed patients may become more depressed.[24]

Supplemental vitamin B_{12} is in a form called cyanocobalamin. The primary reason for taking cyanocobalamin is to compensate for an inability to absorb the vitamin from food. People who lack intrinsic factor, chronic alcoholics, and individuals who have had gastric surgery (including gastric surgery for obesity) may require vitamin B_{12} supplements. Strict vegetarians have an increased risk of vitamin B_{12} deficiency because they do not consume animal products, but they may obtain enough of the vitamin through fortified soy products and cereals. Between 10–30% of adults over age 50 may have stomach acid levels low enough to prevent vitamin B_{12} absorption.[14] The RDA for vitamin B_{12} actually stipulates that people over 51 years of age obtain most of their vitamin B_{12} through fortified foods or supplements.

Vitamin C (Ascorbic Acid)

Vitamin C is the "non-B" water-soluble vitamin.

Functions of Vitamin C

Vitamin C is involved in the production of collagen, the protein that makes up connective tissue. The vitamin C deficiency disease—scurvy—is characterized by breakdown of connective tissue, especially in the skin, the blood vessels, and the gums. Vitamin C also has an important role in:

- Synthesis of neurotransmitters, including norepinephrine, dopamine, bombesin, and cholecystokinin
- Oxidation of fatty acids
- Promoting iron absorption
- Antioxidant activities.

In 1997, researchers discovered that vitamin C reduces oxygen damage to the eyes and may, therefore, be an important factor in the prevention of cataracts.[25] It also prevents damage to the lungs and may prevent cancers in the stomach, the esophagus, the throat, and the mouth.[26] Its function in the prevention of cardiovascular disease is uncertain.

Many have read that vitamin C can prevent colds. Dr. Linus Pauling, a scientist who won the Nobel Prize for Chemistry in 1954 and the Nobel Prize for Peace in 1962, published articles in the early 1970s that suggested vitamin C supplementation might help alleviate the common cold. Subsequent analyses have cast doubt on the capacity of vitamin C to reduce the number of colds people in the general population get each year. But taking more than 200 mg/day of vitamin C for up to six months has reduced the number of colds in populations under physical stress (marathon runners) or exposed to a cold environment (skiers and cold-weather soldiers).[27] Pauling's other conclusions— that taking vitamin C may reduce a cold's duration and symptoms—have been substantiated.[27,28]

Food Sources and Recommended Intakes of Vitamin C

The main source of Vitamin C is from fruits and vegetables, which are listed in Table 6-4. Fruit and vegetable sources of vitamin C are equally bioavailable. After harvest, foods lose a little of their vitamin C, and the vitamin is further lost during long storage and over-cooking. For example, boiling a vitamin C–rich food can reduce its vitamin C content by 50–80%.[29]

Unlike most other animals, humans cannot synthesize vitamin C and must obtain it through the diet. Some vitamin C is stored in the adrenal and pituitary glands as well as the liver, the spleen, the lungs, and the heart. Regular intake of the vitamin is important. The RDA is 90 mg/day for adult men and 75 mg/day for adult women. The Food and Nutrition Board suggests that smokers consume an additional 35 mg/day of vitamin C, raising the recommended intake to 125 mg/day in men and 110 mg/day in women.

TABLE 6-4 Vitamin C–rich Fruits and Vegetables

Food	Amount of vitamin C (mg)
Cantaloupe, medium, ¼	60
Grapefruit, ½ raw	40
Grapefruit juice, ½ c	35
Kiwi, medium, 1 raw	75
Lemon, medium, 1 raw	31
Orange, medium, 1 raw	70
Orange juice, ½ c	50
Papaya, 1 c cubed	85
Strawberries, 1 c sliced	95
Asparagus, cooked, ½ c	10
Broccoli, cooked, ½ c	60
Brussels sprouts, cooked, ½ c	50
Cauliflower, raw or cooked, ½ c	25
Kale, cooked, ½ c	28
Pepper, sweet green or red, raw, ½ c	65
Pepper, sweet green or red, cooked, ½ c	50
Potato, medium, baked	25
Sweet potato, medium, baked	30

The UL of vitamin C is 2,000 mg/day for adults. At 500 mg/day, almost all ingested vitamin C is excreted in the urine.

Vitamin C Supplements

Vitamin C is one of the most commonly supplemented micronutrients in the United States.[14] It is generally found as a supplement in the form of free ascorbic acid, calcium ascorbate, sodium ascorbate, ascorbyl palmitate, or rose hip vitamin C, which is derived from the seed capsule of roses. Vitamin C is absorbed equally well from supplements as from food.[30] About 200 mg of vitamin C taken in divided doses is almost completely absorbed.[31]

People take supplemental vitamin C to prevent colds, cataracts, cardiovascular disease, and cancer. Researchers do not yet know if supplements are as effective as dietary vitamin C in preventing some of these disorders. In addition, whether supplemental vitamin C is as effective as dietary vitamin C in improving iron absorption is uncertain.[32]

Vitamin C is generally considered to be safe at doses several times higher than the RDA. People who take more than 3,000 mg/day may experience diarrhea and nausea, but more severe adverse effects have not been reported because excess vitamin C is not absorbed, metabolized, or retained. Still, vitamin C does alter urinary excretion of calcium oxalate and uric acid, which is linked to kidney stones. People with kidney disease or a predisposition to kidney stones should be advised to limit supplemental vitamin C to 100 mg/day or less. Because vitamin C enhances iron absorption, people with iron-overload diseases should avoid supplements.

What Other Vitamins Should We Be Concerned About?

There are several vitamin-like compounds that do not quite meet the definition of a vitamin but that have known physiological functions. One of these compounds—choline—is important enough to have a DRI.

Choline is needed for maintenance of cell membranes, nerve transmission, and lipid transport. Its name is suggestive of one particular neurotransmitter that it helps to synthesize—acetylcholine. Although the body produces choline, it is considered essential because not enough can be produced to meet our needs. Insufficient choline may cause liver damage. Like folate and vitamin B_{12}, choline helps to normalize plasma homocysteine levels; perhaps it will someday be found to have a role in prevention of cardiovascular disease.

Many foods contain choline. Milk, liver, eggs, and peanuts are especially rich sources. The AI for choline is 550 mg/day for adult men and 425 mg/day for adult women. Excessive choline intake increases perspiration and salivation, causes body odor, impairs growth, lowers blood pressure, and may lead to liver damage. The UL of 3.5 g/day for adults was primarily set to avoid low blood pressure and secondarily to prevent body odor.

Several other "nonvitamins" that have important metabolic functions but are not essential are inositol, carnitine (sometimes called *vitamin B_T*), lipoic acid, para-aminobenzoic acid (PABA), bioflavonoids (vitamin P), and ubiquinone (coenzyme Q).

Supplements of these compounds are not needed because they are probably present in many foods. There is no scientific evidence that supplements of any nonvitamins will improve health or prevent deficiencies in humans.

Case Study. Personal Diet Analysis, Part 3	This case study is a continuation of the one you did in Chapter 5, with yourself as the subject. (If you prefer, and with permission of your instructor, you may select someone else to be your subject. If you select someone else, remember that all information collected for the case study is to be treated with confidentiality).

- Provide a brief description of the subject of this case (you already did this in Chapter 2, unless you are using a different subject, and you may repeat it here).
- Use the Interactive DRI for Healthcare Professionals (http://fnic.nal.usda.gov/interactiveDRI/) to estimate your subject's recommended daily vitamin needs. You will need height, weight, age, and estimated activity level for this.
- Compare vitamin intake, determined from the food record case study in Chapter 5, with recommendations. How close did your subject come to the dietary recommendations presented in this chapter and to the recommendations of the Interactive DRI?
- If your subject uses vitamin or mineral supplements, get a copy of the bottle(s) and estimate additional vitamin consumption. Is your subject at or above the UL for any vitamin?
- Are there any special circumstances that would increase your subject's need for vitamins?
- For each vitamin, list some foods that would be good sources of that vitamin and that are foods your subject would eat.

Water-Soluble Vitamins: Summary

The B vitamins support the body's energy systems and maintain blood health; vitamin C maintains collagen and is an antioxidant. With the exception of niacin, excess intake of water-soluble vitamins does not usually cause adverse effects. Although the likelihood of outright deficiencies of any of these vitamins is small due to food enrichment and fortification, some older adults and people with chronic diseases may benefit from supplements. The role of vitamin supplements in preventing chronic diseases is still under investigation.

BODY WATER

Together, water and minerals make up a substantial portion of body weight—60–70% for water and about 4% for minerals. For a 60-kg (132-lb) woman, this is about 36–42 kg (79–92 lb) of water and 2.4 kg (5 lb) of minerals. Water and minerals are linked: The minerals need water to form electrolytes in the body, and water needs minerals to maintain fluid balance inside and outside of the cells.

Why Is Water Essential?

Water is the most essential nutrient because we can live without it for only a few days. For good health, the average healthy person needs to take in 2–4 liters of water in food and fluids

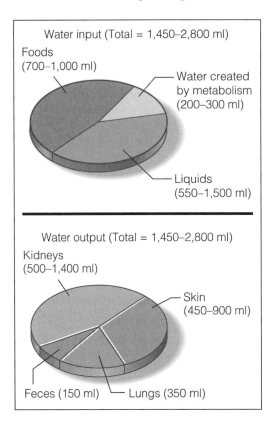

FIGURE 6-3 Water balance.

Source: From Sizer/Whitney, Nutrition, 10E. p. 267. © 2006 Brooks/Cole, a part of Cengage Learning Inc. Reproduced by permission. www.cengage.com/permissions.

each day, which will replace water lost in respiration, perspiration, urination, and elimination. (A liter is 33.8 fluid oz, 4¼ cups, or 1.06 quarts.) Figure 6-3 illustrates water balance.

Functions of Water

You are correct if you surmised that something making up more than half our weight, which we can't live without for very long, must have numerous important functions. Water's functions include:

- Maintenance of body temperature
- Transport of nutrients and other materials needed by the cells
- Elimination of toxic substances or excessive levels of substances by moving them through the kidneys
- Cushioning and protecting organs and tissues, such as a growing fetus, the vital components of the eyes, and the joints

Water is so important that a loss of as little as 3% of body water will reduce blood volume, and loss of 5% will produce noticeable confusion and weakness. A reduction in body water is known as **dehydration** and can be life threatening. Unfortunately, people do not experience increased thirst in proportion to their losses of body water. The best

dehydration
Depletion of body fluids.

way to avoid dehydration is to drink fluids regularly throughout the day. Anyone who exercises has an increased need for water.

Water Distribution in the Body

Most body water is found inside the cells and is called intracellular fluid. About one-third of body water exists outside the cell membranes as extracellular fluid. Blood, lymph, and the water between the cells are examples of extracellular fluids. Because water can move easily across cell membranes, the body has a system for preventing cells from either over-filling or emptying. The major minerals are part of this system.

electrolytes
Substances that become electrical conductors when dissolved in body water; also called ions.

Minerals dissolve in water as salts, which help to regulate the balance between intra-cellular and extracellular fluids. These dissolved salts are called ions, or **electrolytes**, because they have an electrical charge and conduct electricity. "Water follows salt" is a saying in science. Therefore, if the correct concentration of electrolytes is maintained on both sides of a cell membrane, then water balance should also occur. Potassium is the main intracellular ion, and sodium and chloride are the main extracellular ions. Small amounts of magnesium are found both inside and outside of the cells.

When someone is dehydrated or has an extended bout of vomiting and/or diarrhea, fluid and electrolyte imbalances occur that can be very dangerous. In these cases, the initial reduction in extracellular fluid causes water to leave the interior of the cells in an attempt to restore balance. Intracellular and extracellular minerals are also lost, increasing the risk of kidney and heart failure. Restoration of body fluids by drinking water or juice or by intravenous therapy is essential.

Water Sources and Requirements

The best fluid replacer is water. People who do not like the taste of plain water might try adding the juice of lemons and limes. Sport drinks and juices are useful for replacing carbohydrates and electrolytes following prolonged exercise (2 hours or more), but, like soda, they add calories. Anyone trying to reduce caloric intake should stick with water.

The Food and Nutrition Board set an AI for water in 2004. Like the RDAs and AIs for vitamins and minerals, the AI for water varies by gender and life stage. Recommended water intake is presented in Table 6-5.

Physically active people need to consume more water than Table 6-5 recommends. In hot weather, active adults may need two to three times more water than on a cool day. Other conditions that increase the need for fluids include:

- High-protein diet: Recall from Chapter 5 that the kidneys metabolize the nitrogen found in amino acids. Water is needed for this process.
- Low-calorie diet: Ketones, which are produced when fats are burned incompletely, are also processed by the kidneys.
- Diet high in sodium: Sodium not needed for physiological functions is excreted in the urine.
- High-fiber diet: Fiber attracts more water into the large intestine.
- Caffeine and alcohol consumption: Both of these substances are diuretics, which increase urinary output. Diuretic medications also increase the need for water.

TABLE 6-5 DRI Values for Water

Gender and life stage group	Total water (includes drinking water and water in foods and beverages) (L)	Beverages (includes water) (L)
1–3 years	1.3	0.9 (4 cups)
4–8 years	1.7	1.2 (5 cups)
9–13 years		
Males	2.4	1.8 (8 cups)
Females	2.1	1.6 (7 cups)
14–18 years		
Males	3.3	2.6 (11 cups)
Females	2.3	1.8 (8 cups)
19 years and older		
Males	3.7	3.0 (13 cups)
Females	2.7	2.2 (9 cups)
Pregnancy, ages 14–50	3.0	2.3 (10 cups)
Lactation, ages 14–50	3.8	3.1 (13 cups)

Source: Adapted from Otten, J. J., Hellwig, J. P., & Meyers, L. D. (Eds.). (2006). *Dietary reference intakes: The essential guide to nutrient requirements.* Washington, D.C.: The National Academies Press, pp. 156–7.

A good indicator of hydration is the frequency of urination and the appearance of the urine. People who are well hydrated urinate every 2–3 hours (or sooner) and have light-colored, dilute urine. Infrequent urination and urine that is dark and concentrated indicates a need to drink more fluids.

Body Water: Summary

Water and minerals are linked. Minerals dissolve in water as salts, which help to regulate the balance between intracellular and extracellular fluids. Humans cannot survive more than a few days without adequate water intake. More water is needed in hotter environments, by active people, during pregnancy and lactation, and under certain dietary conditions.

THE MINERALS

Minerals are inorganic compounds that make up the "ash" that remains after cremation. There are seven major minerals, which are also called macrominerals because they each make up at least 0.01% of total body weight. There are nine trace minerals, or microminerals, that are found in much smaller amounts in the human body but that are just as important to health and well-being. Figure 6-4 illustrates the proportion of minerals in an adult.

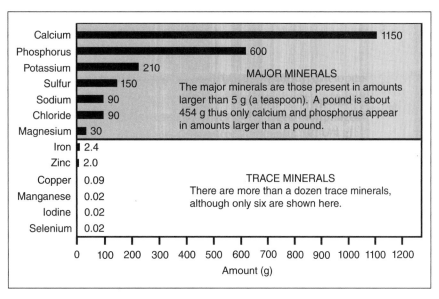

FIGURE 6-4 Minerals in a 60-kg (132-lb) adult.

Calcium, Phosphorus, and Magnesium

Three major minerals—calcium, phosphorus, and magnesium—are primarily stored in bone and share several important functions.

Functions of Calcium, Phosphorus, and Magnesium

Calcium, phosphorus, and magnesium play an important role in skeletal health. During childhood and adolescence, they promote the formation of the skeleton in conjunction with vitamins A, D, and K. Throughout childhood and adolescence, bones grow in length and mass, reaching their maximal length by 20 years old and their peak mass between the ages of 19 and 30 years.[33] During middle age, bone loss begins to exceed bone formation at the rate of 0.3–0.5% per year. At menopause, as women's estrogen production slows and then stops, bone losses of about 3% per year for 5 years are typical.[33] Women whose bone mass was marginal before menopause are at risk of osteoporosis later in life.

Osteoporosis is a disease characterized by a significant loss of bone density, resulting in bone mass more than 2.5 standard deviations below the average for a specific age. A less severe loss of bone mass is called **osteopenia** and increases the likelihood of developing osteoporosis. Slightly over half of men and women age 50 and older in the United States have low bone density.[34] This results in about 1.5 million fractures each year.[14]

Calcium, phosphorus, and magnesium are needed for numerous enzymatic reactions as well as heart and muscle contraction. Calcium and magnesium also appear to work together in maintaining normal blood pressure and preventing spasms in coronary and cerebral blood vessels by keeping smooth muscles in the arteries from contracting. Calcium also modestly lowers the risk of intestinal polyps and colon cancer.[35] Recently, there has been interest in magnesium's role in promoting insulin sensitivity, and one analysis found it to lower the risk of type 2 diabetes.[36]

osteoporosis
Disease characterized by loss of bone mineral density that reduces bone mass and increases susceptibility to fractures.

osteopenia
Condition in which bone mineral density is reduced to a lesser extent than in osteoporosis but that may lead to osteoporosis if untreated.

Food Sources of Calcium, Phosphorus, and Magnesium

Dairy products are the richest sources of calcium, although legumes, canned fish with bones, tofu, several green vegetables, and fortified foods also supply some dietary calcium. The bioavailability of calcium is reduced in foods that contain oxalic acid (spinach, sweet potatoes, rhubarb, and beans) or phytic acid (unleavened bread, seeds, nuts, and some soy products). Calcium in these foods is less well absorbed than calcium from dairy sources.

Phosphorus is found widely in animal products as well as in legumes, cereals, and grains. Magnesium is present in chocolate, nuts, legumes, whole grains, seafood, and leafy green vegetables. Both magnesium and phosphorus have good bioavailability from food sources; however, phosphorus absorption is impaired when large doses of aluminum-containing antacids are consumed, and magnesium is readily lost in food processing and is not routinely added during enrichment. Table 6-6 lists some good food sources of calcium, phosphorus, and magnesium.

Recommended Intakes of Calcium, Phosphorus, and Magnesium

DRIs have been established for calcium, phosphorus, and magnesium:

- The RDA for calcium is 1,000 mg for 19- to 70-year-old men, 1000 mg for 19- to 50-year old women, and 1,200 mg for men over 70 and women over 50. It is highest in adolescence (1,300 mg).
- The RDA for phosphorus is 700 mg/day for adults.

TABLE 6-6 Food Sources of Calcium, Magnesium, and Phosphorus

Food	Calcium (mg)	Magnesium (mg)	Phosphorus (mg)
Navy beans, cooked, 1 c	127	107	286
Tofu, ½ c	130	126	120
Cashews, 1 oz	13	74	137
Sunflower seeds, ¼ c	42	127	253
Whole wheat bread, 1 slice	20	24	64
40% bran flakes, 1 c	21	101	296
Halibut, 4 oz	68	121	323
Pollock, 4 oz	7	83	544
Ground beef, 4 oz	10	28	214
Broccoli, raw, 1 c	94	37	101
Bok choy, cooked, ½ c	79	10	25
Spinach, cooked, ½ c	122	78	50
Sweet potato, baked	32	28	74
Provolone cheese, 1 oz	214	8	140
Nonfat milk, 1 c	301	28	247
Yogurt low fat with fruit, 1 c	369	41	291
Milk chocolate, 1 oz	54	17	61

- The RDA for magnesium is lower for women than for men, 310 mg/day for 19- to 30-year-old women and 400 mg/day for men of the same age; 420 mg for men and 320 mg for women over age 30. The need for magnesium increases in people who absorb it poorly or excrete it readily, such as chronic alcoholics, people with kidney disease, those who take diuretics, and people who have experienced bouts of excessive vomiting and/or diarrhea.

Calcium, Phosphorus, and Magnesium Supplements

Calcium supplementation has increased since the late 1980s as the general public has learned more about osteoporosis. People who are lactose intolerant or allergic to milk often rely on supplements to obtain sufficient calcium. Those who had poor calcium nutrition in childhood or who dislike foods rich in calcium may use supplements to build bone density and slow bone loss. Very lean young women with exercise-induced amenorrhea may take calcium supplements to prevent losses of bone mass.

Is there any value to taking calcium supplements? The answer is yes, but more so at certain ages. Adolescents who receive 1,000 mg of supplemental calcium daily can increase their bone mineral density.[37] Even individuals in their eighties can improve bone mineral density through supplemental calcium.[12] For menopausal women, calcium supplementation is of greater benefit in late menopause than early.

Calcium from supplements (calcium carbonate, acetate, lactate, gluconate, or citrate) is absorbed about as well as the calcium in a glass of milk or the calcium in fortified foods.[37,38] Most people who take a supplement use calcium carbonate because it contains the most elemental calcium per tablet and is inexpensive. Some supplements should be avoided. Oyster shell or bone meal calcium preparations, which are obtained from fossilized limestone quarries, may contain heavy metals, such as aluminum and lead. These preparations could pose a significant danger to children and pregnant women.

To promote maximum absorption, calcium supplements are best taken with meals, at or below a 500 mg/dose.[34] However, because they may inhibit iron absorption, some scientists recommend that calcium supplements should not be taken with a meal by people who have poor iron status, particularly if the meal is one with a low iron content.[39]

Calcium supplements up to 2,000 mg/day have virtually no side effects, other than temporary bloating or constipation in some people. A UL of 2,500 mg/day has been recommended for calcium intake in adults. High calcium intake from supplements probably does not increase the risk of developing calcium-containing kidney stones, as was once thought, but excessive intake can cause calcium deposits in the soft tissues and has no benefit.

Phosphorus supplements are infrequently used because phosphorus deficiencies are rare. Adverse effects from excessive phosphorus intake are equally rare, and the UL is 4,000 mg daily.

Supplemental magnesium is not widely used. While magnesium in food may play a role in preventing type 2 diabetes, supplements are not know to have the same effect. There are few reports of magnesium excesses from high doses of supplements. The UL for magnesium in adults is 350 mg/day above that consumed through food and beverages. Diarrhea is the most commonly reported adverse effect.[33]

Other Major Minerals

The other major minerals are sodium, potassium, chloride, and sulfate. Sodium, potassium, and chloride are important constituents of the fluid inside and outside of the cells. Sulfate is found in conjunction with amino acids in most body tissues.

Functions of the Other Major Minerals

Sulfate is used to make sulfur-containing substances that cannot be obtained through the diet. It is found throughout the body in conjunction with compounds like amino acids, thiamin, and biotin. Sodium and potassium are involved in muscle contraction, transmission of nerve impulses, and maintenance of the balance of fluid inside and outside of the cells. Potassium, the main intracellular ion, has a particularly important role in cardiac activity, and cardiac failure may result when it is depleted. This happens following periods of vomiting and diarrhea or when diuretics ("water pills") are taken to reduce blood pressure or lose weight. Low potassium, called **hypokalemia**, has even killed some individuals with anorexia nervosa and bulimia nervosa. Sodium imbalances may cause water retention, which occurs when water is held in the body to dilute excess sodium, rather than being excreted as urine.

Chloride is also involved in water balance, as a component of extracellular fluid. Chloride is a constituent of hydrochloric acid in the stomach and, like potassium, can be depleted after prolonged vomiting.

Sodium and chloride together comprise table salt, a food additive that has been linked to hypertension. The relationship between dietary sodium and hypertension is direct and a cause for concern. The Institute of Medicine recently released a report, *Strategies to Reduce Sodium Intake in the United States*, to suggest ways to combat this problem. While several other nutrients play a role in regulating blood pressure—calcium, magnesium, potassium—and smoking and obesity undoubtedly contribute to hypertension, sodium intake is a major factor.

hypokalemia
Abnormally low levels of potassium in the bloodstream.

Food Sources of the Other Major Minerals

Thousands of years ago people had to search for salt, but today it is more likely that too much, not too little, will be consumed. Sodium occurs naturally in foods, but most dietary sodium comes from salt added to foods during processing. Table salt, by the way, is also a source of dietary chloride, along with seafood, meat, and eggs. Much of the increase in our caloric intake comes from convenience foods and restaurant meals, which are often loaded with sodium. Table 6-7 illustrates how processed foods have become the leading sources of sodium in the U.S. diet.

Potassium is widespread in foods, particularly fruits, vegetables, legumes, meat, and fish. Highly processed foods are poor choices for obtaining adequate potassium because much potassium is lost in manufacturing.

Recommended Intakes of the Other Major Minerals

When people consume enough sulfur-containing amino acids, no additional sulfate is needed. Thus, there is no DRI for sulfate. Sodium and potassium have AIs for adults of 1.5 g (1,500 mg) and 4.7 g (4,700 mg), respectively.

You might rightly ask how much sodium is too much, rather than how much is too little. The 2010 Dietary Guidelines for Americans will suggest a daily sodium intake below 1,500 mg.[22] The average sodium intake for Americans over age 2 years is 3,614 mg daily.[40] You have already learned that moderating sodium intake can prevent hypertension,

TABLE 6-7 Sources of Sodium: Unprocessed vs. Processed Foods (mg)

Unprocessed foods	Processed foods
Sweet potato, baked in skin, 11 mg	KFC potato wedges, 1 serving, 1,323 mg
Salad (1 c lettuce, ½ c mushrooms, 1 sliced cucumber, 2 T oil & vinegar), 9 mg	Taco Bell taco salad with salsa in shell, 1,500 mg
½ c of most vegetables, cooked, 7–20 mg	Boston Market squash casserole, ¾ c, 1,100 mg
Oatmeal, ½ c, cooked with 1 c milk, 108 mg	Cinnabon classic roll, 801 mg
Ground beef patty, lean, ¼ lb, 101 mg	McDonalds quarter pounder with cheese, 1,310 mg
	Pizza Hut meat lovers stuffed crust pizza, 1 slice, 1,450 mg
Turkey, light meat, 4 oz, 72 mg	Subway Atkins-friendly turkey bacon melt wrap, 1,650 mg
Chicken, light-meat, roasted, 3 oz, 43 mg	KFC hot & spicy chicken breast, 1,230 mg
	Chick Fil-A chicken sandwich, 1,300 mg
	Taco Bell grilled chicken burrito, 1,240 mg
Black beans, cooked from raw, ½ c, 1 mg	Boston Market black beans & rice, ½ c, 525 mg

which is associated with heart attack and stroke, but limiting sodium intake also prevents (1) water retention that might overwork the heart; (2) excess sodium processing by the kidneys; and (3) calcium losses, which are associated with high sodium intake.

Supplements of the Other Major Minerals

Supplements are not needed to obtain adequate sodium, chloride, or sulfate. Sometimes potassium supplements are prescribed for individuals taking diuretics or other drugs associated with potassium depletion. Non-prescription supplementation with potassium is not recommended. Although excess potassium generally induces vomiting, which prevents toxic levels from being ingested, too much potassium could cause heart failure in an unhealthy person.

About 1 million American arthritis-sufferers regularly take sulfate supplements in the form of glucosamine sulfate and chondroitin sulfate.[41] Results on the effectiveness of these in humans is still inconclusive. Some studies have found benefits in decreasing arthritis pain,[41,42] while others have reported little value.[43]

Iron and Zinc

Iron is the most abundant trace mineral, and zinc is the second. These minerals have similar chemical properties and are integral components of several enzyme systems. Insufficient iron or zinc during growth periods (infancy, childhood, early adolescence, and pregnancy) can have permanent effects on development. Iron and zinc also exist in many of the same foods, so an inadequate intake of both may occur when certain foods are avoided.

Functions of Iron and Zinc

Iron is the fourth most abundant element on the earth, but iron deficiency is the most common nutritional deficiency in the world.[44] Most iron in the human body is found in the red blood cells, where it comprises the heme molecule and transports oxygen to the cells. Iron is also a component of myoglobin, a substance similar to hemoglobin that is found in muscle cells, and is a key element in many enzyme systems and physiological processes. For example, cytochromes in the electron transport chain, which was described in Chapter 3, are rich in iron.

Iron deficiency causes a form of anemia, in which the oxygen-carrying capacity of the red blood cells is diminished due to insufficient hemoglobin. This condition is particularly devastating to infants and children and may result in low birth weight, cognitive deficits, and developmental delays. People of any age who are iron deficient may be listless, pale, inattentive, frequently cold, and unable to perform normal activities. They are also frequently ill due to loss of immunity to disease. Of particular concern in childhood is the interaction of iron deficiency with lead poisoning—people who are iron deficient absorb more lead when they are exposed to it.

Zinc is a key component of many enzymes and is found in all organs and tissues, particularly muscle, bone, and liver. Zinc is needed for cell replication, immunity, wound healing, lipid and carbohydrate metabolism, and hormone secretion. A variety of cellular functions are impaired in zinc deficiency. In children, normal growth and development may be interrupted; in adults, immunity may be compromised. At any age, inadequate zinc is associated with poor appetite and skin lesions. Very little zinc is stored in the body, so obtaining regular and adequate amounts of zinc through the diet is particularly important.

Food Sources of Iron and Zinc

Both animal and plant foods provide iron and zinc. These two minerals share a common problem: Although food sources are relatively abundant, the iron and zinc in some foods has low bioavailability. Food sources of iron and zinc are listed in Table 6-8.

Absorption of zinc and iron is enhanced by:

- Composition of the diet: Both minerals are better absorbed from animal products, especially meat and fish, than from plants. The iron in food exists in heme and nonheme forms. Heme iron, which is found in meat, poultry, and fish, is well absorbed. Nonheme iron, found in plant foods, dairy products, and eggs, is more difficult to absorb but makes up a greater proportion of dietary intake. The iron used in food fortification and supplements is also the nonheme type. The terms ferric and ferrous indicate the presence of fortified or supplemental iron.
- Ascorbic acid (vitamin C): Approximately 75 mg of vitamin C promotes iron absorption from plant-based foods that is comparable to the iron available from 3 oz of meat.[45]
- Low iron or zinc stores: People with insufficient stores absorb more.

The absorption of zinc and nonheme iron is inhibited by:

- Polyphenols present in plant foods, coffee, black tea, and red wine: Having tea with or just after a meal can reduce iron absorption by over 60%; coffee reduces absorption by about 40%.
- Oxalic acid: This component of spinach, Swiss chard, berries, chocolate, and tea also limits absorption.
- Phytates: Absorption is impaired by this substance found in whole grains, legumes, nuts, and seeds as well as some soy products.
- The preservative EDTA (ethylenediaminetetraacetate), which promotes color retention in food.

TABLE 6-8 Good Food Sources of Iron and Zinc

Food	Iron (mg)	Zinc (mg)
Beef liver, 3 oz	5.24	4.45
Ground beef, extra lean, 3 oz	2.35	5.47
Chicken leg, 3 oz	1.11	2.43
Oysters, baked or broiled, 3 oz	5.30	72.22
Clams, cooked, 3 oz	23.77	145
Shrimp, cooked, moist heat, 3 oz	2.63	1.33
Tofu, extra firm, 3 oz	1.08	—
Lentils, cooked, ½ c	3.30	1.26
Spinach, chopped, boiled, ½ c	3.21	0.68
Swiss chard, boiled, ½ c	1.98	0.29
Cashews, dry roasted, ¼ c	2.06	1.92
Pecans, raw, ¼ c	0.68	1.22
Sunflower seeds, dried, ¼ c	2.44	1.82
Whole wheat bread, 1 slice	1.43	0.69
All-Bran cereal, 1 c	9.00	3.00
Special K cereal, 1 c	8.70	0.90

- The presence of several other nutrients: For example, calcium from supplements, fortified foods, antacids, or milk reduces iron supplement absorption and may also inhibit zinc absorption.[46] Manganese and nickel, found in nuts, legumes, and leafy vegetables, may also lower iron and zinc absorption.

These observations should not suggest that people who consume a plant-based diet are at risk of iron and zinc deficiency. Although many whole grains and other plant products contain iron inhibitors, the amount of iron and zinc in these foods is generally large enough to ensure some absorption. Also, when the diet is rich in vitamin C, the inhibitory effects are overcome for iron.

Recommended Intakes of Iron and Zinc

The RDA for iron ranges from 8 mg to 18 mg daily, depending upon gender and age, and is highest (27 mg/day) during pregnancy. The reason for the gender-specific RDA is menstrual blood loss in women. Zinc also has an RDA, which is higher in adult men (11 mg/day) than women (8 mg/day).

Iron and zinc deficiencies are not necessarily the result of inadequate intake. Deficiencies mainly occur due to poor mineral absorption; and also result from rapid growth or, in the case of iron, blood loss.

Iron and Zinc Supplements

Iron deficiency anemia is rare in the United States, but iron deficiency is not. Highest rates of iron deficiency are seen in toddlers (prevalence is about 7%) and women (prevalence ranges from 9–16% among 12- to 49-year-old women).[44] People who are especially susceptible to insufficient iron intake and/or absorption and who might require iron supplementation include:

- Individuals restricting caloric intake, especially premenopausal women and adolescents (the typical Western diet provides between 5 and 7 mg of iron per 1,000 kcal)
- Vegetarians who eat no animal products and do not consume a balanced and varied diet
- Infants and young children, whose iron intake often fails to keep up with their growth rate
- Pregnant women, who require more iron to support fetal growth, an enlarged blood volume, and blood loss during childbirth. Iron supplements are often prescribed during pregnancy, but almost three-quarters of women are at least somewhat iron deficient after childbirth.[47]
- People with chronic alcoholism, digestive system disorders, or chronic diseases
- Individuals who have lost a great deal of blood

hemochromatosis *Rare iron metabolism disorder in which iron is deposited in the tissues; treated with blood transfusions.*

When taken, iron supplements are best absorbed with a meal, particularly one rich in vitamin C. Caution is required, as ingestion of too much supplemental iron can damage organs like the heart and liver. About 1 in 250 people is genetically predisposed to a disease called **hemochromatosis**, which causes excessive iron to be absorbed and stored, creating iron overload. These individuals should avoid taking iron supplements without their physicians' knowledge. There is also concern that excess iron intake could increase the risk of cardiovascular disease by increasing oxidation of LDLs, although not all studies have confirmed this.[48] Because iron supplements may inhibit zinc absorption, anytime more than 30 mg of supplemental iron is taken each day, a zinc supplement should also be recommended.

Zinc supplements come in tablets, lozenges, and sprays. They have generated interest because of evidence that zinc inhibits the replication of cold viruses in the laboratory. Studies with humans have been inconsistent but promising. Studies with adults find that taking supplemental zinc, usually as a lozenge, within the first 24 hours of cold symptoms can decrease the duration and severity of the cold.[49] One study with children found that those taking zinc sulfate syrup once a day during cold season had significantly fewer colds than children taking placebo syrup.[50]

Intake of more than 100 mg of supplemental zinc per day can cause severe copper deficiency, a form of anemia, reduced immune response, and a decrease in HDL cholesterol.[51] Even more modest intakes of zinc, 50–75 mg/week, may have similar adverse effects. Side effects of both iron and zinc supplements include gastrointestinal distress, including diarrhea for zinc and constipation for iron. To avoid gastric distress, a UL for iron was set at 45 mg/day for adults. For zinc, a UL of 40 mg/day was established to prevent impeding normal copper metabolism. The best advice for these two nutrients—don't take a supplement if you don't need it.

Selenium and Other Trace Minerals

The other trace minerals are chromium, copper, fluoride, iodine, manganese, molybdenum, and selenium. Selenium is discussed briefly in this section because of its antioxidant role. The other trace minerals are summarized in Table 6-9.

TABLE 6-9   Selected Trace Minerals

Mineral	Function	Food sources	Deficiency problems
Chromium	Formation of glucose tolerance factor, which increases insulin effectiveness	Organ meats, whole grains, cereals, cheese, mushrooms, beer, wine, tea, brewer's yeast	Impaired glucose tolerance; possibly type 2 diabetes
Copper	Various enzymatic reactions	Organ meats, shellfish, nuts, seeds, legumes, dried fruit	Anemia, bone demineralization, loss of immunity
Fluoride	Works with calcium in formation of bones and teeth; prevents dental cavities	Fluoridated water, some fish, brewed tea (also in fluoridated toothpaste)	Tooth decay; may be related to osteoporosis
Iodine	Only known function is formation of thyroid hormones	Seafood, iodized salt; also makes its way into the food supply through food processing and animal feed	Thyroid enlargement; low-sodium diets may increase the risk of iodine deficiency
Manganese	Bone formation; various enzymatic reactions	Grains and vegetables	Unreported in humans
Molybdenum	Various enzymatic reactions	Legumes, cereal grains	Unreported in humans

Functions of Selenium

Selenium is involved in several vital metabolic processes, including DNA repair and immunity. It works with vitamin E as an antioxidant, protecting the cells against oxidative damage. Because of selenium's antioxidant effects and possibly some as yet undefined ability to put a kink in the metabolism of cancerous cells, it may also inhibit the growth of tumors.[52]

Food Sources and Recommended Intakes of Selenium

Selenium is found in animal products, especially seafood and meat, and in a variety of plant foods. Soil content of selenium has an impact on selenium levels in plants and in the animals that eat those plants. Some parts of the world, including many areas in China, lack selenium, so crops grown in those regions have lower selenium content than crops from other parts of the world.

The RDA for selenium is 55 μg for adult men and women. Extremely high doses of selenium can be toxic. The Centers for Disease Control and Prevention reported several cases in China when people regularly consumed about 5 mg of selenium daily as well as one case in the United States when 27 mg of selenium was accidentally ingested daily from a contaminated supplement. In all cases, the result was nausea, nail and hair loss, fatigue, and sour breath.[53] The UL for selenium from food and supplements is 400 μg for adults.

Recommended intakes of the other trace minerals can be found in the table Appendix D.

Body Water and Minerals: Summary

Water makes up 60–70% of body weight, and minerals comprise about 4%. Water is the most essential nutrient, and life cannot be sustained without it. An adequate intake of minerals is needed to promote normal growth and development as well as fluid and electrolyte balance. Minerals also prevent bone loss, anemia, hypertension, and possibly some cancers.

Case Study. Personal Diet Analysis, Part 4	• Use the Interactive DRI for Healthcare Professionals (http://fnic.nal.usda.gov/interactiveDRI/) to estimate your subject's recommended daily mineral needs. You will need height, weight, age, and estimated activity level for this. • Compare mineral intake (determined from the food record case study in Chapter 5) with recommendations. How close did your subject come to the dietary recommendations presented in this chapter and to the recommendations of the Interactive DRI? • If your subject uses vitamin or mineral supplements, get a copy of the bottle(s) and estimate additional mineral consumption. Is your subject at or above the UL for any mineral? • Are there any special circumstances that would increase your subject's need for specific minerals? • For each mineral, list some foods that would be good sources of that mineral and that are foods your subject would eat.

THE ROLE OF MICRONUTRIENTS IN WEIGHT MANAGEMENT

Because they contain no calories, vitamins, minerals, and water do not play a role in weight loss or weight gain in the same way that carbohydrates, fats, and proteins do. Supplementation with micronutrients also provides no benefit in weight gain or loss (although, as you will read later, calcium may aid in weight loss among individuals who do not consume sufficient amounts of the mineral). With the exception of those who are underweight, individuals recovering from bulimia and anorexia, and obese individuals after gastric bypass surgery, vitamin and mineral supplementation at levels higher than recommended for the general public is not needed.

Nevertheless, two aspects of nutrition that involve micronutrients deserve attention in a weight management text: consumption of fruits and vegetables and consumption of dairy. These are briefly discussed in this section.

energy density
The caloric value of a food per unit of weight (kcal/g). Foods with high energy density provide more kcal per food weight and tend to be less satiating than foods with low energy density (people have to eat more to feel full).

How Does Fruit and Vegetable Consumption Relate to Weight Management?

Earlier in this chapter and in Chapter 5 you learned that fruits and vegetables contain several substances that contribute to good health, including fiber, carotenoids, and polyphenols. Because of this, people who consume a diet rich in fruits and vegetables may reduce the risk of cardiovascular disease and some cancers. Fruits and vegetables have additional characteristics that make them especially appealing for individuals trying to lose or maintain weight: they are high in water and low in calories. This makes them low in **energy density**, meaning that they have low caloric value per unit of weight. Water adds weight but not calories.

Why would energy density make a difference? People seem to prefer to eat a consistent weight of food from day to day. Foods that are high in energy density (those that provide lots of kcals but don't weigh much, like potato chips) need to be eaten in greater quantities to produce a feeling of fullness or satisfaction. Foods with higher water content and therefore low energy density, like soup and fruits and vegetables, permit people to consume a desired weight of food but at a lower caloric cost. With their water weight, as well as their fiber content, most fruits and vegetables increase both **satiety** and **satiation**, while lowering caloric intake.[54]

The fruit- and vegetable-rich DASH diet, which you learned about in Chapter 5, not only helps to lower blood pressure but also to lose weight.[55] People on calorie restricted diets who are encouraged to eat more fruits and vegetables may lose more weight than those who simply lower caloric intake.[56,57] Allowing people to consume large quantities of fruits and vegetables might make a "diet" easier to adhere to, even when there are limitations on consuming other foods so that calories are reduced.

satiety *A relatively long-term feeling of fullness or satisfaction that remains after eating, which inhibits further consumption of food.*

satiation *Termination of eating after hunger has been satisfied.*

Do Dairy Products Help People Lose Weight?

You might have read in the popular press that people who consume more calcium, either through supplements or dairy products, weigh less or have an easier time losing excess weight. The research literature is conflicted on this. Some reviews find no weight- or fat-loss benefits of additional calcium,[58,59] while others believe that there is evidence supporting calcium's role in weight management.[60,61]

In the laboratory setting, animals on a low-calcium diet gain fat and those on a high-calcium diet reduce fatness. This effect is not as straightforward with humans. Factors such as gender, age, degree of overweight/obesity, and calcium status may account for some of the differences seen in human studies. The studies with humans that have documented weight loss with increased dietary calcium have noted the following:

- Either dairy or calcium supplements promote weight loss, but dairy may work best.
- There has to be some level of caloric restriction for weight loss to occur.
- People who have lower habitual calcium intake before supplementing with dairy or calcium may experience greater weight loss.
- Obese individuals who are low calcium-consumers may need additional calcium and possibly vitamin D to lose weight on any weight management program.

Why would calcium have any relationship to weight gain and loss? There are several possible explanations:

- Parathyroid hormone and calcitriol (1,25-dihydroxyvitamin D) promote fat storage in adipose tissue. Levels of these hormones increase when calcium intake is low, allowing greater fat storage to take place. And in the presence of adequate calcium, the hormones are suppressed and less fat storage takes place.[62]
- In mice, 1,25-dihydroxyvitamin D also plays a role in adaptive thermogenesis, which was covered in Chapter 3. When 1,25-dihydroxyvitamin D levels are high (which occurs when calcium levels are low), thermogenesis is suppressed. When 1,25-dihydroxyvitamin D levels are low (after adding calcium to the mouse diet), thermogenesis increases.[62] This has not yet been documented in humans.

- Calcium may inhibit fat absorption in the gastrointestinal tract, so less dietary fat is available to be taken up into adipocytes.[60]
- Lack of dietary calcium may stimulate appetite for calcium-containing foods and drive people to eat more.[63]

So, should we add dairy or calcium supplements to our diets to promote weight loss and prevent weight gain? The evidence, while by no means conclusive, does at least partly support the value of attending to one's calcium status when trying to lose weight. The research in this area continues.

Case Study. Personal Diet Analysis, Part 5	Have your subject keep a record of beverage consumption for three typical days (1 weekend day and 2 weekdays). Determine average beverage consumption for those days. • How does your subject's beverage consumption compare with the DRI? • Looking back at food records, does your subject regularly consume foods that provide additional fluid (soups, yogurt)? • Given your subject's activity level and other characteristics, comment on adequacy of fluid consumption. If needed, how could your subject consume more fluid?

The Role of Micronutrients in Weight Management: Summary

Micronutrients and water do not play a role in weight loss or weight gain in the same way that carbohydrates, fats, and proteins do. Individuals who are underweight, recovering from bulimia and anorexia, or who have had gastric bypass surgery may benefit from vitamin and mineral supplementation. Consumption of energy dense, water- and micronutrient-rich fruits and vegetables may help people lose weight while consuming a satisfactory volume of food. And added calcium, preferably through dairy but also through supplementation, has also promoted weight loss in some people.

REFERENCES

1. Voutilainen, S., Nurmi, T., Mursu, J., & Rissanen, T. H. (2006). Carotenoids and cardiovascular health. *American Journal of Clinical Nutrition, 83*(6), 1265–1271.
2. The Alpha-Tocopherol, Beta-Carotene Cancer Prevention Study Group. (1994). The effect of vitamin E and beta-carotene on the incidence of lung cancer and other cancers in male smokers. *New England Journal of Medicine, 330*(15), 1029–1035.
3. Whiting, S. J., & Lemke, B. (1999). Excess retinol intake may explain the high incidence of osteoporosis in northern Europe. *Nutrition Reviews, 57*(6), 192–195.
4. Ford, E. S., & Sowell, A. (1999). Serum alpha-tocopherol status in the United States population:

Findings from the third National Health and Nutrition Examination Survey. *American Journal of Epidemiology, 150*(3), 290–300.
5. Handelman, G. J., Packer, L., & Cross, C. E. (1996). Destruction of tocopherols, carotenoids, and retinol in human plasma by cigarette smoke. *American Journal of Clinical Nutrition, 63*(4), 559–565.
6. Lichtenstein, A. H., & Russell, R. M. (2005). Essential nutrients: Food or supplements? Where should the emphasis be? *JAMA, 294*(3), 351–358.
7. Bischoff-Ferrari, H. A., Willett, W. C., Wong, J. B., Giovannucci, E., Dietrich, T., & Dawson-Hughes, B. (2005). Fracture prevention with vitamin D

supplementation: A meta-analysis of randomized controlled trials. *JAMA, 293*(18), 2257–2264.

8. Wang, S. (2009). Epidemiology of vitamin D in health and disease. *Nutrition Research Reviews, 22*(2), 188–203.

9. Ross, A. C., Taylor, C. L., Yaktine, A. L., & DelValle, H. B. (Eds.). (2011 Pre-Publication Review Copy.) Dietary reference intakes for calcium and vitamin D. Washington, D.C.: National Academies Press.

10. Borradale, D., & Kimlin, M. (2009). Vitamin D in health and disease: An insight into traditional functions and new roles for the "sunshine vitamin". *Nutrition Research Reviews, 22*(2), 118–136.

11. Vieth, R. (1999). Vitamin D supplementation, 25-hydroxyvitamin D concentrations, and safety. *American Journal of Clinical Nutrition, 69*(5), 842–856.

12. Barger-Lux, M. J., Heaney, R. P., Dowell, S., Chen, T. C., & Holick, M. F. (1998). Vitamin D and its major metabolites: Serum levels after graded oral dosing in healthy men. *Osteoporosis International, 8*(3), 222–230.

13. Yoshida, M., Booth, S. L., Meigs, J. B., Saltzman, E., & Jacques, P. F. (2008). Phylloquinone intake, insulin sensitivity, and glycemic status in men and women. *American Journal of Clinical Nutrition, 88*(1), 210–215.

14. Otten, J. J., Hellwig, J. P., & Meyers, L. D. (Eds.). (2006). *Dietary reference intakes: The essential guide to nutrient requirements.* Washington, D.C.: The National Academies Press.

15. Rader, J. I., Calvert, R. J., & Hathcock, J. N. (1992). Hepatic toxicity of unmodified and time-release preparations of niacin. *American Journal of Medicine, 92*(1), 77–81.

16. Institute of Medicine. (1999). *Dietary reference intakes for thiamin, riboflavin, niacin, vitamin B6, folate, vitamin B12, pantothenic acid, biotin, and choline.* Washington, DC: National Academy Press.

17. Heap, L. C., Peters, T. J., & Wessely, S. (1999). Vitamin B status in patients with chronic fatigue syndrome. *Journal of the Royal Society of Medicine, 92*(4), 183–185.

18. CDC. (2008). Quick stats: Spina bifida and anencephaly rates—United States, 1991, 1995, 2000, and 2005. *Morbidity and Mortality Weekly Reports, 57*(1), 15.

19. Pancharuniti, N., Lewis, C. A., Sauberlich, H. E., Perkins, L. L., Go, R. C., Alvarez, J. O., et al. (1994). Plasma homocyst(e)ine, folate, and vitamin B-12 concentrations and risk for early-onset coronary artery disease. *American Journal of Clinical Nutrition, 59*(4), 940–948.

20. Ulrich, C. M. (2005). Nutrigenetics in cancer research—folate metabolism and colorectal cancer. *Journal of Nutrition, 135*(11), 2698–2702.

21. Ericson, U., Sonestedt, E., Gullberg, B., Olsson, H., & Wirfalt, E. (2007). High folate intake is associated with lower breast cancer incidence in postmenopausal women in the Malmo diet and cancer cohort. *American Journal of Clinical Nutrition, 86*(2), 434–443.

22. *Report of the Dietary Guidelines Advisory Committee on the Dietary Guidelines for Americans, 2010.* (2010). Available from http://www.cnpp.usda.gov/DGAs2010-DGACReport.htm. Accessed July 12, 2010.

23. Zimmermann, M. B., & Shane, B. (1993). Supplemental folic acid. *American Journal of Clinical Nutrition, 58*(2), 127–128.

24. Alpert, J. E., & Fava, M. (1997). Nutrition and depression: The role of folate. *Nutrition Reviews, 55*(5), 145–149.

25. Jacques, P. F., Taylor, A., Hankinson, S. E., Willett, W. C., Mahnken, B., Lee, Y., et al. (1997). Long-term vitamin C supplement use and prevalence of early age-related lens opacities. *American Journal of Clinical Nutrition, 66*(4), 911–916.

26. Schorah, C. J., Sobala, G. M., Sanderson, M., Collis, N., & Primrose, J. N. (1991). Gastric juice ascorbic acid: Effects of disease and implications for gastric carcinogenesis. *American Journal of Clinical Nutrition, 53*(1 Suppl), 287S–293S.

27. Douglas, R. M., Hemila, H., D'Souza, R., Chalker, E. B., & Treacy, B. (2004). Vitamin C for preventing and treating the common cold. *Cochrane Database of Systematic Reviews (Online), (4)*(4), CD000980.

28. Hemila, H. (1996). Vitamin C supplementation and common cold symptoms: Problems with inaccurate reviews. *Nutrition, 12*(11–12), 804–809.

29. Ausman, L. M. (1999). Criteria and recommendations for vitamin C intake. *Nutrition Reviews, 57*(7), 222–224.

30. Gregory, J. F., 3rd. (1993). Ascorbic acid bioavailability in foods and supplements. *Nutrition Reviews, 51*(10), 301–303.

31. Levine, M., Rumsey, S. C., Daruwala, R., Park, J. B., & Wang, Y. (1999). Criteria and recommendations for vitamin C intake. *JAMA, 281*(15), 1415–1423.

32. Johnston, C. S. (1999). Biomarkers for establishing a tolerable upper intake level for vitamin C. *Nutrition Reviews, 57*(3), 71–77.

33. Institute of Medicine. (1997). *Dietary reference intakes for calcium, phosphorus, magnesium, vitamin D, and fluoride.* Washington, D.C.: National Academy Press.

34. Nieves, J. W. (2005). Osteoporosis: The role of micronutrients. *American Journal of Clinical Nutrition, 81*(5), 1232S–1239.

35. Chia, V., & Newcomb, P. A. (2004). Calcium and colorectal cancer: Some questions remain. *Nutrition Reviews, 62*(3), 115–120.

36. Larsson, S. C., & Wolk, A. (2007). Magnesium intake and risk of type 2 diabetes: A meta-analysis. *Journal of Internal Medicine, 262*(2), 208–214.

37. Andon, M. B., Lloyd, T., & Matkovic, V. (1994). Supplementation trials with calcium citrate malate: Evidence in favor of increasing the calcium RDA during childhood and adolescence. *The Journal of Nutrition, 124*(8 Suppl), 1412S–1417S.

38. Sheikh, M. S., Santa Ana, C. A., Nicar, M. J., Schiller, L. R., & Fordtran, J. S. (1987). Gastrointestinal absorption of calcium from milk and calcium salts. *New England Journal of Medicine, 317*(9), 532–536.

39. Cook, J. D., Dassenko, S. A., & Whittaker, P. (1991). Calcium supplementation: Effect on iron absorption. *American Journal of Clinical Nutrition, 53*(1), 106–111.

40. Henney, J. E., Taylor, C. L., & Boon, C. S. (Eds.). (2010). *Strategies to reduce sodium intake in the United States*. Washington, D.C.: National Academies Press.

41. Hampton, T. (2007). Efficacy still uncertain for widely used supplements for arthritis. *JAMA, 297*(4), 351–352.

42. Clegg, D. O., Reda, D. J., Harris, C. L., Klein, M. A., O'Dell, J. R., Hooper, M. M., et al. (2006). Glucosamine, chondroitin sulfate, and the two in combination for painful knee osteoarthritis. *New England Journal of Medicine, 354*(8), 795–808.

43. Reichenbach, S., Sterchi, R., Scherer, M., Trelle, S., Burgi, E., Burgi, U., et al. (2007). Meta-analysis: Chondroitin for osteoarthritis of the knee or hip. *Annals of Internal Medicine, 146*(8), 580–590.

44. CDC. (2002). Iron deficiency—United States, 1999–2000. *Morbidity and Mortality Weekly Report, 51*(40), 897–899.

45. Craig, W. J. (1994). Iron status of vegetarians. *American Journal of Clinical Nutrition, 59*(5 Suppl), 1233S–1237S.

46. Fairweather-Tait, S. J., & Teucher, B. (2002). Iron and calcium bioavailability of fortified foods and dietary supplements. *Nutrition Reviews, 60*(11), 360–367.

47. Beard, J. L., Dawson, H., & Pinero, D. J. (1996). Iron metabolism: A comprehensive review. *Nutrition Reviews, 54*(10), 295–317.

48. Sempos, C. T., Looker, A. C., Gillum, R. E., McGee, D. L., Vuong, C. V., & Johnson, C. L. (2000). Serum ferritin and death from all causes and cardiovascular disease: The NHANES II mortality study. National Health and Nutrition Examination study. *Annals of Epidemiology, 10*(7), 441–448.

49. Hulisz, D. (2004). Efficacy of zinc against common cold viruses: An overview. *Journal of the American Pharmacists Association, 44*(5), 594–603.

50. Kurugol, Z., Akilli, M., Bayram, N., & Koturoglu, G. (2006). The prophylactic and therapeutic effectiveness of zinc sulphate on common cold in children. *Acta Paediatrica, 95*(10), 1175–1181.

51. Fosmire, G. J. (1990). Zinc toxicity. *American Journal of Clinical Nutrition, 51*(2), 225–227.

52. Clark, L. C., Combs, G. F., Jr, Turnbull, B. W., Slate, E. H., Chalker, D. K., Chow, J., et al. (1996). Effects of selenium supplementation for cancer prevention in patients with carcinoma of the skin. A randomized controlled trial. Nutritional Prevention of Cancer Study Group. *JAMA, 276*(24), 1957–1963.

53. Centers for Disease Control and Prevention. (1984). Epidemiologic notes and reports selenium intoxication—New York. *Morbidity and Mortality Weekly Reports, 33*(12), 157–158.

54. Rolls, B. J., Ello-Martin, J. A., & Tohill, B. C. (2004). What can intervention studies tell us about the relationship between fruit and vegetable consumption and weight management? *Nutrition Reviews, 62*(1), 1–17.

55. Appel, L. J., Moore, T. J., Obarzanek, E., Vollmer, W. M., Svetkey, L. P., Sacks, F. M., et al. (1997). A clinical trial of the effects of dietary patterns on blood pressure. DASH Collaborative Research Group. *New England Journal of Medicine, 336*, 1117–1124.

56. Singh, R. B., Rastogi, S., Verma, R., Laxmi, B., Singh, R., Ghosh, S., et al. (1992). Randomized controlled trial of cardioprotective diet in patients with recent acute myocardial infarction: Results of one-year follow up. *British Medical Journal, 304*, 1015–1019.

57. Singh, R. B., Dubnov, G., Niaz, M. A., Ghosh, S., Singh, R., Rastogi, S. S., et al. (2002). Effect of an Indo-Mediterranean diet on progression of coronary artery disease in high-risk patients (Indo-Mediterranean Diet Heart Study): A randomized single-blind trial. *Lancet, 360*, 1455–1461.

58. Lanou, A. J., & Barnard, N. D. (2008). Dairy and weight loss hypothesis: An evaluation of the clinical trials. *Nutrition Reviews, 66*(5), 272–279.

59. Trowman, R., Dumville, J. C., Hahn, S., & Torgerson, D. J. (2006). A systematic review of the effects of calcium supplementation on body weight. *British Journal of Nutrition, 95*, 1033–1038.

60. Parikh, S. J., & Yanovski, J. A. (2003). Calcium intake and adiposity. *American Journal of Clinical Nutrition, 77*, 281–287.
61. Zemel, M. B. (2008). Dairy and weight loss hypothesis [letter to the editor]. *Nutrition Reviews, 66*(9), 542–543.
62. Zemel, M. B. (2004). Role of calcium and dairy products in energy partitioning and weight management. *American Journal of Clinical Nutrition, 79*(Suppl), 907S–912S.
63. Major, G. C., Alarie, F. P., Doré, J., & Tremblay, A. (2009). Calcium plus vitamin D supplementation and fat mass loss in female very low-calcium consumers: Potential link with a calcium-specific appetite control. *British Journal of Nutrition, 101*, 659–663.

Physical Activity and Exercise: The Basics

R. Gino Santa Maria/Shutterstock.com

CHAPTER OUTLINE

cappi thompson/Shutterstock.com

Iorga Studio/Shutterstock.com

physical activity
Body movement produced by skeletal muscles that requires energy expenditure and produces overall health benefits.

exercise *Planned, structured, and repetitive physical activity done to improve or maintain one or more components of physical fitness.*

Physical activity is very important in the long-term regulation of body weight and body composition; many experts believe that it is more important than diet. Activity increases caloric expenditure, elevates metabolic rate in some people, and may ultimately help reset the "set point." Even more significant is the role of activity in promoting good health. Active people, whether or not they are overweight/obese, have healthier blood lipid levels, lower blood pressure, and better blood glucose control. They even live longer.

The term **physical activity** encompasses a wide variety of pastimes. It has been defined by the U.S. Department of Health and Human Services as health-enhancing bodily movement produced by contracting skeletal muscles that increases energy expenditure above the basal level.[1] **Exercise** is a type of physical activity defined as "planned, structured, repetitive, and purposive in the sense that the improvement or maintenance of one or more components of physical fitness is the objective."[1] Both physical activity and exercise convey numerous health benefits. This chapter provides a review of the body systems affected by activity; the benefits of activity, including the development of physical fitness; the values of an active lifestyle throughout the lifespan; and the assessment of physical fitness.

BODY SYSTEMS INVOLVED IN PHYSICAL ACTIVITY

When we engage in physical activity, all of the body systems are affected in some way. For example, urinary output is decreased as blood is diverted from the kidneys to the working muscles (excretory system), and sensory information integrated in the brain allows people to fine-tune movements (central nervous system). This section looks closely at two systems that are especially involved during physical activity: the cardiorespiratory system and the musculoskeletal system.

What are the Components of the Cardiorespiratory System?

The cardiorespiratory system provides oxygen and nutrients that make physical activity possible. In return, readers will learn later in this chapter that cardiorespiratory health can be greatly improved by exercise and activity.

Blood

plasma *Liquid component of the blood.*

The average male body contains 5.5 liters of blood, and the average female 5 liters. Blood consists of a liquid component, called **plasma**, which is 90% water, and several types of cells:

- Platelets, which are involved in blood clotting
- Leukocytes, white blood cells that convey immunity
- Erythrocytes, the oxygen-carrying red blood cells

Blood has numerous functions, including regulation of body temperature, acid–base balance, circulation of nutrients and hormones, and oxygen transport. During physical activity, the temperature-regulating function of blood becomes apparent. Heat generated in the working muscles is carried by the blood to the skin, where it is released into the environment. Excessive perspiration reduces blood volume and the body's capacity to cool itself. As learned in Chapter 6, fluid intake might need to increase up to three times the normal amounts during exercise to maintain blood volume and allow temperature regulation.

Several blood tests are useful in assessing overall health and readiness for exercise. Chapter 2 included information about assessment of blood lipids. A person's high-density lipoprotein (HDL) cholesterol level is a good indicator of fitness, as HDL levels tend to be higher in people who exercise regularly. Blood hemoglobin concentration is an indirect indicator of iron status. Poor iron status makes physical activity not only difficult but also unsafe.

The Heart

The heart, pictured in Figure 7-1, is a muscular pump that circulates blood—and the constituents of blood—to the body's tissues and cells. As a pump, the heart also serves to maintain the pressure needed to circulate blood to distant blood vessels and to return blood to the heart.

The heart consists of four chambers. The two upper chambers are the right **atrium** and left atrium, and the two lower chambers are the right and left **ventricles**. The right atrium empties into the right ventricle, and the left atrium empties into the left ventricle. This structure creates a "right heart" and a "left heart," which have coordinated and related functions:

- The right heart is responsible for the **pulmonary circulation**. Deoxygenated blood—blood that has already circulated through the body—returns to the right atrium via the body's largest veins, the superior and inferior vena cavae. This blood empties into the right ventricle and is pumped to the lungs, where it is reoxygenated.
- The left heart is in charge of the **systemic circulation**. Oxygenated blood travels from the lungs to the left atrium, passes into the left ventricle, and is ejected from the heart through the body's largest artery, the aorta.
- Valves between the atria and the ventricles prevent backflow of blood. The tricuspid valve is located between the right atrium and the right ventricle, and the mitral valve is between the left atrium and the left ventricle. Disorders that affect these valves can have serious consequences. Backflow of blood reduces the pumping efficiency of the heart, which can lead to a heart attack during exertion or to heart failure over time.

The heart muscle itself is called the **myocardium**, and it is thickest around the ventricles, especially the left ventricle. The amount of force needed to push blood through the systemic circulation is considerable and, just like large biceps can make lifting heavy objects easier; a strong heart muscle makes the heart's work less taxing.

Contraction of the ventricles is known as systole, and the period between ventricular contractions, when the ventricles are filling with blood, is called diastole. The amount of blood ejected from the left ventricle with each heartbeat is called the **stroke volume (SV)**. The combination of stroke volume and heart rate (HR) produces **cardiac output (Q)**. A healthy adult should have a Q of about 5 l/minute at rest. This could be achieved with an HR of 70 beats per minute and an SV of 70 ml/beat.

$$Q = (70 \text{ beats/minute})(70 \text{ ml}) = 4,900 \text{ ml/minute, or } 4.9 \text{ l/minute.}$$

The heart contains an electrical conduction system that is controlled by the sinoatrial (SA) node (the pacemaker of the heart), located in the right atrium. When the SA node

atria/atrium
Upper chambers of the heart that receive blood from the lungs or systemic circulation and empty into the ventricles.

ventricles Lower, muscular chambers of the heart that pump blood to pulmonary or systemic circulation.

pulmonary circulation
Circulation of deoxygenated blood from the right side of the heart to the lungs and back to the left side of the heart.

systemic circulation
Circulation of oxygenated blood from the left side of the heart to the body tissues and back to the right side of the heart.

myocardium
Heart muscle.

stroke volume (SV) Amount of blood ejected from the left ventricle with each heart beat.

cardiac output (Q) Amount of blood circulated by the heart in 1 minute, calculated by multiplying SV by heart rate; in a healthy adult, Q is approximately 5 l/minute.

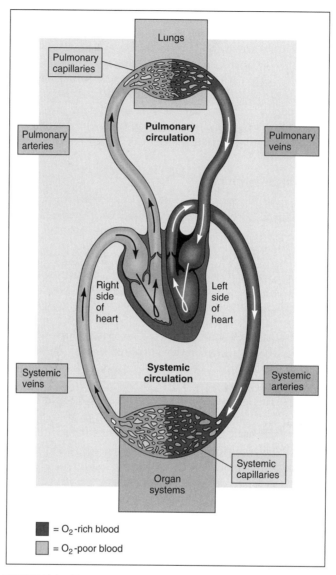

FIGURE 7-1   Basic organization of the cardiovascular system.
The right side of the heart circulates blood to the lungs (the
pulmonary circulation), while the left side of the heart sends
blood throughout the body (systemic circulation). Arteries
branch into arterioles, which further branch into capillaries.
Capillaries rejoin to form venules, which merge to form veins.

is stimulated by signals from the nervous system, it sends impulses to the rest of
the heart via specialized nerve fibers. The muscle cells that comprise the myocardium
are interconnected, so each cell does not need its own connection to these specialized
nerve fibers. When one myocardial cell is stimulated, all of the cells in either the atria or

electrocardio-gram (ECG)
Recording of the electrical events occurring in the myocardium.

the ventricles are stimulated. In this way, both atria contract together as a unit, as do both ventricles. An **electrocardiogram (ECG)** provides a visual representation of the electrical activity of the heart. Each electrical event on the ECG is followed by a mechanical event, such as the filling of the atria or the emptying of the ventricles. An abnormal ECG could indicate serious problems with the heart's capacity to circulate blood, particularly during exertion.

Blood Vessels

When blood leaves the heart, it travels through successively smaller blood vessels, going from arteries to arterioles to capillaries. Oxygenated blood in the systemic circulation loses its oxygen to the tissues while it is in the tiny, thin-walled capillaries. At the same time, carbon dioxide produced in the tissues diffuses into the capillaries and begins its journey back to the heart. As blood returns to the heart, it passes through successively larger blood vessels, going from capillaries to venules to veins.

arteries *Thick-walled, elastic blood vessels that carry oxygenated blood to the tissues.*

arterioles *Small arteries lined with smooth muscle tissue that allows them to narrow (vasoconstriction) or widen (vasodilation).*

veins and venules *Thin-walled blood vessels that carry deoxygenated blood back to the heart.*

capillaries *The smallest blood vessels, through which oxygen and nutrients move into and out of the cells.*

systolic blood pressure *The highest pressure exerted during contraction of the ventricles.*

diastolic blood pressure *Phase of lowest blood pressure in the arteries, which is recorded between ventricular contractions.*

- **Arteries** have thick, muscular walls. When the arteries are healthy, these walls are elastic. The arteries stretch with each heartbeat and then recoil between beats. The arteries maintain blood pressure in conjunction with the arterioles.
- **Arterioles** have thinner walls than arteries, and their walls contain smooth muscle that controls the vessel's diameter. When the smooth muscle contracts, the arteriole narrows (vasoconstriction). When the smooth muscle is relaxed, the arteriole's diameter increases (vasodilation).
- **Venules** are small, thin-walled vessels that carry deoxygenated blood and empty into veins.
- **Veins** have thinner walls than arteries and are relatively inelastic, but they are able to expand to accommodate a significant volume of blood. When you remain on your feet without moving for a long time, blood pools in the leg veins, which reduces SV and limits the transport of oxygen to the brain. The lack of oxygen might make you feel dizzy or cause you to faint.
- **Capillaries** have the smallest diameters of all blood vessels as well as walls that are so thin that fluids, gases, and nutrients may easily flow in and out of them.

Blood pressure is created as blood, pumped by the heart, exerts pressure against the blood vessels. Because arteries are elastic, they "hold" blood pressure and are used to measure it. During systole, the arteries fill with blood, creating **systolic blood pressure**. A systolic blood pressure under 140 mm Hg is desirable. Between ventricular contractions (diastole), no more blood enters the arteries, but blood gradually drains into the arterioles. This phase of lowest pressure in the arteries is called **diastolic blood pressure**. A diastolic blood pressure of 90 mm Hg or less is considered healthy.

High blood pressure ($\geq$ 140/90) may result when the volume of blood is increased, the heart is pumping very forcefully, or the diameter of the blood vessels is reduced. For example, atherosclerotic plaque deposits in the arteries will narrow and harden them, which is a leading cause of hypertension. When the arterioles are constricted, resistance to blood flow (known as peripheral resistance) increases, and blood pressure rises. One cause of peripheral resistance is inadequate intake of calcium, potassium, and magnesium. The measurement of blood pressure is described later in this chapter.

The Lungs

Lung function has a powerful influence on oxygen delivery. Air travels to the lungs through the nose and the mouth, the trachea (popularly called the windpipe), and finally the right and left bronchi. Within the lungs, the bronchi branch into smaller airways called bronchioles, which terminate in alveoli. The thin-walled alveoli are the points at which gas exchange occurs—oxygen from the environment diffuses quickly into the pulmonary capillaries, while carbon dioxide produced in metabolism diffuses from the capillaries into the alveoli. The anatomy of the respiratory system is pictured in Figure 7-2.

A healthy young adult male has a lung capacity of about 5.7 liters of air, and his female counterpart has a lung capacity of about 4.2 liters. Diseases that affect the lungs can impair the individual's ability to exercise. Think of how difficult it is to walk briskly when you have a chest cold, and you will have some idea of the challenges faced by people with asthma or chronic obstructive pulmonary disease.

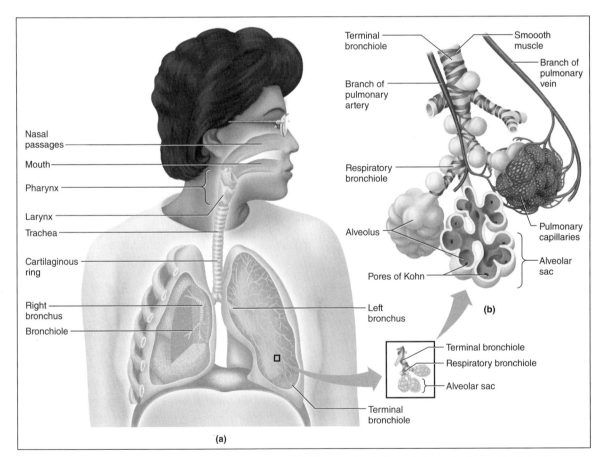

FIGURE 7-2   Basic anatomy of the respiratory system. (a) The trachea brings air into the bronchi of the lungs. Bronchi branch into bronchioles, which terminate in alveoli. (b) Oxygen and carbon dioxide diffuse from the bloodstream into and out of the alveoli and respiratory bronchioles.

What Are the Key Components of the Musculoskeletal System?

The musculoskeletal system consists of skeletal muscles, tendons, ligaments, bones, and cartilage. Together, these elements permit structure, protection, and movement. Regular physical activity strengthens and enhances the functional capacity of the components of the musculoskeletal system.

The Skeletal System

The bones and the cartilage that make up the skeletal system provide structure, support, and protection for underlying tissues and organs. Bone is a dynamic tissue that is replaced every year in infants and every 3–6 years in adults. Two types of bone comprise the skeleton:

cortical bone *Dense bone that makes up about 80% of the skeleton; sometimes called compact bone.*

trabecular bone *Porous bone that makes up the inner part of flat, short, and irregular bones and the shafts of the long bones; this type of bone is especially susceptible to loss of density with age, poor nutritional status, and hormonal imbalance.*

joint *Point at which two or more bones come together.*

ligaments *Slightly elastic, fibrous tissues that hold bones together at a joint and protect the joint.*

cartilage *Strong connective tissue that covers the ends of the bones forming a joint and that absorbs pressure between the bones; over time, cartilage degenerates and becomes the source of pain and stiffness in the joints.*

- **Cortical bone**—The dense bone that makes up about 80% of the skeleton. It is sometimes called compact bone.
- **Trabecular bone**—The porous bone that makes up the inner part of flat, short, and irregular bones and the shafts of the long bones. This type of bone tissue, sometimes called spongy bone, is especially susceptible to loss of density with age, poor nutritional status, and hormonal imbalance.

Bone density increases through childhood and reaches its peak during young adulthood. Bone cells are continually replaced in a process called bone remodeling. If more bone tissue is broken down than replaced, then bone density is lost. The following are important for the development of optimal bone density:

- Good nutrition: Vitamins A, D, and K and the minerals calcium, phosphorus, and fluoride play a part in building healthy bones.
- Physical activity: Weight-bearing activity is especially important in stressing the bones to absorb more bone-building minerals and protein.
- Hormonal balance: Estrogen in women and testosterone in men stimulate bone retention.

Young women with poor calcium intake who engage in little exercise are at risk for the development of osteoporosis later in life. Conversely, as covered in Chapter 1, people who are active and maintain sufficient muscle mass have some protection against osteoporosis. However, even active women may develop osteoporosis if they exercise so much or become so lean that they stop menstruating due to reduced estrogen production.

Ligaments and Cartilage

The point where two or more bones come together is called a **joint**. Examples of joints are the shoulder, elbow, wrist, knee, hip, and ankle. **Ligaments** hold bones together and protect the joints. When ligaments are subjected to great stress, they can become inflamed or tear.

Cartilage is a type of connective tissue that covers the ends of the bones at a joint. It absorbs the pressure between the bones and is so strong that it will return to its original shape even after being severely stressed. Over time, however, cartilage degenerates and becomes the source of pain and stiffness in the joints.

Skeletal Muscle Organization

The three types of muscle found in the human body are skeletal, smooth, and cardiac. All muscle tissue is capable of contracting when it is stimulated by the nervous system. You have already read about the contraction of cardiac muscle (the heartbeat) and smooth muscle (vasoconstriction and vasodilation). The contraction of the third type, skeletal muscle, is needed to maintain posture, move the body, and perform fine-motor activities like writing and painting.

Skeletal muscle is more abundant than cardiac or smooth muscle and comprises between 30% and 45% of body weight in a healthy adult. The more than 400 skeletal muscles in the human body are attached to bones by **tendons** and are held together by several layers of connective tissue. Within each muscle are hundreds to thousands of muscle cells called fibers. **Muscle fibers** are as long as muscles, up to 2.5 feet in length in some cases. Each muscle fiber contains many **myofibrils**, which are contractile proteins made up of various protein filaments. The filaments that appear dark under a microscope are rich in the contractile protein myosin; those that appear light contain more of the protein actin. The alternating bands of light and dark filaments give skeletal muscle its striated appearance. The arrangement of skeletal muscle components is illustrated in Figure 7-3.

Muscle fibers are connected to the nervous system by neurons. Sensory neurons transmit information from the muscles to the central nervous system. Motor neurons carry information from the central nervous system to the muscles. Imagine yourself wading through a stream when your bare foot encounters a sharp rock. The sensory nerves in your foot detect the rock and quickly send this information to the brain, which then transmits impulses through the appropriate motor neurons causing you to jerk your foot up and out of the water.

tendon *Strong, inelastic fibrous tissue that attaches muscle to bone.*

muscle fiber *A muscle cell.*

myofibrils *long strands of contractile protein found in muscle fibers. Myosin and actin are two protein filaments found in myofibrils.*

Muscle Fibers

In Chapter 3, two broad categories of muscle fibers were introduced: **slow twitch (ST)** and **fast twitch (FT)**. ST, or type I, fibers contract relatively slowly, are more aerobic, and tire slowly; FT, or type II, fibers contract rapidly, are more anaerobic, and tire quickly.

Table 7-1 describes some of the differences between ST and FT muscle fibers. Notice that there are actually two categories of FT fibers—fast oxidative glycolytic (FOG) and fast glycolytic (FG). The FOG fibers can behave either aerobically or anaerobically, depending on energy needs of the moment.

Within a single muscle, both types of fibers will typically be found. However, all of the ST muscle fibers will be innervated by ST motor neurons, all of the FT/FOG muscle fibers will be innervated by FT and FOG motor neurons, and all of the FT/FG muscle fibers will be innervated by FT and FG motor neurons.

Most people have about equal numbers of ST and FT fibers. Elite athletes often have an advantageous fiber distribution in specific muscles. For example, some distance runners are made up of 60–90% ST muscle fibers, especially in the muscles of the legs. Up to 75% of muscle fibers of sprinters may be the FT type. Whatever your fiber make-up, you were born that way. Training can enhance the glycolytic or oxidative capacity of each fiber type, but training cannot convert one fiber type to another.

slow-twitch muscle fibers (ST; type I) *Category of muscle fiber characterized by a high aerobic capacity.*

fast-twitch muscle fibers (FT; type II) *Category of muscle fiber characterized by a high anaerobic capacity.*

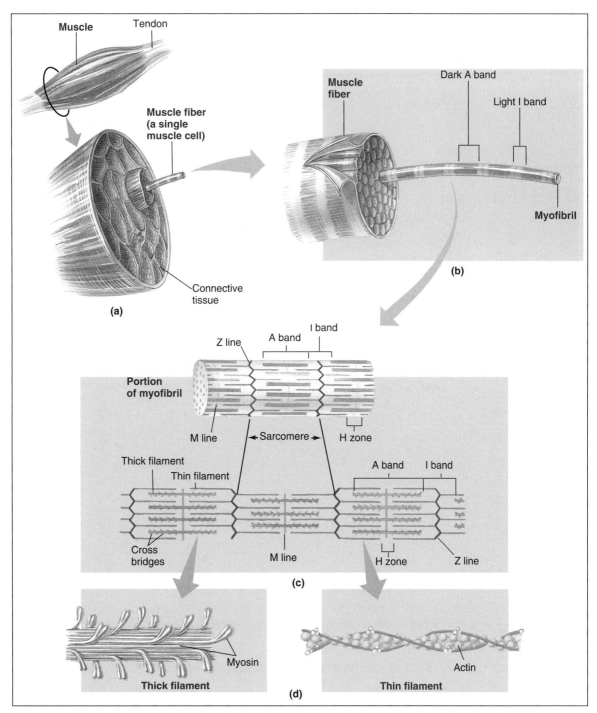

FIGURE 7-3 Organization of skeletal muscle. (a) Cross-section of a whole muscle. (b) Enlargement of one myofibril within a single muscle fiber. (c) Section of a myofibril illustrating the arrangement of actin and myosin filaments. (d) Enlargement of myosin (thick) filament and actin (thin) filament.

TABLE 7-1 Characteristics of Muscle Fibers

| | | Fast twitch (Type II) | |
	Slow twitch (Type I)*	Fast oxidative glycolytic (FOG) (Type IIa)	Fast glycolytic (FG) (Type IIb)
Speed of contraction	Slow	Fast	Fast
Resistance to fatigue	High	Intermediate	Low
Glycolytic enzyme activity	Low	High	High
Capacity for anaerobic metabolism	Low	High	High
Oxidative enzyme activity	High	High	Low
Mitochondria	Many	Many	Few
Capacity for aerobic metabolism	High	High	Low
Color	Red	Red	White
Fiber diameter	Small	Large	Large

*Also called slow oxidative (SO).

dynamic contraction

Muscle contraction in which the muscle changes in length and movement occurs. The concentric phase of a dynamic contraction occurs when the muscle shortens. The eccentric phase occurs as the muscle lengthens.

static contraction

Muscle contraction in which no movement occurs.

isokinetic contraction

Dynamic muscle contraction in which the speed of muscle movement is held constant through the full range of joint motion through the use of special rehabilitation equipment.

Skeletal Muscle Contraction

When a skeletal muscle contracts, the actin and myosin filaments slide across each other along the whole length of the muscle. A muscle contraction in which the muscle changes in length is a **dynamic contraction**. Locomotion, lifting an object above your head, and opening and closing your hand are examples of dynamic muscle contractions. Most dynamic contractions have two phases—**concentric** and **eccentric**. A concentric contraction occurs when the muscle shortens. An eccentric contraction occurs as the muscle lengthens. Consider moving a pile of books from your chair to your desk. When you lift the books, your biceps muscles are contracting concentrically. As you lower the books to the desk, your biceps muscles are still contracting, but this time the contraction is lengthening, or eccentric.

Sometimes a muscle contracts without generating any movement. A diver poised with arms held overhead and a truck driver straining to lift a box that will not move are both demonstrating **static contractions**. There is also a third type of muscle contraction, the **isokinetic contraction**—a dynamic contraction that occurs when the speed of muscle movement is held constant through the full range of joint motion. Special rehabilitation equipment is used for isokinetic muscle contractions.

Body Systems Involved in Physical Activity: Summary

The cardiorespiratory system (blood, heart, blood vessels, and lungs) delivers oxygen and nutrients, which support increasing levels of physical activity. The musculoskeletal system (bones, ligaments, cartilage, tendons, and muscles) provides structure, protection, and movement.

BENEFITS OF AN ACTIVE LIFESTYLE

Early humans have been estimated to consume about 3,000 kcal/day and to expend about 1,000 kcal/day in the physical activities needed to support their survival.[2] Sedentary humans today may consume the same 3,000 kcal/day yet expend only 300 kcal in those non-exercise activities needed to exist in an industrial society. Recall from Chapter 3 that this is not enough to prevent weight gain and certainly not enough to gain the health benefits of physical activity described in this section of the chapter.

One of the best things a person can do to enhance well-being is to become more active. Since 1995 the Centers for Disease Control and Prevention (CDC), American College of Sports Medicine (ACSM), Surgeon General, and National Institutes of Health (NIH) have recommended at least 30 minutes of moderate intensity activity on most if not all days of the week. People who engage in regular exercise or who incorporate physical activity into their daily routines can gain important health benefits: weight loss or maintenance; increased muscle mass; improved mental health; reduced risk of cardiovascular disease, diabetes, and some cancers; less reliance on the health care system; postponement of disability; and even a longer lifespan. The benefits of exercise are summarized in Table 7-2.

What Cardiorespiratory Improvements Result from Activity?

Many experts consider the improvements in cardiorespiratory health that occur among active people to be the most significant benefits of activity. People who engage in regular

TABLE 7-2 Summary of Exercise Benefits

Cardiorespiratory

- Stronger heart muscle
- Improved coronary blood flow
- Lower systolic blood pressure
- Increased cardiorespiratory endurance (VO_{2max})
- Improved respiratory efficiency
- Healthier blood lipid profile
- Increased blood volume

Musculoskeletal

- Increased muscular strength and endurance
- Increased muscle mass
- Increased bone density
- Improved flexibility

Other benefits

- Reduced abdominal fat
- Improved glucose tolerance
- Improved mental health
- Longer life

exercise, particularly vigorous exercise, derive the most benefit, but even individuals who simply try to be more active can gain cardiorespiratory health improvements.

Changes in the Heart, Blood Vessels, and Lungs

Regular physical activity and exercise training cause some notable changes in the heart. Highly trained athletes often experience increased size of the left ventricle, producing an enlarged heart. While some diseases are also associated with a larger than typical heart, in this case it is a healthy adaptation to exercise that helps produce a more efficient cardiovascular system. Even recreational exercisers, whose hearts do not noticeably enlarge, benefit from stronger myocardial muscle fibers. This strength promotes a sometimes impressive increase in SV, which is accompanied by decreases in both resting and exercise HRs. Review the calculation of Q in the preceding section again. What if the resting SV were 80 ml/beat? To circulate the same 4.9 l of blood would require a resting HR of only about 62 beats/minute.

Exercise also improves coronary blood flow—the circulatory system of the heart itself. Because the blood supply to the myocardium may be enhanced, the regular exerciser is less likely to suffer a heart attack during exertion and more likely to survive a heart attack if one occurs.

Resting systolic blood pressure also tends to be lower in exercisers. The reason is that exercise promotes vasodilation and reduces peripheral resistance. Even a moderate program of regular walking can help reduce systolic blood pressure in obese people with hypertension, whether or not weight loss occurs.[3]

Finally, exercise training may improve lung volume and the capacity to move oxygen from the alveoli into the arterial blood quickly. For elite athletes, these improvements can provide significant performance advantages. Recreational exercisers might not notice much difference, unless they have lung disease, but breathing efficiency does make physical activity easier.

Increased Blood Volume

One of the earliest adaptations to regular aerobic exercise training is increased blood volume. Elite endurance athletes may have 20–25% greater blood volumes (mainly plasma) than people who do not exercise. A higher plasma volume reduces the likelihood of dehydration when exercising strenuously or in a hot environment. In addition, oxygen transport is enhanced.

**cardiorespira-
tory endurance**
*Ability of the heart,
the lungs, and the
blood to circulate
sufficient oxygen and
nutrients to sustain
prolonged large-
muscle activity.*

**maximal oxygen
consumption
(VO$_{2max}$)** *Highest
capacity to take in
and use oxygen dur-
ing aerobic activity;
best indicator of
cardiorespiratory
endurance.*

Increased Cardiorespiratory Endurance

Cardiorespiratory endurance is the ability of the heart, the lungs, and the blood to circulate sufficient oxygen and nutrients to sustain prolonged large-muscle activity. The best indicator of cardiorespiratory endurance is **maximal oxygen consumption (VO$_{2max}$),** the highest capacity to take in and use oxygen during aerobic activity. A high VO$_{2max}$ indicates a strong cardiovascular system, increased capacity to carry out daily activities without discomfort, and reduced risk of death from cardiovascular disease.

Aging and inactivity are the leading causes of declines in VO$_{2max}$. The good news is that physical activities, such as walking, running, swimming, and bicycling, increase VO$_{2max}$ even among formerly sedentary older adults. Active obese people may have better cardiorespiratory endurance than sedentary lean people. In one study, previously sedentary obese individuals who walked regularly, participated in some supervised fitness activities, and engaged in recreational activities, like skiing and dancing, lost a small amount of weight (an average of 2.9 kg in men and 1.8 kg in women) over 10 months, but increased their VO_{2max} by 19%.[4] Obese women with low fitness even improved their VO_{2max} following a moderate walking program.[3]

Improved Plasma Lipids

Elevated plasma lipids is a major risk factor for the development of cardiovascular disease, but it can be favorably modified through physical activity. Both intensive physical training and more moderate forms of exercise for periods of 3–14 months have been consistently demonstrated to improve plasma lipoprotein profiles and lower total cholesterol in obese women and men. In addition, people who exercise oxidize fewer low-density lipoprotein (LDL) particles (the bad cholesterol), which reduces the likelihood that oxidized LDLs will be incorporated into atherosclerotic plaque.[4]

Meanwhile, the good cholesterol (HDL) levels generally are higher in lean than overweight individuals. As Chapter 5 pointed out, HDL levels sometimes drop when people on low-fat diets reduce their dietary cholesterol intake. Exercise can counteract this drop and, particularly when weight is lost, increase HDLs.[5]

What causes the beneficial effects of exercise on blood lipids? One theory is that when people follow a prudent diet (less than 30% of calories from fat) while also burning more lipids during intermittent periods of physical activity, they create an optimal state of lipid metabolism. With or without weight loss, this state prevents elevated total cholesterol, triglycerides, and LDL particles, and enhances production of HDL particles from rapidly broken-down chylomicrons and very low-density lipoproteins.

What Musculoskeletal Improvements Result from Activity?

Many changes in the cardiorespiratory system are "invisible," but musculoskeletal improvements are not. Formerly inactive people who engage in strength training are usually pleasantly surprised by obvious gains in strength, balance, and muscle tone.

Improved Muscular Fitness

muscular strength *Ability of the muscles to exert forceful, maximal contractions.*

muscular endurance *Capacity of the muscles for repeated movement or to sustain a prolonged contraction in one position.*

hypertrophy *Increased muscle fiber size resulting from greater protein uptake after strength training.*

Muscular fitness encompasses both strength and endurance. **Muscular strength** is the ability of the muscles to exert forceful, maximal contractions. **Muscular endurance** is the muscles' capacity for repeated movement or to sustain a prolonged contraction in one position, such as a gymnast's ability to hold a position for several seconds.

Strength is developed through high-intensity resistance training. Weight lifting is one example of resistance training that builds strength. Endurance is developed through lower-intensity repetitive activities, like running and calisthenics. The two elements of fitness are related in that improvements in one are usually accompanied by improvements in the other.

When muscles are subjected to repeated stress, such as during weight training, muscle fibers take up more protein and become enlarged, a phenomenon known as **hypertrophy**. The muscles become stronger and more resistant to injury. Hypertrophy also enhances the exerciser's functional capacity:

- The ability to rise from a chair or get out of the bathtub, preventing the need for assisted living

- The distance walked without tiring, allowing participation in activities that develop cardiorespiratory endurance
- The capacity to support more stress around a joint, helping to cope with arthritis of the lower extremities
- Retention of balance and coordination, preventing falls and debilitating injuries

Individuals who engage in mostly endurance training, such as distance running, may experience hypertrophy primarily in the ST muscle fibers. More intense activities, such as weight training, generally build greater muscle size, which occurs through hypertrophy of the FT muscle fibers.

Strength improves as muscles get larger. Neuromuscular factors also improve strength. Muscles that are trained recruit more motor units faster and, therefore, generate more powerful contractions. Strength improvements of 25–30% in middle-aged adults doing resistance training are typical.[6] Older adults, who lose considerable strength due to aging, may improve even more dramatically.

Increased Bone Density

Bone strength is also enhanced by physical activity. Exercise is a key component in the prevention and treatment of osteoporosis because it helps to maintain, and even build, bone. Under the mechanical stress of exercise, bone formation outpaces bone loss. Weight-bearing aerobic activities and strength training are excellent bone-builders. Even walking about 1 mile each day will maintain or increase bone density.

In addition, people who lose weight may benefit from the bone-preserving effects of physical activity. Diet-induced weight loss is associated with bone losses ranging from 4.0% to 6.9%, whereas exercise-induced weight loss decreases bone mass by 1% or less.[7]

Improved Flexibility

flexibility Range of motion available at a joint.

Flexibility refers to the range of motion available at a joint. Flexibility is determined by the shape of the joint as well as by the muscles, tendons, and ligaments that cross the joint. Some extremely flexible people say that they are double-jointed. They don't really have two joints at one point on the body, but they do have very loose connections holding the joint together, so they can often assume contortionist-like positions.

People don't need to be as flexible as contortionists, but lack of flexibility can increase the risk of muscle injury, limit the ability to engage in activities of daily living, and produce poor posture. Although poor flexibility is unlikely to shorten your life, attaining good flexibility can have many health-related benefits:

- Increased respiratory capacity because the chest can expand to its fullest potential
- Reduced frequency of low back and neck pain
- Lessened chance of injury during exercise and while carrying out activities of daily living

Golgi tendon organs Receptors in tendons that promote muscle relaxation and, after training, contribute to flexibility.

Active people are more flexible than inactive people, and those who engage in stretching exercises for specific areas of the body are the most flexible. Stretching elongates the muscles and the tendons, keeping them elastic. It also activates receptors in the tendons, called **Golgi tendon organs**, which promote muscle relaxation. Continued stimulation of Golgi tendon organs through stretching allows greater elongation of the muscles without injury.

How Does Physical Activity Affect Body Fatness?

Chapter 3 introduced the concept that nonresting energy expenditure—the energy expended in exercise and daily physical activity—contributes to maintenance of a healthy body weight. Chapter 4 stressed that too much body fat can be harmful to health, particularly when fat is stored primarily in the abdominal fat cells. Upper-body obesity is far more likely than lower-body obesity to be associated with elevated blood lipids, hypertension, and type 2 diabetes—all risk factors for cardiovascular disease.

Fortunately, internal visceral abdominal fat is very responsive to exercise. People who regularly engage in vigorous leisure activities tend to have smaller abdominal adipocytes and, therefore, reduced cardiovascular disease risk.[3,8]

Activity is generally more effective in reducing the size of the abdominal fat deposits than the gluteofemoral fat. One probable reason is that abdominal adipocytes have higher sensitivity to catecholamines, which are released during activity. In addition, for reasons as yet unknown, physical activity may decrease lipoprotein lipase levels in selective areas, which would allow the fat cells in those areas to mobilize fat rather than to store it.

Age, gender, level of fatness, and heredity affect the capacity of activity to reduce abdominal fat:

* People who have higher central fat stores before they become active generally lose the most fat during exercise.
* Because men tend to store more fat centrally than women, men respond better than women to the fat-burning properties of physical activity.
* The few studies on the effects of exercise on reducing abdominal fat in children suggest that activity has a positive effect on central fat in children, too.[9]
* Because research findings vary, genetic factors are no doubt involved in regulating the capacity of physical activity to affect fat distribution.

There is an interaction between abdominal fat and other cardiovascular risk factors. Insulin-resistant, obese individuals who deposit fat abdominally are most likely to have elevated cholesterol and triglyceride levels. When these individuals become more active, abdominal fat shrinks, insulin sensitivity improves, and harmful blood lipid levels drop. Gluteofemoral fat cells also shrink under the influence of exercise, just not as much as internal abdominal cells. And, for fat cells in any area, exercise prevents further filling, assuming that caloric intake and output are in balance.

What Are Some Other Health Benefits of Physical Activity?

In addition to improved fitness and functional capacity, active people derive other benefits from physical activity that reduce the risk of chronic disease and premature death.

Improved Glucose Tolerance

Even when minimal or no weight is lost, physical activity and exercise improve insulin sensitivity and glucose tolerance. Every time we move about, and especially when we exercise, we use glycogen, which must be replaced by circulating glucose. This process alone helps to normalize the blood glucose level and improve insulin sensitivity, possibly also due to increased liver uptake of insulin as a result of exercise. Activity level is directly related to glucose tolerance, so obese type 2 diabetics who engage in higher levels of activity generally show more improvements in glucose tolerance and glycosylated hemoglobin, a marker of long-term glucose control. Even nonvigorous physical activity—as long as it is ongoing—improves insulin sensitivity.[10]

Improved Mental Health

The physiological and psychological effects of physical activity can enhance mental health by lowering stress, improving mood, reducing anxiety, and lessening depression—even in clinically depressed people. The effect of physical activity on mental health has a biological basis. Researchers theorize that brain serotonin and β-endorphin levels rise during aerobic activities and strength training, which makes people feel better and perhaps even sleep better. In addition, being active has psychological effects, such as gaining a sense of mastery or control over one's life; focusing attention away from weight or guilt about eating; improving self-esteem as fitness increases; working off anger and hostility; and enjoying food without worrying about weight gain. Ample evidence exists to conclude that physical activity enhances mental health.

Reduced Mortality

Active people may add an average of 2–7 years to the length of their lives. The Harvard Alumni Study is a long-term investigation of chronic disease risk that has involved nearly 20,000 Harvard alumni followed since the 1960s. Several published reports from the study note that men who expend more energy every week have lower death rates at all levels of body mass index.[11] Vigorous exercise conveys the greatest longevity.[12,13]

Other studies of men and women support these conclusions. Researchers from the Institute for Aerobics Research in Dallas followed more than 30,000 men and women from 1970 to 1989. All were classified by fitness level on the basis of a maximal exercise test. Moderately fit men and women had lower death rates than low-fitness individuals, and those with the highest fitness had the lowest mortality rates, whether they were obese or nonobese.[14] A longitudinal study of more than 21,000 men at the Institute confirmed that lean, unfit men had twice the death rates from all causes as obese, fit men, and fit lean and fit obese men had similar mortality rates.[15]

Lower-intensity activities like walking can offer longevity benefits, too. Nonsmoking, retired men of Japanese ancestry in the Honolulu Heart Study lived longer when they walked more. After 12 years, 21% of men who walked more than 2 miles/day had died versus 43% of men who walked less than 1 mile daily.[16]

Benefits of an Active Lifestyle: Summary

Physical activity improves several aspects of physical and psychological functioning, including improved cardiorespiratory endurance, strength, and flexibility. In addition, metabolic effects—like improved blood lipids, normalized blood glucose, and less abdominal fat—lower the risk of disease and may even extend life.

Case Study. Married and Thinking About Children, Part I	Michael and Kari are a married couple who have become sedentary and gained weight since their wedding 5 years ago. They have recently been talking about starting a family. Big fans of reality television shows, they were inspired after watching morbidly obese people on one show lose significant amounts of weight and gain impressive levels of fitness after working with exercise trainers. "Let's join that health club that keeps sending us coupons," Kari said to Michael one evening. "We can work with a personal trainer and get buff again. And I'll be healthier when I get pregnant."

Michael is 32 years old, 5′ 11″ tall, and weighs 195 lb. At a recent health fair, his blood pressure was measured as 141/82. Kari is 30 years old, 5′ 5″ tall, and weighs 175 lb. Her blood pressure is 130/68.

- Determine Michael's and Kari's BMIs. In what health classification does their BMI place each of them.
- Discuss some general health benefits that Michael and Kari could expect from an exercise program.
- What are health benefits that would be particularly important to Kari in preparation for pregnancy?

BENEFITS OF AN ACTIVE LIFESTYLE THROUGHOUT THE LIFESPAN

Physical activity has value for people of all ages, fitness levels, and physical conditions. There are three specific points in the lifespan when activity and exercise can have great benefits: childhood, pregnancy, and late adulthood.

What Are Exercise Benefits in Childhood?

Although the diagnosis of chronic diseases associated with inactivity, such as cardiovascular disease and type 2 diabetes, most commonly occurs in adulthood, the foundations of chronic disease are built in childhood. Chronic disease risk factors begin to accumulate at a young age, and autopsy studies of young children have found pre-atherosclerotic fatty streaks in the arteries.[17]

Like adults, children can improve cardiorespiratory fitness through endurance training, strength through strength training, and flexibility through stretching exercises. Other exercise benefits that have been observed in children include:

- Improved sports performance
- Prevention of injury
- Reductions in blood pressure in hypertensive children
- Improved body composition
- Development of exercise habits that may carry over into adulthood[18]

Unfortunately, American children, like adults, have become more sedentary in recent years. Although children have a natural inclination toward physical activity, low-energy pursuits like watching television, working on computers, and playing video games are attractive alternatives. Parents and caregivers need to encourage, model, and facilitate physical activity in children.

Appropriate Exercise Regimens for Children

Active, informal play is ideal for children because it involves many different muscle groups, promotes socialization, and, most important, is fun. When children participate in a variety of physical activities and games, they are less likely to become bored or to suffer from repetitive or overuse injuries.

Children can also benefit from more formal exercise programs, such as running, swimming, and strength training. When it comes to exercise, however, children are not miniature adults. Their exercise regimens must be tailored to their physical immaturity.

Children have a lower Q than adults at all workloads. Therefore, endurance training usually needs to be at a lower intensity for a child than an adult. Children may not be able to keep up with adult runners, swimmers, or skiers without becoming dangerously overtired. In addition, children have a lower blood volume and perspire less than adults, so they are more likely to experience heat illness. Special precautions must be taken for children participating in endurance activities in a warm environment.

Pre-pubertal boys and girls of any age will not increase muscle size as a result of resistance training, but they can improve strength. An appropriate strength-training program for children is closely supervised and prohibits maximal lifts. These precautions prevent injury, particularly to the **growth plates** of the bones. The growth plates, or epiphyseal plates, are areas at the ends of the long bones where growth in length and width occurs. Because epiphyseal plates do not ossify or mineralize until puberty (or later in some cases), excessive stress to this area could cause injury, malformation, or even stunted growth. Maximal lifts, repetitive motions like pitching or stress from long-distance running, could adversely affect the growth plate. The American Academy of Pediatrics (AAP) recommends low- to moderate-intensity resistance training for the development of strength in pre-pubertal children and avoidance of maximal and competitive weight lifting, power lifting, and body building.[19] Before participating in strength training, children should have achieved balance and postural control (usually by ages 7 to 8 years).

Although overweight children may look big, they are not necessarily strong. They should follow the same precautions for resistance training as non-overweight children. Children and adolescents with cardiomyopathy, pulmonary hypertension, or Marfan syndrome in which the elastic fibers of the aorta are fragmented should not participate in resistance training.

growth plates (epiphyseal plates) *Areas at the ends of the long bones where growth in length and width occurs. When epiphyseal plates ossify, peak height and bone length has been achieved.*

What Are Exercise Benefits During Pregnancy?

Between 30–40% of women in Western countries exercise during pregnancy.[20, 21] Recent research has found that women who are active before pregnancy and remain active during pregnancy have a lower risk of pre-eclampsia.[22] Additional benefits of exercise during pregnancy include:

- Prevention of excessive maternal weight gain, which is especially important for overweight women
- Improvement of physical fitness, which can help with postural stability and make the later months of pregnancy more comfortable
- In some women, promotion of an easier, shorter labor and reduced likelihood of cesarean delivery
- Improvement in sleep

Appropriate Exercise Regimens During Pregnancy

Women who were physically active before becoming pregnant can generally continue to be active during pregnancy. Expending up to 2,000 kcal/week in exercise has not been found to be harmful to mother or fetus.[20] This would be equivalent to walking about an hour each day or doing weight training for an hour five times a week.

However, too much exercise can prevent appropriate maternal weight gain, retard growth of the fetus, and even subject the fetus to excessively high temperatures. Some modification of the pre-pregnancy exercise regimen may be necessary to protect fetal and maternal health. Women who were sedentary before becoming pregnant can benefit from exercise but should start slowly—by walking at a modest pace, for example—and work up to participation in supervised exercise.

Most mild to moderate aerobic activities are safe and enjoyable during pregnancy. Low-impact aerobic dance, brisk walking, swimming, or using a stair-stepper or exercise cycle for 30–40 minutes three to five times a week are excellent activities with no adverse effects that might harm the mother or the developing fetus.[23] Even strength training is acceptable for women who were engaged in this activity before pregnancy. The only modifications advised are avoiding heavy resistance and an emphasis on correct breathing while lifting weights. (Proper breathing during exercise is discussed in Chapter 8). For safety, exercise machines are preferable to free weights. Because exercise machines limit range of motion, they are less likely than free weights to cause muscle or joint injuries.

Stretching exercises are also appropriate during pregnancy. Hormonal changes during pregnancy make the tissues around the pelvis very pliable in preparation for childbirth, so any stretching should be passive and mild.

As pregnancy progresses, once-enjoyable exercises may become less so. Pedaling an exercise cycle could become mechanically impossible, and a changing sense of balance could make an aerobics class too difficult. Walking and swimming are usually well tolerated throughout pregnancy.

Exercise Precautions

While mild to moderate exercise is both safe and effective for most pregnant women, any woman with a high-risk pregnancy should avoid formal exercise programs. For example, women who develop high blood pressure or diabetes during pregnancy are at risk of even more serious conditions that may harm them or the fetus. Activities more strenuous than walking may not be advised. Women with bleeding from the fourth month onward or who have had preterm labor with this or any pregnancy should not exercise. In addition,

Swimming is an excellent form of exercise during pregnancy.

Sergey Chirkov, 2010/Shutterstock.com

women carrying multiple fetuses must use caution to avoid activities that might compromise the blood and nutrient supply to more than one baby or induce premature labor.

All pregnant women should observe the following exercise precautions:

- Follow an exercise schedule (every other day, for example), rather than exercising intermittently.
- Keep well hydrated so body temperature does not rise during exercise.
- Check HR frequently and maintain an exercise HR that does not exceed 150 beats/ minute.
- After the fourth month of pregnancy, avoid doing exercises lying flat on the back.
- Avoid any activities where abdominal trauma might occur, such as softball, karate, or team sports.
- Don't overexert. Any activity that causes fatigue should be avoided for the entire pregnancy.

Women who have any concerns about the health and safety of exercise during pregnancy should be urged to speak with their physician.

Exercise During Lactation

Exercise can also be beneficial to breast-feeding women. Milk production increases in lactating women who exercise, energy expenditure increases, and body fat decreases more than in breast-feeding women who do not exercise.[24] At least 500 additional kcal/day are needed to support exercise and milk production.

What Are Exercise Benefits for Older Adults?

The growing population of adults age 65 and older has the potential to gain significant benefits through increased physical activity. Chronic disease, functional impairment, and disability need not occur with increasing age and, in many instances, may be prevented or postponed by adopting a higher level of activity. Other benefits of physical activity for older adults include:

- Reduced risk factors for chronic disease, such as improved glucose tolerance and insulin sensitivity, normalized blood lipid levels, reduced intrahepatic fat content, and reduced blood pressure
- Lower incidence of strokes
- Reduced risk of premature death
- Less need for medications, many of which have side effects
- Increased bone density
- Increased range of motion of the joints used in activities of daily living
- Fewer falls, which is a significant benefit for the frail elderly, among whom falls are a leading cause of fatal injury
- Improved strength (researchers estimate that FT muscle fiber size decreases by about 30% between ages 20 and 80 and fiber number declines dramatically after age 50, but resistance training can halt this)[25]
- Reductions in body fat (one study found older men and women who engaged in strength training 3 days/week for 12 weeks gained 3 lb of muscle and lost 4 lb of fat while eating slightly more each day)[26]

- Improved mobility, even among older adults who already had mobility disabilities.[27]
- Modest effects on depression

The benefits of exercise may be less obvious for older than younger adults, especially for individuals who cannot exercise above light intensity. For example, indicators of fitness like body composition, muscular strength, and VO_{2max} do not measurably improve without a stimulus of moderate to high intensity. Still, the health effects of physical activity accrue, even without significant fitness gains.

Barriers to Involving Older Adults in Physical Activity

More so than for people of younger ages, there are considerable barriers to getting older adults more active. Older adults may have health problems and may never have been encouraged to exercise. At this point in their lives, they may not see a value to becoming active. Health professionals, especially physicians, do their patients a great service by encouraging even small increments of activity.

People who were never encouraged to exercise may have misconceptions about it. Some may equate the term exercise with a specific activity like jogging or with high-intensity activity. Older adults may simply not know how to get started, or may lack social support to do so. Exercise programs for older adults must include components of education, support, and safe, appropriate activity.

Appropriate Exercise Regimens for Older Adults

According to the CDC, about 80% of older adults have one chronic condition and half have two or more.[28] The most common chronic diseases reported in older adults—osteoarthritis, hypertension, heart disease, and type 2 diabetes mellitus—affect exercise capacity.

The optimal program includes elements of strength, endurance, and flexibility.

- Strength training: Supervised resistance training may need to be the first physical activity introduced to an older client. This is because of the loss of muscle mass that typically accompanies aging in the inactive person. Some older adults will not be able to start a walking program without first building leg strength. Although resistance equipment builds strength most rapidly, elastic exercise bands and even light ankle and hand weights can be used in the home for resistance training.
- Endurance activities: Large-muscle aerobic activities, such as walking, swimming, and cycling, are excellent activities for older adults. Because SV usually declines with age, the intensity of these activities should be reduced and may be gradually increased as exercise capacity improves. Individuals with osteoarthritis, especially those who are obese, must avoid vigorous activities that put excess stress on the joints. Many communities have walking clubs at shopping malls, and senior citizen centers often sponsor low-impact activities like water aerobics.
- Flexibility training: Mild stretching exercises can be performed almost anywhere and contribute to ease of joint movement. When flexibility exercises are combined with strength training, gait and balance may improve. Yoga is an excellent activity that combines elements of strength, endurance, and flexibility and that can be enjoyed throughout the lifespan.

Exercise Precautions

Any exercise regimen should avoid aggravating existing diseases or conditions and prevent injury. Strategies for providing a safe exercise program include:

- Obtaining medical clearance from the individual's physician before beginning an exercise program. (This topic is discussed further in Chapter 8.)
- Having water available to ensure that participants are well hydrated.
- Walking on sidewalks, preferably on streets with less automobile traffic. (When possible, outdoor walking should take place in daylight hours.)
- Wearing good-fitting shoes with cotton sport socks to avoid blisters that might become infected in the person with poor circulation.
- Avoiding exercise when the person is ill or, in the case of people with arthritis, during acute episodes of joint inflammation.
- Providing competent supervision for group exercise programs, particularly for people with dementia.

Benefits of an Active Lifestyle Throughout the Lifespan: Summary

Physical activity provides benefits to people in all life stage groups. There is no "one size fits all" for exercise. A safe and appropriate exercise regimen needs to be considered for the young, old, and pregnant.

Case Study. Married and Thinking About Children, Part 2	While waiting to see a fitness professional at the health club, Kari picked up some brochures on exercise programs during pregnancy. "I thought that women should take it easy when they're pregnant," she said to Michael, "but it looks like I can keep exercising." • What is the American Congress of Obstetricians and Gynecologists position statement on exercise during pregnancy? • What kinds of exercise would be safe during pregnancy? • What are appropriate exercise precautions that Kari should observe during her pregnancy?

PHYSICAL FITNESS ASSESSMENT

physical fitness
Ability to carry out everyday tasks without undue fatigue and with energy left over to enjoy leisure activities and meet unforeseen emergencies.

Physical activity contributes to improved physical fitness at all ages. **Physical fitness** is the ability to carry out everyday tasks without undue fatigue and with energy left over to enjoy leisure activities and meet unforeseen emergencies.[29] It includes sports-skill related fitness (balance, agility, power, and coordination) and aspects of fitness that contribute to health—cardiorespiratory endurance, body composition, muscular strength and endurance, and flexibility—collectively known as **health-related physical fitness**. A third aspect of physical fitness is **physiologic fitness**, which includes metabolic fitness, the attainment of metabolic improvements, such as normalized blood glucose level, reduced abdominal fat, and healthy bone mineral density, that lower the risk of chronic disease.

health-related physical fitness *Aspects of fitness that contribute to health—cardio-respiratory endurance, body composition, muscular strength and endurance, flexibility, and metabolic fitness.*

physiologic fitness *Attainment of metabolic improvements, such as normalized blood glucose level and reduced abdominal fat, that lower the risk of chronic disease.*

Why Assess Physical Fitness?

Physical fitness assessment has three broad purposes for the client and the professional: (1) To provide baseline information about physical fitness status, (2) to assist in devising realistic goals for change and developing an activity program that can meet those goals, and (3) to document changes that occur as a result of increasing physical activity. You have already learned about the use of nutrition assessment in weight-management programs. The purposes of these two types of assessment are very similar.

A comprehensive physical fitness assessment measures all of the elements of health-related physical fitness: cardiorespiratory endurance, body composition, muscular strength and endurance, and flexibility. Several techniques for assessing each are presented here. The nutritionist or exercise professional can decide which is most appropriate for each client. Assessment of physiologic fitness includes several measures presented in Chapter 2, such as waist circumference and blood glucose.

If several fitness tests are to be administered in one session, the ACSM suggests the following test sequence to ensure that participating in one test does not adversely affect the results of another test:[30]

1. Any resting measurements (HR, blood pressure)
2. Body composition
3. Cardiorespiratory endurance
4. Muscular fitness
5. Flexibility

To prepare for fitness testing, the client should be instructed to drink plenty of water during the day before testing; to refrain from eating, smoking cigarettes, or using caffeine or alcohol for 3 hours before testing; to come to the testing session well rested; and to avoid strenuous exercise on the test day. During fitness testing, the client should wear comfortable, non-constricting clothing. All individuals should be screened for health problems before taking a fitness test. Chapter 8 provides a simple medical clearance form called the PAR-Q that can be used for both exercise and fitness testing clearance.

How Is Cardiorespiratory Fitness Assessed?

Measures of cardiorespiratory fitness range from the simple to the sophisticated. Blood pressure and resting HR are always obtained first, followed by determination or estimation of VO_{2max}.

sphygmoman-ometer *Device used to measure blood pressure, which consists of an inflatable arm band, a bulb and valve to regulate air pressure in the cuff, and a mercury or aneroid manometer.*

Measurement of Blood Pressure and Heart Rate

Blood pressure and HR are two very basic indicators of the status of the cardiovascular system. Blood pressure is measured with a device called a **sphygmomanometer**, shown in Figure 7-4. With the client seated comfortably, an inflatable cuff is wrapped around the upper arm and aligned with the brachial artery. The cuff is inflated until the pressure gauge on the sphygmomanometer is at about 200 mm Hg. A stethoscope is placed over the brachial artery near the bend of the arm, just below the blood pressure cuff. Air is gradually released from the cuff at the rate of 2–3 mm Hg/second as the technician listens for the beginning of heart sounds. The first sound heard, as the pressure of blood in the cardiovascular system overcomes pressure in the cuff, signifies the systolic

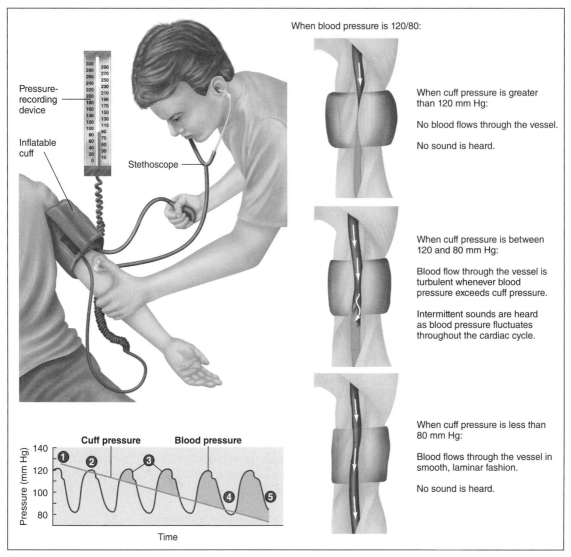

When blood pressure is 120/80:

Pressure-recording device

Inflatable cuff

Stethoscope

When cuff pressure is greater than 120 mm Hg:

No blood flows through the vessel.

No sound is heard.

When cuff pressure is between 120 and 80 mm Hg:

Blood flow through the vessel is turbulent whenever blood pressure exceeds cuff pressure.

Intermittent sounds are heard as blood pressure fluctuates throughout the cardiac cycle.

When cuff pressure is less than 80 mm Hg:

Blood flows through the vessel in smooth, laminar fashion.

No sound is heard.

Cuff pressure   Blood pressure

Pressure (mm Hg)

Time

FIGURE 7-4 Blood pressure is measured using a sphygmomanometer.

blood pressure. While air continues to be released from the cuff, additional heart sounds are heard. The point where no further sounds are detected is the diastolic blood pressure.

To minimize the risk of measurement error, proper cuff fit is important. Large cuffs are available for individuals with big upper arms. Smaller cuffs are available for children and adults with small arms.

Table 7-3 provides blood pressure classifications from the National Heart, Lung, and Blood Institute. Hypertension is indicated by a systolic blood pressure that is 140 mm Hg or above or a diastolic pressure that is 90 mm Hg or above. Cardiorespiratory endurance and muscular fitness testing should be avoided in individuals whose systolic blood pressure exceeds 200 mm Hg or whose diastolic blood pressure is above 115 mm Hg. These

TABLE 7-3 Classification of Blood Pressure for Adults

Blood pressure classification	Systolic blood pressure (mm Hg)	Diastolic blood pressure (mm Hg)
Normal	<120	and < 80
Prehypertension	120–139	or 80–89
Stage I Hypertension	140–159	or 90–99
Stage 2 Hypertension	≥160	or ≥100

Source: National Heart, Lung, and Blood Institute. (2004). The seventh report of the Joint National Committee on Prevention, Detection, Evaluation, and Treatment of High Blood Pressure (JNC7). Bethesda, MD: National Institutes of Health. Available from: http://www.nhlbi.nih.gov/guidelines/hypertension. Accessed March 8, 2011.

individuals should be referred to their physician for medical clearance before testing the cardiorespiratory or musculoskeletal systems.

HR is measured at either the radial or the carotid pulse. The radial artery is found on the thumb side of the inner wrist. Either wrist can be used for the measurement. The carotid arteries are located just under the jawline, either to the right or to the left of the midline of the neck. The fingertips are placed over the artery and gentle pressure is exerted. The beats are counted for 15 seconds and then multiplied by four to calculate HR in beats per minute.

An irregular or racing pulse might be indicative of cardiovascular disease but, because HR varies so much from person to person, there is no "normal" HR. Measuring exercise HR is necessary for one of the field tests of cardiorespiratory endurance described below; in Chapter 8, readers will use resting HR to develop exercise prescriptions.

Measurement of VO_{2max}

The most accurate way to determine cardiovascular endurance is to measure VO_{2max}. The measurement of VO_{2max} is usually performed in a laboratory or a clinic and rarely in a classroom or a nonclinical setting. This is typically done using open-circuit spirometry. The individual in the following photograph is inspiring (inhaling) room air and expiring (exhaling) through a mouthpiece into a gas-sampling system. He/she is wearing a nose clip to prevent air from entering or escaping through the nose. Although the individual in the photograph is on a motorized treadmill, a cycle ergometer may also be used.

© Philip Gould/CORBIS

Open-circuit spirometry with a treadmill or cycle ergometer is used to measure VO_{2max}.

In a true VO_{2max} test, the individual is subjected to increasing levels of difficulty by changing the grade of the treadmill or increasing the resistance of the cycle ergometer. Initially, the individual responds to increases in exercise difficulty by inspiring more oxygen and expiring more carbon dioxide. The point at which oxygen intake levels off, despite increased exercise intensity, is the VO_{2max}.

Most of the time, VO_{2max} is not directly measured but is estimated from a submaximal exercise test. The individual pedals a cycle ergometer or walks on a motorized treadmill at a designated workload for several minutes. HR is measured at specific intervals. Workload may be increased once or twice during the test until a predetermined HR is achieved. Because there is a direct relationship between HR and oxygen consumption, VO_{2max} can be calculated from HR during exercise.

Field Tests of Cardiorespiratory Endurance

Several simple field tests of cardiorespiratory endurance have also been developed. One of these, the Rockport One-Mile Fitness Walking Test, requires no special equipment or facilities—only a stopwatch, a good pair of walking shoes, and a 1-mile walking course—although use of a pulse rate monitor is desirable. A quarter-mile track works very well for this test. Individuals walk 1 mile as quickly as possible without stopping. Heart rate is obtained during the final minute of the walk; use of a pulse rate monitor makes this much easier than trying to find and count a pulse. Best results are obtained when people are motivated enough to maintain a brisk pace for the duration of the test. Some researchers have found the Rockport Test to overestimate VO_{2max} in low to moderately fit people and to underestimate it in very fit people.[31] Still, this test is appropriate for people of low to average fitness or in situations in which an accurate measure of VO_{2max} is not needed.

VO_{2max} is expressed as milliliters of oxygen consumed per kilogram of body weight per minute (ml/kg/min). This equation is calculated mathematically from the gas samples obtained during a maximal exercise test, from HR in a submaximal test, or from several variables in the Rockport test. The formulas in Figure 7-5 are used to calculate VO_{2max} from the One-Mile Walk Test results. Comparing the test results with Tables 7-4 and 7-5 gives an indicator of cardiorespiratory fitness level.

$VO_{2max} = 132.853 - (0.0769 \times WT) - (0.3877 \times AGE) + (6.3150 \times SEX) - (3.2649 \times T)$
$- (0.1565 \times HR)$

WT = Weight in pounds

AGE = Age in years

SEX = 0 for female; 1 for male

T = Walk time in minutes and seconds, to the nearest tenth of a minute (seconds divided by 60 = tenths of a minute)

HR = heart rate in beats/minute, taken in the final minute of the walk

---

FIGURE 7-5 VO_{2max} can be calculated using One-Mile Walk Test results.

Source: Kline, G. M., Porcari, J. P., Hintermeister, R., Freedson, P. S., Ward, A., McCarron, R. F., et al. (1987). Estimation of VO_{2max} from a one-mile track walk, gender, age, and body weight. *Medicine and Science in Sports and Exercise, 19*(3), 253.

TABLE 7-4 Estimation of Cardiorespiratory Fitness from VO_{2max} (ml/kg/min) for Men

Age (years)	Superior ≥ 95th percentile	Excellent 80th–90th percentile	Good 60th–75th percentile	Fair 40th–55th percentile	Poor 20th–35th percentile	Very Poor ≤ 15th percentile
20–29	55.5–60.5	51.1–54.0	45.6–48.5	41.7–44.8	38.0–41.0	26.5–36.7
30–39	54.1–58.3	48.3–51.7	44.1–47.0	40.7–43.9	36.7–39.5	26.5–35.2
40–49	52.5–56.1	46.4–49.6	42.4–44.9	38.4–41.0	34.8–37.6	25.1–33.8
50–59	49.0–54.0	43.3–46.8	39.0–41.8	35.5–38.1	32.0–34.8	22.8–30.9
60–69	45.7–51.1	39.6–42.7	35.6–38.3	32.3–34.9	28.7–31.6	19.7–27.3
70–79	43.9–49.6	36.7–39.5	32.4–35.2	29.4–31.6	25.7–28.4	18.2–24.6

Source: Data reprinted with permission from The Cooper Institute, Dallas, Texas, from a book called *Physical Fitness Assessments and Norms for Adults and Law Enforcement,* pp. 27–29. Available from: http://www.cooperinstitute.org. Accessed March 8, 2011.

TABLE 7-5 Estimation of Cardiorespiratory Fitness from VO_{2max} (ml/kg/min) for Women

Age (years)	Superior ≥ 95th percentile	Excellent 80th–90th percentile	Good 60th–75th percentile	Fair 40th–55th percentile	Poor 20th–35th percentile	Very Poor ≤ 15th percentile
20–29	49.6–54.5	43.9–46.8	39.5–42.4	36.1–38.5	32.3–35.2	23.7–30.9
30–39	47.4–52.0	42.4–45.3	37.7–41.0	34.2–36.9	30.9–33.8	22.9–29.4
40–49	45.3–51.1	39.6–43.1	35.9–38.6	32.8–35.2	29.4–32.3	22.2–28.2
50–59	41.0–46.1	36.7–38.8	32.6–35.2	29.9–32.3	26.8–29.4	20.1–25.8
60–69	37.8–42.4	32.7–35.9	29.7–32.3	27.3–29.4	24.6–26.6	19.5–23.9
70–79	37.2–42.4	30.6–32.5	28.1–29.8	25.9–28.0	23.5–25.3	16.8–22.2

Source: Data reprinted with permission from The Cooper Institute, Dallas, Texas, from a book called *Physical Fitness Assessments and Norms for Adults and Law Enforcement,* pp. 34–36. Available from: http://www.cooperinstitute.org. Accessed March 8, 2011.

How Is Body Composition Assessed?

body composition
Proportion of body weight composed of fat and lean tissue. Obesity, an excess of body fat, occurs when total fat makes up 32% or more of body weight in women and 25% in men.

Body composition refers to the makeup of the body, specifically the amount of fat and lean tissue. Ideally, the greatest proportion of body weight is composed of nonfat tissue—water, muscle, and bone. Still, for good health, each individual requires a certain minimal level of fat, called essential fat. Females need more essential fat than males (12% of body weight versus 3% in males), primarily to support reproductive functions. Storage fat is fat accumulated over and above physiological need. Although storage fat is not required for physiological functions, it will support health during times of famine and is harmful only when accumulated in excessive amounts.

For women, an excess of body fat—obesity—occurs when total fat makes up 32% or more of body weight. For men, 25% fat is considered to be unhealthy. As the proportion of body fat increases, so does the risk of cardiovascular disease, diabetes, certain cancers, sleep apnea, and other health problems, as presented in Chapter 1.

Although many obese individuals are also overweight, weight is not a good indicator of an unhealthy body composition. Total body weight consists of fat, bone, water, muscle, and other organs and tissues. Overweight individuals may have an excess of any or all of these components. Some overweight people have excessive body fat, but others do not. Bodybuilders, for example, carry a lot of muscle weight and may weigh more than their ideal body weight, but they are not usually overfat or even unhealthy.

Body composition is most accurately measured by techniques that specifically target fat rather than weight. Skinfold measurements, bioelectrical impedance analysis, and underwater weighing are among such assessment techniques. Less accurate but still useful in body composition determination are circumference measures and the body mass index. Finally, height-weight tables may be used, but they are subject to much misinterpretation. (Refer to Chapter 2 for details about assessing body composition.)

How Is Muscular Fitness Assessed?

absolute strength *Maximal amount of weight that can be lifted.*

relative strength *Maximal weight lifted divided by body weight; also called the strength-to-weight ratio.*

Tests of muscular fitness can measure either relative or absolute levels of strength and endurance. An individual's **absolute strength** is indicated by the maximum amount of weight he or she can lift. Because women have less muscle mass than men, their absolute strength is usually 60–85% that of men. When **relative strength** is calculated, however, women and men may have similar levels of strength. Relative strength, sometimes called the strength-to-weight ratio, is calculated by dividing weight lifted by body weight.

Both muscular strength and endurance are specific to each muscle group. Ideally, several tests of strength and endurance will be performed, using different muscles, to obtain an accurate picture of muscular fitness. The bench press is a measure of strength often used in a fitness center or a clinical setting.

Upper Body Strength Test: Bench Press

The bench press is a good test of upper body strength. It may be done using a seated press, a bench press machine, or a free weight. For the seated press, the arm handles should be directly in front of the chest, and the individual should push the handles forward as the arms are fully extended. For the bench press machine or free weight, the arm

David Madison/Stone/Getty Images

The bench press is one tool for measuring upper-body strength.

handles or bar should be just in front of the shoulders, and the individual should lift the handles up as the arms are fully extended (without locking the elbows). To prevent back injury, the individual should keep the back as flat as possible on the bench during the lift.

The weight that the individual can lift with one maximal contraction is determined by trial and error, starting with very low weights for less fit people. The individual should warm up by performing lifts with a lower amount of weight than will be used for the test. During each lift, the individual must exhale to prevent a rise in blood pressure.

Tables 7-6 and 7-7 provide estimates of muscular strength based on test results. To use the tables, divide the weight lifted by the individual's body weight using either pounds or kilograms.

The bench press test is not suitable for the frail elderly, people recovering from a recent heart attack, or anyone with musculoskeletal problems that could be aggravated by weight lifting.

Field Tests of Strength

When resistance-training equipment is not available, a simple field test of strength is the sit-and-rise test.[32] The individual stands with the knees slightly bent, back pressed

TABLE 7-6 Estimation of Muscular Strength from the Bench Press (strength-to-weight ratio weight lifted/body weight) for Men

Age (years)	Superior ≥ 95th percentile	Excellent 80th–90th percentile	Good 60th–75th percentile	Fair 40th–55th percentile	Poor 20th–35th percentile	Very Poor ≤ 15th percentile
20–29	≥1.63	1.32–1.48	1.14–1.26	0.99–1.10	0.88–0.96	≤ 0.84
30–39	≥1.35	1.12–1.24	0.98–1.08	0.88–0.96	0.78–0.86	≤0.75
40–49	≥1.20	1.00–1.10	0.88–0.96	0.80–0.86	0.72–0.78	≤0.69
50–59	≥1.05	0.90–0.97	0.79–0.87	0.71–0.77	0.63–0.70	≤0.60
60 +	≥0.94	0.82–0.89	0.72–0.79	0.66–0.70	0.57–0.65	≤0.56

Source: Reprinted with permission from The Cooper Institute, Dallas, Texas, from a book called *Physical Fitness Assessments and Norms for Adults and Law Enforcement*, p. 33. Available from: http://www.cooperinstitute.org. Accessed March 8, 2011.

TABLE 7-7 Estimation of Muscular Strength from the Bench Press (strength-to-weight ratio = weight lifted/body weight) for Women

Age (years)	Superior ≥ 95th percentile	Excellent 80th–90th percentile	Good 60th–75th percentile	Fair 40th–55th percentile	Poor 20th–35th percentile	Very Poor ≤ 15th percentile
20–29	≥ 1.01	0.80–0.90	0.70–0.77	0.59–0.68	0.51–0.58	≤0.50
30–39	≥0.82	0.70–0.76	0.60–0.65	0.53–0.58	0.47–0.52	≤0.45
40–49	≥0.77	0.62–0.71	0.54–0.60	0.50–0.53	0.43–0.48	≤0.42
50–59	≥0.68	0.55–0.61	0.48–0.53	0.44–0.47	0.39–0.43	≤0.38
60 +	≥0.72	0.54–0.64	0.47–0.53	0.43–0.46	0.38–0.41	≤0.36

Source: Reprinted with permission from The Cooper Institute, Dallas, Texas, from a book called *Physical Fitness Assessments and Norms for Adults and Law Enforcement*, p. 41. Available from: http://www.cooperinstitute.org. Accessed March 8, 2011.

Spencer Grant/PhotoEdit

A grip dynamometer is used to get an overall assessment of strength.

against a door or a wall with nonskid surface. She slowly lowers her body as far as possible—but not so far that the knees flex below 90°. This test approximates functional strength and can be used to assess progress after beginning an exercise program.

A **grip dynamometer** can also be used to get an overall assessment of strength. This portable device, shown in the photo above, measures static strength. The individual being assessed squeezes the device as hard as possible to obtain a measure of strength in kilograms of force exerted. Some evidence suggests that hand grip strength is related to the strength of other muscles. In one study of more than 3,000 healthy men between 45 and 68 years of age, grip strength predicted functional limitations 25 years later.[33]

grip dynamometer
Device used to measure hand grip strength in kilograms of force exerted.

Muscular Endurance Tests

Muscular endurance is measured by having the individual perform as many repetitions of a particular exercise as possible. Common tests are pull-ups or push-ups to estimate upper body endurance and sit-ups or curl-ups to estimate abdominal muscle endurance. In a typical test, the individual performs as many repetitions of the exercise as possible in the time allowed (usually 1 minute). Young, healthy, fit people rarely have difficulty with these tests. Older adults with lower levels of fitness, musculoskeletal disorders, or frailty should be evaluated on a case-by-case basis before undergoing muscular endurance testing.

The most common muscular endurance tests are administered as follows:

- Push-up test for males: The individual positions himself on hands and toes, with the hands pointing forward and shoulder-width apart and the back straight. He lowers himself almost to the floor and then pushes back up to a straight-arm position as many times as possible in 1 minute. Use Table 7-8 to interpret push-up test results.
- Modified push-up test for females: The individual positions herself on hands and knees, with hands shoulder-width apart, ankles in plantar flexion, knees bent at 90°, and back straight. She lowers herself almost to the floor and then pushes back up to a straight-arm position as many times as possible in 1 minute. Use Table 7-9 to interpret modified push-up test results.

The push-up test has been modified for women, who tend to have less upper-body strength and endurance than men.

TABLE 7-8 Estimation of Muscular Endurance from the Push-Up for Men

Age (years)	Superior ≥ 95th percentile	Excellent 80th–90th percentile	Good 60th–75th percentile	Fair 40th–55th percentile	Poor 20th–35th percentile	Very poor ≤ 15th percentile
20–29	≥ 62	47–57	37–44	29–35	22–27	≤ 19
30–39	≥ 52	39–46	30–36	24–29	17–21	≤ 15
40–49	≥ 40	30–36	24–29	18–22	11–16	≤ 10
50–59	≥ 39	25–30	19–24	13–17	9–11	≤ 7
60 +	≥ 28	23–26	18–22	10–16	6–9	≤ 5

Source: Reprinted with permission from The Cooper Institute, Dallas, Texas, from a book called *Physical Fitness Assessments and Norms for Adults and Law Enforcement*, p. 32. Available from: http://www.cooperinstitute.org. Accessed March 8, 2011.

TABLE 7-9 Estimation of Muscular Endurance from the Modified Push-Up for Women

Age (years)	Superior ≥ 95th percentile	Excellent 80th–90th percentile	Good 60th–75th percentile	Fair 40th–55th percentile	Poor 20th–35th percentile	Very poor ≤ 15th percentile
20–29	≥ 45	36–42	30–34	23–29	17–22	≤ 15
30–39	≥ 39	31–36	24–29	19–23	11–17	≤ 9
40–49	≥ 33	24–28	18–21	13–17	6–11	≤ 4
50–59	≥ 28	21–25	17–20	12–15	6–10	≤ 4
60 +	≥ 20	15–17	12–15	5–12	2–4	≤ 1

Source: Reprinted with permission from The Cooper Institute, Dallas, Texas, from a book called *Physical Fitness Assessments and Norms for Adults and Law Enforcement*, p. 39. Available from: http://www.cooperinstitute.org. Accessed March 8, 2011.

- Curl-up: The curl-up test, also called a crunch, is considered to be less likely to cause muscle soreness or injury than the traditional sit-up test. Before testing, lay two strips of tape about 4 inches apart across a mat. The individual lies on his back on the mat with the arms extended at the sides and the palms down so that the fingertips touch the tape mark closest to the shoulders. The knees should be bent at least 90° so the feet are flat on the floor or the mat (feet must not be held or anchored). To perform a

TABLE 7-10 Estimation of Muscular Endurance from Curl-Up Test, Men (number of curl-ups performed in one minute)

Age (years)	Excellent	Very good	Good	Fair	Needs improvement
15–19	25	23–24	21–22	16–20	≤ 15
20–29	25	21–24	16–20	11–15	≤ 10
30–39	25	18–24	15–17	11–14	≤ 10
40–49	25	18–24	13–17	6–12	≤ 5
50–59	25	17–24	11–16	8–10	≤ 7
60–69	25	16–24	11–15	6–10	≤ 5

Source: Canadian Physical Activity, Fitness & Lifestyle Approach: CSEP-Health & Fitness Program's Appraisal and Counselling Strategy, 3rd edition, © 2003. Reprinted with permission from the Canadian Society for Exercise Physiology.

TABLE 7-11 Estimation of Muscular Endurance from Curl-Up Test, Women (number of curl-ups performed in one minute)

Age (years)	Excellent	Very good	Good	Fair	Needs improvement
15–19	25	22–24	17–21	12–16	≤ 11
20–29	25	18–24	14–17	5–13	≤ 4
30–39	25	19–24	10–18	6–9	≤ 5
40–49	25	19–24	11–18	4–10	≤ 3
50–59	25	19–24	10–18	6–9	≤ 5
60–69	25	17–24	8–16	3–7	≤ 2

Source: Canadian Physical Activity, Fitness & Lifestyle Approach: CSEP-Health & Fitness Program's Appraisal and Counselling Strategy, 3rd edition, © 2003. Reprinted with permission from the Canadian Society for Exercise Physiology.

curl-up, the individual curls his head and shoulders forward, sliding the palms along the mat until they touch the tape mark closest to the feet. He then returns to the beginning position such that fingertips touch the beginning tape mark. The individual performs as many crunches as possible, to a maximum of 25, in one minute. The test administrator can help the person by setting a metronome at 50 beats per minute and instructing the client to curl up on one beat, down on the next. The test is stopped when the individual cannot keep up the test pace. Use Tables 7-10 and 7-11 for interpreting curl-up test results.

How Is Flexibility Assessed?

No single test gives an overall assessment of flexibility, because flexibility, like muscular strength and endurance, is specific to each joint.

Goniometer Measurements

goniometer

Instrument used to measure range of motion at a joint.

Very accurate measures of joint flexibility can be obtained with a **goniometer**. Using a goniometer is similar to using a protractor to determine the size of an angle. When measuring joint flexibility, the center of the goniometer is placed at a joint, such as the elbow. The stationary arm of the goniometer is placed on the portion of the limb that will remain stable during the measurement. For the elbow joint, this would be along the upper arm. The movable arm of the goniometer is placed on the portion of the limb that will move—in this example, the forearm. Elbow joint range of motion is measured by first extending the forearm as far as possible and then flexing it as much as possible. The difference between the two measurements is the range of motion of the elbow joint.

Field Tests of Flexibility

The most widely used field test of flexibility is the sit-and-reach test. This test measures flexibility of the hamstring muscles, which are on the back of the thighs. For many years, the sit-and-reach test was also thought to be a good measure of lower back flexibility and, therefore, an indicator of the tendency to develop low back pain or a lower back injury. Today, few exercise physiologists believe that the sit-and-reach test is a valid measure of lower back flexibility.

The sit-and-reach test is most easily performed using a sit-and-reach box. After warming up the hamstring muscles, the client sits with her back against a wall and the bottom of the feet, without shoes, pressed against the box. With the knees extended, the individual slowly leans forward, pushing the hands along the measuring tape or the yardstick affixed to the top of the box. Dropping the head between the arms while reaching improves the sit-and-reach score. The score is the farthest point reached with the fingertips. The best of three trials represents the sit-and-reach score. Use Tables 7-12 and 7-13 to interpret sit-and-reach test results. These results are based on the feet being placed at the 10-inch mark on the measuring tape.

TABLE 7-12 Estimation of Hamstring Flexibility from Sit and Reach Test, Men (in inches)

Age (years)	Excellent	Very good	Good	Fair	Needs improvement
15–19	≥ 39	34–38	29–33	24–28	≤ 23
20–29	≥ 40	34–39	30–33	25–29	≤ 24
30–39	≥ 38	33–37	28–32	23–27	≤ 22
40–49	≥ 35	29–34	24–28	18–23	≤ 17
50–59	≥ 35	28–34	24–27	16–23	≤ 15
60–69	≥ 33	25–32	20–24	15–19	≤ 14

Source: Canadian Physical Activity, Fitness & Lifestyle Approach: CSEP-Health & Fitness Program's Appraisal and Counselling Strategy, 3rd edition, © 2003. Reprinted with permission from the Canadian Society for Exercise Physiology.

Note: the feet are placed at the 10-inch mark on the measuring tape

TABLE 7-13   Estimation of Hamstring Flexibility from Sit and Reach Test, Women (in inches)

Age (years)	Excellent	Very good	Good	Fair	Needs improvement
15–19	≥ 43	38–42	34–37	29–33	≤ 28
20–29	≥ 41	37–40	33–36	28–32	≤ 27
30–39	≥ 41	36–40	32–35	27–31	≤ 26
40–49	≥ 38	34–37	30–33	25–29	≤ 24
50–59	≥ 39	33–38	30–32	25–29	≤ 24
60–69	≥ 35	31–34	27–30	23–26	≤ 22

Source: Canadian Physical Activity, Fitness & Lifestyle Approach: CSEP-Health & Fitness Program's Appraisal and Counselling Strategy, 3rd edition, © 2003. Reprinted with permission from the Canadian Society for Exercise Physiology.

Note: the feet are placed at the 10-inch mark on the measuring tape

Physical Fitness Assessment: Summary

Measurement of health-related physical fitness provides clients with baseline information to set realistic goals for improvement and to document fitness gains. A thorough fitness assessment includes measures of cardiorespiratory endurance, body composition, muscular fitness, and flexibility.

Case Study. Married and Thinking About Children, Part 3

Lou, a fitness specialist at the health club, met with Michael and Kari and explained that fitness assessment always preceded development of an exercise regimen. "But we're healthy," Michael protested. "All we need is some help jump-starting an exercise program." Lou explained why fitness assessment was an important part of designing an exercise program, and Michael and Kari reluctantly complied. The following are results of those assessments.

Assessment results	Kari	Michael
Blood pressure	130/68	141/82
Resting heart rate	75 bpm	80 bpm
HR at completion of Rockport test	125 bpm	127
Time to completion of Rockport test	20 minutes	15 minutes
Estimated VO_{2max}		
Estimated cardiovascular fitness level		
Percent fat	30%	17%
Max. bench press	88 lb	179 lb
Push-ups completed	8	15
Curl-ups completed	15	8
Sit-and-Reach results	35 cm	22 cm

- Why would Lou insist on performing fitness assessments before designing an exercise program for Michael and Kari?
- You already calculated Kari's and Michael's BMI and determined each one's BMI classification. What additional information does their percent fat provide?

(Continued)

**Case Study.
(Continued)**

- Using the appropriate formula in Figure 7-5, estimate Kari's and Michael's VO$_{2max}$ and cardio-vascular fitness levels (Note that you will have to perform some additional calculations to use Tables 7-4 and 7-5).
- What do the rest of their assessment results tell you about their levels of muscular strength, muscular endurance, and flexibility.
- Given these results, what additional benefits might an exercise program offer them?

REFERENCES

1. U.S. Department of Health and Human Services. (2008). *2008 physical activity guidelines for Americans.* Washington, D.C.: U.S. Department of Health and Human Services.
2. Saris, W. H., Blair, S. N., van Baak, M. A., Eaton, S. B., Davies, P. S., Di Pietro, L., et al. (2003). How much physical activity is enough to prevent unhealthy weight gain? Outcome of the IASO 1st Stock conference and consensus statement. *Obesity Reviews, 4*(2), 101–114.
3. Roussel, M., Garnier, S., Lemoine, S., Gaubert, I., Charbonnier, L., Auneau, G., et al. (2009). Influence of a walking program on the metabolic risk profile of obese postmenopausal women. *Menopause, 16*(3), 566–575.
4. Vasankari, T. J., Kujala, U. M., Vasankari, T. M., & Ahotupa, M. (1998). Reduced oxidized LDL levels after a 10-month exercise program. *Medicine and Science in Sports and Exercise, 30*(10), 1496–1501.
5. Katzmarzyk, P. T., Leon, A. S., Rankinen, T., Gagnon, J., Skinner, J. S., Wilmore, J. H., et al. (2001). Changes in blood lipids consequent to aerobic exercise training related to changes in body fatness and aerobic fitness. *Metabolism: Clinical and Experimental, 50*(7), 841–848.
6. Pollock, M. L., Gaesser, G. A., Butcher, J. D., Despres, J. P., Dishman, R. K., Franklin, B. A., et al. (1998). The recommended quantity and quality of exercise for developing and maintaining cardiorespiratory and muscular fitness, and flexibility in healthy adults: American College of Sports Medicine position stand. *Medicine and Science in Sports and Exercise, 30*(6), 975–991.
7. Pritchard, J. E., Nowson, C. A., & Wark, J. D. (1996). Bone loss accompanying diet-induced or exercise-induced weight loss: A randomised controlled study. *International Journal of Obesity and Related Metabolic Disorders, 20*(6), 513–520.
8. You, T., Murphy, K. M., Lyles, M. F., Demons, J. L., Lenchik, L., & Nicklas, B. J. (2006). Addition of aerobic exercise to dietary weight loss preferentially reduces abdominal adipocyte size. *International Journal of Obesity, 30*(8), 1211–1216.
9. Suliga, E. (2009). Visceral adipose tissue in children and adolescents: A review. *Nutrition Research Reviews, 22*(2), 137–147.
10. Mayer-Davis, E. J., D'Agostino, R., Jr, Karter, A. J., Haffner, S. M., Rewers, M. J., Saad, M., et al. (1998). Intensity and amount of physical activity in relation to insulin sensitivity: The insulin resistance atherosclerosis study. *JAMA, 279*(9), 669–674.
11. Paffenbarger, R. S., Jr, Hyde, R. T., Wing, A. L., & Hsieh, C. C. (1986). Physical activity, all-cause mortality, and longevity of college alumni. *New England Journal of Medicine, 314*(10), 605–613.
12. Lee, I. M., Hsieh, C. C., & Paffenbarger, R. S. Jr. (1995). Exercise intensity and longevity in men: The Harvard alumni health study. *JAMA, 273*(15), 1179–1184.
13. Lee, I., & Paffenbarger, R. S., Jr. (2000). Associations of light, moderate, and vigorous intensity physical activity with longevity: The Harvard alumni health study. *American Journal of Epidemiology, 151*(3), 293–299.
14. Blair, S. N., Kampert, J. B., Kohl, H. W., 3rd, Barlow, C. E., Macera, C. A., Paffenbarger, R. S., Jr, et al. (1996). Influences of cardiorespiratory fitness and other precursors on cardiovascular disease and all-cause mortality in men and women. *JAMA, 276*(3), 205–210.
15. Lee, C. D., Blair, S. N., & Jackson, A. S. (1999). Cardiorespiratory fitness, body composition, and all-cause and cardiovascular disease mortality in men. *American Journal of Clinical Nutrition, 69*(3), 373–380.
16. Hakim, A. A., Petrovitch, H., Burchfiel, C. M., Ross, G. W., Rodriguez, B. L., White, L. R., et al. (1998). Effects of walking on mortality among nonsmoking retired men. *New England Journal of Medicine, 338*(2), 94–99.

17. Rowland, T. W. (1990). *Exercise and children's health.* Champaign, IL: Human Kinetics.

18. Raitakari, O. T., Porkka, K. V., Taimela, S., Telama, R., Rasanen, L., & Viikari, J. S. (1994). Effects of persistent physical activity and inactivity on coronary risk factors in children and young adults. The cardiovascular risk in young Finns study. *American Journal of Epidemiology, 140*(3), 195–205.

19. Council on Sports Medicine and Fitness, American Academy of Pediatrics. (2008). Strength training by children and adolescents. *Pediatrics, 121*(4), 835–840.

20. Hatch, M., Levin, B., Shu, X. O., & Susser, M. (1998). Maternal leisure-time exercise and timely delivery. *American Journal of Public Health, 88*(10), 1528–1533.

21. Juhl, M., Andersen, P. K., Olsen, J., Madsen, M., Jorgensen, T., Nohr, E. A., et al. (2008). Physical exercise during pregnancy and the risk of preterm birth: A study within the Danish national birth cohort. *American Journal of Epidemiology, 167*(7), 859–866.

22. Shaikh, H., Robinson, S., & Teoh, T. G. (2009). Management of maternal obesity prior to and during pregnancy. *Seminars in Fetal & Neonatal Medicine,* doi:10.1016/j.siny.2009.10.003.

23. McMurray, R. G., Hackney, A. C., Guion, W. K., & Katz, V. L. (1996). Metabolic and hormonal responses to low-impact aerobic dance during pregnancy. *Medicine and Science in Sports and Exercise, 28*(1), 41–46.

24. Lovelady, C. A., Lonnerdal, B., & Dewey, K. G. (1990). Lactation performance of exercising women. *American Journal of Clinical Nutrition, 52*(1), 103–109.

25. Wright, V. J., & Perricelli, B. C. (2008). Age-related rates of decline in performance among elite senior athletes. *American Journal of Sports Medicine, 36*(3), 443–450.

26. Campbell, W. W., Crim, M. C., Young, V. R., & Evans, W. J. (1994). Increased energy requirements and changes in body composition with resistance training in older adults. *American Journal of Clinical Nutrition, 60*(2), 167–175.

27. Manini, T. M., Newman, A. B., Fielding, R., Blair, S. N., Perri, M. G., Anton, S. D., et al. (2009). Effects of exercise on mobility in obese and nonobese older adults. *Obesity,* doi:10.1038/oby.2009.317.

28. CDC. (2009). *Healthy aging—improving and extending quality of life among older Americans: At a glance, 2010.* Available from http://www.cdc.gov/chronicdisease/resources/publications/AAG/aging.htm.. Accessed September 13, 2010.

29. U.S. Department of Health and Human Services. (1996). *Physical activity and health: A report of the Surgeon General.* Atlanta, GA: Centers for Disease Control and Prevention.

30. American College of Sports Medicine. (2010). *ACSM's guidelines for exercise testing and prescription* (8th ed.). Philadelphia: Lippincott Williams & Wilkins.

31. Kline, G. M., Porcari, J. P., Hintermeister, R., Freedson, P. S., Ward, A., McCarron, R. F., et al. (1987). Estimation of VO_{2max} from a one-mile track walk, gender, age, and body weight. *Medicine and Science in Sports and Exercise, 19*(3), 253–259.

32. Kligman, E. W., Hewitt, M. J., & Crowell, D. L. (1999). Recommending exercise to healthy older adults. *Physician and Sportsmedicine, 27*(11), 42–44, 49, 52–56, 61–62.

33. Rantanen, T., Guralnik, J. M., Foley, D., Masaki, K., Leveille, S., Curb, J. D., et al. (1999). Midlife hand grip strength as a predictor of old age disability. *JAMA, 281*(6), 558–560.

Physical Activity, Health, and Weight Management

R. Gino Santa Maria/Shutterstock.com

CHAPTER OUTLINE

cappi thompson/Shutterstock.com

Iorga Studio/Shutterstock.com

P hysical activity offers many benefits to the overweight/obese person—improved muscular fitness, reduced body fat stores (especially abdominal fat), prevention of further weight gain, and improved health. Both increased physical activity and participation in formal exercise programs can help people lose or maintain weight.

ROLE OF PHYSICAL ACTIVITY IN WEIGHT MANAGEMENT

Physical activity is an important component of weight management. Physical activity metabolizes stored fat, which reduces the size of the fat stores, while increasing lean body mass. Both of these effects help the body to achieve and/or maintain a normal metabolic rate. Weight loss cannot be predicted or even guaranteed as a result of physical activity, but the metabolic effects of physical activity are important for preventing further fat gain and, in most people, will promote at least some fat losses.

What Are the Fat-Burning Effects of Low- to Moderate-Intensity Activity?

Recall from Chapter 3 that most of the energy needed to support metabolism is generated from fats and carbohydrates in aerobic metabolic processes known as the citric acid cycle and electron transport. In the first minutes of moderate exercise, before aerobic metabolism can fully power up, carbohydrates, creatine phosphate, and stored ATP are used for fuel. After a few minutes, lipid oxidation progressively increases, gradually surpassing use of carbohydrate as an aerobic exercise fuel as fatty acids are released from adipocytes. Lipids make their greatest contribution to metabolism during exercise at about 65% of maximal oxygen consumption (VO_{2max}). Figure 8-1 illustrates that when exercise intensity exceeds 65% of VO_{2max} muscle glycogen use exceeds that of lipids.

People who engage in regular activity and improve their physical fitness use more lipids for fuel and begin burning fat earlier during exercise. The adipose tissue of trained individuals appears to become more sensitive to catecholamines, which are released during exercise. This advantage is lost when people stop exercising.

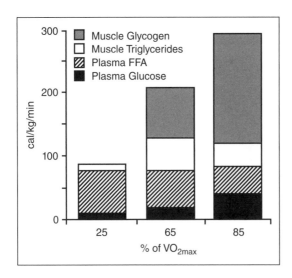

FIGURE 8-1 Both carbohydrate and fat contribute to energy expenditure during exercise. At 25% through 65% VO_{2max} free fatty acids (FFA) and muscle triglycerides make their greatest contribution.

Source: Adapted with permission from Romijn, J. A., Coyle, E. F., & Sidossis, L. S. (1993). Regulation of endogenous fat and carbohydrate metabolism in relation to exercise intensity and duration. *American Journal of Physiology, 265,* E380–E389.

Lipids are not just used as fuels during exercise; they also make a substantial contribution to energy expenditure after exercise through excess post-exercise oxygen consumption (EPOC), a concept introduced in Chapter 3. An important reason for fat-fueled EPOC is to allow carbohydrate to be used for rebuilding glycogen stores. One study that used a calorimetry chamber to measure nutrient oxidation during 3 hours of exercise and for several hours thereafter determined that lipid oxidation was still elevated 17 hours after exercise.[1]

What Are the Fat-Burning Effects of High-Intensity Activity?

More carbohydrate than fat is burned in activities that exceed 65% of VO_{2max}. However, because high-intensity exercise has a higher caloric cost than moderate-intensity exercise, even a lower proportion of fat use may expend substantial fat calories. Consider the example presented in Figure 8-2. A 65-kg person with a VO_{2max} of 40 ml/kg/minute exercises for 40 minutes at 50% of VO_{2max} one day and another day at 70% of VO_{2max}. Almost the same amount of fat is burned in each exercise, and the higher-intensity exercise uses over 100 more kcal. Fat can be lost through higher intensity exercise.

High-intensity exercise may have its greatest effect on adaptations that occur as a result of exercise:

skeletal muscle 3-hydroxyacyl CoA dehydrogenase (HADH) *Enzyme involved in beta-oxidation, in which fats are prepared for the citric acid cycle.*

- **Skeletal muscle 3-hydroxyacyl CoA dehydrogenase (HADH)** levels increase after high-intensity interval training but not after endurance training.[2] HADH is an enzyme involved in β-oxidation, the metabolic process that strips away carbons to prepare fats to enter the citric acid cycle.
- Norepinephrine in the bloodstream is increased as a result of higher-intensity exercise, such as strength training, and may be partly responsible for the shift to greater fat oxidation after exercise.
- Greater glycogen depletion after high-intensity exercise could cause even more fat to be oxidized in the hours following exercise, as carbohydrate is retained to restore glycogen.

Kcal expended in 40 minutes	261 kcal	366 kcal
	130.5 carbohydrate kcal	256 carbohydrate kcal
	130.5 fat kcal	110 fat kcal
Exercise intensity	50% VO_{2max} "moderate"	70% VO_{2max} "strenuous"
Examples of activities at this intensity	Brisk walking, tennis, swimming laps	Racquetball, handball, running 10-minute mile
	Moderate	**Strenuous**

FIGURE 8-2 Proportionately more fat is burned in moderate-intensity activities, but strenuous activities burn more total kilocalories.

- High and sustained EPOC has been reported following high-intensity exercise, which can significantly increase energy expenditure.

A note of caution is necessary. Many obese people, older adults, and people with health problems cannot safely engage in activities above a moderate level of intensity. The key to using physical activity for fat burning is to select a level of activity that is safe and well tolerated. The next sections of this chapter describe how to develop such a program.

What Are the Advantages of Preserving Lean Body Mass?

Because muscle tissue is a major contributor to resting metabolic rate and is responsible for muscular strength, endurance, and functional abilities, maintaining or increasing lean body mass is important for good health. Aging and inactivity cause losses of lean tissue, and these losses are accelerated in people who lose weight through calorie restriction. Lean tissue is always lost when calorie-restricted diets are followed, whereas fat accounts for the majority of weight lost through physical activity.[3]

weight-bearing activity Physical activity, such as walking and running, in which the individual supports his or her body weight on the legs.

Weight-bearing activity is needed to preserve muscle mass; resistance exercise, such as strength training, is needed to increase it. Because exercise increases or maintains lean tissue, exercisers often lose less weight than dieters (remember that muscle tissue is heavy due to its high water content). Exercisers almost always lose more fat, however. People who maintain a higher percentage of lean body mass after weight loss have a tremendous advantage—they seem to be better at maintaining lost weight.[4]

The kind of exercise may have different effects on lean tissue. One example, provided by a study of overweight women who expended 300 kcal on a treadmill at either moderate intensity or high intensity four times a week for 12 weeks, shows why focusing on "weight" should be avoided. Both groups lost 5 lb of fat, but the high-intensity exercisers gained 4.3 pounds of lean tissue, so their actual weight loss was only 0.7 lb.[5] Although weight loss may be smaller than expected as a result of exercise, both fitness and functional capacity will increase. In addition, because lean tissue is more compact than fat, even without significant weight loss, there may be a decrease in girth measurements.[6] When waist girth decreases, not only does health risk drop but possibly clothing size, too.

How Does Activity Affect Appetite?

Some individuals are afraid that if they become more active they will want to eat more and the calorie-burning effects of exercise could be neutralized. There is a scientific basis for this fear. Think back to Chapter 4 where you learned about homeostasis. If homeostatic mechanisms like set point do exist—and there is evidence that they do—then an increase in physical activity could upset the balance, perhaps driving the individual to eat more or to engage in less non-exercise energy expenditure (NEAT).

Here is what is known about food intake and physical activity, based on several recent reviews:[7–9]

- Very active individuals tend to be leaner than those who are sedentary, yet they sometimes eat a great deal.

- Preference for dietary carbohydrate is increased in some people who exercise, dietary fat in others.
- Exercise may suppress hunger over the short-term and, in some studies, reduces energy intake among regular exercisers a small but significant amount over the long-term.
- High-intensity physical activity suppresses food consumption to a better degree than low-intensity physical activity.
- There is a great deal of individual variation in appetite-response to physical activity.

orexigenic
Promoting food intake.

Hormones involved in regulating hunger and satiety are undoubtedly implicated in the hunger and appetite responses to physical activity. Ghrelin, an **orexigenic** hormone that was covered in Chapter 4, has been studied in this regard. Recall that ghrelin levels increase during periods of food restriction (including between meals) and decrease after eating. Researchers have found that ghrelin levels increase as weight is lost after exercise programs lasting from 12 weeks to a year.[10, 11] There may be gender differences in hormonal response to exercise that explain why men and women often lose different amounts of weight after exercise. Overweight/obese women in one study had higher levels of ghrelin than overweight/obese men after a program of exercise training.[12] Thus, women may be more inclined to increase calorie consumption following exercise (perhaps restoring set point to defend fat stores needed for reproduction), while men do not.

Even if exercise does increase appetite in some people, weight *gain* will not occur as long as energy balance is maintained. Also, if people only partially compensate for increased activity by eating a bit more, the cost of exercise is still greater than the caloric value of food consumed, and weight loss can occur. This emphasizes two important factors for physically active people:

- Avoiding extreme caloric restriction
- Focusing on manageable dietary change as well as physical activity for weight management.

Does Exercise Prevent Weight Regain?

One of the most reliable predictors of long-term success in weight maintenance is physical activity. Whatever the method of weight loss, men and women who regularly engage in some form of exercise are more successful in preventing regain of lost weight. Even when weight is initially lost through severe caloric restriction (a method that is almost never effective for long-term weight loss), exercise can help prevent weight regain.

Higher amounts of activity are needed to prevent weight regain than the U.S. government recommends for health. Most government agencies and sports medicine groups recommend 30 minutes of moderate intensity activity every day (equivalent to about 150–200 kcal/day). Studies that focus on the amount of activity needed to prevent weight gain and regain suggest much higher caloric expenditure. The Institute of Medicine recommends 60 minutes of moderate intensity activity on most days.[13] Studies of successful weight-loss maintainers suggest that up to 90 minutes of moderate activity daily may be needed.[3,14] Even with personal trainers and high amounts of physical activity, most people who lose weight—whether in exercise or NEAT—regain some of it.[15] However, those who remain very active regain less weight than those who are either sedentary or only lightly active.

Role of Physical Activity in Weight Management: Summary

Physical activity is essential for weight management because it burns fat and increases muscle mass—two conditions that are metabolically favorable for weight loss, weight maintenance, or prevention of weight regain. The effects of exercise on appetite are variable, with some eating more and others not. But even among those individuals who partially compensate for increased physical activity by eating more, if energy out is greater than energy in, weight loss can occur.

DEVELOPING ACTIVITY PROGRAMS THAT WORK

As discussed in the previous section, most government agencies and fitness organizations recommend that every adult should accumulate at least 30 minutes of moderate-intensity physical activity on most—and preferably all—days of the week. These guidelines are appropriate for promoting health and fitness but may not be enough to prevent weight gain and probably are not sufficient to prevent weight regain in the previously obese. This section will explore the kinds of activities effective in preventing and treating overweight/obesity, as well as the quantity of activity recommended.

Which Types of Activity Are Most Effective for Weight Management?

Doing anything safely is better than doing nothing. This statement leaves little room for debate, but you should know that every activity does not have the same principal effect. Generally, aerobic activities are recommended to improve cardiorespiratory fitness, reduce body fat, and lower the risk of chronic disease, whereas anaerobic activities are recommended to increase lean body mass, improve strength, and develop sports-specific fitness.

Aerobic Activities

high-impact activities *Physical activity, such as running and jumping, in which the legs are subjected to intensive, repetitive contact with the exercise surface.*

Aerobic activities are large-muscle, continuous activities that comfortably raise the heart rate and promote endurance. Carbohydrates, fats, and proteins are the fuels used in aerobic activities. When pastimes like brisk walking, biking, jogging, and cross-country skiing are done for a prolonged period, substantial calories are expended. Any aerobic activity will improve cardiorespiratory fitness as long as it provides enough of an overload to stimulate the cardiovascular system. Even walking, an activity that most people can do every day, may reduce body fat, increase high-density lipoprotein cholesterol (HDL-C), and improve cardiorespiratory fitness. Moderate-intensity activities (discussed later) are well-tolerated by most people. **High-impact activities**, like running and jumping, are not recommended for previously sedentary or overweight/obese people because of their association with injuries. During any activity, fat is burned from deposits throughout the body; fat stores are not lost at a greater rate from deposits near the muscles involved in the exercise. So, there is no reason to choose activities that specifically work the fatter parts of the body.

Heavier people generally expend more energy in weight-bearing aerobic activities (for example, walking, jogging, and cross-country skiing) than in non-weight-bearing activities (such as cycling and swimming). However, weight and fat losses have been found to be approximately comparable in obese women who either walk or ride an exercise cycle. Swimming and other water-based exercise may not be the best activities for fat loss, but they improve cardiorespiratory fitness to the same degree as other aerobic activities.[16, 17]

Anaerobic Activities

Anaerobic activities are more intense (greater than 70% of VO_{2max}) and, therefore, cannot be engaged in for as long a time as aerobic activities. Short-distance cycling, running, or stair climbing at a high intensity are anaerobic activities, as are weight lifting and other forms of resistance training. Less fat is burned as exercise intensity increases, and, during maximal exercise, carbohydrates are the only nutrients that can be used for energy. But, because high-intensity exercise has a higher caloric cost than moderate intensity exercise, even a lower proportion of fat use may expend substantial fat calories (see Figure 8-2).

Although aerobic activities are most often recommended for weight management, there is some physiological basis for promoting more widespread use of anaerobic activities. High-intensity activities create greater EPOC than more moderate activities, and considerable fat oxidation occurs following anaerobic exercise. It is probably not a coincidence that people who regularly exercise at a high intensity are leaner than those who do not. High-intensity activities also have a higher caloric cost and may increase fitness more rapidly than lower-intensity activities. Resistance training has the added benefit of increasing lean tissue and, while it is not as effective in reducing body fat as aerobic activity, has been shown to prevent gains in intraabdominal fat.[18]

Most people cannot tolerate exercise at 70% VO_{2max}, although this would substantially increase energy expenditure. Even young, healthy people find high-intensity training to be difficult. An activity so intense that it injures or wears participants out for the rest of the day is of no value, even if it does burn considerable calories. Two types of physical activity that combine elements of aerobic and anaerobic activity and may be tolerated well are interval training and circuit training. They may be particularly effective in weight management and fitness.

Combining Aerobic and Anaerobic Activities

interval training
Type of exercise in which short periods of high-intensity activity are alternated with longer periods of low-intensity activity.

Interval training is a type of exercise that alternates shorter periods of more intense activity with longer periods of moderate activity. In a walking program, for instance, the individual who has gradually increased walking distance to 1 mile a day might try inserting a 15-second brisk pace every 5 minutes. Or a person using a stair-stepper might do two 30-second intervals of slightly more intense stepping in each 30-minute workout. To avoid injury, high-impact intervals (running, for example) should be avoided by sedentary, older, and heavier individuals.

Competitive athletes who want to accomplish a greater volume of training in a limited workout time have used interval training for years. The technique can also be adopted by people with low fitness to increase the vigor of exercise in small bursts, to expend more energy in the same amount of exercise time, and to avoid exercise boredom. Obviously, any higher-intensity intervals must be compatible with the individual's fitness level.

circuit weight training *Type of weight training in which the individual moves rapidly among weight stations, increasing not only muscular strength and endurance but also cardiorespiratory endurance.*

Circuit weight training is weight training in which no more than 30 seconds elapses between each exercise. This strategy not only increases strength and endurance, but it also improves VO_{2max}. One example of a circuit program uses ten weight resistance machines (squat press, leg curl, leg extension, chest press, seated row, shoulder press, pull down, triceps press, biceps curl, and abdominal crunch) and an exercise cycle. Individuals perform one resistance exercise for 60 seconds, then do 60 seconds on the exercise cycle, moving through the entire circuit of weight machines and alternating with the cycle.[6] So in the course of 25 minutes, participants do a substantial amount of training. Combining different forms of exercise in this manner promotes lipid oxidization (from the aerobic component) and increased fat-free mass (from the anaerobic component).

How Is the Appropriate Intensity of Exercise Determined?

Engaging in 30 minutes of moderate intensity activity every day promotes health, but what, exactly, is "moderate"? How do you know when a moderate intensity activity becomes vigorous?

The three most common indicators of exercise intensity are heart rate, metabolic equivalents, and perceived exertion. Using these indicators, activities may be broadly classified by intensity, as indicated in Table 8-1. In practice, these methods are used as follows:

- Heart rate (HR): used when health status makes it imperative that a specific intensity not be exceeded or to periodically determine whether a particular intensity is being maintained

TABLE 8-1   Classification of Physical Activity Intensity for Endurance Activity Lasting up to 60 Minutes

	Relative intensity			Absolute intensity (METs) in healthy adults[*]			
Intensity	VO_2R[†] (%) HRR[‡](%)	Maximal heart rate (%)	RPE[§]	Young (20–39 yr)	Middle age (40–64 yr)	Old (65–79 yr)	Very old (80 +yr)
Very light	<20	<35	<10	<2.4	<2.0	<1.6	≤1.0
Light	20–39	35–54	10–11	2.4–4.7	2.0–3.9	1.6–3.1	1.1–1.9
Moderate	40–59	55–69	12–13	4.8–7.1	4.0–5.9	3.2–4.7	2.0–2.9
Hard	60–84	70–89	14–16	7.2–10.1	6.0–8.4	4.8–6.7	3.0–4.25
Very hard	≥85	≥90	17–19	≥10.2	≥8.5	≥6.8	≥4.25
Maximal[ǁ]	100	100	20	12.0	10.0	8.0	5.0

Borg, G.A.V. (1982). Psychophysical bases of perceived exertion. *Medicine and Science in Sports and Exercise, 14*, 377–381, adapted by permission.
[*]Absolute intensity (METs) values are approximate mean values for men. Mean values for women are approximately 1–2 METs lower than those for men.
[†]VO_2R = oxygen uptake reserve, which is the difference between resting and maximal oxygen consumption.
[‡]HRR = heart rate reserve, which is the difference between resting and maximal heart rate.
[§]Borg rating of perceived exertion on a 6–20 scale from Borg, G.A.V. (1982). Psychophysical bases of perceived exertion. *Medicine and Science in Sports and Exercise, 14*, 377–381, adapted by permission.
[ǁ]Maximal values are mean values achieved during maximal exercise by healthy adults.

- Metabolic Equivalent (MET): used to quickly select activities in an intensity range, to keep a Physical Activity Record, or to estimate the caloric value of an activity
- Perceived exertion: for everyday use in helping a client find a comfortable activity intensity

Heart Rate

The best indicator of how hard people are working is their oxygen consumption, which increases with workload until the individual can work no harder. That point is the VO_{2max}. Unfortunately, VO_{2max} can only be measured in a laboratory, making it impractical for everyday use as an indicator of exercise intensity. However, HR increases proportionately with oxygen consumption, so HR can be used to gauge intensity in place of VO_{2max}.

Exercise HR is usually calculated using the heart rate reserve method. The **heart rate reserve (HRR)** is the difference between maximal HR and resting HR, where maximal HR is estimated from a formula and resting HR is measured by taking the client's pulse. Think of HRR as what you have to work with when you want to increase effort. The appropriate exercise intensity for you is calculated by using a percentage of your HRR. If you had access to a laboratory, you could determine **reserve oxygen consumption (VO_2R)** by subtracting resting oxygen consumption (VO_{2rest}) from VO_{2max}. But you do not need to measure this directly because the percentage of VO_2R that represents desired exercise intensity corresponds with the same percentage of HRR.

To determine exercise HR using the HRR concept, follow the directions below. Figure 8-3 provides an example of calculating an exercise HR range.

1. Estimate maximal HR using the formula:
   a. maximal HR for nonobese = 220 − age in years
   b. maximal HR for obese = 200 − [(0.5)(age in years)]

heart rate reserve (HRR) *Difference between maximal HR and resting HR, which is manipulated to create the desired exercise intensity.*

reserve oxygen consumption (VO_2R) *Difference between VO_{2max} and resting oxygen consumption.*

Maximum HR	= 200 − [1/2 age in years] = 200 − 21 = 179 beats/minute
HRR	= Maximum HR − Resting HR = 179 − 82 = 97 beats/minute
Light intensity	= 20% − 39% of HRR (see Table 8-1) = (0.20)(97); (0.39)(97) = 19.4 (round down to 19); 37.83 (round up to 38)
Exercise heart rate	= Intensity + resting HR = 19 + 82; 38 + 82 = 101−120 beats/minute

FIGURE 8-3 You can calculate resting HR using HRR. In this example, Myrah is 42 years old and has a resting heart rate of 82 beats/minute. Her body mass index is 30 kg/m². She has been sedentary for many years and wants to begin exercising at a light intensity.

2. Determine HRR:

   HRR = maximal HR − resting HR

3. Determine the desired intensity of the activity. Note that, in Table 8-1, moderate intensity corresponds with 40–59% of HRR. Take 40% of HRR to get the lower limit of exercise heart rate [(HRR)(0.40)], and take 59% of HRR to get the upper limit [(HRR)(0.59)].

4. Add back the resting HR to the HR ranges calculated in Step 3.

beta blockers

Class of medications prescribed for hypertension, angina, and irregular heart beat; these drugs block beta-adrenergic receptors, which decreases HR and the strength of heart contraction.

To use HR as an indicator of exercise intensity, clients need to wear a heart rate monitor or learn to check their pulse periodically during exercise. Before exercising, clients can calculate a 10-second HR range by dividing the exercise HR range by 6. Then, to monitor intensity, they need only pause for a 10-second pulse count during exercise.

Calculated HR is not a perfect indicator of exercise intensity. Formulas that predict maximal HR may be less accurate for adults over age 40. An alternative formula has been suggested for this age group, but has not yet been widely adopted: 208 − (0.7 × age).[19] People who are taking medications that lower HR, such as **beta blockers**, cannot use calculated HR as an indicator of exercise intensity. They need to have their exercise HR determined from a physician-supervised exercise test. Smokers, even young people in good health, may have a blunted HR response to exercise.[20] Exertion, indicated by HR, is masked in some smokers, greatly increasing the risk of heart attack and death during exercise. Smokers should be counseled to quit the habit and to lower their exercise intensity.

Metabolic Equivalents (METs)

Another way to express the intensity of physical activity is by MET value. As discussed in Chapter 2, 1 MET is approximately equal to an individual's resting metabolic rate: 3.5 ml oxygen consumption per kilogram body weight per minute, or about 1 kcal/kg/hour. A 3 MET activity requires three times the energy expenditure of a 1 MET activity.

Although the concept of METs may be difficult for non-professionals to comprehend, physical activities can be classified by their MET values, providing an easy reference for a client who wants or needs to exercise at a particular intensity. Table 3 in Appendix B lists the MET value of several dozen activities. These are organized into three categories by intensity: light (less than 3 METs), moderate (3—5.9 METs), and vigorous (6 METs and higher). Table 3 is derived from a much larger Compendium of Physical Activities. The authors of the compendium caution users that the MET values of these activities were derived for "average" individuals and may not be as accurate for obese individuals.[21] For example, a weight-bearing activity may actually have a higher intensity in a person carrying excess weight.

Individuals who have a physician-supervised exercise test often bring a MET-based exercise prescription to the nutrition or fitness professional. Notice in Table 8-1 that, as people get older (and presumably less fit), the MET value that corresponds with "moderate intensity" progressively decreases. So, at age 25, moderate-intensity activities have a MET value of 4.8–7.1, whereas, at age 75, moderate intensity corresponds with 3.2–4.7 METs. A fit 75-year-old could keep doing what she did in her 20s, although at a slightly lower intensity; but an unfit 75-year-old should be advised to select something at a lower intensity.

METs can be fairly easily converted to kilocalories to get an idea of the energy cost of a particular activity. Let's say your 70-kg client does 30 minutes of stretching each day. According to Table 3 in Appendix B, stretching is a moderate-intensity activity with a value

rating of perceived exertion (RPE)
System of subjectively assessing exercise intensity by asking oneself, "How hard am I working?" Also known as the Borg scale.

of about 2.5 METs. Its caloric value is, therefore, approximately 2.5 kcal/kg/hour. For a 70-kg person, this is equal to (2.5)(70) = 175 kcal/hour, or about 90 kcal/half hour.

Perceived Exertion

In the 1950s, the Swedish exercise physiologist Gunnar Borg devised a system that is based on subjective assessment of exercise intensity (how you feel during exercise; how out of breath you are; how much you are sweating). The original Borg scale, or **rating of perceived exertion (RPE)**, pictured in Figure 8-4(a), is a scale from 6 to 20 that allows clients to answer the question, "How hard am I working?" Numbers on the scale correspond to HR when they are multiplied by 10. So, a rating of 12 (between "fairly light" and "somewhat hard") is roughly equivalent to an HR of 120 (12 × 10). A rating of 12–16 ("somewhat hard" to "hard") on the original Borg scale is appropriate for apparently healthy people who do not have cardiovascular disease but who may have cardiovascular risk factors.[22]

The revised Borg scale, pictured in Figure 8-4(b), ranges from 0 to 10 and may be easier for people to use in quantifying their feelings of exertion. In this scale, 10 corresponds with "very, very strong," while 3 is "moderate." Borg suggests that the revised scale may allow people to more accurately express their exertion symptoms. The Borg scale was developed from studies with young men, so it may not be as accurate for women, older adults, or the obese.

Another indicator of perceived exertion that some people prefer is the "talk test"—the ability to talk while exercising. Although this method is not terribly scientific, many

(a) Original Borg Scale		(b) Revised Borg Scale		
6		0	Nothing at all	
7	Very, very light	0.5	Very, very weak	(just noticeable)
8		1	Very weak	
9	Very light	2	Weak	(light)
10		3	Moderate	
11	Fairly light	4	Somewhat strong	
12		5	Strong	(heavy)
13	Somewhat hard	6		
14		7	Very strong	
15	Hard	8		
16		9		
17	Very hard	10	Very, very strong	(almost maximum)
18				
19	Very, very hard			
20				

FIGURE 8-4 Ratings of perceived exertion. In the original Borg scale (a) numbers on the scale correspond to HR when they are multiplied by 10. The revised Borg scale (b) may be easier for people to use when expressing how exerting a physical activity is.

Source: Borg, G.A.V. (1982). Psychophysical bases of perceived exertion. *Medicine and Science in Sports and Exercise, 14,* 377–381; adapted by permission.

people report that they know intensity is just right when it is easy to carry on a conversation while exercising. If they are too out of breath to talk comfortably, they slow down, no matter what their exercise HR may be. An important point in determining appropriate intensity is listening to yourself. Because estimation of maximal HR is inexact, a calculated exercise HR may be too high. If the client's body says, "Slow down," she should listen.

How Much Exercise Is Enough?

volume of work
Combination of exercise duration and intensity.

The intensity of physical activity needed to improve fitness and health or to promote weight loss is directly related to the activity's duration. This combination of intensity and duration creates what exercise physiologists call the **volume of work**. Lower-intensity exercise for a longer duration is roughly comparable in volume to higher-intensity exercise for a shorter duration.

Volume of Activity Needed for Health and Cardiorespiratory Fitness

More exercise is needed to improve fitness than to improve health. For health, people should accumulate 30 minutes of moderate-intensity aerobic activity at least five days a week.[23] Walking at 3 mph is a moderate-intensity activity (about 3.3 METs). A 200-lb person engaging in this volume of work expends about 150 kcal. To improve cardiorespiratory endurance and attain metabolic fitness, an exercise intensity starting at 40–50% VO_2R/HRR for 20–60 minutes at least three times per week is needed. The more fit individual will need to exercise at a higher intensity to further improve cardiorespiratory endurance. A caloric expenditure of about 2,000 kcal/week offers the greatest protection against premature death from cardiovascular disease.[24]

Physical activity does not have to be continuous to be effective. According to the American College of Sports Medicine (ACSM), several 10-minute activity segments can be just as effective as one continuous exercise session.[25] Several researchers have documented equivalent improvements in VO_{2max}, blood pressure, and caloric expenditure when people either walked continuously for 30–40 minutes or for three or four 10-minute segments separated by several hours. This is good news for people who find continuous bouts of activity physically difficult or who need to work physical activity into an already-full day. Table 8-2 summarizes recommendations for aerobic exercise programs.

Volume of Activity Needed for Muscular Strength and Endurance

Muscular strength and endurance are developed by overloading the muscles through a combination of resistance (the weight lifted) and repetitions (the number of times the weight is lifted). Although both strength and endurance are improved through weight training, strength gains occur when there is more resistance and fewer repetitions, and endurance improves with less resistance and more repetitions.

The ACSM recommends the following for adults as a safe and effective exercise volume that improves both strength and endurance.[6]

- 8 to 10 exercises for the major muscle groups
- 8 to 12 repetitions (also called "reps") of each exercise
- Use moderate speed (about 6 seconds per repetition)
- Train 2 or 3 times a week, but not on consecutive days

TABLE 8-2 Aerobic Exercise Program Guidelines

	Goal	
	Cardiorespiratory fitness	**Physiologic fitness and weight management**
Type of activity	Large-muscle activities: low-impact aerobic dance, bicycling, cross-country skiing, hiking, inline skating, jogging, racquetball (singles), rope skipping, rowing, stair-climbing, swimming, walking	Large-muscle activities: low-impact aerobic dance, bicycling, hiking, inline skating, jogging, rowing, stair-stepping, swimming, walking, water aerobics
Intensity	For more fit people, 60–84% HRR; >6 METs; for less fit people, 40–59% HRR; 5–6 METs	For more fit people, 40–59% HRR; 5 METs; for less fit people, 20–39% HRR; 3–4 METs
Duration and frequency	20–60 minutes, either continuous or in 10-minute segments; longer duration at lower intensity or shorter duration at higher intensity three times/week	30 minutes of moderate-intensity physical activity on most, preferably all, days of the week; may be continuous or intermittent
Volume	250–300 kcal/day	200 kcal/day
Warm-up	5 minutes of chosen activity at a lower intensity before exercise begins	5–10 minutes of chosen activity at a lower intensity before exercise begins
Warm-down	Remain standing after exercise and continue activity at a lower intensity for 5 minutes or walk	Remain standing after exercise and continue activity at a lower intensity for 5–10 minutes or walk

Andresr, 2010/Shutterstock.com

The seated row is a resistance training exercise that will build muscular strength and endurance

TABLE 8-3   Muscular Strength and Endurance Program Guidelines

	Goal	
	To primarily develop strength	**For musculoskeletal health and weight management**
Type of activity	Dynamic weight-training exercises using weight machines: • Bench press (for chest, shoulders, upper arms) • Shoulder press (for shoulders, upper arms) • Pull-down or seated row (for chest, back, shoulders, arms) • Leg press (for buttocks, front and back of thighs) • Leg extension (for front of thighs) • Leg curl (for back of thighs) • Back extension (for back)	
Volume	8–12 repetitions of each exercise (select weight by trial and error, starting with low weight) two times/week	10–15 repetitions of each exercise (select weight by trial and error, starting with low weight) two times/week
Warm-up	Stretch first, or do low-intensity lifts, to increase blood circulation in the muscles to be stressed	Stretch first, or do low-intensity lifts, to increase blood circulation in the muscles to be stressed
Precautions	Breathe while lifting	Breathe while lifting; older and more frail individuals should lift lighter weights

resistance training *Exercises that use weights of varying resistance to build muscular strength and endurance.*

People with heart disease, who are frail, or over age 60 should decrease the resistance and increase the number of repetitions to 10 to 15. Eventually, the initial weight selected will be easily lifted, and the individual may move to a higher weight. **Resistance training** should not be done more than three times a week or on consecutive days. Beyond that, few additional gains occur and the risk of injury increases. Weight machines are safer than free weights for beginners. Alternatively, light resistance exercise can be done with elastic bands or light hand and ankle weights. Table 8-3 summarizes recommendations for muscular strength and endurance programs.

Volume of Activity Needed for Weight Loss and Maintenance

In its position statement, *Appropriate Physical Activity Intervention Strategies for Weight Loss and Prevention of Weight Regain in Adults*, the ACSM recommends the following duration of physical activity:[26]

- For minimal weight loss, at least 150 minutes/week
- For moderate weight loss (4 to 7 lbs), 150 to 250 minutes/week
- For weight losses of over 10 lbs, 225 to 420 minutes/week

Of course, the volume of activity needed for weight loss depends on caloric intake. Assuming that caloric intake is not excessive, the individual who can engage in a volume of activity approaching 2,000 kcal/week will generally lose more fat weight than the individual who can tolerate only a low volume of activity. This could be achieved by engaging in a 3.3 MET walk for an hour each day; or exercising to a Slimnastics exercise tape

(5 METs) for an hour five days a week; or swimming laps at a moderate pace for an hour (about 7 METs) 3 times a week. Dividing 60 minutes of activity into shorter, 10-minute bouts has also been found successful in promoting weight loss.[27]

Keeping off lost weight may require considerably higher energy expenditure, especially when the individual is attempting to maintain a body weight below her biological set point. One researcher found that men needed to exercise at least three times per week, expending a total of 1,500 kcal, to maintain weight losses averaging 12 kg for 3 years.[28] Overweight women who engaged in 275 minutes of exercise weekly, expending about 1,500 kcal, were able to maintain losses of 10% of body weight for a year.[29] In another study, women trying to maintain weight loss needed to engage in moderate exercise for 80 minutes/day or vigorous physical activity 35 minutes daily.[14]

Is Physical Activity Advisable for the Severely Obese?

Most people with a body mass index (BMI) up to 30 can safely and comfortably engage in physical activity, including exercise. People with a BMI of 40 and above tend to exercise with great difficulty and even find increasing daily physical activity a challenge. Those with BMIs between 30 and 40 have great variability in their capacity for physical activity and exercise.

Approximately six percent of the adult U.S. population is severely obese –100 lb over ideal body weight, or having a BMI at or above 40.[30] According to the Centers for Disease Control's (CDC) National Center for Health Statistics, this rate has doubled in the past 15 years. Severely obese people need to begin an activity program very slowly and only after being cleared by a physician for increased physical activity. The initial exercise prescription might be walking to the corner and back or making two extra trips up the stairs each day and then gradually adding slightly more distance. Eventually, the individual might consider a very low-impact formal exercise program, such as water aerobics or walking in a pool. Water reduces stress on the joints, making it an ideal exercise environment for heavier people. Unfortunately, there are few places for very large individuals to exercise in water without feeling embarrassed or uncomfortable.

An additional challenge in recommending exercise for very large people is that it may not promote weight loss to the same extent as in the more moderately obese. There are two reasons for this: First, a severely obese individual may not be able to expend enough energy to achieve caloric deficit, even though moving a larger body requires more energy. Second, several experts have pointed out that exercise may be less effective in people with hyperplastic obesity (an excessive number of fat cells) than with hypertrophy (a normal number of enlarged fat cells). Recall from Chapter 4 that people with more severe obesity, particularly beginning early in life, are more likely to have hyperplasia of the fat cells. Exercise should still be encouraged because metabolic benefits can be derived without significant weight loss.

How Can Physical Activity Be Incorporated into Daily Life?

For people who are reluctant or unable to engage in formal exercise, finding a way to incorporate activity into everyday life offers the greatest potential for a lifelong intervention that works. People in industrialized nations used to spend a great deal of time and energy in daily chores. Modernization has freed us from chopping wood, gathering food, climbing stairs, and even walking to the next cubicle at work to deliver a message. The daily savings in caloric expenditure that results from modern conveniences could add up to several hundred calories each day, more than enough to account for a 5- to 10-lb annual weight gain.

Recently scientists have become interested in exploring the benefits of simply increasing daily energy expenditure in everyday activities, known as NEAT. Considering most people's reluctance to engage in exercise, this could offer the potential for small weight losses—or at least no weight gain. Consider the following example of activities that could become part of a 70-kg person's daily routine:

- Fifteen minutes of sweeping (the kitchen floor, the patio, the front walk, and the deck) expends about 44 kcal.
- Walking at a slow pace to the mailbox (5 minutes), around the block (15 minutes), and to the convenience store (10 minutes) expends 70 kcal.
- Walking up and down ten flights of stairs over the course of a day (5 minutes) expends 20–25 kcal.

No one would suggest that these three examples could result in sufficient caloric expenditure for significant weight loss, but they are initial steps toward beginning a more active lifestyle. Additional ideas for incorporating activity into everyday life are presented in Table 8-4.

TABLE 8-4 Incorporating Physical Activity into Everyday Life

Encourage clients to:

At home...
- Play with their children.
- Always use the bathroom on another floor.
- Stop using the television remote control, or get up when using it.
- Walk in place during television commercials.
- Use a cordless telephone or cell phone, and walk around while talking.
- Sweep the floors, patio, front walk, and/or deck every day.
- Keep a pair of walking shoes near the front door.
- Walk their dog, a neighbor's dog, a friend's dog, or their "inner dog" every day.

At work...
- Wear comfortable shoes, or keep a comfortable pair stashed in the office.
- When possible, deliver messages in person rather than by e-mail or telephone.
- Walk on breaks.
- Take "recess."
- Get off the bus or the subway one stop early.
- Park farther away from the workplace.

In general...
- Sit less.
- Take the stairs more often.
- Avoid using drive-thrus; park the car and get out.
- Move whenever they can.

Public health approaches:
- Put prompts next to elevators, reminding people to take the stairs.
- Promote community design that allows and encourages walking and biking.

TABLE 8-5 No Equipment, No Frills Movement Program

Throughout the day, periodically move your body.
- Flex and extend your feet.
- Shake your hands and flex and extend your hands and fingers.
- Inhale and exhale fully.
- Swing your right arm across your body and extend it out to the side; repeat with left arm.
- Sitting in a chair, extend your left leg at the knee; repeat with right leg.
- While standing, rise up on toes. With both feet flat on the floor, balance on one leg at a time.

Stretching exercises for the major muscle groups should be done 2–3 days a week. Do a slow, sustained stretch at the point of mild discomfort and hold each stretch for 10–30 seconds.
- Turn your head to the right and hold; then to the left.
- Pull your right arm across the front of your body and hold with left hand; repeat with left arm.
- Rotate your trunk around to the right, then the left.
- Seated on the floor with legs in a "V" and knees slightly bent, stretch forward over one leg at a time.
- Lie on your back. Pull right knee to your chest; repeat with left knee.

Weight-management professionals need to encourage clients to expend energy. Eventually, the person who gets into the "activity habit" may be ready to adopt a regular exercise program. A very low-impact form of exercise that people might start with is simply body movement, which requires no special equipment or skill and can keep people limber. When combined with stretching, within each individual's physical capabilities, flexibility may improve and chronic pain, like lower back pain, may diminish. A sample movement and stretching program is presented in Table 8-5.

Developing Activity Programs that Work: Summary

Activities of various types, intensity, duration, and frequency can contribute to health, fitness, and weight management. Before beginning a formal exercise program, individuals should explore ways to incorporate more activity into their everyday routines. For weight loss, a more substantial volume of activity may be needed than that required to promote health.

Case Study. The Novice Exerciser, Part I

Janii is a 58-year-old single woman living in a major metropolitan area. She has never exercised in her life, but she enjoys walking. She uses a bus stop farther from her home most work days, which means that she walks about three miles each day ("My most peaceful time," Janii calls it). On Saturdays she walks to a nearby Farmer's Market to buy vegetables. She is 5' 5" tall and weighs 156 lb. Her blood pressure is around 142/90, and her doctor has encouraged her to keep walking but to try to adopt an exercise program. At a recent medical appointment, her resting heart rate was 79 beats per minute.

On Saturday Janii stopped in at the community recreation center in her neighborhood and inquired about exercise programs. The young man on duty at the information desk provided her with a brochure listing the center's exercise programs. "Several of these would fit into my schedule,"

(Continued)

Case Study.
(Continued)

Janii said, "but I don't know what some of them are. Spinning? Tae Bo? And should I come once a week? Will the instructor know if I'm overdoing it?"

The young man at the desk happened to be one of the center's instructors and a certified fitness trainer. He escorted her to his office and got information about her exercise goals ("not get heavier and stay off blood pressure medicine") and her health.

What will he tell Janii about:

1. A good exercise heart rate for her.
2. Exercises that meet her goals.
3. How long she should exercise on each occasion.
4. A potentially beneficial exercise program for her.

KEEPING PHYSICAL ACTIVITY SAFE

Exercise presents a bit of a paradox: On the one hand, some people die while engaging in it, especially vigorous exercise. On the other hand, exercise offers a number of health benefits, including a lower risk of having a cardiac arrest during exercise. In addition, exercising incorrectly or too strenuously can cause musculoskeletal injuries, sometimes severe enough to require medical attention. Nevertheless, medical and fitness organizations, including the American Heart Association and ACSM, believe that the benefits of exercise far outweigh the risks.

What Is the Risk of Sudden Death During Exercise?

When people under age 40 die during exercise, the cause is generally undetected congenital heart disease. When people over 40 die during exercise, the cause is usually coronary artery disease. Available data on the incidence of death during physical exertion suggest that the risk is low. Among the general population, the rate of heart attack is about 1 per 375,000 to 888,000 person-hours of exercise; and this is 7 times higher than the cardiac death rate during exercise.[22] In medically supervised cardiac rehabilitation programs the heart attack rate is about 1 per 116,906 patient-hours, and the fatality rate is very low.[31]

Factors That Increase the Risk of Cardiac Events During Exercise

The following factors seem to be especially associated with fatal or non-fatal heart attacks during exercise:

- Intensity of exercise: higher intensity activity (> 6 METs) is associated with increased risk of cardiovascular events in both young people and adults and in those with diagnosed or undiagnosed heart disease.[22,31]
- Previously sedentary lifestyle: the least active individuals have as much as a 50 times greater chance of suffering a heart attack during vigorous activity than the most active. Most cardiac deaths occur in people who habitually exercise less than one day/week.

- Clinical cardiovascular disease: risk of sudden death increases among people who have been diagnosed with cardiovascular disease. Among persons with CVD, jogging has been associated with the highest incidence of cardiac arrest.[22]
- Cardiovascular risk factors: the presence of hypertension, elevated blood lipids, diabetes, cigarette smoking, obesity, and family history of early cardiovascular disease may also increase the risk of heart attack or death during exertion.
- Age: perhaps because the incidence of cardiovascular disease and the presence of risk factors increase with age, the likelihood of exercise-related death is higher among older adults.
- Environmental factors: The risk of cardiac death during exercise may increase during hot and cold temperature extremes.
- Time of day: The likelihood of having a heart attack or dying from a heart attack is greatest first thing in the morning.[31]

Underlying Mechanisms of Exercise-Related Cardiac Death

hypertrophic cardiomyopathy
Heart condition characterized by abnormal thickening of the muscles in the left ventricular wall, which reduces the blood-pumping capacity of the heart and can cause disturbances in electrical transmission.

Exercise probably causes heart attacks in susceptible people for a few reasons. Platelet aggregation is more common in sedentary people than active people, resulting in an increased tendency to form atherosclerotic plaque and clots. (Recall from Chapter 5 that there are also dietary factors that promote platelet aggregation.) During exertion, diseased coronary arteries may go into spasm rather than dilating as healthy coronary arteries would. Such a spasm might not only reduce blood (and oxygen) flow to the affected area but could break loose a piece of atherosclerotic plaque, which might lodge in a smaller artery. Plaque may also be broken away when systolic blood pressure rises, which is a natural occurrence during exercise. There may be previously undetected structural abnormalities, such as narrowed or missing coronary arteries, weakness in the aorta, or **hypertrophic cardiomyopathy**, that are apparent only during a bout of strenuous activity.

How Should People Be Screened for Participation in Physical Activity?

The risk of having a heart attack or stroke during exercise—or of dying from a cardiac event—can be partly minimized by identifying the people at greatest risk. There is, however, no screening instrument that can reliably identify everyone at risk. Screening instruments can:

- Identify many people who have cardiovascular disease or increased risk of cardiovascular disease and lead to their referral to medical personnel (People with major cardiovascular risk factors, as listed in Table 8-6, are most likely to have cardiovascular disease and to benefit from exercise testing.)
- Suggest a safe and appropriate exercise prescription that offers the lowest risk to the client
- Indicate which individuals would be better served by a medically supervised exercise program
- Spare people who do not need an exercise stress test from the time, discomfort, and expense of getting one

TABLE 8-6   Who Should Undergo Exercise Testing?

American College of Sports Medicine recommends exercise testing for:
- Men ≥ 45 years and women ≥ 55 years who are initiating vigorous exercise (≥60% VO_{2max})
- Anyone with 2 or more major cardiac risk factors (cigarette smoking, high blood cholesterol, hypertension, physical inactivity, obesity and overweight, diabetes)
- People with signs or symptoms of heart disease
- Individuals with diagnosed cardiac, pulmonary, or metabolic disease

American College of Cardiology and American Heart Association recommend:
- No routine exercise testing for asymptomatic persons who engage in moderate intensity activity
- Consultation with a physician by symptomatic individuals or those with cardiovascular disease, diabetes, other active chronic disease, or medical concerns before increasing physical activity or engaging in vigorous physical activity

Source: Haskell, W. L., Lee, I., Pate, R. R., Powell, K. E., Blair, S. N., Franklin, B. A., et al. (2007). Physical activity and public health: Updated recommendation for adults from the American College of Sports Medicine and the American Heart Association. *Circulation, 116*(9), p. 1089.

Cardiac Screening Questionnaires

The American Heart Association (AHA) and the ACSM make the following recommendations about exercise screening:[32]

- Before exercise, all adults should be screened to estimate their risk of cardiovascular disease.
- Screening questionnaires should be interpreted by qualified staff.
- Individuals with medical risks should be referred to their physician to obtain a medical evaluation.
- Individuals with medical risks who do not obtain a medical evaluation should be required to sign an assumption of risk or release before participating in a formal exercise program. These individuals should be advised to reduce the intensity of exercise and be alert to signs and symptoms of heart attack or stroke.

The PAR-Q (Physical Activity Readiness Questionnaire) is a widely used, practical questionnaire for screening people for cardiovascular disease prior to participating in exercise. The PAR-Q, presented in Figure 8-5, also includes a question about musculoskeletal problems. Its seven yes/no questions can be answered in moments, and the scoring is simple: If one or more questions is answered "yes," then the client should contact his physician or nurse practitioner. The PAR-Q has a **sensitivity** of about 100%, which means that almost everyone who has a medical problem needing a physician's referral responds "yes" to one or more questions. The PAR-Q has a **specificity** of about 80%, indicating that approximately 80% of individuals who do not have a medical reason to see a physician before exercise respond "no" to all questions. Like most screening questionnaires, the sensitivity of PAR-Q for predicting abnormal exercise **electrocardiogram (ECG)** is low.[33]

The AHA and the ACSM also have a pre-participation screening questionnaire that is a bit longer than the PAR-Q but is based on the same idea—to identify people who may be at high risk for exercise and direct them to medical personnel for evaluation. This questionnaire, shown in Figure 8-6, asks more direct questions about blood pressure,

sensitivity
Capacity of an assessment to correctly detect people who have a disease or condition.

specificity
Capacity of an assessment to correctly detect people who are free of disease.

electrocardiogram (ECG)
Noninvasive test that measures the electrical activity of the heart.

Physical Activity Readiness
Questionnaire - PAR-Q
(revised 2002)

PAR-Q & YOU

(A Questionnaire for People Aged 15 to 69)

Regular physical activity is fun and healthy, and increasingly more people are starting to become more active every day. Being more active is very safe for most people. However, some people should check with their doctor before they start becoming much more physically active.

If you are planning to become much more physically active than you are now, start by answering the seven questions in the box below. If you are between the ages of 15 and 69, the PAR-Q will tell you if you should check with your doctor before you start. If you are over 69 years of age, and you are not used to being very active, check with your doctor.

Common sense is your best guide when you answer these questions. Please read the questions carefully and answer each one honestly: check YES or NO.

YES	NO		
☐	☐	1.	Has your doctor ever said that you have a heart condition <u>and</u> that you should only do physical activity recommended by a doctor?
☐	☐	2.	Do you feel pain in your chest when you do physical activity?
☐	☐	3.	In the past month, have you had chest pain when you were not doing physical activity?
☐	☐	4.	Do you lose your balance because of dizziness or do you ever lose consciousness?
☐	☐	5.	Do you have a bone or joint problem (for example, back, knee or hip) that could be made worse by a change in your physical activity?
☐	☐	6.	Is your doctor currently prescribing drugs (for example, water pills) for your blood pressure or heart condition?
☐	☐	7.	Do you know of <u>any other reason</u> why you should not do physical activity?

If

you

answered

YES to one or more questions

Talk with your doctor by phone or in person BEFORE you start becoming much more physically active or BEFORE you have a fitness appraisal. Tell your doctor about the PAR-Q and which questions you answered YES.

- You may be able to do any activity you want — as long as you start slowly and build up gradually. Or, you may need to restrict your activities to those which are safe for you. Talk with your doctor about the kinds of activities you wish to participate in and follow his/her advice.
- Find out which community programs are safe and helpful for you.

NO to all questions

If you answered NO honestly to <u>all</u> PAR-Q questions, you can be reasonably sure that you can:
- start becoming much more physically active — begin slowly and build up gradually. This is the safest and easiest way to go.
- take part in a fitness appraisal — this is an excellent way to determine your basic fitness so that you can plan the best way for you to live actively. It is also highly recommended that you have your blood pressure evaluated. If your reading is over 144/94, talk with your doctor before you start becoming much more physically active.

DELAY BECOMING MUCH MORE ACTIVE:
- if you are not feeling well because of a temporary illness such as a cold or a fever — wait until you feel better; or
- if you are or may be pregnant — talk to your doctor before you start becoming more active.

PLEASE NOTE: If your health changes so that you then answer YES to any of the above questions, tell your fitness or health professional. Ask whether you should change your physical activity plan.

<u>Informed Use of the PAR-Q</u>: The Canadian Society for Exercise Physiology, Health Canada, and their agents assume no liability for persons who undertake physical activity, and if in doubt after completing this questionnaire, consult your doctor prior to physical activity.

No changes permitted. You are encouraged to photocopy the PAR-Q but only if you use the entire form.

NOTE: If the PAR-Q is being given to a person before he or she participates in a physical activity program or a fitness appraisal, this section may be used for legal or administrative purposes.

"I have read, understood and completed this questionnaire. Any questions I had were answered to my full satisfaction."

NAME _____

SIGNATURE _____   DATE_____

SIGNATURE OF PARENT _____   WITNESS _____
or GUARDIAN (for participants under the age of majority)

Note: This physical activity clearance is valid for a maximum of 12 months from the date it is completed and becomes invalid if your condition changes so that you would answer YES to any of the seven questions.

CSEP
SCPE © Canadian Society for Exercise Physiology   Supported by: 🍁   Health   Santé
                                                                      Canada Canada

FIGURE 8-5 The Physical Activity Readiness Questionnaire (PAR-Q) is a screening tool for cardiovascular disease risk that might preclude exercise.

Source: Physical Activity Readiness Questionnaire (PAR-Q) © 2002. Used with permission from the Canadian Society for Exercise Physiology www.csep.ca.

Assess your health needs by marking all *true* statements.

History
You have had:

_____ a heart attack

_____ heart surgery

_____ cardiac catheterization

_____ coronary angioplasty (PTCA)

_____ pacemaker/implantable cardiac
          defibrillator/rhythm disturbance

_____ heart valve disease

_____ heart failure

_____ heart transplantation

_____ congenital heart disease

If you marked any of the statements in this section, consult your health care provider before engaging in exercise. You may need to use a facility with *medically qualified staff.*

Symptoms

_____ You experience chest discomfort with exertion.

_____ You experience unreasonable breathlessness.

_____ You experience dizziness, fainting, or blackouts.

_____ You take heart medications.

Other health issues:

_____ You have musculoskeletal problems.

_____ You have concerns about the safety of exercise.

_____ You take prescription medication(s).

_____ You are pregnant.

Cardiovascular risk factors:

_____ You are a man older than 45 years.

_____ You are a woman older than 55 years or you have
          had a hysterectomy or you are postmenopausal.

_____ You smoke.

_____ Your blood pressure is greater than 140/90.

_____ You don't know your blood pressure.

_____ You take blood pressure medication.

_____ Your blood cholesterol level is >240 mg/dl.

_____ You don't know your cholesterol level.

_____ You have a close blood relative who had a heart attack before age 55
          (father or brother) or age 65 (mother or sister).

_____ You are diabetic and take medicine to control your blood sugar.

_____ You are physically inactive (i.e., you get less than 30 minutes of
          physical activity on at least 3 days/week).

_____ You are more than 20 lb overweight.

If you marked two or more of the statements in this section, you should consult your health care provider before engaging in exercise. You might benefit by using a facility with *professionally qualified exercise staff* to guide your exercise program.

_____ None of the above is true.

You should be able to exercise safely without consulting your health care provider in almost any facility that meets your exercise program needs.

FIGURE 8-6 The AHA/ACSM Health/Fitness Preparticipation Screening Questionnaire identifies people who may be at high risk if they exercise.

Source: From American College of Sports Medicine & American Heart Association. (1998). Recommendations for cardiovascular screening, staffing, and emergency policies at health/fitness facilities. Joint position statement. *Medicine and Science in Sports and Exercise, 30*(6), 1009–1018; reprinted by permission.

Physician Referral Form*

Dear Dr._____

Your patient, [name], would like to begin a program of exercise and/or sports activity at [name of health/fitness facility]. After reviewing his/her responses to our cardiovascular screening questionnaire, we would appreciate your medical opinion and recommendations concerning his/her participation in exercise/sports activity. *Please provide the following information and return this form to [name, address, telephone, fax]:*

    1. Are there specific concerns or conditions our staff should be aware of before this individual engages in exercise/sports activity at our facility? Yes/No
       If yes, please specify:

    2. If this individual has completed an exercise test, please provide the following:
       a. Date of test:_____
       b. A copy of the final exercise test report and interpretation
       c. Your specific recommendation for exercise training, including heart rate limits during exercise:

    3. Please provide the following information so that we may contact you if we have any further questions:
    _____I AGREE to the participation of this individual in exercise/sports activity at your health/fitness facility.
    _____I DO NOT AGREE that this individual is a candidate to exercise at your health/fitness facility
       because_____

Physician's signature:_____
Physician's name:_____
Address:_____
Telephone:_____
Fax:_____

Thank you for your help.

*Must be accompanied by a medical release form.

FIGURE 8-7   This sample Physician Referral Form may be used to notify physicians that their patients are about to begin an exercise program and to obtain permission to start an exercise program.

Source: From American College of Sports Medicine & American Heart Association. (1998). Recommendations for cardiovascular screening, staffing, and emergency policies at health/fitness facilities. Joint position statement. *Medicine and Science in Sports and Exercise, 30*(6), 1009–1018; reprinted by permission.

---

stress test

Cardiorespiratory exercise test supervised by a physician in which the individual exercises on a treadmill or a cycle to a particular HR or percentage of VO_{2max} while blood pressure and ECG are monitored.

---

lipids, activity habits, family history, and pregnancy, which might make it more predictive of exercise complications.

An exercise screening questionnaire may not be needed if a complete nutritional assessment has already been done. If the client was referred by her physician, then the weight-management professional need not send the client back for another evaluation before beginning exercise, but the physician should be told that the weight-management intervention will include exercise. Figures 8-7 and 8-8 provide sample physician referral and medical release forms.

Cardiorespiratory Exercise Tests

Cardiorespiratory exercise tests, sometimes called **stress tests**, are used in the diagnosis and treatment of cardiovascular disease. A stress test is conducted on a treadmill or a cycle ergometer while blood pressure and ECG activity are monitored. Of particular interest are HR and systolic blood pressure responses to exercise, exercise capacity,

Form for Release of Medical Information

1. I hereby authorize_____to release the following information
   from the medical record of
Patient's name:_____
Address:_____
Telephone:_____
Date of birth:_____
2. Information to be released
(If specific treatment dates are not indicated, then information from the most recent
visit will be released.)
_____Exercise test   _____Most recent history and physical exam
_____Most recent clinic visit   _____Consultations
_____Laboratory results (specify)_____
_____Other (specify):_____
3. Information to be released to
Name of person/organization:_____
Address:_____
Telephone:_____
4. Purpose of disclosure of information:_____
5. I do not give permission for disclosure of this information other than that specified above.
6. I request that this consent become invalid 90 days from the date I sign it or _____
_____.
   I understand that this consent can be revoked at any time except to the extent that disclosure made in good
   faith has already occurred in reliance of this consent.
7. Patient's signature:_____
   Date:_____
   Witness:_____
          (please print)
   Signature:_____

FIGURE 8-8   This sample form for Release of Medical Information is used to obtain your client's permission for getting information pertinent to safe participation in an exercise program.

Source: From American College of Sports Medicine & American Heart Association. (1998). Recommendations for cardiovascular screening, staffing, and emergency policies at health/fitness facilities. Joint position statement. *Medicine and Science in Sports and Exercise, 30*(6), 1009–1018; reprinted by permission.

asymptomatic

Free of symptoms of a disease or condition, although the disease or condition may still be present.

and ECG findings. Abnormalities in any of these indicators would represent a positive exercise test.

The U.S. Preventive Services Task Force, American Heart Association, and American College of Cardiology do not recommend routine use of exercise testing. Among the evidence that they cite to support this are: (1) among **asymptomatic** people, most ECGs that suggest heart disease may be present are seen in people who do not have a cardiac event in the next 5–10 years; and (2) a study in which 71% of adults who had an abnormal exercise treadmill test had normal findings on angiography.[34]

Stress tests are predictive of cardiac events 3 to 12 years in the future between 40% and 62% of the time.[34] They are less predictive in women than men, simply because they have not been studied as widely in females. Stress tests are of their greatest value in: men over 45 and women over 55 who plan to engage in vigorous exercise; people with diabetes; and people with multiple cardiovascular risk factors.[35]

Obesity in the absence of other major cardiovascular risk factors may not be cause for heightened concern. Data collected for over 20 years from the Aerobics Center Longitudinal Study suggest that low cardiorespiratory fitness is as great a predictor of CVD death as smoking, diabetes, high cholesterol, and hypertension in overweight and obese men.[36] Others report similar findings in women. Asymptomatic obese women with an exercise capacity of under 5 METs had three times the death rates of those with an exercise capacity of 8 METs or higher. Simple cardiorespiratory fitness tests like the Rockport test described in Chapter 7 may be useful in identifying obese individuals with low exercise capacity and suggest the need for a stress test. Habitual participation in physical activity, determined by physical activity assessment such as the example in Chapter 2, may also serve as an indirect determinant of a person's exercise capacity.

Another way to estimate risk of cardiac complications from exercise is to measure HR recovery after exercise. HR should decline rapidly after physical activity is terminated, by 12 or more beats/minute. During the warm down period, HR decline that is 12 beats/minute or less one minute after exercise termination is cause for concern.[35]

What Are Other Considerations for Safe Exercise?

Sudden death during exercise is a rare complication of exercise. Far more likely are musculoskeletal injuries and dehydration.

Musculoskeletal Injuries

muscle strain
Common injury characterized by microscopic tears in muscle fibers, which results in pain, swelling, and inflammation.

People of any age who overdo exercise may develop a musculoskeletal injury. The most common injury associated with exercise is **muscle strain**, in which microscopic muscle tears leak fluid, become inflamed, and cause pain. Strains can result from either endurance exercise—walking, riding, or swimming at too high an intensity for too long a time—or from resistance exercise—lifting too much weight too many times and using incorrect form. Some soreness is inevitable when inactive people become more active or move from physical activity to formal exercise. Soreness should not last for more than 1 week or get worse. The sentiment "No pain, no gain" is inaccurate and can be dangerous.

Most musculoskeletal injuries occur because of two factors: (1) exercising at too high an intensity, and/or (2) engaging in high-impact activities.[22] The previously sedentary or infrequent exerciser should progress slowly to minimize soreness and prevent strains. People who have an old muscle injury need to be particularly careful not to overdo exercise. Table 8-7 suggests some ways to reduce the risk of musculoskeletal injury.

Warming Up and Warming Down to Prevent Injury

For people who are engaging in formal exercise, time should be allotted for a warm-up before and warm-down after exercise. Warming up increases muscle temperature and pliability, preventing muscle strains. A warm-up also increases blood flow to the heart muscle, preventing spikes in systolic blood pressure and reducing the probability of abnormal electrical rhythms and heart attack. Warming down at the end of activity slows blood flow into the working muscles, moves metabolic byproducts like lactic acid out of the muscles, and prevents blood from pooling in the legs. If someone were to abruptly stop

TABLE 8-7 Reducing the Risk of Musculoskeletal Injury in Inactive People

- Teach clients the correct techniques for using aerobic or strength-training equipment.
- Supervise initial uses of equipment.
- Avoid the use of free weights for strength training until the client has increased strength and developed some skill in resistance training. Weight-training machines generally require less skill and prevent overextending range of motion.
- Teach strength-training exercises that can be performed in a slow, controlled manner to avoid explosive, jerky movements.
- Provide an individualized exercise prescription that stipulates exercise intensity, duration, and frequency.
- Advise clients to start slowly, with one or two exercise sessions per week of 5–10 minutes duration and to gradually increase duration and frequency of training.
- Recommend that clients not train the same muscle groups every day. It is safer to take a day off between exercise sessions and perform low-intensity activities on the "off" day.
- Be sure clients know the signs of overexertion and are advised to stop exercising if they feel out of breath, faint, or are in pain. No one should exercise on days when they have musculoskeletal pain, although low-intensity walking may be permitted.
- Warm up thoroughly before exercising.

exercising and stand still, the lack of adequate blood flow to the brain could cause dizziness and even fainting. Stretching done after exercise when the muscles are warm further promotes flexibility and reduces the chance of injury.

Older sedentary people and those who have cardiovascular disease risk factors need to do a long warm-up—at least 10 minutes of very low-intensity activity. As they become more accustomed to activity and see fitness improvements, a 5-minute warm-up is usually sufficient to get blood flowing into the working muscles and the heart. By the end of the warm-up, heart rate should be about 20 beats per minute below the target HR range. The warm-down should be about as long as the warm-up and can simply be a lower-intensity version of the chosen physical activity or walking.

Good Hydration

When exercising in warm weather, high humidity, or a combination of both, adequate fluid intake is essential to prevent heat stress. People who are obese, diabetic, or have cardiovascular disease are even more susceptible to temperature extremes. The higher the intensity of exercise, the greater the need for adequate hydration. Because thirst is not a good indicator of the need for fluids, professionals need to advise clients to adopt a regular schedule of hydration during and after exercise.

Competitive athletes are often advised to increase their endurance by using sports drinks that are no more than 8% carbohydrate. People engaging in noncompetitive moderate activity can replace lost fluids with plain water. Sports drinks provide unneeded calories for the exerciser who is trying to lose or maintain weight.

Who Can Safely Exercise?

Most people can increase their daily physical activity or participate in a formal exercise program with minimal risk. For people with existing musculoskeletal conditions, such as arthritis or osteoporosis, medical clearance should be obtained before increasing physical activity level. In most cases, low-impact and low-intensity activities (walking, water aerobics, and exercise cycling) are completely safe. Even adults over age 45 who have cardiovascular risk factors can participate in low- to moderate-intensity activities. As age and cardiovascular risk factors or severity of musculoskeletal conditions increase, the risk of exercise increases when the individual moves from moderate to high intensity.

Closer exercise supervision is needed for people who have cardiovascular disease or type 2 diabetes. Individuals with stable cardiovascular disease who can be taught to regulate exercise intensity (by reading a heart rate monitor or taking their pulse) may participate in low-intensity exercise without supervision. Any exercise of moderate or high intensity should be supervised—in a fitness center that serves clinical populations or in a clinical exercise program. Individuals with cardiovascular disease that puts them at higher risk during exercise and/or who cannot regulate exercise intensity should perform all exercise in clinically supervised settings. Individuals with unstable cardiovascular disease, hypertrophic cardiomyopathy, very high resting blood pressure (systolic >200 mm Hg, diastolic >110 mm Hg), **aneurysm**, or **thrombophlebitis** should restrict their physical activity.

People with type 2 diabetes need to monitor their blood glucose before and after exercise. Those who have vascular complications should pay close attention to the condition of their feet, checking for blisters and ulcers after exercise. The dose of sulfonylurea medications may need to be adjusted as insulin sensitivity improves.

aneurysm *Pouch that forms in the wall of a blood vessel, usually from hypertension or atherosclerosis, resulting in weakening of the blood vessel and the possibility of rupture.*

thrombophlebitis *Inflammation of a vein, which can lead to formation of blood clots.*

Keeping Physical Activity Safe: Summary

Physical activity is an important contributor to good health, and most people can safely engage in low to moderate levels of activity. For those with cardiovascular disease or risk factors, we can minimize any dangers of exercise by proper screening, warm-up and warm-down, hydration, and appropriate levels of physical activity.

MOTIVATING PEOPLE TO ADOPT AN ACTIVE LIFESTYLE

"You need to get more exercise." These six words are easily spoken by fitness and health professionals every day. In fact, nothing is less likely to happen than a sedentary person readily translating these words into action. Consider what a person must do to become more active: (1) select an activity, (2) determine when to engage in the activity, (3) purchase any equipment needed to begin the activity, (4) find a place to engage in the activity, (5) become motivated enough to start, and (6) keep it up day after day.

This section of the chapter looks at how weight management professionals can facilitate the transition from inactivity or low activity to a more active lifestyle, perhaps even one that includes exercise. Because continuing to be active is as important as starting an activity regimen, ways to maintain physical activity are also reviewed.

Who Is—and Isn't—Active?

CDC collects information about American health habits, including physical activity, through its Behavioral Risk Factor Surveillance System (BRFSS), National Health

Interview Surveys, and Youth Risk Behavior Surveys (YRBS). Data from these surveys indicate that few of us meet government recommendations for physical activity. About 39% of adults engage in no leisure-time physical activity, 30% in some activity, and 31% regularly participate in physical activity.[37] While about one-fourth of those 18 and older engage in vigorous activity three times/week, approximately 60% have never engaged in vigorous physical activity for more than 10 minutes.[37] Among 9th to 12th graders, 35% participated in moderate to vigorous physical activity for an hour a day, five days a week, and one-fourth participated in no physical activity during the week before the YRBS survey.[38]

Race and gender are factors that influence participation in physical activity at all ages. Most likely to be inactive among adults and adolescents are females and Hispanics. Education level and place of residence also affect activity level. Inactivity is more common among people with less than a high school education and those living in major metropolitan areas, where people might have concerns about safety, and less common among college graduates and those living in non-metropolitan areas.[37]

How Do We Motivate People to Initiate Physical Activity?

These physical activity trends suggest some of the barriers and difficulties inherent in encouraging people to become more physically active. People who are not by nature active and who have little or no experience with formal exercise programs cannot be expected to jump into action just because we inform them that there is something to be gained from it.

Adults who are completely inactive are probably not yet thinking about becoming more active. Adults who are already active on most days represent people who need to be encouraged to maintain their active lifestyle. The remainder of the adult population is not completely inactive but is less active than they should be. The strategies in this section of the chapter are particularly relevant to this last group.

Active People Feel Better and Are in More Control

Weight-management professionals need to communicate the benefits of activity with an emphasis on well-being. Certain benefits may be particularly motivating:

- Psychological effects: For some people, feelings of strength and confidence, improved mood, and reduced anxiety and stress may be more compelling reasons to start—and keep—exercising than improved blood pressure and lipid levels.[6] Often people do not realize that their exercise program, not weight loss, is helping them to feel better, and you can point out this fact to them.
- Weight maintenance: By itself, increased physical activity may not cause significant weight loss, especially in people who have a limited capacity for movement. Still, a program of regular physical activity can prevent additional weight gain as well as weight regain.

Weight-management professionals should put the emphasis on health, over which active people have some control, rather than weight loss, over which they have limited control. Clients are then freed from worrying whether they lost weight this week and they can focus on genuine and measurable health benefits.

Many Benefits Result from Few Lifestyle Changes

The first lifestyle change that should be explored is simply reducing the amount of time spent in sedentary behavior. Uninterrupted sitting time at work or at home in front of the television or computer has health implications that are just like those seen in people on prolonged bed rest. Research in the Australian Diabetes, Obesity, and Lifestyle Study verifies that taking activity breaks from sedentary time can improve triglyceride levels, fasting plasma glucose, and abdominal fat levels.[39] Standing during television commercials, getting up from one's desk or computer every 20 minutes and walking around the office may be the baby steps toward a more active lifestyle. At work, this does not have to be a random walk—people might use these activity breaks to deliver information instead of relying on email or telephone calls.

People often overestimate the time needed to derive benefits from physical activity. Naturally, more exercise will yield more dramatic results, but fitness and body composition improvements have been seen with as little as two strength training sessions each week or three 20-minute higher-level aerobic exercise sessions weekly.[6,23] Small changes, like exercising two days a week for 20 minutes, are less disruptive than exercising 3 days a week for an hour, and these changes still offer tangible health benefits. It is important, however, for previously sedentary people not to do too much when making the transition to an active lifestyle. A modest activity program may yield fewer gains at first, but in the long run be adhered to better.

People vary greatly in the kinds of activities they enjoy, so it is fortunate that many types of activities are beneficial. The key is to try a variety of activities until one that is enjoyable and meets one's goals is found. Some people like to join a health club because the array of exercise machines builds in variety. More self-motivated people may prefer having exercise equipment at home to reduce embarrassment and remove excuses for not exercising ("I don't feel like driving to the gym" or "I forgot my sneakers today"). Many people are happy to stick to walking, an activity that can be done anywhere and anytime. People who are uncomfortable in high-and even moderate-intensity activity and who cannot adhere to a 30-minute continuous program may be reassured to know that lower-intensity activity offers many health benefits.

Think of activity as being on a continuum, with inactivity at one end and high-intensity exercise at the other. The weight-management professional's goal is to move people along the continuum to their highest point of enjoyment, comfort, and safety. Figure 8-9 illustrates this progression from inactivity to an active lifestyle.

Clear, Written, or Verbal Instructions Help People to Get Started

We would not send a person on a low-sodium diet home without written guidelines for shopping, preparing meals, and eating in restaurants. The same consideration should be given to the person for whom physical activity is recommended. Useful information includes:

- Health clubs and gyms that are size friendly; area malls that have walking programs
- Types of activity that other clients have enjoyed
- Recommended frequency, intensity, and duration of various types of activity

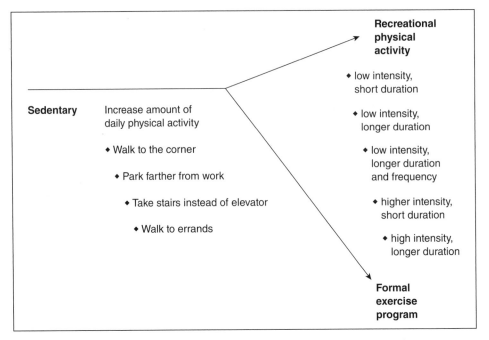

Sedentary

Increase amount of
daily physical activity

♦ Walk to the corner

♦ Park farther from work

♦ Take stairs instead of elevator

♦ Walk to errands

**Recreational
physical
activity**

♦ low intensity,
   short duration

♦ low intensity,
   longer duration

♦ low intensity,
   longer duration
   and frequency

♦ higher intensity,
   short duration

♦ high intensity,
   longer duration

**Formal
exercise
program**

FIGURE 8-9 Individuals should be encouraged to move along the continuum from an inactive to an active lifestyle to the extent of their capabilities.

- Specific instructions for using weight machines, stair-steppers, or other exercise equipment
- Methods for self-regulating exercise intensity, including perceived exertion and exercise HR

Setting up exercise groups can also help people to get started. Exercising with a "buddy" provides socialization and positive reinforcement for becoming more active. It also builds confidence by offering opportunities for verbal encouragement and sharing successful physical activity experiences.

Barriers to Becoming Active Should Be Addressed

A number of barriers prevent people from becoming more active. Some barriers are tangible, like safety issues for those who live in unsafe neighborhoods, or the cost of joining an exercise facility or purchasing home equipment for low-income clients. Some barriers are physiological, such as low functional capacity or joint deterioration. Many obese individuals also contend with psychological barriers. A person who has been teased about her weight may be reluctant to put herself on display while exercising. The person who had to run laps as punishment for being "fat and lazy" may not see any type of physical activity as fun.

Weight-management professionals who discuss barriers to becoming more active can assist their clients in finding solutions. Professionals can also maintain lists of weight-sensitive fitness centers, gyms, and personal trainers (yes, even public recreation centers may have personal trainers on staff). We cannot ignore very real barriers to adopting an active lifestyle and assume the client will work it out. And, if a barrier cannot be overcome, acknowledge it.

How Do We Motivate People to Adhere to Physical Activity?

When people return to an inactive lifestyle, they lose most of the beneficial effects of physical activity rather quickly. Within two weeks of inactivity, for example, cardiorespiratory fitness gains begin to disappear. Weight lost through exercise is likely to be regained if activity is not maintained.

relapse *Complete breakdown in behavior, return to old behaviors, and loss of confidence in one's ability to sustain change.*

lapse *Temporary setback during the process of change.*

Obese individuals who become more physically active tend to find that activity is easier to adhere to than diet. Still, most reports on adherence to exercise programs observe that about half of nonobese persons in formal exercise programs will drop out within a year.[40] Although there is very little research on exercise adherence with obese people, there are some indications that adherence rates may be similar to the nonobese. The term **relapse** refers to a relatively permanent reversion to old behaviors after a **lapse** or series of lapses. An exercise lapse is not necessarily a bad thing, and clients should be prepared for lapses to minimize the probability of relapse. These suggestions may help keep people motivated and prevent relapse.

Set Realistic Goals

People base their willingness to continue a behavior on the extent to which the behavior meets their expectations. Therefore, setting realistic and attainable activity goals is critical to adherence. Activity goals should be based on the client's current activity level (determined by a physical activity assessment) and readiness to change: Does the client know that activity has health benefits? Does she believe that she needs to be more active? What have his past experiences been with trying to become more active? What barriers to change exist?

Health-related goals are more reliably met than weight goals. Although people who are able to engage in substantial amounts of exercise can lose 12% or more of body weight from exercise alone, for many obese persons, engaging in 60 or more minutes of vigorous activity every day is not only impractical but could also be unsafe. Most studies of moderate exercise programs (1–3 hours/week) do not show 12% weight losses. It is far better to set a predictable health goal—and perhaps see weight loss as the icing on the cake—than to set an unpredictable weight goal and see clients give up because it cannot be met.

Even people who are willing to increase their daily physical activity considerably need to be realistic about the amount of fat that can be lost and the time required to lose it. In most cases, weight gain has occurred over a period of years. Why should substantial weight loss happen in a month? The professional needs to continually remind clients that the energy deficit created by moderate exercise is small and therefore weight loss will occur slowly and gradually. Also, if muscle tissue is gained at the same time fat is lost, there may be metabolic benefit but little weight change to show for it.

Initially, clients should keep activity records to record their progress (distance walked, minutes on the treadmill, exercise HR attained) and how they feel (perceived exertion, muscle aches). This approach is extremely motivating to most people. Keeping accurate self-reports is also helpful in looking realistically at goal attainment. Obesity researchers estimate that 40–60% of overweight women overstate the amount of exercise they perform, and these women appear to lose less weight than women who report exercise with more accuracy.[41]

For individuals who are trying to reduce time spent in inactivity or who are following a walking regimen, pedometers offer an effective way to not only set goals but to monitor progress. A meta-analysis of data from 26 studies on the effectiveness of pedometers found that motivation was increased in participants who set step goals.[42]

Build in Rewards

Few people have the capacity to engage in a behavior that does not produce at least some reward. If we want people to remain active, then we need to build in reinforcement (and remind clients of all of the unseen benefits from activity).

- When appropriate, recommend group exercise, which offers socialization and group reinforcement.
- Provide periodic individual counseling/booster sessions.
- Remind clients to try new activities or to alter activity intensity and duration for variety. At least initially, adherence rates are higher among people who exercise intermittently rather than continuously, and varying one's routine can reinforce continued participation.[40]
- Prevent injuries and complications of exercise by appropriately screening clients for increased activity.

Prepare for Weight-Loss Plateaus

Eventually even the most adherent person will experience a slowing of weight loss. Researchers who look at the metabolic effects of exercise note that fat oxidation steadily decreases as fat is lost until the individual reaches a point at which additional fat losses do not occur. People who are prepared for this plateau, who have learned to enjoy being more active, and who have not exclusively focused on weight loss as a physical activity goal will be better able to look to the health and weight-maintenance effects of exercise. People who are not prepared may adopt the attitude of "Why bother?" and return to inactivity.

There is a great deal of individual variation in response to physical activity. Trainers sometimes call people high-responders if they show rapid and dramatic responses to exercise and low-responders if they do not appear to derive as much benefit from activity. The exercise physiology literature contains many examples of differences in energy nutrients oxidized, skeletal muscle metabolism, lipoprotein lipase activity, cardiovascular capacity, and other factors that influence the response to exercise. Individual variation is affected by:

- Genetic differences: Studies of identical twins illustrate that the tendency of exercise to reduce overall body fat and abdominal fat stores is at least partly controlled by heredity.[43]
- Fat pattern: Abdominal fat is more readily mobilized during exercise than gluteofemoral fat.
- Number of fat cells: People with hyperplasia are not as responsive to weight reduction with exercise.
- Mechanical efficiency: Because individuals perform the same activity with different levels of efficiency, estimated energy expenditure in a given activity may vary among people by as much as 25%.[44]

Because of these factors, people will vary in the weight-loss potential of exercise. So, mathematically predicting weight loss based on 1 pound of fat having a caloric equivalent of 3,500 to 4,000 kcal is simply not be a good idea. People do not always lose expected amounts of weight based on exercise performed, and we do our clients a disservice by predicting weight loss.

Combine Physical Activity with Diet Change

Keeping in mind that substantial individual variation exists, there is a dose effect with respect to physical activity and weight loss. As physical activity volume increases, so does weight loss. People who are particularly motivated by weight loss may find that combining physical activity with moderate dietary changes will promote greater weight loss and better adherence. The ACSM suggests that a 500- to 700-calorie energy restriction combined with physical activity will increase weight loss.[26] Extremes in either diet or activity should be actively discouraged. As covered in Chapter 4, any combination of reduced caloric intake and increased energy expenditure through physical activity should never exceed 1,000 kcal/day. A low-fat, low-calorie diet combined with 30–60 minutes of vigorous exercise three times a week may initially promote dramatic weight loss; however, adherence to such an extreme program is unlikely.

Motivating People to Adopt an Active Lifestyle: Summary

Most Americans do not exercise, and over one-third of overweight adults engage in no leisure-time physical activity. People thinking about becoming more active may be helped in making the transition to an active lifestyle by learning that significant change is not needed to get benefits from physical activity, addressing the barriers to physical activity, and receiving specific instructions for exercise. Setting realistic goals, finding rewards for activity, and preparing for weight plateaus and lapses may help promote adherence.

IMPORTANCE OF ACTIVITY IN HEALTH AND WEIGHT MANAGEMENT

Physical activity offers benefits that go far beyond weight loss or maintenance. Whether or not a person is overweight, adopting an active lifestyle improves fitness, reduces the risk of cardiovascular disease, and promotes overall better health. Individuals who develop cardiorespiratory fitness through exercise tend to live longer, even if they are overweight.

Ideally, weight management includes a variety of behavioral changes for healthy diet and activity. People who want to lose weight will find that, in the long run, increasing physical activity and adopting a regular exercise program are more beneficial than making extreme dietary changes. An active lifestyle can do many things that caloric restriction cannot: reduce abdominal fat, maintain lean body mass, improve fitness, enhance psychological well-being, and improve glucose tolerance and blood lipid levels. Adhering to an active lifestyle is also easier than following a severe calorie- or fat-restricted diet. For people who want to maintain their weight, becoming more physically active allows them to continue to enjoy food without constant worry about caloric intake while they improve their health.

Well over half of the U.S. adult population has never engaged in vigorous physical activity for more than 10 minutes. Among those who do initiate an exercise program, adherence is low. To get people off the couch, weight-management professionals need to clearly promote the message that activity offers many of the same health benefits as exercise. Knowing that one does not have to participate in formal exercise to gain health benefits may improve adherence to physical activity.

The new National Physical Activity Plan is a collaboration between the public and private sectors which aims to create a national culture that supports physically active lifestyles.[45] The plan offers strategies at local, state, federal, and institutional levels for promoting activity. The importance of this collaborative approach cannot be overstated. Individual efforts are commendable, but broader efforts are needed to make physical activity accessible to all.

Case Study. The Novice Exerciser, Part 2	Today, at age 65, Janii can't even walk a half-mile to the post office without feeling winded. She now has full-blown high blood pressure, in addition to asthma. At 5′ 5″ tall and 200 pounds, she is also now obese.

Janii shakes her head in dismay at what has happened to her. "Six years ago I was 50 pounds lighter and had my blood pressure under control with medication. I walked a lot and I ate regular, healthy meals. I went to the community center two times a week for exercise. But when I left my job downtown to stay home and take care of my mother, everything went to pot." Instead of being able to go to exercise class on her way home from work, Janii's schedule revolved around her mother's needs. "And now that my mother is gone," Janii says, "I want to be active again, but I just don't have the energy." Janii recently stopped in at the community center to inquire about exercise classes. "They had me fill out something called a PAR-Q, and they didn't like my answers to questions 2, 4, and 6. So now I have to go to my doctor again."

1. Discuss the barriers to becoming more physically active that Janii faces.
2. What are some potential solutions to these barriers?
3. What type of activity/exercise program would you suggest for Janii (and her doctor's approval)? Use Appendix B Table 3 to help pull your ideas together for this program.

REFERENCES

1. Bielinski, R., Schutz, Y., & Jequier, E. (1985). Energy metabolism during the postexercise recovery in man. *American Journal of Clinical Nutrition, 42*(1), 69–82.
2. Tremblay, A., Simoneau, J. A., & Bouchard, C. (1994). Impact of exercise intensity on body fatness and skeletal muscle metabolism. *Metabolism: Clinical and Experimental, 43*(7), 814–818.
3. Saris, W. H., Blair, S. N., van Baak, M. A., Eaton, S. B., Davies, P. S., Di Pietro, L., et al. (2003). How much

physical activity is enough to prevent unhealthy weight gain? Outcome of the IASO 1st Stock Conference and consensus statement. *Obesity Reviews, 4*(2), 101–114.
4. Horton, T. J., & Hill, J. O. (1998). Exercise and obesity. *Proceedings of the Nutrition Society, 57*, 85–91.
5. Grediagin, A., Cody, M., Rupp, J., Benardot, D., & Shern, R. (1995). Exercise intensity does not effect body composition change in untrained, moderately

overfat women. *Journal of the American Dietetic Association*, 95(6), 661–665.

6. Westcott, W. (2009). ACSM strength training guidelines: Role in body composition and health enhancement. *ACSM's Health & Fitness Journal*, 13(4), 14–22.

7. Elder, S. J., & Roberts, S. B. (2007). The effects of exercise on food intake and body fatness: A summary of published studies. *Nutrition Reviews*, 65(1), 1–19.

8. Martins, C., Morgan, L., & Truby, H. (2008). A review of the effects of exercise on appetite regulation: An obesity perspective. *International Journal of Obesity*, 32(9), 1337–1347.

9. King, N. A., Caudwell, P., Hopkins, M., Byrne, N. M., Colley, R., Hills, A. P., et al. (2007). Metabolic and behavioral compensatory responses to exercise interventions: Barriers to weight loss. *Obesity*, 15(6), 1373–1383.

10. Martins, C., Kulseng, B., King, N. A., Holst, J. J., & Blundell, J. E. (2010). The effects of exercise-induced weight loss on appetite-related peptides and motivation to eat. *Journal of Clinical Endocrinology and Metabolism*, 95(4), 1609–1616.

11. Foster-Schubert, K. E., McTiernan, A., Frayo, R. S., Schwartz, R. S., Rajan, K. B., Yasui, Y., et al. (2005). Human plasma ghrelin levels increase during a one-year exercise program. *Journal of Clinical Endocrinology and Metabolism*, 90(2), 820–825.

12. Hagobian, T. A., Sharoff, C. G., Stephens, B. R., Wade, G. N., Silva, J. E., Chipkin, S. R., et al. (2009). Effects of exercise on energy-regulating hormones and appetite in men and women. *American Journal of Physiology- Regulatory, Integrative, and Comparative Physiology*, 296, R233–R242.

13. Otten, J. J., Hellwig, J. P., & Meyers, L. D. (Eds.). (2006). *Dietary reference intakes: The essential guide to nutrient requirements*. Washington, D.C.: The National Academies Press.

14. Schoeller, D. A., Shay, K., & Kushner, R. F. (1997). How much physical activity is needed to minimize weight gain in previously obese women? *American Journal of Clinical Nutrition*, 66(3), 551–556.

15. Tate, D. F., Jeffery, R. W., Sherwood, N. E., & Wing, R. R. (2007). Long-term weight losses associated with prescription of higher physical activity goals are higher levels of physical activity protective against weight regain? *American Journal of Clinical Nutrition*, 85(4), 954–959.

16. Takeshima, N., Rogers, M. E., Watanabe, E., Brechue, W. F., Okada, A., Yamada, T., et al. (2002). Water-based exercise improves health-related aspects of

fitness in older women. *Medicine and Science in Sports and Exercise*, 34(3), 544–551.

17. Gwinup, G. (1987). Weight loss without dietary restriction: Efficacy of different forms of aerobic exercise. *American Journal of Sports Medicine*, 15(3), 275–279.

18. Schmitz, K. H., Hannan, P. J., Stovitz, S. D., Bryan, C. J., Warren, M., & Jensen, M. D. (2007). Strength training and adiposity in premenopausal women: Strong, healthy, and empowered study. *American Journal of Clinical Nutrition*, 86(3), 566–572.

19. Tanaka, H., Monahan, K. D., & Seals, D. R. (2001). Age-predicted maximal heart rate revisited. *Journal of the American College of Cardiology*, 37(1), 153–156.

20. Bernaards, C. M., Twisk, J. W., Van Mechelen, W., Snel, J., & Kemper, H. C. (2003). A longitudinal study on smoking in relationship to fitness and heart rate response. *Medicine and Science in Sports and Exercise*, 35(5), 793–800.

21. Ainsworth, B. E., Haskell, W. L. Whitt, M. C., Irwin, M. L., Swartz, A. M., Strath, S. J., et al. (2000). Compendium of physical activities: An update of activity codes and MET intensities. *Medicine and Science in Sports and Exercise*, 32(9), S498–504.

22. Fletcher, G. F., Balady, G. J., Amsterdam, E. A., Chaitman, B., Eckel, R., Fleg, J., et al. (2001). Exercise standards for testing and training: A statement for healthcare professionals from the American Heart Association. *Circulation*, 104(14), 1694–1740.

23. Haskell, W. L., Lee, I. M., Pate, R. R., Powell, K. E., Blair, S. N., Franklin, B. A., et al. (2007) Physical activity and public health: Updated recommendations for adults from the American College of Sports Medicine and the American Heart Association. *Circulation*, 116, 1081–1093.

24. Lee, I. M., Hsieh, C. C., & Paffenbarger, R. S., Jr. (1995). Exercise intensity and longevity in men. The Harvard alumni health study. *JAMA*, 273(15), 1179–1184.

25. American College of Sports Medicine. (2010). *ACSM's guidelines for exercise testing and prescription* (8th ed.). Philadelphia: Lippincott Williams & Wilkins.

26. Donnelly, J. E., Blair, S. N., Jakicic, J. M., Manore, M. M., Rankin, J. W., Smith, B. K., et al. (2009). American College of Sports Medicine position stand. Appropriate physical activity intervention strategies for weight loss and prevention of weight regain for adults. *Medicine and Science in Sports and Exercise*, 41(2), 459–471.

27. Jakicic, J. M., Winters, C., Lang, W., & Wing, R. R. (1999). Effects of intermittent exercise and use of

home exercise equipment on adherence, weight loss, and fitness in overweight women: A randomized trial. *JAMA, 282*(16), 1554–1560.

28. Pavlou, K. N., Krey, S., & Steffee, W. P. (1989). Exercise as an adjunct to weight loss and maintenance in moderately obese subjects. *American Journal of Clinical Nutrition, 49*(5 Suppl), 1115–1123.

29. Jakicic, J. M., Marcus, B. H., Lang, W., & Janney, C. (2008). Effect of exercise on 24-month weight loss maintenance in overweight women. *Archives of Internal Medicine, 168*(14), 1550–1559.

30. Flegal, K. M., Carroll, M. D., Ogden, C. L., & Curtin, L. R. (2010). Prevalence and trends in obesity among US adults, 1999–2008. *JAMA, 303*(3), p. 235–241.

31. Thompson, P. D., Franklin, B. A., Balady, G. J., Blair, S. N., Corrado, D., Estes, N. A., 3rd, et al. (2007). Exercise and acute cardiovascular events placing the risks into perspective: A scientific statement from the American Heart Association Council on Nutrition, Physical Activity, and Metabolism and the Council on Clinical Cardiology. *Circulation, 115*(17), 2358–2368.

32. Balady, G. J., Chaitman, B., Driscoll, D., Foster, C., Froelicher, E., Gordon, N., et al. (1998). Recommendations for cardiovascular screening, staffing, and emergency policies at health/fitness facilities. *Circulation, 97*(22), 2283–2293.

33. Gordon, N. F. (1998). Pre-participation in health appraisal in the nonmedical setting. *ACSM's resource manual for guidelines for exercise testing and prescription* (3rd ed., pp. 341–346). Baltimore: Williams & Wilkins.

34. U.S. Preventive Services Task Force. (2004 February). *Screening for coronary heart disease, topic page.* Rockville, MD: Agency for Healthcare Research and Quality. Available from http://www.ahrq.gov/clinic/uspstf/uspsacad.htm. Accessed July 16, 2010.

35. Lauer, M., Froelicher, E. S., Williams, M., Kligfield, P., & American Heart Association Council on Clinical Cardiology, Subcommittee on Exercise, Cardiac Rehabilitation, and Prevention. (2005). Exercise testing in asymptomatic adults: A statement for professionals from the American Heart Association Council on Clinical Cardiology, Subcommittee on Exercise, Cardiac Rehabilitation, and Prevention. *Circulation, 112*(5), 771–776.

36. Wei, M., Kampert, J. B., Barlow, C. E., Nichaman, M. Z., Gibbons, L. W., Paffenbarger, R. S., Jr., et al. (1999). Relationship between low cardiorespiratory fitness and mortality in normal-weight, overweight, and obese men. *JAMA, 282*(16), 1547–1553.

37. *National health interview surveys: Summary health statistics for U.S. adults.* (2009). Available from www.cdc.gov/nchs/nhis.htm. Accessed July 15, 2010.

38. Eaton, D. K., Kann, L., Kinchen, S., Shanklin, S., Ross, J., Hawkins, J., et al. (2008). Youth risk behavior surveillance—United States, 2007. *Surveillance Summaries, MMWR, 57*(SS–4), 1–136.

39. Healy, G. N., Dunstan, D. W., Salmon, J., Cerin, E., Shaw, J. E., Zimmet, P. Z., et al. (2008). Breaks in sedentary time: Beneficial associations with metabolic risk. *Diabetes Care, 31*(4), 661–666.

40. Dishman, R. K. (Ed.). (1988). *Exercise adherence: Its impact on public health.* Champaign, IL: Human Kinetics.

41. Jakicic, J. M., Polley, B. A., & Wing, R. R. (1998). Accuracy of self-reported exercise and the relationship with weight loss in overweight women. *Medicine and Science in Sports and Exercise, 30*(4), 634–638.

42. Bravata, D. M., Smith-Spangler, C., Sundaram, V., Gienger, A. L., Lin, N., Lewis, R., et al. (2007). Using pedometers to increase physical activity and improve health: A systematic review. *JAMA, 298*(19), 2296–2304.

43. Bouchard, C., Tremblay, A., Despres, J. P., Theriault, G., Nadeau, A., Lupien, P. J., et al. (1994). The response to exercise with constant energy intake in identical twins. *Obesity Research, 2*(5), 400–410.

44. Heitmann, B. L., Kaprio, J., Harris, J. R., Rissanen, A., Korkeila, M., & Koskenvuo, M. (1997). Are genetic determinants of weight gain modified by leisure-time physical activity? A prospective study of Finnish twins. *American Journal of Clinical Nutrition, 66*(3), 672–678.

45. National Physical Activity Plan. Available from http://www.physicalactivityplan.org. Accessed July 15, 2010.

Behavioral Approaches to Weight Management

R. Gino Santa Maria/Shutterstock.com

cappi thompson/Shutterstock.com

lorga Studio/Shutterstock.com

Behavioral approaches to weight management focus on the role of inactivity and an unhealthy diet in the development and maintenance of obesity. The emphasis is on improved diet, normalized food habits, increased physical activity, and prevention of further weight gain, not necessarily weight loss. This statement may seem contradictory in a text on weight management, but consider:

- Only tobacco causes more deaths each year than inactivity and poor diet.[1]
- Chronic disease risk can be reduced through a program of diet and activity modification.
- Biological factors exert a strong influence on body weight that may be difficult to overcome.

Some obese people need to lose weight to regain health. Fortunately, however, their weight losses do not have to be large to improve blood cholesterol, blood pressure, and glucose control. Because weight loss is not a behavior that an individual can control, behavioral approaches focus instead on things that are within the individual's command—poor diet and inactivity. With these behaviors as targets, weight loss may or may not occur, but health is sure to improve.

Behavior change is a process that takes time. People who successfully make a change progress through several stages, from denial and defensiveness to willingness and readiness for action. The culture(s) with which people identify exert a powerful influence on weight-management behaviors. This chapter explores the process of change and factors that influence it, with an emphasis on diet, activity, and eating behaviors. We will also look at the reasons behind recommendations for dietary change and suggest dietary strategies that will improve overall health, promote compliance, and attend to individual needs.

HOW PEOPLE CHANGE

transtheoretical model *Theory of behavior change, also known as the*

stages of change *Theory, that describes change as progressing through six stages: precontemplation, contemplation, preparation, action, maintenance, and termination.*

In 1959, a review of obesity treatment programs found that very few obese individuals lose significant amounts of weight in treatment and that 95% of those who lose weight regain it within 5 years.[2] More recent estimates suggest that about one-third of dieters regain lost weight within a year and return to their initial weights in 3–5 years.[3] Some experts believe that biological controls of body weight are so powerful that no treatment can overcome them. A major factor behind the anti-dieting movement is this perceived failure of most weight-loss programs.

Interventions based on sound principles of behavior change, particularly those beginning in childhood, have much higher success rates than non-behavioral programs in helping people to keep off lost weight. Even people using pharmacological treatments or liquid meal replacements experience greater success when linked with behavior change.[4,5] Behavior-change programs are built on our understanding of why and how people change. An excellent model for this concept is the **transtheoretical model** of change, sometimes called the **stages of change theory**. This theory is derived from several other theories that attempt to explain health behavior (this is the reason that it is called "transtheoretical"), such as the health belief model, social cognitive theory, and the theory of reasoned action. Stages of change theory can help us in designing individualized weight-management approaches that promote success and minimize failure.

TABLE 9-1   The Initial Stages of Change

Stage	Characteristic comment about diet	Characteristic comment about activity
Precontemplation	"My fat intake is just fine."	"My grandmother never exercised a day in her life, and she lived to 85."
Contemplation	"I think I could stand to cut out some fats from my diet."	"Maybe I wouldn't get so out of breath if I walked a bit more."
Preparation	"Two percent milk isn't as bad as I thought. Maybe I will be able to reduce my dietary fat."	"I enjoyed walking in the pool when we went to Florida. Maybe I can find a pool with a walking program."

Stages of Change Theory

Many theories have attempted to explain why people change behavior but, until development of the transtheoretical model, no theory considered that change occurs over time.[6,7] Change is not a single event—as in "I will reduce my fat intake to 20% of total calories." Change is a process—as in "I will think about looking for low-fat recipes, I will start reading food labels more closely, I will buy a magazine and try some recipes, and, eventually, I will begin asking for low-fat items on restaurant menus."

The transtheoretical model describes change as progressing through six stages: precontemplation, contemplation, preparation, action, maintenance, and termination. Table 9-1 illustrates characteristic comments of people in the early stages of change. Most obese individuals are thought to be in these early stages for behaviors related to weight loss, such as increasing physical activity, utilizing portion control, and decreasing fat intake.[8]

Identifying a person's stage of readiness for change and developing an intervention that matches readiness can have two important effects: (1) It increases the likelihood that the individual will progress to at least the next stage of change, and (2) it reduces the likelihood that the individual will drop out of a behavior-change program or discontinue the new behavior. The transtheoretical model has been used in the treatment of addictions and anxiety disorders, in AIDS prevention and cancer screening, for smoking cessation, and in nutrition and exercise programs.

Matching Intervention to Individual

The stages-of-change model is useful because it matches the intervention with the client's readiness for change. Making this match is not an exact science, but determining a person's readiness for change can maximize success and minimize failure. Studies of smokers have found that when stage-matched programs are used and people progress through only one stage during treatment, morale and ultimately long-term success are greatly enhanced.[9] Change takes time. During a 10-week or 15-week program, not everyone will take action but, if clients feel successful at having moved from contemplation to preparation, they will be more likely to eventually move to the action stage.

The transtheoretical model has been applied to numerous nutritional and exercise interventions.[10–13] From these we have learned several lessons:

- People may know that immediate weight loss or a change in diet or physical activity is needed to improve health, but this knowledge is rarely enough to spur them to action.
- Weight loss that occurs before people are ready for action is unlikely to be maintained.

• People in earlier stages of change may benefit from taking the time to talk and express concerns, build rapport with a potential change agent, and gain nutrition information. They may not benefit from an intervention imposed too soon.

Table 9-2 summarizes weight-management approaches for people in each stage of change. Using the detailed information in this chapter on the six stages of change should help you match clients to their stage of change and recommend stage-tailored interventions.

How People Change: Summary

The transtheoretical model describes change as progressing through six stages: precontemplation, contemplation, preparation, action, maintenance, and termination. Identifying a person's stage of readiness for change, developing a weight-management intervention

TABLE 9-2 Weight-Management Approaches for People at Each Stage of Change

Precontemplation stage	Contemplation and preparation stages	Action stage	Maintenance stage
Conduct assessments.	Clarify feelings, values, and attitudes about behavior change (self-reevaluation).	Set reasonable and attainable goals for change. Diet and activity goals are more appropriate than weight-loss goals.	Develop problem-solving skills to cope with situations that promote relapse.
Determine client's perceptions of severity of the problem, susceptibility to the problem, benefits of action, and barriers to action (health belief model).	Build a support system.	Use contingency contracting and rewards.	Identify support systems and groups.
Provide information about the problem, particularly about benefits to change (consciousness raising).	Counteract barriers to change with salient benefits.	Self-monitor diet, activity, and other behaviors.	Plan for relapse.
Convey acceptance of client.		Identify and control environmental stimuli that promote unhealthy eating and inactivity.	Promote restructuring of social environment.
		Modify eating styles that promote unhealthy eating.	
		Replace negative self-statements with positive self-statements (cognitive restructuring).	

TABLE 9-3 Characteristics of the Precontemplation Stage

- No intent to change within the next 6 months (for weight management, no intent to reduce portion sizes, amount of food consumed, or dietary fat; or to increase fruit and vegetable consumption or physical activity level)
- May deny that a problem exists
- Lack confidence in ability to change behavior
- Most common stage for program dropouts

that matches readiness, and addressing problem behaviors will increase the likelihood that the individual will progress to at least the next stage. This approach also reduces the likelihood that the individual will drop out. Determining readiness for change is important for maximizing success and minimizing failure.

BEHAVIORAL STRATEGIES FOR PEOPLE THINKING ABOUT CHANGE

According to the stages of change theory, people who are thinking about change are precontemplators. In reality, the only reason many precontemplators are thinking about change is because their doctor said they'd better lose some weight, get some exercise, or alter their diet. Precontemplators have no immediate plans to change their behavior.

What Characterizes the Precontemplation Stage?

precontemplation stage First stage of change; individuals in this stage do not intend to change during the next 6 months, may lack confidence in their ability to make a change, and often deny that a problem exists.

People in the **precontemplation stage** do not intend to change during the next 6 months. They may lack information about the need for change, or they may have failed in the past and now lack confidence in their ability to make a change. They may spend considerable energy denying that a problem exists and take pains to avoid receiving information about their health problem. Characteristics of precontemplators are summarized in Table 9-3.

If people in this stage were coerced into coming to see you, you would probably characterize them as unmotivated and unwilling to change. However, because you would be concerned about their health, you might refer them to a particular program or you might design a diet plan for them. They will not go to the program more than once or twice, nor will they follow your diet plan, because they are not ready. Almost all health program drop-outs are precontemplators.[14]

You can recognize precontemplators by their denial that there is a problem and their lack of intent to alter their behavior. Reaching precontemplators is challenging. Weight-management professionals either do not even try, or they convince themselves that the resistant client will come around in a few weeks, see the benefits of a change in diet and/or activity, and make a permanent change. This belief sets up both the client and the professional for failure.

How Does the Health Belief Model Explain Early Stages of Change?

health belief model Idea that health behavior depends on perceived susceptibility to a health problem, perceived severity of the problem, and perceived benefits of and barriers to health action.

People in the early stages of change use several processes as they move from resistance to contemplation to action. The **health belief model** offers an explanation for these processes and proposes that an individual's health behavior depends on four factors:

1. Perceived susceptibility to a health problem: If Carla's physician advises weight loss to avoid hypertension, how susceptible does Carla believe she actually is to developing hypertension.

2. Perceived severity of the health problem: Does Carla believe that hypertension is a serious condition?

3. Perceived benefits of health action: Is weight loss to prevent hypertension enough of a reason for Carla to try to lose weight?

4. Perceived barriers to health action: What does Carla see as obstacles to change (for example, bland food, no time to exercise), and does she believe that she can overcome these obstacles?

For people in the earliest stages of change, perceived severity of the health problem and perceived benefits of action are key factors in helping to initiate change.[15]

What Are Some Strategies for Precontemplators?

A full-blown weight-management intervention will not help a person in the precontemplation stage of readiness, but things can be done to move precontemplators along. At this stage, individuals are examining their feelings about change and becoming aware of the pros and the cons (the benefits and the costs) of changing their behavior. Without directly seeking information, they are forming an opinion about the pros ("I'll feel better"; "I won't need blood pressure medication") and the cons ("I'll have to eat less"; "My food will taste bland").

consciousness-raising *Building awareness about a behavior that needs to be changed.*

To move into the contemplation stage, the perceived benefits of change must increase.[9] So, in the precontemplation stage, the most helpful interventions are: (1) to determine the client's perceptions of the problem and (2) to provide information and give feedback about the advantages of changing, sometimes called **consciousness raising**. Approaches for precontemplators are summarized in the first column of Table 9-2.

Determining Perceptions of the Problem

Weight-management professionals must assess perceived susceptibility, threat, benefits, and barriers because these are the factors that influence health behavior. Assessment should include:

- Measuring indicators of the health problem and determining the client's understanding of the relationship of the problem to weight ("What is your blood pressure? Has it gotten higher since you've gained weight?")

- Gauging the client's perceptions of the severity of the problem ("What did your doctor tell you about high blood pressure? Have you done any reading about it?")

- Learning about past attempts at weight loss to address the problem and the outcomes of those attempts ("Have you ever tried to lose weight to reduce your blood pressure? What happened?")

- Identifying the client's attitude toward changing diet and activity habits ("How do you feel about reducing the salt in your diet?" "Are you interested in increasing your activity level?")

- Determining the client's perceptions of the difficulties involved in making needed changes ("What do you think would be hardest for someone who wants to increase their physical activity?" "Do you think people can follow a lower salt diet in today's world?")

- Learning whether the client is in denial about a problem ("Do you think that your blood pressure is too high?" "Do you think that you need to lose weight?")

As a result of your conversations with Carla, you might be able to convince her to read an article or a blog about how others like her increased their physical activity relatively easily by using the stairs at work or walking at the mall on weekend mornings. Learning that she could benefit from engaging in an activity that she can imagine herself doing might nudge Carla toward contemplating change.

Consciousness Raising

Consciousness-raising interventions build awareness. One way to increase awareness is by providing information. Some people are unaware of the problems associated with a particular health condition or the relationship between their body weight and their health. Their physician may have failed to make this clear or had insufficient time to discuss it thoroughly. For these people, you might offer brochures or fact sheets that describe the health problem and its long-term effects. You might direct them to a Web site on the Internet where they could find answers to frequently asked questions without having to interact with anyone.

Keep in mind that helping someone to perceive weight-related behaviors as a problem can be difficult. Whereas smoking has some social stigma, very little about eating does. Cigarette sales are restricted, sale of French fries is not; parents try to keep their children away from cigarettes yet hold birthday parties at fast-food restaurants. Cigarettes are no longer advertised on American television, but food advertising exceeds that of any other U.S. industry.[16] Although society seems to worship thinness, weight-management professionals work in an eating-promoting environment.

Some precontemplators are aware of the health problems associated with their weight but are defensive and angry. They may have had contact with medical professionals who demeaned them, believing that weight gain is a sign of weakness. They may have a relative who weighed 300 lb and lived to be 90. By the time these individuals come to you, their level of defensiveness and denial is quite high. You will need to convey a sense of acceptance. Moving the conversation away from weight to some specific aspect of behavior is helpful. Scare tactics should be avoided. If perceived severity and susceptibility are too great, most people simply deny that the particular health condition could affect them.

Behavioral Strategies for People Thinking About Change: Summary

Precontemplators do not intend to change their behavior in the next 6 months. They make up a large number of weight-management program drop-outs. The health belief model shows how people move from resistance to contemplation to action as they assess their perceived susceptibility to a health problem and its severity along with perceive benefits and barriers to change. Interventions at this stage of readiness focus on ing the client's perceptions of the problem, raising consciousness about and conveying acceptance.

| Case Study. The Ex-Football Player, Part I | Luis is a 28-year-old who works for an engineering firm in California. His parents moved to the United States from Mexico when he was 4 years old. He was active and athletic through high school but became less active in college and now finds himself about 20 lb above his healthy weight. He is thinking about his tenth high school reunion in June, three months away, and worries what people he hasn't seen in a long time will think about his increased girth.

"Man, I was a real popular guy in school—football player, track star—I looked GOOD. But I'm not looking so good right now," Luis told his wife one day.

"Why don't you try this diet book," his wife said. "It could really help you."

Luis followed the book's suggestions for a lower-fat diet, and he even visited a health club several times each week. He stopped eating his usual Mexican meals and gave up beer, and by June he had lost almost all of the excess weight. The week after the reunion, Luis only went to the gym once, and resumed drinking a beer with dinner. His mother came for a visit, and he ate her traditional Mexican dishes every night. By the end of the summer his lost weight had been regained.

• What are some possible reasons for Luis's weight regain?
• Suggest some strategies, based upon health behavior theory that might help Luis move to the next stage of change and perhaps experience long-term success in managing his weight. |

BEHAVIORAL STRATEGIES FOR PEOPLE READY TO ACT

People who are getting ready to act are in the contemplation or preparation stage. Although moving into these stages from precontemplation implies that action is imminent, people can remain stalled in contemplation for some time.

What Characterizes the Contemplation and Preparation Stages?

contemplation stage *Stage of change in which individuals are preparing for change in the next 6 months; sometimes people get stuck in this stage for quite a long time.*

preparation stage *Stage of change in which an individual is ready for change within the next month and has typically already made some small change in behavior.*

People in the **contemplation stage** are preparing for change during the next 6 months. If you were to ask them about their readiness to change, they would tell you that they are seriously considering change. They are examining the pros and the cons of changed behavior more critically than they did as precontemplators and may have already tried to make a change. Still, sometimes people who plan to make a change within 6 months get stuck and are contemplators for quite a long time.

Once an individual reaches the point of making change in the next month, he has entered the **preparation stage**. Preparation is characterized by making some change, even if it is small. The individual may switch from whole milk to 2%, may get a salad instead of French fries, or may order broiled fish instead of fried. People in the preparation stage are excellent candidates for weight-management interventions. Table 9-4 summarizes characteristics of those in the contemplation and preparation stages.

How Does the Theory of Planned Behavior Help Explain Readiness for Change?

The **theory of planned behavior** helps to explain some of the factors that influence change in the preparation stage. The theory states that a person's intention to perform a

TABLE 9-4 Characteristics of the Contemplation and Preparation Stages

- Preparing for change within the next 6 months
- May have made a small behavior change already (for weight management, may have started eating less at a meal to compensate for earlier splurges; avoiding fast foods; eating fruit as a dessert; and looking for ways to add more physical activity every day)
- May become stalled in contemplation

theory of planned behavior *Theory that suggests that intention to perform a behavior is the best determinant of whether the behavior will be adopted. Three factors influence intent: general attitudes toward the behavior, belief in the importance of the behavior to others, and perceived behavioral control.*

specific behavior is the best determinant of whether the person will adopt the behavior. Three factors influence intent:[17]

1. General attitudes toward the behavior: We would want to know what value Carla places on weight loss for blood pressure control.
2. Belief in the importance of the behavior to others: Has Carla's family expressed concern about her weight or blood pressure?
3. Perceived behavioral control: To what extent does Carla believe that her weight, diet, or activity is under her control, or do obstacles to change put the health behavior out of her reach?

The theory of planned behavior evolved from an earlier theory by the same author, which was called the theory of reasoned action.[18] The original theory was relevant only to behaviors under voluntary control; adding the concept of perceived behavioral control extends the theory to behaviors that are not solely voluntary, such as weight management behaviors. Thus the theory of planned behavior tells you that Carla is unlikely to move toward change when she is told to "lose 15 lb" (especially if she has already tried and failed). Carla is more likely to move toward action when advised to "get 30 minutes of physical activity three times a week." Carla can perceive activity as being under her control to a greater extent than weight loss is, even if she doesn't express this. The theory of planned behavior is the model that has been used most frequently to explain why people exercise.[19]

What Are Some Strategies for People Ready to Act?

People who have overcome denial and defensiveness will benefit from weight-management approaches that prepare them for action. These include further clarification of beliefs and attitudes, developing a support system, and counteracting barriers with benefits. Strategies for people ready to act are summarized in the second column of Table 9-2.

Self-Reevaluation

Self-reevaluation is a process of further clarifying values and determining feelings about the problem behavior and the notion of behavior change. Self-reevaluation includes:

- Increasing perceived vulnerability to the health problems associated with excess weight, a lack of physical activity, or an unhealthy diet: This helps build motivation for change. ("How did you feel when you read that people with your cholesterol level have 20 times the risk of heart attack?")
- Examining personal capabilities for change: This helps build confidence in the ability to change. ("What are some of your ideas for starting a walking program?")

Social Support

Mobilizing social support for behavior change usually involves working with significant others (spouse, parent, partner, and close friend) to encourage both obvious and not-so-obvious kinds of assistance: food preparation, exercise buddy, creating a climate of encouragement and acceptance, and happily eating new foods. Support within the family is particularly important for obese children, who generally rely on others for meals and transportation. In addition, family and community support may be essential for low-income people who may have difficulty getting low-fat foods or access to opportunities for physical activity without the help of others.

Weight-management professionals can help clients to identify the level of support available and to locate sources of additional social support. When support within the household is lacking, the individual might be directed to classes, clubs, or community organizations for people with similar goals. Many larger communities now have size-friendly gyms where overweight individuals may feel more comfortable and accepted. Some churches have become sites for community health promotion programs. The Internet also offers online social support.

Evaluating Pros and Cons of Change

For individuals to move from preparation to action, the costs of change (the cons) must be perceived as decreasing, which is why interventions at this stage target the cons. The cons have to decrease only half as much as the pros must increase to get people moving toward change.[14]

Nonetheless, the cost of change may be great, particularly for dietary habits. The American Dietetic Association's 2008 National Trends Survey reported that Americans perceive the following as major barriers to healthy eating:[20]

- Not wanting to give up favorite or preferred foods (73%)
- Satisfied with the way they currently eat (79%)
- Perception that healthy eating takes too much time (54%)
- Lack of understanding of nutritional guidelines (41%)

People will not readily change their dietary behaviors just because you tell them to. After perceived barriers have been identified, the professional must help the client to see benefits that are not incompatible with her personal beliefs. For example, a person with high cholesterol who fears giving up favorite foods may be reminded that reductions in blood cholesterol can also be accomplished through portion control and different cooking methods.

Behavioral Strategies for People Ready to Act: Summary

People ready to act are either in the contemplation stage and are preparing for change in the next 6 months or in the preparation stage, where some change—however small—has already been attempted. The theories of reasoned action and planned behavior state that a person's intention to perform a specific behavior is the best determinant of whether the behavior will be adopted. Intent is influenced by attitudes toward the behavior, beliefs in the importance of the behavior to others, and perceived control over the behavior. Interventions at this stage of change help people to move toward action by clarifying beliefs about the problem behavior (self-reevaluation), mobilizing social support for change, and finding benefits to change that outweigh costs of change.

BEHAVIORAL STRATEGIES FOR PEOPLE TAKING ACTION

Weight-management professionals are most comfortable with the action stage—the stage in which clients are actively adopting healthier behaviors. Professionals can help clients achieve success by not moving them to action before they are ready and by helping them to adhere to action interventions.

action stage *Stage of change in which the individual has adopted a new behavior; this stage lasts from the day that the goal behavior is adopted for about 6 months.*

What Characterizes the Action Stage?

During the **action stage**, steps are taken to change the environment or the behavior in a way that facilitates change in diet and/or activity and, ultimately, weight management behaviors. The action stage lasts about 6 months, beginning the day that the goal behavior is adopted. Some goal behaviors are more difficult to adopt than others. For example, switching from ice cream to low-fat yogurt may happen fairly easily, assuming that acceptable flavors of low-fat yogurt are available. In contrast, reducing dietary sodium may take many more steps and much more preparation. Weight management usually involves changing more than one behavior, and people tend to underestimate the length of time needed for this to happen.

social-cognitive theory *People adopt new behaviors that are reinforced, promote self-efficacy, and result from interactions between factors within the person and the person's environment.*

How Does Social-Cognitive Theory Explain Behavior Change?

Social-cognitive theory states that people adopt new behaviors that they feel confident about performing and that are reinforced. The interaction between factors inside of the person and the person's environment influences behavior change.[21] The three key features of this theory are:

reinforcement *Expected outcome resulting from performance of a behavior; may be an external reward or an internal feeling.*

1. **Reinforcement**: People perform a behavior because they expect that a particular outcome will occur. The outcome, which is often outside of the person, reinforces continued performance of the behavior.
2. **Self-efficacy**: People continue performing only behaviors with which they feel successful. In other words, there is confidence that change can occur.
3. **Reciprocal determinism**: Reciprocal interactions among personal, behavioral, and environmental factors affect behavior change.

self-efficacy *Sense of confidence in one's abilities.*

reciprocal determinism *Interactions among personal, behavioral, and environmental factors that determine the course of behavior.*

The word reciprocal means that one factor affects another factor. For example, Michael has been successful at reducing dietary fat by eating breakfast every day instead of munching potato chips and other snacks at his desk. He particularly enjoys toasted bagels with jelly. He starts bringing a bagel to breakfast meetings where only doughnuts are usually served. At first, this behavior is difficult because Michael is the only person not eating doughnuts. Shortly, however, the person who orders the food gets calls from others who want bagels at meetings. Gradually, some doughnuts are replaced with bagels, bread, and fruit. Seeing others eat healthier foods reinforces Michael's behavior and increases his confidence in being able to eat tasty, healthy food at meetings. So, he initiates a new behavior—eating an additional serving of fruit every day.

Social-cognitive theory is the most frequently used model in weight-management programs.[12] It is particularly relevant to people who are actively engaged in change. We know from social-cognitive theory that people will perform behaviors if they expect a certain outcome, that people's actions are influenced by others, and that the environment and individual interact in determining behavior. Self-efficacy is part of this reciprocal process.

In studies that have measured the self-efficacy of people in different stages of change, self-efficacy has been found to increase as people move from contemplation to action and maintenance. Therefore, interventions for people in the action stage focus heavily on building self-efficacy. Self-efficacy is specific to each behavior. A person who feels confident about taking a walk every evening after work may have less belief in his ability to reduce dietary fat, especially if family members refuse to eat certain kinds of foods.

What Are Some Strategies for Action?

For people to change, they need to feel that their current pattern of behavior is harmful, believe that the change undertaken will be beneficial, and have confidence that they can implement the change. Action strategies based on social-cognitive theory use goal setting, provide reinforcement for new behaviors, and include skill-building activities and activities that increase self-efficacy. These strategies are summarized in the third column of Table 9-2.

Goal Setting

Assessment and self-reevaluation activities in the contemplation and preparation stages should have helped to identify perceived barriers and suggested some ways to overcome them. Goal setting involves deciding on reasonable and biologically attainable directions for change. Progressing slowly will increase the likelihood of success and build self-efficacy.

Goal setting is most helpful when the goals are related to behaviors that are under the person's control. As you discovered from Chapters 3 and 4, weight loss is not solely under volitional control. So, a goal of substituting two servings of fruit for two high-fat snacks is more attainable than a goal of losing 3 lb this week. Other reasonable, attainable goals include participating in an enjoyable movement activity every day and eating three meals and two snacks daily.

If clients insist on having weight-loss goals, then the weight-management professional must help them settle on a realistic goal weight. A 6-month weight loss goal of 10% of current weight is biologically and genetically reasonable for most people and can significantly improve health. Unfortunately, many people select unattainable goal weights. A study of women with an average weight of 218 lb found that most clients wanted to lose 32% of their current weight (almost 70 lb), would be disappointed with a 17% loss (37 lb), and would consider a 55-lb loss acceptable but would not be particularly happy with it.[22] Actual weight loss after a program that lasted almost a year averaged 35 lb. A 35-lb weight loss is admirable but was disappointing to the women in this study.

In discussing weight goals, consider assessment results indicating the client's fat cell size and fat distribution. The individual who has been obese for a long time and who has had difficulty losing weight may have an excessive number of enlarged fat cells (hypertrophic hyperplastic obesity), making weight loss more difficult. The individual with lower-body obesity has fewer health complications but may have more difficulty losing weight than the individual with abdominal obesity. Goal weight must reflect these realities.

Setting weight goals in the absence of healthy nutrition and activity goals should be avoided. Although weight loss is one possible outcome of changes in diet and activity, people who focus exclusively on changes in weight may overlook the less obvious health benefits of dietary and activity behavior changes.

Contingency Contracting and Reinforcement

An effective tool for motivating individuals to comply with a new dietary and activity regimen is a behavioral contract with agreed-upon rewards for specific behaviors called **contingency contracting**. The contingencies might be attending weight-management educational or counseling sessions, participating in a group exercise program three times per week, or eating a specified number of meals/snacks each day.

contingency contract

Behavioral contract with agreed-upon rewards for specific behaviors.

The reinforcement, or reward, should be specified in advance and must be meaningful to the individual. Rewards that have been used in weight-management programs include earning back money deposited in a bank account or getting tangible items, like cookbooks, clothing, or movie tickets.

Self-Monitoring

Self-monitoring is a critical activity in the action stage of change. Two examples of self-monitoring have already been discussed in Chapter 2: the 3-Day Diet Record and the Physical Activity Record. Regularly recording agreed-upon aspects of behavior has the following values:

- Increases awareness of behaviors, such as eating and activity, or of physiological parameters, such as blood glucose level or blood pressure
- Reinforces new behaviors by providing a visual reminder of those behaviors
- Helps in stimulus control by illustrating circumstances that may promote (or inhibit) unwanted behaviors
- Develops analytical and problem-solving skills as unwanted stimuli are examined and eliminated or modified
- Provides evidence that setbacks in behavior are controllable

The last value of self-monitoring is particularly relevant to maintenance of behavior. When individuals are recording fat grams, for example, they begin to see that eating an occasional high-fat food does not defeat an entire diet program. The fat in a bag of potato chips can be dispassionately recorded and accounted for by making small dietary changes over the rest of a week. This normal eating behavior does not have to be interpreted as failure and lead to relapse.

The diet and physical activity records described in Chapter 2 and Appendix B can be used for continued self-monitoring. Depending on the client's goal, the forms may be modified. For example, if reducing fat intake is the primary dietary goal, then the client might be given a fat gram counter and taught to record the fat content of each food and beverage consumed, rather than waiting for the nutritionist to analyze the diet and provide feedback.

Self-monitoring promotes dietary compliance and even weight loss in some people during high-risk situations, such as the annual holiday season.[23] People who stick to their dietary monitoring begin to discover that foods aren't really "good" or "bad" and that overeating once in a while can be compensated for by moderation most of the time.

Weighing is another form of self-monitoring. While some researchers worry that weighing may increase body dissatisfaction, anxiety, and depression in individuals trying to lose weight,[24] numerous studies confirm the value of self-weighing in promoting weight loss and weight maintenance.[25–27] Doing this in a group, such as Weight Watchers, may even desensitize individuals to any potential negative effects of weighing.[28]

A few cautions about monitoring body weight: overweight individuals need to be advised that short-term weight fluctuations are likely due to changes in body water rather than changes in fat stores; that reduced caloric intake may result in dramatic weight loss, as reduced carbohydrate stores are accompanied by lost water; and that some women experience considerable water-weight fluctuations during the menstrual cycle. Individuals who experience significant loss of body fat and weight initially typically see weight loss slowing, even when eating and exercise habits are maintained. If weight is overemphasized as a criterion for success, failure to continue to lose weight at the same rate may cause some people to become frustrated and fall back on old habits.

Stimulus Control

The social and physical environment in which most of us live and work promotes obesity. Environmental factors are powerful counters to changing—and maintaining change in—diet and activity. For example, most vending machines are loaded with high-fat, high-calorie snacks, and soda is often cheaper than bottled water; the food served at social events tends to be abundant, rich, and very tasty; many neighborhoods are car friendly and pedestrian unfriendly; and television commercials urge us to eat, eat, eat!

Weight-management professionals must help clients to identify all of the stimuli promoting inappropriate eating and inactivity so that change can be encouraged. Assessment of diet and physical activity will identify the individual's typical pattern of eating and activity and situations associated with lack of control. Stimulus control involves rearranging the physical and social environment to minimize undesirable behaviors and maximize desired behaviors:

- Avoiding situations that promote problem behaviors:
  - Do not dine out for the first few weeks of a new dietary regimen.
  - Decline invitations to parties.
  - Store food out of sight.
  - Have others do the grocery shopping.
- Modifying the environment to promote desirable behaviors:
  - Leave walking shoes beside the front door.
  - Put the exercise equipment near the television set.
  - Pack a work-out bag the night before and leave it near the front door.
  - Store food only in the kitchen or the pantry.
  - Freeze leftovers immediately, or bring a portion for lunch the next day.
- Altering behavior within the environment:
  - Eat only in the dining room or kitchen table.

- Schedule meals and snacks rather than spontaneously eating whatever is available.
- Use a shopping list at the grocery store.
- Do grocery shopping after eating.
- Plan ahead for eating in restaurants and at parties.
- Eat desserts or high-risk foods only in restaurants, where a single portion can be ordered and perhaps shared with a companion.
- Stretch or walk around the house during every television commercial.

Modifying Eating Style

Some people have eating styles that promote excess food consumption. If these people are lean and active, they probably will not become obese. However, if they have a family history of obesity, are sedentary, and are already overweight, modification of eating style is necessary to prevent further weight gain. Eating too fast, eating when preoccupied with other things, and feeling compelled to clean their plates are common eating problems that may need to be addressed.

For people who are rapid eaters, excessive food may be consumed before the brain receives satiety signals. Strategies to reduce eating speed include:

- Chew food a specific number of times before swallowing.
- Lay down the eating utensil after each third bite of food.
- Put a quantity of food on the plate at the kitchen counter, go to the kitchen or dining room table to eat, and do not return to the kitchen counter for more food for at least 15 minutes.
- Eat everything from a plate or a bowl, even bagged snack foods, cookies, and candy.

Pairing eating with some other activity may be a problem for some people. If they are preoccupied with reading or watching television while eating, then they lose track of how much food has been consumed. Effective strategies for these individuals include:

- Avoid reading, watching television, or engaging in any behavior other than enjoying food while eating.
- Eat only at the dining room or kitchen table so that eating becomes paired with an appropriate location.

Many people were taught to clean their plates as children and, as adults, find it difficult to leave food on the plate after a meal, even when they are full. These individuals may be helped if they:

- Use smaller plates, bowls, and glasses so that less food is presented at one time.
- Make a conscious effort to leave a bit of food on the plate after a meal.

Normalization of Eating

An important first step in establishing healthy eating patterns is the normalization of eating. Skipping breakfast and other meals and overreliance on fast food are behaviors associated with weight gain. Before we ask clients to change their food choices, we may need to help them reestablish normal eating patterns. This has been called healthy, unrestrained eating because it emphasizes an end to calorie- and food-restricted dieting and helps people learn what it feels like to be hungry versus full. When a consistent pattern of meals and snacks is followed, there is less chance of rapidly eating excessive calories due to extreme hunger.

Clients should be encouraged to eat three meals and two or three snacks every day and—at first—to record what they eat. Initially, some weight gain may occur because the emphasis at this point is normalizing eating, not choosing lower-fat or lower-calorie foods. This approach is absolutely essential to help people establish a pattern of eating based on the physiological need for food and recognition of hunger and satiety. People who snack all the time may never feel hungry, and those who regulate eating by finishing the whole bag or cleaning their plate may be unable to recognize satiety. Physiologically, this is also important because people who follow a consistent pattern of meals and snacks every day have better blood glucose control.

An important part of normalizing eating is stressing that all foods are acceptable. The notion of being "good" when eating a salad and "bad" when eating brownies must stop. A diet that works needs to have a place for all kinds of foods, without guilt or self-recrimination.

This initial phase of normalizing eating might last as long as 6 months. Behaviors described in this section of the chapter—social support, self-monitoring, stimulus control, and modifying eating style—should be emphasized as part of normalized eating. When a normal pattern of eating has been established, dietary changes that respect food preferences and practices may be gradually introduced. A slow approach to change will not result in rapid weight loss or dramatic changes in blood pressure or blood lipids but ultimately will promote long-term compliance.

Cognitive Restructuring

We assume that our thoughts and perceptions represent reality. Obese individuals often have cognitions (thoughts) that increase the difficulty of behavior change.[29] They may see themselves in a negative and critical way, tying self-esteem to weight; and they may see foods as being good or bad, which is sometimes called dichotomous thinking. When a bad food is eaten, this confirms the individual's low self-opinion and frequently leads to overeating.

cognitive restructuring
Behavior change technique in which people develop rational and positive self-statements to counteract negative self-thoughts.

Cognitive restructuring is a technique that helps people develop rational and positive self-statements to counteract negative self-thoughts. Substitute behaviors (positive thoughts) replace problem behaviors (negative thoughts). Cognitive restructuring involves more than simply parroting positive statements. To develop counterarguments, the client must be helped to examine negative statements critically and seek their meaning. Many times, negative thoughts are the product of old assumptions and circumstances that no longer exist.

For example, just before walking into the conference room, Rosa says to herself, "I look so fat today. I'll never be thin so I might as well have a doughnut instead of a bagel" (negative thought). In exploring this negative self-statement with Rosa, the weight-management professional helps Rosa identify the meaning of the negative thought. Often, the meaning is something like this: "I look bad because I'm fat; other people also think I look bad, and this makes them see me in a very negative way; therefore, my fatness makes me an undesirable person (lazy, weak, stupid, etc.); so I might as well live up to everyone's expectations of me and eat something high-fat." Or Rosa might think, "I've already eaten one doughnut today. I'll just blow it and have another" (dichotomous thinking). The professional will help Rosa see that she is defining doughnuts as bad and (probably) making herself feel worse by eating a "bad" food.

TABLE 9-5   Cognitive Restructuring

Negative self-talk	Possible meaning or purpose of statement	Realistic self-talk
"I can't do this exercise program. I never stick with anything."	Overgeneralizing	"Exercising regularly has always been a problem for me. Being flexible in my activity program should help."
"I got a performance bonus, but that's only because none of the skinny people did much this quarter."	Minimizing self-value and overemphasizing value of others based on personal attributes	"I've really improved my sales strategy. I deserved that bonus."
"I ate so much last night only because my husband wanted to go out to eat."	Putting responsibility for own behaviors on other people	"I really didn't need to have dessert last night. Next time I'll bring home dessert and have it the next day."
"Now I've blown it. I ate six cookies. I might as well finish the package."	Dichotomous thinking: Foods can be classified as good and bad.	"I enjoyed those cookies. I'll cut back on my snacking this week."

deep breathing
Consciously practicing deep, slow breathing at times when stress level increases. Involves sitting comfortably, inhaling deeply so the abdomen rises, and then exhaling from the abdomen, visualizing stress and tension flowing outward.

imagery *Visualizing oneself in a relaxing situation so fully that the body responds as if one is in that situation.*

meditation *Individual sits in a comfortable chair in a place without distractions with eyes closed. Individual focuses on an unchanging stimulus in the environment (clock ticking, sound of waves), discarding all thoughts and concentrating on the stimulus. Practice for 10–20 minutes each day.*

To counteract negative thoughts, Rosa needs to critically examine her beliefs about herself, her eating, and how others see her. When cognitive restructuring is coupled with relaxation techniques, it can reduce the stress of situations that elicit negative thoughts. Rosa can learn to replace "I look fat today" with "I made an important contribution to our last meeting, and I have some good ideas today."

Several other examples of the technique of replacing negative self-statements with positive ones are listed in Table 9-5.

Stress Management

Stress-management techniques allow us to replace negative behaviors (nail biting, drug abuse, insomnia, and overeating) with positive ones (time management and relaxation). Exercise is such an effective stress-management technique that it is recommended in many addiction-treatment programs. In weight management, exercise has a dual value: energy expenditure and relief of tension. A daily walk at lunchtime or after work contributes to weight management and helps to promote relaxation. Developing a heightened sense of fitness can itself help to relieve stress.

Reducing stress through relaxation exercises can also address some of the psychological issues surrounding weight. Learning to lower one's physiological responses to stressful situations produces a greater sense of control. The ability to control stress response may increase self-efficacy and help a person feel more confident about approaching eating and activity behaviors.

Several excellent relaxation techniques can be performed almost everywhere (except driving a car) and require no special equipment. **Deep breathing**, **imagery**, and **meditation** are widely accepted as effective strategies for stress reduction.

Pete Saloutos, 2010/Shutterstock.com

Yoga is an excellent stress reducer that also burns calories.

Skill Building: Diet and Activity

It is easy and fun to give advice to others, but acknowledging such advice by changing behavior is very difficult. Previously sedentary people and those who know little about nutrition need a lot of help to develop the motivation and skills to begin incorporating dietary change and physical activity into their lives. Weight-management professionals can make it easier for clients to accept advice by carefully tailoring interventions to each individual. The starting point must be an initial assessment. Once the assessment suggests the target of an intervention, skill-building activities can be recommended. These are discussed further later in this chapter.

Behavioral Strategies for People Taking Action: Summary

The action stage lasts for about 6 months, beginning the day that the goal behavior is adopted. Social-cognitive theory explains how interactions among personal, behavioral, and environmental factors help people change. Interventions that provide reinforcement and build self-efficacy facilitate change. These include goal setting, contingency contracting and reinforcement, self-monitoring, stimulus control, modifying eating style, normalizing eating, cognitive restructuring, stress management, and building diet and activity skills.

Case Study. The Ex-Football Player, Part 2	A year after his high school reunion, Luis was required by his employer to get a physical examination so he could qualify for a substantial health insurance plan. Luis's physician expressed concern that his BMI was 29 and suggested that he get some exercise. "Exercise will help keep your blood pressure down, too," the doctor remarked. "Your pressure is a bit high for someone of your age."

Luis started walking with his wife after work and after about 2 months decided to pay another visit to the health club that he had stopped using. He was enjoying the walking and thought that he was ready for more variety in his exercise routine. He was encouraged that he had been able to work physical activity into his schedule and felt confident that he could keep it up. He made an appointment with a personal trainer at the club.

- If you were the personal trainer, what kinds of goals would you talk about with Luis?
- How would you address self-efficacy?
- What might be different for Luis this time, and how can you keep him in "action?"

BEHAVIORAL STRATEGIES FOR MAINTENANCE AND TERMINATION

Some strategies needed in the maintenance stage to continue the change process and/or prevent regaining weight are the same as those used in the action stage to alter diet and activity habits and perhaps lose weight. Self-monitoring, for example, can continue to promote regular physical activity and prevent overeating. Dietary skills developed in the action stage are modified and continued for maintenance. Exercise behaviors are particularly important for maintenance of lost weight and for health reasons.[30,31]

What Characterizes the Maintenance and Termination Stages?

maintenance stage *Stage of change reached by people who have continued to practice a new behavior for at least 6 months.*

termination stage *Final stage of change, at which point the individual is considered to have successfully terminated an undesirable behavior and adopted a goal behavior. Weight-management behaviors do not always have a clearly defined termination point.*

People in the **maintenance stage** have continued to practice the goal behavior for at least 6 months. For some behaviors, there is a point at which the individual is considered to have successfully terminated the undesirable behavior, adopted the goal behavior, and is at no risk of relapse. In smoking cessation, for example, a person who has not smoked for 5 years is considered to have reached termination.[9] For weight-management behaviors, there is no clearly defined **termination stage** but rather a lifetime of change maintenance. In weight management—where so many different behaviors are involved—people may be in the action stage for some behaviors and the maintenance stage for others. A goal of eating one serving of fruit each day might be easily met while, at the same time, a goal of gradually reducing fat intake to 25% of total calories might take a year of effort and continued vigilance.

Researchers have noticed that people who reach the maintenance stage tend to have greater self-efficacy than those still in the action stage.[32] This observation supports the idea that self-efficacy varies among different behaviors. The client might have considerably more confidence at having fruit every day than further reducing fat intake. It also supports the value of reasonable goals. Goals more likely to be met have the potential to

TABLE 9-6   Characteristics of the Maintenance and Termination Stages

- Adherence to goal behavior for at least 6 months
- Increased self-efficacy
- Continued possibility of lapses (maintenance)
- No risk of lapses (termination)

increase self-efficacy and the likelihood of continuing successful behaviors. There may also be a maintenance advantage of gradually incorporating new diet and activity behaviors into one's lifestyle. Such behaviors may ultimately become routine and easier to continue than an extreme weight-loss behavior like following a liquid diet. Characteristics of the maintenance and termination stages are summarized in Table 9-6.

What Are Some Strategies for Maintenance?

relapse Complete breakdown in behavior, return to old behaviors, and loss of confidence in one's ability to sustain change.

lapse Temporary setback during the process of change.

Although behavior maintenance uses some of the same skills suggested for people taking action, several strategies are particularly necessary for promoting maintenance. These strategies focus on coping with lapses in behavior and preventing relapse. The term **relapse** refers to a relatively permanent return to old behaviors after a **lapse** or series of lapses. The difference between a lapse and a relapse is not quantifiable. In other words, experts don't say that three lapses equals a relapse. If a person has three lapses but continues to self-monitor and practice healthy behaviors, then she has not had a relapse. In contrast, if she has one lapse, gives up hope, and reverts to unhealthy behaviors, then she has had a relapse.

Earlier in this chapter you learned that most people who lose weight through dieting regain about one-third of lost weight within a year and return to initial weight in 3 to 5 years. Behavioral programs can help people to keep off much of lost weight by focusing on strategies for maintaining diet and activity behaviors.

Development of Problem-Solving Skills

Various clinical studies involving dietary changes have identified two factors that account for lapses:[15]

- Intrapersonal factors: Factors within the person, such as mood and emotional state, that often precipitate overeating and divergence from a diet plan
- Situational factors: Social situations in which food is served, eating in restaurants, and other occasions where people are exposed to food

Therefore, developing ways to cope with life circumstances and with emotional upset is important. Rather than falling back on emotion-focused or escape-avoidance methods of coping—such as overeating and staying in bed while hoping that the problem will "go away"—confrontational problem-solving skills should be used. These include:

- Identifying existing or potential problems
- Considering alternative behaviors when confronted with problems
- Determining the costs and the benefits of various behavior alternatives
- Choosing the best alternative

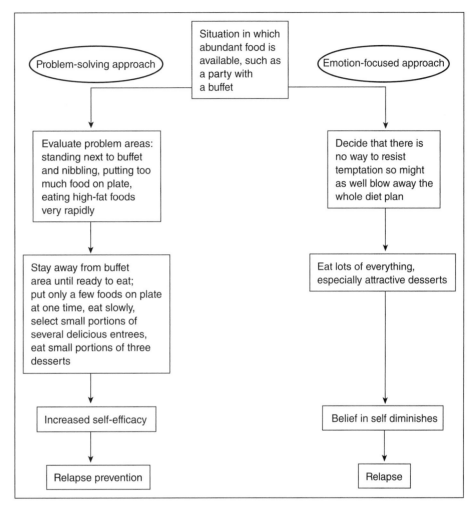

FIGURE 9-1 Relapse can be prevented through problem solving.

Source: This model is based on a model of the process of lapse and relapse developed by Brownell, K.D., & Rodin, J. (1990). *The weight maintenance survival guide.* Dallas: American Health Publishing.

Figure 9-1 presents an example of applying a problem-solving approach to one high-risk eating situation. The individual using this approach has identified the problem (tempting foods in a social situation, the tendency to graze at a buffet and eat too much) and determined a course of action that will minimize excess consumption while allowing pleasure (not standing next to the buffet table, putting only a few foods on the plate at one time, taking small portions of favorite desserts). Increased belief in self is the result. A person who uses a more emotional approach, in contrast, decides that such a tempting array of foods could never be resisted, overeats, loses self-efficacy, and relapses.

In fact, the client has been using problem-solving skills all along: when the client and weight-management professional worked together to set goals, when strategies for modifying diet and activity were adopted, and when new ways to control stimuli were tried. In the maintenance stage, a problem-solving approach includes helping clients to identify situations in which they believe problems are likely to occur and devising approaches to these situations.

Support Groups

People often need a boost to continue diet and exercise behaviors, even when they appear to have successfully adopted them. Support group meetings or "booster" sessions with a nutritionist or an exercise physiologist can be very helpful to people trying to stick with change. Some people enjoy the companionship and support of others who have similar goals, and some need guidance or additional information. Web-based interactive technology offers additional ways to provide support. Several recent studies using email reminders and interactive computer-tailored interventions have found at least modest success in minimizing weight regain and maintaining diet and exercise behaviors.[10,26,33]

Planning for Relapse

Most people who are making a behavior change experience temporary setbacks during the process. When people have not been prepared for these temporary setbacks, they may assume they lack the character and willpower to continue. Their response to a lapse may be complete breakdown in behavior and a return to old behaviors. In contrast, people who are prepared for imperfection and know that they can dust themselves off and proceed with behavior change have higher self-efficacy and are less likely to relapse. Lapses, and even relapse, can be important opportunities for reflection and learning.

Relapse is more likely when change is initiated in the precontemplation or contemplation stages. One way we minimize relapse is by moving people into the action stage only when they are ready. In addition, people who use medications or surgery for weight loss without attending to behavior are at great risk of relapse. For these individuals, behavior-change strategies must be initiated before or during the pharmacological/surgical intervention.

It is also important to clarify clients' expectations about weight loss periodically. Someone who expects to lose 70 lb and does not may have a greater risk of relapse. The client who thinks his problems at work or in social situations will go away with weight loss needs to confront this belief realistically.

In addition to the previously discussed problem-solving strategies, people can plan for relapse by role-playing situations in which lapses might occur. Working through reactions to a high-risk situation and practicing responses can improve clients' repertoire of behaviors. Some people are helped if they are encouraged to lapse. Seeing that a lapse can be controlled and need not turn into relapse can be particularly empowering.

Restructuring the Environment

A public health view of maintaining healthy diet and activity behaviors is to change the social and policy environment, instead of the person. The federal government could approach this problem like other public health issues, such as failure to wear seatbelts, alcohol use by minors, and smoking. For nutrition, legislation might limit advertising of high-fat foods and fast-food chains, keep high-fat foods out of schools and vending machines, or tax high-fat foods at a higher rate. For physical activity, some ideas include giving tax incentives to businesses and communities that provide fitness facilities or requiring physical activity time in every school day from kindergarten through graduation.

Although such sweeping changes are not likely to occur, weight-management professionals can have an impact at the community and the local levels. We help many people when we model normal eating and exercise behaviors, encourage administrators to stock vending machines with moderately priced healthy food choices, negotiate low-cost health club memberships for co-workers, offer to teach people to use fitness equipment, organize a lunchtime walking group, encourage the cafeteria to post the nutrient content of meals, and have healthy food choices available at meetings that we organize. (The public health approach is discussed further in Chapter 12.)

Do People Who Keep Off Lost Weight Have Secrets of Success?

Keeping off lost weight is difficult. Recently, there has been renewed interest in people who maintain weight loss for a prolonged period of time. The "secrets" of these weight suppressors might help others who need or want to lose weight.

What Is a Weight Suppressor?

The term weight suppressor has been used to describe people who lose a significant amount of weight and keep off the weight for at least a year.[34] Data from organized weight-loss programs suggest that there are very few weight suppressors in the adult population. However, most people who attempt weight loss are not in organized programs and, therefore, may not be identified. In addition, people who enter treatment programs may be different from those who lose weight on their own. The number of weight suppressors in the population is probably not as low as estimated in the 1959 review mentioned at the beginning of this chapter (5%) but is not higher than 15–20%.

The National Weight Control Registry (NWCR) is the largest ongoing study of weight suppressors in the United States. The registry was established in 1994 and is maintained at two locations—Brown University Medical School and the University of Colorado Health Sciences Center. Individuals are eligible for enrollment in the registry if they are at least 18 years of age, have lost at least 30 lb, and have maintained a 30-lb weight loss for a year or more. In fact, most people in the registry have lost over 70 lb and maintained that loss for over 5 years.[35] About 80% are women, and most are Caucasians. Read more about the NWCR at their Web site: www.nwcr.ws.

Weight suppressors initiate weight-loss attempts for many reasons. Over three-quarters of individuals in the NWCR report that a particular event precipitated their weight loss.[30] For women, this was usually emotional in nature (husband left; someone said she was fat) or related to a life event (important anniversary). For men, the trigger was far more likely to be related to medical causes (back pain, fatigue, varicose veins) or to be spontaneous ("just decided to do it").

Although most registry participants had previously tried to lose weight, this time there was a more compelling reason and a greater commitment. Others have noticed similar resolve in successful weight suppressors.[36] Whether the triggering event is what kicks weight suppressors from preparation to action or whether the trigger simply strengthens the capability of people ready for action is not known.

Successful Weight Suppression Behaviors

With or without professional assistance, you will notice several behaviors common to almost all weight loss and maintenance success stories:

- Dietary fat is reduced: Participants in the NWCR reported their fat intake as 29% of total calories, far less than the average American's fat intake.[37]
- Portion control is practiced: Many weight suppressors reported eating fewer calories as a strategy for keeping off weight.
- Almost everyone exercises: Becoming more physically active appears to be a requirement for maintaining weight loss.
- About a third weigh themselves daily: those that weighed themselves more frequently had lower BMIs and less weight regain.[38]

Exercise is a key behavior for maintaining weight loss. Self-reported physical activity levels of registry participants are impressive, although there is considerable variability. About 35% report expending more than 3,000 kcal/week, while 25% report less than 1,000 kcal/week.[39]

The Challenge of Weight Suppression

Very little information is available from people who have tried to lose weight about their perceptions of the difficulties of weight loss. The NWCR asked its participants to rate the difficulty of weight loss and weight maintenance and found that most people reported both processes to be difficult. Men were more likely than women to say that weight loss was easy. Despite the common notion that weight maintenance is harder than weight loss, 42% of these individuals said that weight loss was the more difficult task. Only 25% reported that weight maintenance was more difficult.[30] A recent NWCR study that compared individuals who had lost weight with bariatric surgery to those who had used non-surgical methods found that the non-surgical weight suppressors had to work harder than the surgical patients to maintain lost weight.[40]

Even NWCR participants experience lapses leading to weight regain. Unfortunately, researchers still do not know a great deal about how to keep weight management lapses from turning into full-blown relapses. However, registry researchers have found that when lapses lead to weight gains as minor as 1–2 kg (under 5 lb), recovery is difficult. Those in the registry who react quickly to reverse small lapses are most successful in preventing relapse.[41]

Quality of Life of Weight Suppressors

Chapter 1 discussed some of the difficulties in determining whether obesity is associated with a lower quality of life and whether weight loss makes people happier. The National Weight Control Registry attempted to assess this by asking participants to rate how weight loss affected their general health and well-being, interactions with others, self-confidence, and time spent thinking about food and weight.

The majority of people (more than 85%) said that weight loss had improved their quality of life, level of energy, mobility, mood, self-confidence, and physical health.[42] Some perceived that their interactions were improved with individuals of the opposite sex (65%), the same sex (50%), strangers (69%), and their spouse (56%). For the most part, registry participants did not exhibit signs of psychological distress.

Unfortunately, a small proportion of registry participants reported that they now thought more about food (14%) or weight (20%). This finding highlights a potentially negative outcome of weight loss: If people are thinking more about food, then they may be feeling deprived; and, if people are thinking more about weight, then they may be dissatisfied with their appearance. Healthy dietary and activity behaviors should not make people feel deprived or dissatisfied.

"Secrets" to Weight-Loss and Weight-Maintenance Success

Popular diet books, videos, and commercial weight-loss programs would like us to believe that there is a special dietary regimen that will ensure weight-loss success: People who follow this special regimen will succeed, and those who do not lose weight—or fail to keep it off—have failed to follow the regimen correctly.

The experiences of weight suppressors have shown that successful strategies are built around long-term changes in dietary and exercise practices that are appropriate for everyone, not just people seeking weight loss. These strategies really aren't secrets at all:

- Readiness to change
- Awareness of what you are eating
- Portion control and less fat, at least some of the time
- Physical activity
- Development of problem-solving skills for reducing stress and managing life's challenges
- Quick response to lapses
- Long-term commitment to change

Behavioral Strategies for Maintenance and Termination: Summary

People in the maintenance stage have continued to practice the goal behavior for at least 6 months. Maintenance may be a relatively permanent stage without a clearly defined termination. Maintenance strategies focus on coping with lapses in behavior and preventing relapse. Problem-solving skills, support groups, planning for relapse, and, when possible, restructuring the environment all are maintenance strategies. The National Weight Control Registry offers some important lessons for eating and exercise behaviors from people who have kept lost weight off.

SKILL BUILDING FOR MODIFYING DIET AND ACTIVITY

Remember that we want individuals to build self-efficacy. Offering a client a list of exercises or a list of foods that should be eaten or avoided does not build self-efficacy. Developing diet-management skills or skills for becoming more active does build self-efficacy.

What Physical Activity Skills Are Needed?

Obese individuals who become more physically active tend to find that activity is easier to adhere to than diet.[43] Particularly when individuals are counseled to select activities that are the "best fit" for them, a regular exercise program becomes everything that a restrictive diet is not: energizing, self-directed, unrestrained, and fun.

Physical activity interventions, which were covered in Chapter 8, are based on several general principles:

- Participating in an activity that is safe and enjoyable builds self-confidence and promotes adherence.
- Health benefits occur when people become more physically active, and the benefits are greatest when people gradually adopt a formal exercise program.
- Some people may need to learn to be physically active.
- Skills in self-regulating exercise intensity (such as target heart rate and rate of perceived exertion) will help people stick with an exercise program and avoid injury.

What Dietary Skills Are Needed?

Dietary interventions for weight management are based on several general principles:

- The best way to build self-efficacy is to experience success. Therefore, we start with easy tasks and gradually progress to the more difficult. For example, switching from whole milk to 2% milk may be easier than modifying favorite recipes to lower their fat content.
- Food-focused interventions work better than interventions based on broad nutritional recommendations. Advising someone to "eat a diet low in fat" is not as easy to translate into practice as advising him to bake chicken instead of frying it and to have a slice of whole grain bread instead of a biscuit.
- Certain skills can empower individuals to create a diet of pleasurable and acceptable foods. These include reading food labels, modifying the nutrient content of recipes, planning meals, and developing a shopping list.
- Before people can change, they may need to normalize their eating by planning three meals and two or three snacks each day. Doing this can ultimately help prevent relapse and promote weight maintenance.
- Skills in grocery shopping, cooking, and eating in restaurants may need to be developed to facilitate dietary behavior change.

What Are Some Strategies for Modifying Dietary Fat Intake?

Most of us can benefit from a diet that is less than 30% fat, contains more monounsaturated and polyunsaturated fats than saturated fats, and provides less than 300 mg of cholesterol per day. A diet with these fat characteristics is heart healthy and, when combined with portion control and physical activity, may help some people to lose weight.

Lowering Fat Consumption

Most Americans consume more than 30% of calories in the form of fat. Even if weight loss is not a goal, the average person could improve cardiovascular health by modifying fat intake. A person eating 2,000 kcal/day who goes from 30% to 25% fat will save about

100 kcal/day—enough to lose up to 10 lb in a year if energy expenditure is maintained. People who have successfully lowered their fat consumption recommend several strategies:

- Reduce intake of high-fat milk products, high-fat meats, and added fats and oils.
- Limit consumption of sweets that have hidden fats.
- Use lower fat cooking methods, such as baking and broiling meats, trimming the fat from meat, and limiting fried foods.

If people are willing to modify a few food choices, then sweeping changes in fat consumption are generally not needed. Once individuals realize the sources of their dietary fat, they can identify the fat that they can't imagine doing without and determine ways to reduce hidden fats that they can live without. For example, a person may be unwilling to use skim milk on cereal but may not object to cooking with skim milk. Or a person who usually sautés vegetables in butter might taste vegetables sautéed in lemon juice and a little Worcestershire sauce and find that the new flavor is an acceptable substitute for 200 fat calories. Table 9-7 presents additional strategies for lowering dietary fat.

Fat-Modified Diet Plans

To reduce the risk of cardiovascular disease, both the National Cholesterol Education Program and the American Heart Association recommend lowering total dietary fat, saturated fat, and cholesterol. This can be accomplished in many ways, including consuming less whole milk and meat and replacing saturated fats with grains, vegetables, fruits, and legumes. Individuals not accustomed to cooking with legumes will be more likely to try

TABLE 9-7   Strategies for Reducing Dietary Fat Content

Cook with low-fat alternatives	• Use skim milk instead of whole milk in sauces. • When a recipe calls for more than one egg, try using just the white of every other egg.
Use fat in moderation for flavor	• Use butter or margarine in small quantities on bread or potatoes. • Substitute low-fat sour cream for butter. • Try herbal substitutes, nonfat sauces, or low-fat chicken broth to add flavor.
Reduce or eliminate meat	• Experiment with texturized vegetable protein in casseroles. • Use tofu in small amounts. • Have fish instead of meat. • Limit meat intake to no more than 6 oz/day.
Control portions	• Eat a smaller quantity of high-fat foods.
Switch to lower-fat varieties of preferred foods	• Try tuna in water instead of oil. • Switch to low-fat salad dressing, or make your own dressing with more vinegar, lemon juice, and water and less oil. • Use low-fat milk and cheeses instead of whole milk.
Reduce or eliminate alcohol consumption	

them if they are given recipes and cooking tips. Replacing saturated fats and *trans* fatty acids with canola or olive oil is also heart healthy. When dietary fats are reduced, the logical replacement (if calories are being maintained) is dietary carbohydrate. However, weight loss cannot occur if more carbohydrate is eaten than fat is removed. If calories are being reduced, then the type of carbohydrate consumed must be chosen carefully. Empty-calorie carbohydrate foods like sweets and snack foods should be reserved for occasional consumption.

Whatever changes are made, they must be maintainable. Some people find that after a period on a lower-fat diet they lose their appetite for high-fat foods like whole milk and fried foods.[44] This certainly helps maintain the low-fat habit. Others may need a more moderate approach, first switching to healthier fats and then gradually reducing total fat intake.

A person trying to stay on a 1,800 kcal/day diet that is 20% fat has a fat budget of 360 kcal, or 40 grams, of fat. This challenge is enormous and may require continued education, skill development, and social support. Barriers to maintaining a low-fat diet increase with the severity of fat restriction. The individual may need to budget more money as well as time to prepare food. New shopping and preparation techniques may need to be learned. The dietetics professional must acknowledge and address these barriers and help clients to develop skills for overcoming them.

What Are Some Strategies for Controlling Calories?

The combination of excess dietary fat and calories share the responsibility for weight gain in our population. Ongoing nutrition studies suggest that some people have simply added low-calorie foods to a high-fat/high-calorie diet, rather than substituting low-calorie foods for high-calorie foods. And if we eat a large quantity of low-calorie foods, we can still get a substantial number of calories.

For weight loss to occur either caloric intake must drop or caloric expenditure must increase. People who reduce caloric intake to 1,000–1,200 kcal/day do lose weight. An initially rapid loss of weight is due to depleted glycogen stores, which are water heavy. Commercial diet programs and hospital-based, very-low-calorie diets can guarantee substantial weight losses this way; however, low-calorie diets are almost never effective for long-term weight loss. Unless rapid weight loss is absolutely necessary to preserve health, there is no benefit to extreme caloric restriction and there are numerous and severe health detriments:

- Caloric restriction may precipitate a powerful drive to eat.[45] Feelings of hunger, deprivation, and food craving often terminate in binge eating, which can become chronic.
- Weight loss of more than 0.3 lb/day is associated with unacceptable reductions in lean body mass. When people lose 1 lb/day, 85% of the weight lost is lean tissue.[46] If this continues, then protein depletion results.
- Low-calorie diets may increase susceptibility to developing gallstones.
- Low-calorie diets are poorly adhered to, resulting in greater likelihood of weight regain.

For individuals who want or need to lose weight or who are trying to maintain weight loss, calorie control is a necessary part of the diet plan. Calorie control is not synonymous with low-calorie, however.

Diet recommendations that emphasize moderation, rather than restriction, are more likely to succeed over the long term. In addition, food guides are useful for moderating caloric intake.

Portion Size as a Calorie-Control Tool

Simply reducing portion sizes can help people reduce caloric intake without the bother of counting calories. Many people respond better when they can eat a little of what they like rather than eliminating favorite foods altogether. Portion control also builds self-efficacy by allowing the client to eat in a variety of settings, to use foods that are culturally acceptable, and to satisfy their hunger and appetite for particular foods. Reducing portion sizes makes physiological sense, too. There is evidence that more frequent, smaller meals may reduce serum total cholesterol and low-density lipoprotein cholesterol levels.[47]

The best way to teach people about portion sizes is to pull out the measuring cups, teaspoons, and food scale. There is no substitute for putting cooked macaroni into a cup and emptying it onto a dinner plate to demonstrate what 1 cup of cooked macaroni looks like. An excellent alternative to the cups and spoons method is to use the human hand as a guide to portion size. Using the average woman's hand, a 3-oz portion of meat, fish, or poultry is approximately the size of the palm; a closed fist is about 1 cup; the tip of the thumb is the size of 1 teaspoon; and the thumb itself is about 1 tablespoon.

It is also instructive for people to give themselves what they consider to be a serving, to measure it, and then to compute the caloric value. Did they think they were getting ½ cup of ice cream when it was really 1 cup? Would they be satisfied with a little less today?

Using Food Labels

Food labels include useful information for people trying to adopt a healthier diet, and most people report at least glancing at the label when making food purchasing decisions.[48] However, whether people actually understand what they are reading is open to question. Instructions in using a food label for diet planning should emphasize that:

- Nutrition information on the label pertains to an amount that has been designated as one serving, which may or may not be the entire container. If the label indicates that the bag of chips contains two servings and the client has eaten a whole bag, the client needs to multiply "calories per serving" by 2 to determine the number of calories consumed. Sodium, fat, and other nutrients also must be doubled.
- The "% daily value" listing indicates what percentage of the recommended amount for each nutrient listed is obtained in a serving. This shows how each food fits into the overall diet.

Let's put these two concepts together. A cookie label states that two cookies equal one serving, which provides 25% of the daily value for total fat. The client who eats four cookies knows that she will be using half of her recommended daily fat intake in two servings. She can also compare similar products to determine which one best fits her dietary goals. The percent daily value is based on a 2,000 kcal/day diet that includes 65 g of total fat. People who consume fewer kilocalories need to keep in mind that a product providing 25% of the daily value for total fat will be giving them an even higher percentage of their daily fat intake.

Skills for weight management can be learned.

Eating Away from Home

In 1977, 16% of all meals and snacks were eaten away from home—in restaurants, fast-food places, schools, cafeterias, and bars and from vending machines. In 1995, 27% of meals were eaten away from home.[49] Today, the Economic Research Service of the U.S. Department of Agriculture estimates that spending on food away from home accounts for almost half of U.S. food expenditures.[50] If you conclude that people consume more calories when eating outside of the home than when in the home, then you are correct. People tend to eat larger portions in restaurants, and restaurant foods are often higher in calories than home-cooked versions.

People who frequently eat foods prepared by others outside of the home can benefit from developing skills in food selection and portion control:

- Learn to pick lower-fat entrées from available menu choices.
- Request sauces and dressings on the side.
- Order an appetizer as an entrée to get a smaller portion.
- Avoid French fries and onion rings; have a salad or baked potato instead.
- Split the entrée with a companion and each get a full-size salad.
- Ask for a "doggy bag" immediately upon being served, and take half the meal home for lunch the next day.
- Limit alcohol consumption.

There is no reason to automatically exclude alcohol from the diet of a person who is trying to lose or maintain weight. In people not at risk of alcoholism or liver disease, alcohol can increase the enjoyment of a meal and help promote relaxation. Some studies indicate that moderate alcohol consumption lowers the risk of cardiovascular disease.

However, alcohol contains 7 kcal/g, and those calories count just like carbohydrate, fat, and protein. In addition, alcohol tends to be oxidized itself, rather than promoting oxidation of other nutrients, and can thus reduce fat oxidation. Too much alcohol may also increase blood pressure and triglycerides. Therefore, good advice for including alcohol in the diet is to regard it as a food to be used in moderation, not exceeding one or two alcoholic drinks per day. Additional strategies for dietary change are presented in Table 9-8.

Skill Building for Modifying Diet and Activity: Summary

Fat, carbohydrate, and protein intake can be modified to reduce the risk of chronic disease and promote weight loss or maintenance. A more plant-based diet has characteristics

TABLE 9-8 Strategies for Dietary Change

Precontempla-tion stage	Contemplation and preparation stages	Action stage	Maintenance stage
Provide information about benefits of reducing dietary fat and matching energy in with energy out	Provide information about benefits of reducing dietary fat and matching energy in with energy out. Normalize eating. Set goals for dietary change: • Modifying fat intake • Modifying caloric intake (portion sizes) • Planning for eating in restaurants and social situations	Implement goals using skills that empower clients to create a diet of pleasurable and acceptable foods: • Modify nutrient content of recipes • Modify food preparation methods • Incorporate new foods into the diet • Use diet planning tools • Eat in restaurants and social situations • Read food labels	Monitor intake periodically. Weigh self periodically. Attend support groups, or use email or Internet support. Take cooking classes on using low-fat recipes.

that make it particularly useful for weight management. Moderate reduction in caloric intake may promote weight loss, but low-calorie diets in general are ineffective for long-term control of weight. Learning to use information from food labels, eat in restaurants, and practice portion control are important skills for weight management.

CULTURE AND BEHAVIOR CHANGE

Chapter 1 presented information about the disproportionate occurrence of overweight and obesity among racial and ethnic minorities in the United States, especially among women. Recall from Table 1-3 that over half of African American women and over 40% of Mexican American women are classified as obese, compared with about one-third of Caucasian women. Also recall from Chapter 4 that obesity affects a significant proportion of several Native American and Pacific Islander populations. Diabetes, hypertension, and cardiovascular disease are more prevalent in several minority populations due, at least in part, to high rates of overweight and obesity. Minorities who are poor have been especially hard hit by obesity and obesity-related chronic diseases.

Because of weight-related health problems, weight management is particularly critical for non-Caucasian populations. This section of the chapter explores the effect of culture on behavioral approaches to weight management.

What Is Culture?

culture *Patterns of thinking, feeling, and behaving that are learned and shared by a group of people.*

Culture has been defined as "learned patterns of thinking, feeling, and behaving that are shared by a particular group of people."[51] Ethnicity is one aspect of culture. People might also consider themselves to be part of a particular culture due to other shared characteristics, such as age, sexual orientation, religion, or leisure pursuits.

The notion of America as a melting pot presumes that people who arrived in this country would lose at least some of their cultural identity in favor of adopting beliefs and practices of the dominant culture. A better analogy is to think of America as a pot of soup, where each ingredient maintains its distinctive flavor but combines in a way that enhances the flavor and the value of the whole. Professionals who appreciate clients' cultural identity and diversity are better equipped to help clients acquire and adhere to weight-management behaviors.

How Does Culture Influence Weight Management?

All weight-related attitudes and behaviors arise from the culture with which we identify. For example, a 40-year-old woman from a rural community in the Southeast, where girls marry young and become grandmothers at an early age, may be accepting of middle-age weight gain and unwilling to engage in physical activities that she associates with young people. That woman's daughter, who goes to college in California on an athletic scholarship, may reject identification with her rural culture and adopt the values of one of her new cultures—Californian, college student, or athlete, for example. Several factors influenced by culture will affect a client's response to weight-management interventions: notions about acceptable body size, acceptable foods and food preparation methods, and physical activity.

Attitudes Toward Body Size

The supposedly ideal female who has a slim waist, hips, and thighs and wears a size 4 is primarily a product of Caucasian, Western cultures. Many non-Caucasian cultures do not stigmatize overweight and obesity, and they may consider large size to be attractive, an indicator of good health, and, sometimes, preferable.[52,53] Since not everyone values leanness to the same extent, the lure of cosmetic improvement through weight loss may be less compelling than health benefits associated with changes in eating and exercise behaviors.

Weight-management professionals should not assume that all members of a given ethnic group have the same attitude about obesity. While culture certainly influences many of our attitudes and values, income, occupation, and educational level also interact to influence attitudes toward body size.

Foods and Dietary Practices

Culture is an important influence on food selection, preparation, and pattern of consumption. Ethnicity and income are the components of culture that influence food choice the most. Dietary guidelines that fail to acknowledge food acceptability and affordability are unlikely to be followed. An excellent example is provided by a study of Cherokee Indians attending an outpatient diabetes clinic.[54] For 1 month, all type 2 diabetics were interviewed by a nutritionist. Of the 90 people visiting the clinic, 77 (86%) had failed to comply with their recommended dietary treatment; they cited reasons such as these:

- Felt hungry, weak, or dizzy on the diet
- Didn't understand how to use the Exchange System diet
- Couldn't get to the store and were dependent on foods brought by family and friends

- Didn't like to be told what to eat
- Preferred own foods and cooking methods (which often involved adding fat and frying)

Another study used focus groups with African American men and women in North Central Florida to examine difficulties in making dietary changes. Many participants believed that they would have to sacrifice part of their cultural heritage to conform to government dietary recommendations.[55] One participant likened it to "eating like white people."

Dietary change strategies must respect cultural differences if people with a strong cultural orientation are to adhere to recommendations. This can involve providing food lists that contain culturally specific foods, using computerized diet-analysis programs that have ethnic foods in the database, and recommending cooking methods and recipes that are compatible with food preferences and aversions.

Sometimes differences are less cultural and more socioeconomic. When we consider culture, are we seeing dietary preferences dictated by ethnicity/religion or by poverty? People on a limited income may not be able to afford certain kinds of fruits and vegetables, or they may be limited to food choices available at a local market. People eligible for U.S. Department of Agriculture surplus foods often find that, despite weight-management recommendations, many surplus foods are high in fat. Dietetics professionals can help low-income clients add fruits and vegetables to their diets by providing practical information about:

- Purchasing fresh fruits and vegetables in season (some farm markets, even in urban areas, accept food stamps)
- Substituting canned fruits and vegetables for fresh ones
- Determining the quantity needed to serve varying numbers of people (to prevent waste)
- Comparing prices between bulk items and smaller quantities

Physical Activity

Caucasian Americans have higher levels of physical activity than minority Americans. Surveys indicate that African, Hispanic, and other non-Caucasian Americans are more likely to report no participation in leisure-time physical activity and less likely to exercise five times per week than Caucasians.[56]

Differences in physical activity level may result from the value placed on activity by people in different cultures. Some evidence suggests that the family is a more important influence on participation in physical activity among minority women than among Caucasian women.[57] Socioeconomic status also explains ethnic differences in physical activity level. For example, people in minority cultures who live in communities where safety is a concern cannot be expected to conform to an activity plan that recommends an evening walk after work. Likewise, those with limited incomes will not be able to join a health club, and people working several jobs with inflexible hours may not have the time to exercise.

Weight-management professionals must acknowledge cultural attitudes toward activity as well as situational factors related to income that limit opportunities for leisure-time physical activity. Recommendations for activities that can be done in the home and as a family may be better adhered to than standard recommendations to get more exercise.

How Can Behaviorally Based Interventions Become Culturally Relevant?

Most nutritionists and others in the health fields are non-Hispanic Caucasians.[58,59] Because professionals tend to be more comfortable with approaches that support their own cultural values, most weight-management interventions probably reflect Caucasian, middle-class values. Some of the behavior-change approaches discussed in this chapter, such as goal setting, assertive problem solving, and taking responsibility for one's own health, are more compatible with the values of the dominant culture than minority cultures.

Weight-management interventions can be adapted to meet the needs of diverse populations. Although most research in this area is new and much more is needed, behavioral interventions that include minority populations have shown success in altering weight management behaviors.[11,33,60,61] Several strategies for adapting weight-management approaches to non-Caucasian cultural groups are summarized in Table 9-9. When considering these strategies, keep in mind the following general points:

TABLE 9-9 Culturally Relevant Approaches to Weight Management

Client characteristic	Intervention approach
Low literacy, low educational level, primary language other than English	• Provide materials and instruction in multiple languages. • Minimize use of written materials. • Develop materials at lower-grade reading levels. • Use interviews rather than questionnaires. • Translate food labels and lists.
Perception that obesity is not a problem	• Stress health benefits of change.
Finances and living situation limit opportunities for physical activity	• Emphasize activities that can be done at home (stretching and endurance activities requiring no equipment). • Provide examples of daily activities that burn calories (housework, walking to a farther bus stop, taking stairs). • Provide list of shopping centers with walking programs.
Finances and living situation limit food choices	• Provide recipes that use surplus foods as ingredients. • Emphasize portion control with high-fat foods. • Become familiar with foods available at market in client's neighborhood and use these in designing diet changes. • Provide a list of nearby farm markets or food cooperatives
Strong culturally related food preferences and food preparation methods	• Use ethnic foods in food demonstrations. • Modify recipes with lower-fat ingredients. • Modify cooking methods to reduce fat and calories.
No social or family support for exercise	• Emphasize physical activity rather than formal exercise. • Help client identify opportunities for family participation. • Develop a community intervention through a church or other social center.

- Cultural bias is normal. We tend to interpret interactions with others through our own cultural filter, but there is not a right and wrong culture.
- Gaining an understanding of the client's cultural orientation is important, but we must avoid stereotyping. Not everyone in the same racial, ethnic, age, or other cultural group has similar beliefs and values or is from the same socio-economic class.
- Western values about weight and health are not shared by all cultures.
- About half of Americans have low literacy levels.[62] This means that they read at or below the eighth-grade level. In addition, some minority clients new to the United States may be able to communicate verbally in English but may be unable to read food lists, food labels, and other materials. For educational materials to be of any value, they must be sensitive to the comprehension level and language preference of clients.

Cultural insensitivity may doom interventions to failure before they begin. By assessing—and incorporating into the treatment program—poverty, literacy, food preferences, family customs, and health beliefs, weight-management professionals acknowledge the powerful influence of culture on behavior.

Culture and Behavior Change: Summary

Because of weight-related health problems, weight management may be particularly critical for non-Caucasian populations. Culture influences our attitudes about weight, food, and activity. Culturally relevant weight-management approaches reflect differences in attitudes toward obesity, food preferences, dietary practices, physical activity beliefs, and socioeconomic status.

EXPECTED OUTCOMES FROM A BEHAVIORAL APPROACH TO WEIGHT MANAGEMENT

A behavioral approach to weight management considers clients' readiness for change, is based on assessment results, and is culturally sensitive. Modest changes in diet and activity can be expected to have the following outcomes after approximately 6 months of action:

1. Improved blood lipids, with people having the poorest lipid values showing the greatest improvement
2. Improved glucose control
3. Modest reductions in blood pressure if the person was initially hypertensive
4. Improved stress management
5. Increased cardiovascular fitness
6. Greater ease in moving around
7. Possibly a 5–10% reduction in body weight, with maintenance of muscle mass

Individuals who make dietary and activity improvements a part of their lives can expect to see these positive outcomes maintained over the long term, which is really our primary goal. Studies that have examined the outcomes of weight-loss treatment programs involving entire families have found better results with obese children than with their obese parents. In follow-up studies lasting as long as 10 years, weight regain occurs in both parents and children, but there is generally better maintenance of lost weight in children.[63]

Part of the explanation is biological. When healthy diet and activity behaviors begin early for children, accumulation of additional fat cells may be prevented. As you learned in Chapter 4, it is easier to shrink a normal number of fat cells than to contend with excess fat accumulation in an increased number of cells. Children have the further advantage of growing in height. If additional fat accumulation can be prevented, then weight eventually catches up with height.

An additional part of the explanation is that children seem to be better able to increase their physical activity levels than adults. Perhaps this is because children think of activity as recess, whereas adults think of activity as work. The lesson is that people stick with behaviors that are fun and have pleasurable consequences. Weight-management behaviors regarded as punitive and restrictive are unlikely to be followed for long. The focus must be on behaviors that can be continued for a lifetime.

| **Case Study. The Ex-Football Player, Part 3** | Luis has gradually lost fat and gained muscle as a result of his fitness regimen with a personal trainer. But both he and the trainer agree that he would make more progress if he could modify his diet. While most days Luis is happy with the meals that his wife prepares, he cannot pass up his mother's traditional Mexican dishes (enchiladas, pork tacos, fried beans, stuffed peppers), and he manages to find a reason to visit his mother's house at least twice a week for her food. He also has had trouble staying away from the fast food served at lunch meetings at work.

Luis makes an appointment with a sports nutritionist who consults at the health club. The nutritionist provides information about the Exchange System and suggests that Luis try to follow a 2,000 kcal/day diet for several weeks. She provides some sample eating plans for him based on this caloric intake.

• Develop two different eating plans for Luis, based on a 2,000 kcal/day diet. Provide ideas for breakfast, lunch, dinner, and two snacks.
• Now, modify one of those plans to include one beer and traditional Mexican foods, staying within the 2,000 kcal/day. Was this difficult? How could Luis get some of his favorite foods and continue to moderate his intake? |

REFERENCES

1. McGinnis, J. J., & Foegl, W. H. (1993). Actual causes of death in the U.S. *JAMA, 270,* 2207–2212.

2. Stunkard, A., & McLaren-Hume, M. (1959). The results of treatment for obesity: A review of literature and report of a series. *Archives of Internal Medicine, 103,* 79–85.

3. Wadden, T. A., & Phelan, S. (2002). Behavioral assessment of the obese patient. In T. A. Wadden, & A. J. Stunkard (Eds.), *Handbook of obesity treatment* (pp. 186–226). New York: Guilford Press.

4. LeCheminant, J. D., Jacobsen, D. J., Hall, M. A., & Donnelly, J. E. (2005). A comparison of meal replacements and medication in weight maintenance after weight loss. *Journal of the American College of Nutrition, 24*(5), 347–353.

5. Wadden, T. A., Berkowitz, R. I., Womble, L. G., Sarwer, D. B., Phelan, S., Cato, R. K., et al. (2005). Randomized trial of lifestyle modification and pharmacotherapy for obesity. *New England Journal of Medicine, 353*(20), 2111–2120.

6. DiClemente, C. C., & Prochaska, J. O. (1982). Self change and therapy change of smoking behavior: A comparison of processes of change in cessation and maintenance. *Addictive Behaviors, 7,* 133–142.

7. Prochaska, J. O., & DiClemente, C. C. (1983). Stages and processes of self-change of smoking: Toward an integrative model of change. *Journal of Consulting and Clinical Psychology, 51,* 390–395.

8. Sutton, K., Logue, E., Jarjoura, D., Baughman, K., Smucker, W., & Capers, C. (2003). Assessing dietary and exercise stage of change to optimize weight loss interventions. *Obesity Research, 11*(5), 641–652.

9. Prochaska, J. O., & Velicer, W. F. (1997). The transtheoretical model of health behavior change. *American Journal of Health Promotion, 12*(1), 38–48.

10. Vandelanotte, C., De Bourdeaudhuij, I., Sallis, J. F., Spittaels, H., & Brug, J. (2005). Efficacy of sequential or simultaneous interactive computer-tailored interventions for increasing physical activity and decreasing fat intake. *Annals of Behavioral Medicine, 29*(2), 138–146.

11. Suris, A. M., Trapp, M. C., DiClemente, C. C., & Cousins, J. (1998). Application of the transtheoretical model of behavior change for obesity in Mexican American women. *Addictive Behaviors, 23*(5), 655–668.

12. Palmeira, A. L., Teixeira, P. J., Branco, T. L., Martins, S. S., Minderico, C. S., Barata, J. T., et al. (2007). Predicting short-term weight loss using four leading health behavior change theories. *International Journal of Behavioral Nutrition and Physical Activity, 4,* 14.

13. Nelson, M. S., Robbins, A. S., & Thornton, J. A. (2006). An intervention to reduce excess body weight in adults with or at risk for type 2 diabetes. *Military Medicine, 171*(5), 409–414.

14. Prochaska, J. O. (1993, May/June). Working in harmony with how people change naturally. *Weight Control Digest, 3*(3), 251–254.

15. Brownell, K. D., & Cohen, L. R. (1995). Adherence to dietary regimens. 2: Components of effective interventions. *Behavioral Medicine, 20*(4), 155–164.

16. Harrison, K., & Marske, A. L. (2005). Nutritional content of foods advertised during the television programs children watch most. *American Journal of Public Health, 95*(9), 1568–1574.

17. Ajzen, I. (1991). The theory of planned behavior. *Organizational Behavior and Human Decision Processes, 50,* 179–211.

18. Ajzen, I., & Fishbein, M. (1980). *Understanding attitudes and predicting social behavior.* Englewood Cliffs, NJ: Prentice-Hall.

19. Courneya, K. S., & Bobick, T. (2000). Integrating the theory of planned behavior with the processes and states of change in the exercise domain. *Psychology of Sport and Exercise, 1,* 41–56.

20. American Diabetes Association. (2006). Standards of medical care in diabetes—2006. *Diabetes Care, 29* (Suppl 1), S4–S42.

21. Bandura, A. (1986). *Social cognitive theory: Social foundations of thought and action.* Englewood Cliffs, NJ: Prentice-Hall.

22. Foster, G. D., Wadden, T. A., Vogt, R. A., & Brewer, G. (1997). What is a reasonable weight loss? patients' expectations and evaluations of obesity treatment outcomes. *Journal of Consulting and Clinical Psychology, 65*(1), 79–85.

23. Baker, R. C., & Kirschenbaum, D. S. (1998). Weight control during the holidays: Highly consistent self-monitoring as a potentially useful coping mechanism. *Health Psychology, 17*(4), 367–370.

24. Dionne, M. M., & Yeudall, F. (2005). Monitoring of weight in weight loss programs: A double-edged sword? *Journal of Nutrition Education and Behavior, 37*(6), 315–318.

25. Raynor, H. A., Jeffery, R. W., Ruggiero, A. M., Clark, J. M., Delahanty, L. M., & Look AHEAD (Action for Health in Diabetes) Research Group. (2008). Weight loss strategies associated with BMI in overweight adults with type 2 diabetes at entry into the look AHEAD (Action for Health in Diabetes) trial. *Diabetes Care, 31*(7), 1299–1304.

26. Wing, R. R., Tate, D. F., Gorin, A. A., Raynor, H. A., & Fava, J. L. (2006). A self-regulation program for maintenance of weight loss. *New England Journal of Medicine, 355*(15), 1563–1571.

27. Linde, J. A., Jeffery, R. W., French, S. A., Pronk, N. P., & Boyle, R. G. (2005). Self-weighing in weight gain prevention and weight loss trials. *Annals of Behavioral Medicine, 30*(3), 210–216.

28. O'Neil, P. M., & Brown, J. D. (2005). Weighing the evidence: Benefits of regular weight monitoring for weight control. *Journal of Nutrition Education and Behavior, 37*(6), 319–322.

29. Wadden, T. A., & Foster, G. D. (1992). Behavioral assessment and treatment of markedly obese patients. In T. A. Wadden, & T. B. VanItallie (Eds.), *Treatment of the seriously obese patient* (pp. 290–330). New York: Guilford Press.

30. Klem, M. L., Wing, R. R., McGuire, M. T., Seagle, H. M., & Hill, J. O. (1997). A descriptive study of individuals successful at long-term maintenance of

substantial weight loss. *American Journal of Clinical Nutrition, 66*(2), 239–246.

31. Anderson, J. W., Konz, E. C., Frederich, R. C., & Wood, C. L. (2001). Long-term weight-loss maintenance: A meta-analysis of US studies. *American Journal of Clinical Nutrition, 74*(5), 579–584.

32. Prochaska, J. O., Norcross, J. C., Fowler, J. L., Follick, M. J., & Abrams, D. B. (1992). Attendance and outcome in a work site weight control program: Processes and stages of change as process and predictor variables. *Addictive Behaviors, 17*(1), 35–45.

33. Svetkey, L. P., Stevens, V. J., Brantley, P. J., Appel, L. J., Hollis, J. F., Loria, C. M., et al. (2008). Comparison of strategies for sustaining weight loss: The weight loss maintenance randomized controlled trial. *JAMA, 299*(10), 1139–1148.

34. French, S. A., & Jeffery, R. W. (1997). Current dieting, weight loss history, and weight suppression: Behavioral correlates of three dimensions of dieting. *Addictive Behaviors, 22*(1), 31–44.

35. Hill, J. O., Wyatt, H., Phelan, S., & Wing, R. (2005). The National Weight Control Registry: Is it useful in helping deal with our obesity epidemic? *Journal of Nutrition Education and Behavior, 37*(4), 206–210.

36. Ferguson, K. J., Brink, P. J., Wood, M., & Koop, P. M. (1992). Characteristics of successful dieters as measured by guided interview responses and restraint scale scores. *Journal of the American Dietetic Association, 92*(9), 1119–1121.

37. Phelan, S., Wyatt, H. R., Hill, J. O., & Wing, R. R. (2006). Are the eating and exercise habits of successful weight losers changing? *Obesity, 14*(4), 710–716.

38. Butryn, M. L., Phelan, S., Hill, J. O., & Wing, R. R. (2007). Consistent self-monitoring of weight: A key component of successful weight loss maintenance. *Obesity, 15*(12), 3091–3096.

39. Catenacci, V. A., Ogden, L. G., Stuht, J., Phelan, S., Wing, R. R., Hill, J. O., et al. (2008). Physical activity patterns in the National Weight Control Registry. *Obesity, 16*(1), 153–161.

40. Bond, D. S., Phelan, S., Leahey, T. M., Hill, J. O., & Wing, R. R. (2009). Weight-loss maintenance in successful weight losers: Surgical vs. non-surgical methods. *International Journal of Obesity (2005), 33*(1), 173–180.

41. Phelan, S., Hill, J. O., Lang, W., Dibello, J. R., & Wing, R. R. (2003). Recovery from relapse among successful weight maintainers. *American Journal of Clinical Nutrition, 78*(6), 1079–1084.

42. Klem, M. L., Wing, R. R., McGuire, M. T., Seagle, H. M., & Hill, J. O. (1998). Psychological symptoms

in individuals successful at long-term maintenance of weight loss. *Health Psychology, 17*(4), 336–345.

43. Skender, M. L., Goodrick, G. K., Del Junco, D. J., Reeves, R. S., Darnell, L., Gotto, A. M., et al. (1996). Comparison of 2-year weight loss trends in behavioral treatments of obesity: Diet, exercise, and combination interventions. *Journal of the American Dietetic Association, 96*(4), 342–346.

44. Gorbach, S. L., Morrill-LaBrode, A., Woods, M. N., Dwyer, J. T., Selles, W. D., Henderson, M., et al. (1990). Changes in food patterns during a low-fat dietary intervention in women. *Journal of the American Dietetic Association, 90*(6), 802–809.

45. Dulloo, A. G., Jacquet, J., & Girardier, L. (1997). Poststarvation hyperphagia and body fat overshooting in humans: A role for feedback signals from lean and fat tissues. *American Journal of Clinical Nutrition, 65*(3), 717–723.

46. Yang, M. U., & Van Itallie, T. B. (1992). Effect of energy restriction on body composition and nitrogen balance in obese individuals. In T. A. Wadden, & T. B. Van Itallie (Eds.), *Treatment of the seriously obese patient* (pp. 83–106). New York: Guilford Press.

47. Jenkins, D. J., Wolever, T. M., Vuksan, V., Brighenti, F., Cunnane, S. C., Rao, A. V., et al. (1989). Nibbling versus gorging: Metabolic advantages of increased meal frequency. *New England Journal of Medicine, 321*(14), 929–934.

48. Cowburn, G., & Stockley, L. (2005). Consumer understanding and use of nutrition labeling: A systematic review. *Public Health Nutrition, 8*(1), 21–28.

49. Lin, B. H., Guthrie, J., & Frazão, E. (1999). *Away-from-home foods increasingly important to quality of American diet.* Washington, D.C.: Economic Research Service, U.S. Department of Agriculture.

50. Economic Research Service, U.S. Department of Agriculture. (2010). *Food CPI and expenditures.* Available from http://www.ers.usda.gov/Briefing/CPIFoodand Expenditures. Accessed July 13, 2010.

51. Curry, K. R., & Jaffe, A. (1998). *Nutrition counseling and communication skills.* Philadelphia: WB Saunders.

52. National Heart, Lung, and Blood Institute. (1998). *Clinical guidelines on the identification, evaluation, and treatment for overweight and obesity in adults: The evidence report.* Bethesda, MD: National Institutes of Health.

53. Davis, D. S., Sbrocco, T., Odoms-Young, A., & Smith, D. M. (2010). Attractiveness in African American and Caucasian women: Is beauty in the eyes of the observer? *Eating Behaviors, 11*(1), 25–32.

54. Broussard, B. A., Bass, M. A., & Jackson, M. Y. (1982). Reasons for diabetic diet noncompliance

among Cherokee Indians. *Journal of Nutrition Education, 14*(2), 56–57.

55. James, D. C. S. (1998). Improving the diets of African Americans: Challenges and opportunities. *Health Educator, 30*(1), 29–37.

56. U.S. Department of Health and Human Services. (1996). *Physical activity and health: A report of the Surgeon General.* Atlanta, GA: Centers for Disease Control and Prevention.

57. Ransdell, L. B., & Wells, C. L. (1998). Physical activity in urban white, African-American, and Mexican-American women. *Medicine and Science in Sports and Exercise, 30*(11), 1608–1615.

58. Rogers, D., American Dietetic Association, American Dietetic Association Foundation, & Commission on Dietetic Registration. (2005). Report on the American Dietetic Association/ADA Foundation/Commission on dietetic registration 2004 dietetics professionals needs assessment. *Journal of the American Dietetic Association, 105*(9), 1348–1355.

59. Committee on Institutional and Policy-Level Strategies for Increasing the Diversity of the U.S. Health Care Workforce. (2004). *In the nation's compelling interest: Ensuring diversity in the health care workforce.* Washington, D.C.: National Academies Press.

60. Vasquez, I. M., Millen, B., Bissett, L., Levenson, S. M., & Chipkin, S. R. (1998). Buena alimentacion, buena salud: A preventive nutrition intervention in Caribbean latinos with type 2 diabetes. *American Journal of Health Promotion, 13*(2), 116–119.

61. Raymond, N. R., & D'Eramo-Melkus, G. (1993). Non-insulin-dependent diabetes and obesity in the black and hispanic population: Culturally sensitive management. *The Diabetes Educator, 19*(4), 313–317.

62. Sum, A., Kirsch, I., & Traggart, R. (2002). *The twin challenges of mediocrity and inequality: Literacy in the U.S. from an international perspective.* Princeton, NJ: Educational Testing Service.

63. Epstein, L. H., Valoski, A. M., Kalarchian, M. A., & McCurley, J. (1995). Do children lose and maintain weight easier than adults: A comparison of child and parent weight changes from six months to ten years. *Obesity Research, 3*(5), 411–417.

Non-Behavioral Approaches to Weight Management

CHAPTER OUTLINE

R. Gino Santa Maria/Shutterstock.com

cappi thompson/Shutterstock.com

lorga Studio/Shutterstock.com

M oderate changes in diet and physical activity are effective in improving blood pressure, blood lipids, and glucose tolerance. They can also lower the risk of cardiovascular disease, diabetes, and other chronic diseases. While diet and activity may help people to lose weight and maintain lost weight, their inability to dramatically alter body weight has led many individuals—and in some instances their health-care providers—to try more extreme approaches to weight loss. These measures include surgery, drugs, and low-calorie diets. Do these methods work? What are their long-term effects in weight control? This chapter reviews the benefits and risks of commonly used weight-loss drugs, dietary supplements, low-calorie diets, bariatric surgery, and localized fat reduction.

WEIGHT-LOSS DRUGS

amphetamines
Classification of drugs that stimulate the central nervous system and have high potential for abuse.

Food and Drug Administration (FDA) *Federal agency responsible for overseeing the safety of food, drugs, nutritional supplements, cosmetics, and medical devices.*

The first widely used weight-loss drugs were **amphetamines**, or "uppers," which were popular in the 1950s and 1960s. Because amphetamines have a high potential for abuse and only modest success in promoting weight loss, their use as diet drugs was short-lived, and the **U.S. Food and Drug Administration (FDA)** did not approve any new weight-loss medications between 1973 and 1996. Since 1996, several new drugs for obesity have been marketed, other new pharmacological agents are in the pipeline, and weight-loss drugs have been prescribed with increasing frequency. The development of these approaches reflects a change in the way people view obesity—as a chronic condition that can be treated with medications.

What Characterizes a Good Weight-Loss Drug?

Regarding obesity as a condition that might respond to medication, like hypertension and diabetes, is a fairly recent approach. As with other chronic conditions, there will probably never be a single drug that is effective or safe for all people.

The federal Food, Drug, and Cosmetic Act requires the FDA to ensure the safety and effectiveness of new drugs. To make this determination about weight-loss drugs, the FDA considers:[1]

- Risk versus benefit: How much risk is there in using the drug over a long period of time, perhaps a lifetime? Many obese people are basically healthy, so it would be inappropriate to harm their health just for the sake of weight loss.
- Potential for abuse: What is the addictive potential of the drug? Does it offer any psychoactive effects that would promote abuse?
- Weight-loss potential: A "successful" weight-loss drug promotes a weight reduction of at least 5% more than placebo; or twice as much weight as placebo; or at least 35% of users lose 5% or more of baseline weight.
- Potential to reduce comorbidities: Does the drug also limit or treat conditions that often occur with obesity, such as type 2 diabetes, hypertension, and sleep apnea?

Today's anti-obesity drugs fall into two broad categories: appetite suppressants, which reduce energy intake, and lipase inhibitors, which reduce fat absorption. In addition, some other drugs, primarily intended for treating depression or smoking, may promote weight loss and are appropriate in certain circumstances. Weight-loss medications are generally recommended for people with a body mass index (BMI) of at least 30 or people with risk factors or chronic disease whose BMI is at least 27.

Which Drugs Reduce Energy Intake?

anorectic, or anorexiant, drugs *Drugs that suppress appetite and reduce food intake.*

catecholaminergic drugs *Category of stimulant anorectic drugs that suppress appetite by elevating levels of norepinephrine and dopamine in the central nervous system.*

serotonergic drugs *Category of anorectic drugs that suppress appetite by elevating levels of serotonin in the central nervous system.*

Medications that reduce food intake by suppressing appetite and promoting satiety are known as **anorectic**, or **anorexiant, drugs**. Some of these drugs (amphetamines, phentermine, diethylproprion, and buproprion) affect the catecholaminergic system. These **catecholaminergic drugs** act somewhat like stimulants but, with the exception of amphetamines, have little potential for abuse. They increase norepinephrine and dopamine levels in the brain, which act on the ventromedial hypothalamic nucleus and other areas of the brain to inhibit eating. Other drugs with weight-loss potential (fenfluramine, dexfenfluramine, fluoxetine, and sertraline) affect the serotonergic system. **Serotonergic drugs** have no stimulant effects but, by either increasing serotonin levels in the brain or inhibiting the reuptake of serotonin, they reduce appetite. Some drugs, like sibutramine, have both catecholaminergic and serotonergic properties. And the drug rimonabant, not approved for use in the United States but used in Europe and South America, inhibits receptors in the brain that stimulate intake of high-fat and sweet foods. The weight-loss drugs discussed in this section are summarized in Table 10-1.

Phentermine and Diethylproprion

Phentermine and diethylproprion hydrochloride are appetite suppressants that help most users lose weight. The U.S. Drug Enforcement Agency classifies these as schedule IV drugs. This means that they are currently accepted for medical use and have low potential for abuse, but if abused may cause physical or psychological dependence. They therefore tend to be taken for a short time, generally about 12 weeks. When use of the drugs stops, weight regain is typical. The adverse effects of phentermine and diethylproprion are relatively mild and related to their stimulant properties—headache, insomnia, dizziness, anxiety, and irritability. As stimulants, they could have adverse effects on the cardiovascular system and are not recommended for people who have moderate to severe hypertension. There have been reports of stroke resulting from phentermine use for weight loss.[2]

Phentermine, the phen of fen-phen, was once widely prescribed in combination with fenfluramine or dexfenfluramine. The drug combination increased weight loss more than phentermine alone. When obese individuals taking fen-phen were found to have an unexpectedly high incidence of cardiac valve disease and **primary pulmonary hypertension**, fenfluramine and dexfenfluramine were withdrawn from the market. Although phentermine was not implicated in either disorder, people who have heart disease, especially heart valve disease, should not take it. The safety of taking phentermine in combination with other weight-loss drugs has not been determined, so this practice is not advised.

primary pulmonary hypertension *Rare and life-threatening condition in which blood pressure in the pulmonary artery is elevated; signs include shortness of breath, fatigue, dizziness, and fainting.*

Sibutramine

Sold under the name of Meridia or Reductil, sibutramine has catecholaminergic and serotonergic properties. It inhibits the reuptake of dopamine, norepinephrine, and serotonin, so it enhances satiety, reduces appetite, and may also stimulate thermogenesis. One study reported the thermic effect of 30 mg of sibutramine each day to be equivalent to that induced by ephedrine or the caffeine in a cup of coffee.[3]

TABLE 10-1 Common Weight-Loss Drugs and Their Effects

Generic name	Trade name(s)	Average weight loss expected	Side effects
Appetite suppressants			
Phentermine	Adipex-P®, Fastin®, Obenix®, Obephen®, Obermine®, Obestin®, Phentamine®, Phentride®, T-Diet®, Zantryl®	8 lb*	headache, insomnia, dizziness, anxiety, irritability, elevated heart rate and blood pressure
Diethylproprion hydrochloride	Tenuate®, Tepanil, Tenuate Dospan®	6.6 lb*	
Sibutramine	Meridia, Reductil	Varies, but at least 5% of body weight	elevated heart rate and blood pressure; higher risk of heart attack and stroke (no longer sold in the U.S.)
Rimonabant	Zimulti, Acomplia	Not approved in United States	depression, anxiety, agitation, sleep disorders, and attempted suicide or suicidal ideation
Fat absorption inhibitor			
Orlistat	Xenical	18 lb at 1 year*	gas, bloating, rectal discharge, fecal urgency, fatty/oily bowel movements; vitamin supplementation needed to avoid deficiency of fat soluble vitamins
Medications not primarily intended for weight loss but that promote weight loss			
Fluoxetine	Prozac	Wide range	lack of energy, fatigue, headache, insomnia, nausea, diarrhea, dry mouth, anxiety, sexual dysfunction
Sertraline	Zoloft	Few studies	
Buproprion	Wellbutrin; Zyban	9 lb*	dry mouth, insomnia
Topiramate	TOPAMAX	6.5% of initial weight*	fatigue, nervousness, difficulty with concentration or attention, confusion, depression, anorexia, language problems, anxiety, mood problems, changes in taste
Zonisamide		6% of initial weight*	
Over-the-counter medications			
Orlistat (low-dose)	alli	5% of initial weight	Similar to prescription orlistat, but milder; vitamin supplementation recommended

*These values were derived from pooled analysis of studies of the medications, reported in Li, Z., Maglione, M., Tu, W., Mojica, W., Arterburn, D., Shugarman, L. R., et al. (2005). Meta-analysis: Pharmacologic treatment of obesity. *Annals of Internal Medicine*, 142(7), 532–546.

Based on the problems experienced with fen-phen, many experts publicly questioned the safety and effectiveness of sibutramine. The FDA's own advisory council recommended rejection of sibutramine because it did not believe the drug's safety had been proved.[4] Nevertheless, the FDA approved the drug.

Clinical trials of sibutramine demonstrated average weight losses at 1 year of up to 10 lb more than individuals taking a placebo, when sibutramine or placebo was combined with lifestyle modification.[2] For studies extending into a second year, sibutramine users gradually gained weight but still had average weight losses of about 14 lb.[5] In one year-long study, sibutramine users who also had lifestyle modification (weekly group meetings, food and physical activity records) lost an average of 26 lb, compared with those taking sibutramine alone (average loss of 11 lb) or having lifestyle modification alone (average loss of 15 lb).[6] In addition, some improvements of glycemic control, waist circumference, and blood lipid levels were reported among sibutramine users.[2,5]

Sibutramine has always been known to have side effects, including dry mouth, headache, constipation, and insomnia. Increases in systolic and diastolic blood pressure and resting heart rate were reported in clinical trials. Recent reports from the Sibutramine Cardiovascular Outcome Trial (SCOUT), which began in 2003, found more serious cardiovascular complications. Obese individuals in SCOUT lost an average of 5.7 lb over 6 weeks, and those who were randomly assigned to continue taking sibutramine for up to 3 years lost and maintained an additional 4 lb, compared with a placebo group.[7] But the sibutramine group had a significantly higher rate of heart attack and stroke. This prompted the FDA to issue a warning in November 2009 and in October 2010 to request the drug's manufacturer to voluntarily withdraw it from the U.S. market, which the company did.

Rimonabant

Tetrahydrocannabinol (THC) is the active chemical in the plant *Cannabis sativa*, the source of marijuana. THC stimulates one of the cannabinoid receptors in the brain (CB-1), which leads to overeating of high fat and sweet foods. Rimonabant is a CB-1 antagonist, so it inhibits intake of these same foods.

Several clinical trials of rimonabant in Europe and North America demonstrated that significant weight loss can occur in individuals taking the drug. However, serious psychological effects were also reported: depression, anxiety, agitation, sleep disorders, and attempted suicide or suicidal ideation. Although rimonabant has been approved in several European and South American nations, in 2008 the European Medicines Agency (EMEA) recommended suspension of the marketing authorization for rimonabant, which means that no more prescriptions can be written.[8] The FDA has already unanimously rejected the drug application due to concerns about its safety and effectiveness.

Other Appetite-Suppressing Drugs

There are a number of medications approved for other uses that also promote weight loss. While these may be prescribed for obese individuals or those who are overweight with additional comorbidities, they should generally only be used to treat the condition for which they were developed.

Fluoxetine and Sertraline

Fluoxetine (sold under the brand name Prozac) and sertraline (Zoloft) are serotonergic drugs that increase serotonin levels in the brain by selectively inhibiting its reuptake. The result is both improved mood and, sometimes, increased satiety. Fluoxetine is used to treat depression, obsessive-compulsive disorder, and bulimia; in addition to treating depression, sertraline is also prescribed for panic and posttraumatic stress disorders.

More information is available about the weight-reducing effects of fluoxetine than sertraline. After taking fluoxetine for 6 months, weight losses of up to 20 lb have been reported, but at 1 year only about half of studies report weight loss.[2] So this is not a medication that should be considered for long-term use, if weight loss is a primary goal. However, since many antidepressant drugs also promote weight gain, fluoxetine and sertraline may be helpful for depressed overweight/obese individuals who do not want to gain weight.

The side effects of fluoxetine and sertraline include lack of energy, fatigue, headache, insomnia, nausea, diarrhea, dry mouth, anxiety, and sexual dysfunction. Fluoxetine should be used with caution in people who have heart disease, diabetes, or other chronic health problems. Because fluoxetine increases serotonin levels in the brain, it should not be taken with dietary supplements that also may affect serotonin or act on the central nervous system, such as St. John's Wort and tryptophan.

Buproprion

Buproprion hydrochloride, also known as Wellbutrin and Zyban, inhibits the reuptake of norepinephrine and dopamine and is used for smoking cessation and the treatment of depression. Studies of weight loss among buproprion users report a wide range of weight lost, but about 9 lb seems to be average.[2] For an overweight/obese individual trying to quit smoking, buproprion could be very helpful in preventing the weight gain sometimes seen during smoking cessation. Dry mouth and insomnia are the most common side effects.

Anti-Seizure Drugs

Topiramate and zonisamide are used to treat seizure disorders and migraines. Clinical trials of both drugs revealed that appetite suppression and weight loss occurred in most people. While weight loss varies based on drug dose, losses average about 6.5% of original body weight.[2,9] Potential adverse effects of these medications include fatigue, nervousness, anxiety, difficulty with concentration or attention, confusion, depression, tingling sensation on the skin, constipation, dry mouth, and changes in taste.

A medication that combines phentermine and topiramate is under development. Early clinical trials of this medication (called Qnexa) with individuals having an average BMI of 38 found it to be more effective than placebo or phentermine or topiramate alone in promoting losses of at least 10% of body weight over 6 months.[10] In addition, during its phase 2 clinical trial the drug reduced incidence of sleep apnea by 69%, according to DrugWatch.com. The FDA rejected the manufacturer's application for Qnexa in October 2010 and asked for more data on possible birth defects and cardiovascular complications prior to making a final decision in 2011. However, the FDA recently gave preliminary approval to another new drug, Contrave, which combines buproprion and naltrexone (used in treatment of addictions) and has shown modest weight-loss success.

Can Drugs Reduce Fat Absorption?

The food additive olestra (Olean), discussed in Chapter 5, is a nonabsorbable fat substitute added to snack foods to reduce their caloric content. Orlistat (sold under the name Xenical) is a drug that also promises to reduce caloric intake, in this case by blocking

(45) absorption of about 30% of dietary fat from the small intestine. Orlistat is approved by the FDA for long-term treatment of obesity.

Studies of orlistat document weight losses of about 5½ lb at 6 months and 18 lb at 1 year.[2] A 4-year study that included lifestyle change plus either orlistat or placebo reported average weight losses of 13 lb in the orlistat group and 7 lb in the placebo group.[11] The orlistat group also had a reduced incidence of type 2 diabetes.

Orlistat offers additional health benefits in some users, including lower systolic blood pressure, waist circumference, low-density lipoprotein cholesterol levels, fasting serum glucose, and insulin levels. Unfortunately, it has several unpleasant gastrointestinal side effects that are similar to the side effects of olestra—gas, bloating, rectal discharge, fecal urgency, (46) and fatty/oily bowel movements. Orlistat users who overconsume fat are especially susceptible to gastrointestinal side effects, such as diarrhea and rectal leaking, which actually may help users to follow a lower fat diet. Because orlistat reduces absorption of the fat-soluble vitamins (A, D, E, and K), vitamin supplementation is recommended.

The FDA has reported at its Web site 32 reports of liver injury in orlistat users between 1999 and 2008. In six cases, users experienced liver failure. This remains under investigation. The long-term effects of orlistat use (beyond 4 years) are unknown.

What About Over-the-Counter Drugs?

over-the-counter (OTC) drug *Drug sold without a prescription.*

Over-the-counter (OTC) drugs offer several benefits to both consumers and health care professionals. First, health care costs are lower when individuals can obtain medications without having a medical office visit. This also saves the consumer money and time. Second, the individual may actually obtain treatment for a condition that would have been ignored, had a visit to the health care provider been necessary. Third, direct access to OTC drugs may make consumers better educated, as they see (on television) or read (in magazines and drug labels) about the particular drug being used. Nevertheless, there are concerns about use of OTC drugs. The consumer may incorrectly self-diagnose a condition and delay treatment that could have been effective. In addition, even though there is evidence that consumers do read drug labels—and the FDA requires that OTC drug manufacturers prove that consumers can comprehend the drug label—there is no way to assure that every OTC drug user actually follows dosage instructions and understands what a particular drug does.[12]

OTC Orlistat

The first weight loss drug approved by the FDA for over-the-counter use was alli, which contains orlistat at a lower dose than the prescription drug Xenical. Alli, coupled with lifestyle changes (portion-controlled diet, use of a food diary, increased physical activity), promotes weight losses of 5–10% in overweight individuals over 2 to 4 months.[13,14] A study that looked at actual consumer use of alli found that almost all users followed dosage instructions correctly, took advantage of the educational materials provided in the starter package, and reduced fat and calorie intake. More than half increased their physical activity.[14] Gastrointestinal side effects are similar to those for prescription orlistat but are milder due to the lower dose.

Phenylpropanolamine (PPA)

Phenylpropanolamine (PPA) has catecholaminergic properties (which makes it useful for appetite control) and causes blood vessels to constrict (which reduces nasal congestion).

Regarded as safe and effective for relief of nasal congestion and weight control for over 30 years, the FDA asked manufacturers to voluntarily withdraw PPA-containing medications from the market in 2000 because of its links to hypertension and increased risk of hemorrhagic stroke. In 2005 the FDA reclassified PPA as "not generally recognized as safe and effective." Cold remedies and weight-loss drugs that contained PPA have either been reformulated or removed.

Are New Drugs on the Horizon?

Some of the peptides and hormones produced by the body (and discussed in Chapter 4) may ultimately offer promise as obesity drugs. Given the wide range of known causes of obesity, a number of pharmacological solutions are likely.

Peptide YY (PYY)

Recall from Chapter 4 that peptide YY (PYY) is a peptide secreted in the small intestine in response to food intake. It inhibits NPY (neuropeptide Y secretion) secretion in the brain, which slows and then stops eating. Obese individuals may secrete less PYY than lean, which puts them at a biological disadvantage in being able to respond to physiological signals of satiety.[10]

In one study, intravenous administration of PYY reduced caloric intake at a buffet by about a third in lean and obese subjects.[15] A PYY nasal spray is currently being tested. Early indications were that it reduced caloric intake and promoted weight loss without side effects,[10] but the pharmaceutical company called MDRNA did not report successful weight loss outcomes in its phase 2 clinical trials.

Another peptide secreted in the small intestine in proportion to caloric intake is oxyntomodulin. Obese volunteers who injected themselves with oxyntomodulin three times a day for 4 weeks experienced small weight reductions (0.45 kg/week), compared to a control group.[16,17] What is most interesting about this is that at the same time the research subjects were reducing caloric intake (by about 128 kcal/meal), they also increased energy expenditure and physical activity levels by about 143 kcal/day.[17]

Leptin

As covered in Chapter 4, leptin is a hormone produced in the fat cells. The hypothalamus responds to leptin by suppressing appetite and increasing energy expenditure. Although some rodents do not produce leptin and gain significant amounts of weight, obese humans apparently produce normal amounts. However, the small number of obese individuals who are truly leptin deficient do respond to the drug by reducing caloric intake and losing weight; in addition, improvements in triglyceride levels and glycemic control occur.[18]

Chapter 4 also pointed out that some overweight/obese individuals may produce sufficient leptin but have lost sensitivity to it. Amylin may help restore that sensitivity. Like insulin, amylin is secreted by the β-cells of the pancreas in response to a meal. And, like insulin, amylin promotes satiety. A synthetic form of amylin called proamlintide is used in diabetes when insulin alone is ineffective. When obese diabetic and nondiabetic men were given proamlintide after fasting overnight, both groups reduced caloric intake at a buffet meal.[19] Additional studies have documented that significant weight loss can occur with proamlintide even without lifestyle change, but greater weight loss and maintenance of weight loss takes place when lifestyle change is included.[20] Rats that received both leptin and amylin reduced food intake and increased their metabolic rates, with resultant weight loss.[5]

Melanin-Concentrating Hormone Antagonists

Overproduction of melanin-concentrating hormone (MCH), which is synthesized in the hypothalamus, causes overeating and weight gain. MCH antagonist would work against such overproduction. Studies with animals show that this is exactly what happens, and the result is a decrease in food intake.[10] No studies have yet been conducted with humans.

β_3-agonists

Recall that both white and brown adipocytes are rich in β_3-adrenergic receptors. When these receptors are stimulated by catecholamines, they turn on thermogenesis. Weight gain is likely when β_3-adrenergic receptors are absent or ineffective. Studies in three countries have identified mutations in the β_3-adrenergic receptor gene—in Pima Indians, Finns, and French hospital patients—that increased the tendency to gain weight.[21–23]

β_3-**agonists** *Drugs that stimulate the β_3-adrenergic receptors and increase thermogenesis.*

β_3-**agonists** are drugs that mimic catecholamines to stimulate the β_3-adrenergic receptors. Several β_3-agonists have increased brown adipose tissue mass in rodents, and scientists thought that it was just a matter of time before similar effects would be seen in humans. Unfortunately, human studies have not yielded beneficial results, and research in this area has stalled. Eventually, drugs may be developed that target PRDM16, a protein in brown adipose tissue (BAT) that was introduced in Chapter 4. If this protein could be activated without causing irreparable harm, BAT might become more active in humans, with resultant increases in energy expenditure.[24]

How Effective Is Drug Therapy?

Most obese individuals who take weight-reducing drugs lose between 5–10% of initial body weight (2 to 10 kg) during the course of drug therapy. Some weight-loss drugs may also lower harmful blood lipids and blood pressure and improve insulin sensitivity, probably at least partly due to the weight loss. Generally, most weight loss occurs in the first 6 months, followed by weight maintenance or regain, even when drug use is continued.

Keep in mind that the effectiveness of a particular drug taken during a clinical trial may not be replicated when the medication is used in real life. In a research study, there are often intensive lifestyle interventions, with individual and group supports not available to individuals outside of clinical trials. In clinical trials, both orlistat and sibutramine achieved their effectiveness in combination with lifestyle change. Studies of both drugs outside of clinical trials reveal that the majority of users stop taking the drugs at 6 months, and only 6–8% continues to use them at 1 year.[25]

Although most weight-loss drugs in use today are not addictive, some people think of them as easy fixes that spare them from making changes in diet and activity. Behavior change is still needed. Research subjects taking a placebo also lose weight as a result of these lifestyle prescriptions—they just don't lose as much weight as those taking the drug. Realistic goals are needed. With average weight losses of 4½ lb (10 kg) or less, most people, unless their goals are modest, will not achieve their desired weight from drug therapy alone. Severely obese individuals, who may wish to lose more than 14 lb (30 kg), or 15% of body weight, will not achieve this goal using currently available drugs.

Who should take weight-loss drugs? There is no way to know, based on current information, what type of person will be most successful with weight-loss medications. The following criteria should be observed when determining an individual's appropriateness for drug therapy for obesity:

- BMI at or greater than 30 kg/m^2; or 27 kg/m^2 with at least one health risk factor
- Inability to lose weight or maintain weight loss with behavioral treatment
- Presence of comorbidities or the need for surgery, where immediate weight loss is required
- Family history of obesity (people with a genetic predisposition toward obesity may have difficulty losing weight through behavioral programs)
- Readiness for change. Drugs are most effective when accompanied by calorie control and physical activity

Because of potential short-term side effects and lack of information about the long-term consequences of taking weight-loss drugs, anorectic drugs are not recommended for cosmetic weight loss. They should be reserved for moderately and severely obese people who have a medical need to lose weight. Individuals with medical contraindications to drug therapy, including symptomatic cardiovascular disease, cardiac arrhythmias, liver or kidney disease, and a history of psychiatric illness, will need close consultation with a health care provider to determine the safest weight-loss drug for them. Weight-loss drugs should not be used during pregnancy.

Drug therapy should be accompanied by changes in eating and activity behaviors because there is currently no drug that is completely safe for prolonged use. Failure to lose about 1 lb/week (0.45 kg/wk) over an initial period of 1 month should prompt a reassessment of the value of drug therapy for obesity.[26] What can be expected from weight loss drugs is summarized in Figure 10-1.

Weight-Loss Drugs: Summary

Presently there are few weight-loss drugs on the market, and only one that is available over the counter. Weight-loss drugs offer some potential for more rapid weight loss than diet and physical activity but, in the long run, are not much more effective than behavior change in long-term weight loss and maintenance of lower weight. The potential side

---

Weight Loss/Health
- Loss of 5–10% of initial weight
- After 6 months, some regain is typical
- Improvements in blood lipids, glycemic control, sleep apnea with some medications

Complications/Side Effects
- Headache, irritability, dry mouth
- Increased blood pressure, constipation (sibutramine)
- Gas, rectal discharge, fecal urgency (orlistat)
- Effects of long-term use unknown

FIGURE 10-1 What to expect from weight-loss drugs.

effects of drugs are of some concern, as increased risk of heart attack and stroke (sibutramine), gastrointestinal distress (orlistat), and chronic insomnia and headaches (phentermine) may impair health. No weight-loss drug is currently known to be safe and effective for long-term use. Investigations of new drugs and continued evaluation of existing drugs offer the promise of improved pharmacological treatments for obesity.

Case Study. The Traveling Man, Part I	John is a 34-year-old accountant with a small firm that specializes in tax law. He works long days, typically 6 days a week, and travels 5 or 6 days every month. He likes to cook and to eat, enjoying the social aspects of a meal. But, as a busy single man, John finds himself frequently dining out and relying on convenience food. He is 6 feet tall and weighs 250 lb. Both John and his doctor want him to lose weight to get his blood glucose under better control and to avoid going on blood pressure medication. "If you don't start exercising," his doctor recently observed, "you will find yourself taking medication for hypertension and type 2 diabetes." For a few weeks John complied with his doctor's recommendation. He used the fitness facilities when staying in hotels and tried out a few health clubs in his neighborhood, even though exercise made his knees hurt. But as tax season approached, he got even busier at work, and the exercise diminished. John asked his doctor about taking Xenical, which he heard about from a client. His doctor discussed the medication with him and agreed to write him a prescription. • Is John an appropriate candidate to take a weight-loss drug? • What information should John's physician give him about Xeincal? • What kind of weight loss might John expect from taking this medication? How could he improve the chance of losing more weight?

DIETARY SUPPLEMENTS

dietary supplement

Product (other than tobacco) that contains a vitamin, mineral, amino acid, herb or other botanical, any extract, or combination of these. Dietary supplements are intended for ingestion in pill, capsule, tablet, or liquid form and are not considered to be conventional foods or the sole items consumed in a meal.

Most of us think that vitamins and minerals are the only **dietary supplements**. Dietary supplements also include 1,500 to 1,800 herbs, botanicals, metabolites, extracts, and combinations of these. A dietary supplement is defined by the FDA as a product (other than tobacco) intended to supplement the diet that contains a vitamin, mineral, amino acid, herb or other botanical, any extract, or combination of these. By definition, dietary supplements are ingested in pill, capsule, tablet, or liquid form; are not considered to be conventional foods; and are not the sole items consumed in a meal.

Dietary supplements are widely available through grocery stores, health food stores, and pharmacies as well as through the mail and over the Internet. About half of the adult U.S. population uses dietary supplements at an estimated annual cost of over $20 billion.[27,28] A growing dietary supplement market is aimed at individuals trying to lose weight.

How Are Dietary Supplements Regulated?

The FDA is charged with regulating drugs, cosmetics, dietary supplements, certain kinds of foods, and various medical products. Before passage of the **Dietary Supplement Health and Education Act (DSHEA)** in 1994, the FDA regulated dietary supplements in the same way it regulates foods, which includes evaluating the safety of ingredients. Since 1994, the FDA's oversight of the supplement industry has been constrained. The DSHEA puts the legal burden on the FDA to show that a dietary supplement is *unsafe*, rather than on the manufacturer to show that a product is *safe* before it is marketed. Supplement manufacturers are required to have scientific evidence substantiating claims made about a product but are not required to provide the FDA with such evidence before marketing the product.

One effect of this casual regulation of dietary supplements is that U.S.-manufactured dietary supplements do not necessarily follow a standard formula and may vary in potency and ingredients from company to company and even from batch to batch. For example, FDA analysis of more than 125 dietary supplements labeled as containing ephedrine alkaloids found the amount of ephedrine to vary from a trace to 110 mg in a single dose.[29] Laboratory analysis of supplements that were supposedly 71% chitosan found that capsules contained only 42% chitosan.[30] Many products marketed as "natural" or "herbal" actually contain harmful ingredients. The FDA reports at its Web site that it has found some to contain sibutramine, rimonabant, anti-seizure medications, diuretics, and even phenolphthalein, a cancer-causing chemical.

U.S. laws contrast with the strict regulation of dietary supplements in Germany, where herbal supplements are based on identical formulas. The FDA does have the authority to establish "good manufacturing practices" for dietary supplements, which would ensure that adequate manufacturing conditions exist to produce safe and properly labeled products. Some trade groups already recommend good manufacturing practices that are followed by their member companies.

Drugs Versus Dietary Supplements

Dietary supplements are not drugs. The FDA defines a drug as a product that is intended to diagnose, cure, mitigate, treat, or prevent disease. Extensive clinical study and premarket review of drugs is required to ensure that they do what they claim without causing harmful side effects. Because dietary supplements are not subject to such rigorous review, they cannot claim to treat or cure specific diseases and conditions. New labeling requirements for dietary supplements took effect in March 1999 and affect what can be claimed on a label. Figure 10-2 shows a dietary supplement label.

Dietary Supplement Claims

The DSHEA permits dietary supplement manufacturers to make three types of claims about their products:

1. Nutrient-content claims: Supplements that contain a specified amount of a nutrient may claim to be "high in" or an "excellent source of" the nutrient.
2. Health claims: If scientific evidence has established a link between a dietary supplement and a disease, then this link may be stated. Examples of scientifically valid supplement–health links include folic acid and neural tube defects, calcium and osteoporosis, and psyllium seed husk and heart disease.
3. Structure/function claims: Structure/function claims relate to the role of an ingredient in affecting normal human structure or function. For example, fiber maintains bowel regularity; calcium builds strong bones. These claims may also refer to deficiency diseases that result from lack of a nutrient in the diet, such as scurvy and vitamin C. All structure-function claims must be true, based on the manufacturer's interpretation of

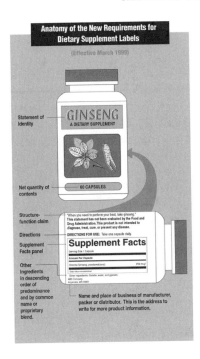

FIGURE 10-2 Dietary supplement labels cannot claim to treat or cure specific diseases and conditions.

Source: Kurtzweil, P. (1998, September–October; revised 1999, January). An FDA guide to dietary supplements. FDA Consumer Magazine. [Available online at http://www.fda.gov.]

the scientific literature. The FDA recently broadened its interpretation of structure-function claims and narrowed the definition of disease, which will permit supplement manufacturers to make more health claims. The disclaimer "This statement has not been evaluated by the Food and Drug Administration. This product is not intended to diagnose, treat, cure, or prevent any disease" must accompany structure-function claims.

Which Supplements Claim to Promote Weight Loss?

A person searching the Internet or the aisles of a health food store will find dozens of dietary supplements that claim to help people lose weight. There are over 50 single supplements and 125 combination products that are marketed as promoting weight loss.[31] These may be packaged as "fat burners," "dieter's teas," "metabolic boosters," "fat blockers," or in a variety of other ways suggesting that they accelerate metabolism and rid the body of fat. Use of these supplements is appealing to consumers because they are readily available, nonprescription, promise an easier solution than changes in diet or exercise, and create the illusion of being safe and effective. A telephone survey in 2002 estimated that 15% of adults had ever used a weight-loss supplement, and 18- to 34-year-old women were the highest users (about 17%).[32] The most common of these dietary supplements are discussed in this section and summarized in Table 10-2.

Ephedra/Ma Huang

Ephedrine is a central nervous system stimulant that can decrease appetite, increase metabolic rate, and mobilize fat from the fat cells. It is derived from plants of the genus *Ephedra sinica*. Although ephedrine is available only by prescription, other plant

TABLE 10-2 Dietary Supplements Claiming to Promote Weight Loss

Supplement	Weight-loss claim	Weight-loss facts	Adverse effects
Ephedra/Ma huang	Speeds up metabolism, burns fat	Small effect on weight loss	Gastrointestinal, psychiatric, and cardiovascular complications. Risk of heart attack and stroke prompted FDA to ban these products
Caffeine	Speeds up metabolism, burns fat	Effects on fat loss inconclusive	Dizziness, headache, nausea, addiction
Chromium	Reduces fat, increases lean body mass	Effects on lean body mass inconclusive; no effect on fat loss	Two cases of renal insufficiency with high doses
Soluble fibers	Increases satiety, reduces appetite	Increases satiety, improves glucose and lipid levels; no effect on fat loss	At high doses could cause micronutrient losses; abdominal discomfort
Hydroxycitric acid/ Garcinia cambogia	Blocks fat storage, suppresses appetite	No effect on fat loss	None reported
Chitosan	Reduces dietary fat absorption	No effect on fat loss	Some gastrointestinal symptoms reported
Pyruvate	Speeds up oxidation of carbohydrates	Small effects on weight loss when combined with diet or exercise	None reported
Dieter's teas	Prevents calorie absorption	No effect on fat loss	Powerful laxative effect: diarrhea, vomiting, dehydration, laxative dependency; 4 reported deaths
Green tea	Increased metabolic rate	No effect on fat loss	None reported

derivatives of *Ephedra sinica*, also known simply as ephedra or Ma huang, are available in dietary supplements. Ma huang/ephedra supplements appeal to people who want to boost their energy and lose weight. The telephone survey mentioned earlier found that 74% of adult users of weight-loss supplements had used products containing ephedra, bitter orange (a related stimulant), or caffeine.[32] Ephedra can have serious gastrointestinal, psychiatric, and cardiovascular effects, including increased blood pressure, irregular heart beat, and more forceful heart contraction.

Not surprisingly, Ma huang/ephedra does not live up to its claims as a weight-loss miracle. Although ephedrine increases energy expenditure, clinical trials in which obese individuals followed a low-calorie diet while taking either ephedrine or a placebo found only small differences in weight loss between the two groups (about 2 lb per month).[33] There are no long-term studies. People taking ephedra as an energy booster find that tolerance develops fairly quickly, so a higher dose is needed to get the same effect. When people stop taking ephedra, appetite may return with a vengeance, energy may drop, and some people even experience temporary depression.

To increase ephedra's weight-loss potential, it may be added to supplements in combination with other stimulants. Some evidence supports this tactic. When obese individuals following a low-calorie diet took ephedrine plus caffeine, they lost an average of 36½ lb (16.6 kg) over 24 weeks, 7½ lb more than people taking a placebo.[34] The practice of taking ephedra–stimulant combinations is so dangerous that it should be vigorously discouraged. Dietary supplements containing ephedra/Ma huang and kola nut or other potent sources of caffeine generated so many adverse-reaction complaints to the FDA (including heart attack, stroke, and seizures) that the agency banned the sale of ephedra products in 2004.

Caffeine

One hundred to 300 mg of caffeine per day (the equivalent of one to three cups of coffee) is not harmful to most adults. More than 300 mg/day may cause elevated heart rate and blood pressure and promote gastric distress, insomnia, nervousness, and muscle twitches. More than 500 mg/day has been associated with extreme anxiety, heart palpitations, paranoia, and even death.[35] Caffeine is also addictive and produces withdrawal symptoms—headache, fatigue, muscle pain, and nausea—when people stop consuming it.

The interest in caffeine as a dietary supplement for weight loss is based on its thermogenic and fat-metabolizing effects. The amount of caffeine in a cup of coffee (about 100 mg) can produce 3–4% elevations in resting metabolic rate for a short time, and the metabolic response increases with higher doses. Studies of endurance athletes indicate that the working muscles burn more fat when caffeine is consumed about an hour before exercise. Whether these effects are beneficial for weight management is debatable.

One study found that taking caffeine with food improved defective diet-induced thermogenesis in formerly obese individuals.[36] The study's authors estimated that even an increased energy expenditure of 75–110 kcal/day could help normalize body weight. Yet, in another study, obese individuals taking 200 mg of caffeine a day for 6 months did not lose more weight than individuals taking a placebo.[34]

The International Olympic Committee bans caffeine at levels of 12 mg/ml in the urine. This level could result from the consumption of six to eight cups of strong coffee, four Vivarin, eight No-Doz, or one cup of guarana tea. Because high levels of caffeine are harmful, particularly when combined with other stimulants, dietary supplements that contain caffeine are not recommended.

Chromium

Chromium is a mineral needed in small amounts for normal carbohydrate and lipid metabolism as well as insulin function. Chromium deficiency causes neurological disorders and increases in blood glucose, insulin, cholesterol, and triglycerides. Fortunately, such an occurrence is rare, and most people get plenty of chromium through their diets. Trivalent chromium is the form found in food and supplements, and it is usually combined with either picolinate or nicotinate to increase gastrointestinal absorption.

The FDA examined claims that chromium improves insulin sensitivity and concluded in 2005 that "the existence of ... a relationship between chromium picolinate and either insulin resistance or type 2 diabetes is highly uncertain."[37] Nevertheless, because chromium reduces circulating insulin and because insulin promotes fat storage, it would be logical to presume that chromium might positively affect body composition. A review of the effects of chromium on body composition reached these conclusions:[38]

- Over half of studies of chromium supplementation during weight training failed to find an effect of chromium on lean body mass.
- Chromium picolinate supplements of at least 400 mg per day for a minimum of 12 weeks may be needed before an effect on muscle mass is seen.

Studies of chromium's effects on obesity have not found it to promote weight loss in the obese.[5,31] In weight-control products, chromium is often combined with stimulants, such as Ma huang and kola nut. There are no published studies evaluating the short- or long-term effects of supplements that contain these chromium combinations.

The estimated safe and adequate daily dietary intake for chromium is 200 mg/day.[38] Although trivalent chromium is thought to be safe at higher doses, in two reported cases, individuals taking chromium picolinate developed renal insufficiency. In one instance, a woman taking 600 mg/day for 6 weeks developed chronic kidney problems.[39] In the other case, the individual who had taken 1,200–2,400 mg/day for 4–5 months was eventually able to recover kidney function.[40]

Fiber

Chapter 5 mentioned that fiber draws water into the intestines and creates a feeling of fullness. It also helps remove cholesterol and may promote better glycemic control. Products containing fiber—psyllium, guar gum, and glucomannan—have found their way onto the weight-loss supplement shelves, claiming to improve health and control appetite. These products are safe and may improve glucose and lipid levels. While there is not widespread evidence of fiber's value for weight loss, recall from Chapter 5 that at least one study did find small reductions in caloric intake and weight losses of about 4½ lb over 4 months. These effects can occur from consumption of either whole food fiber or viscous fibers added to drinks.

Hydroxycitric Acid/Garcinia Cambogia

Hydroxycitric acid is the active ingredient in the herb Garcinia cambogia, which is extracted from the rind of an orange-like fruit known as Malabar tamarind or Brindall berry. The rationale for using garcinia for weight loss is that hydroxycitric acid blocks the action of an enzyme needed to store fat (adenosine triphosphate-citrate [pro-3S]-lyase) and suppresses food intake. Hydroxycitric acid is an ingredient in many weight-control products.

Garcinia slows weight gain in studies with animals. Studies with humans have been poorly controlled and short in duration, with the exception of a 1998 report published in JAMA (Journal of the American Medical Association).[41] Researchers evaluated the effects of either a placebo or 1,500 mg of hydroxycitric acid per day on 180 moderately overweight men and women who followed a 1,200 kcal diet (high in fiber and 20% of calories as fat) for 12 weeks. Individuals in both groups lost a significant amount of weight (between 6 and 9 lb over 12 weeks), but there were no differences in weight or fat loss between the two groups. No harmful effects of hydroxycitric acid were noted.

Other Weight-Loss Supplements

A number of additional dietary supplements claim to affect fat metabolism or absorption. To date, none of these has been found effective in promoting significant weight loss.

Chitosan: A component of many diet preparations, chitosan is derived from the shells of crustaceans to increase the transit time of food through the gastrointestinal tract and reduce intestinal absorption of fat. Studies with mice and chickens find that chitosan supplements prevent weight gain on a high-fat diet, reduce blood cholesterol levels, and reduce blood glucose levels in lean (but not obese) animals with type 2 diabetes. Most studies of this product with obese humans have paired chitosan with a low-calorie diet. The one study that gave either chitosan or a placebo to people who maintained their usual caloric intake found no differences in body weight between the groups after 28 days.[30] The product appears to be safe.

Pyruvate: Recall from Chapter 3 that pyruvate is the three-carbon byproduct of glycolysis that enters the citric acid cycle. In theory, excess pyruvate might accelerate aerobic metabolism, oxidizing blood glucose and preventing conversion of excess carbohydrate to fat. Few controlled studies of pyruvate have been published. In one, overweight individuals taking 6 g of pyruvate in combination with exercise training had weight losses of about 2.5 lb over 6 weeks.[42]

Dieter's teas: Various so-called dieter's teas claim to enhance weight loss by speeding food through the digestive tract so calories cannot be absorbed. These are marketed as "herbal" products, suggesting that they are natural and therefore safe. Generally these teas contain plant-derived laxatives, such as senna, aloe, dandelion, rhubarb root, cascara, castor oil, and buckthorn. They are usually powerful laxatives that can cause diarrhea, nausea, vomiting, stomach cramps, fainting, dehydration, and—in four reported cases—death.[43] People who abuse these products or use them for a year or more may ultimately pay a high price. Laxative dependency sometimes leads to chronic constipation and, in at least one case reported to the FDA, loss of function necessitating removal of the colon. The FDA has recommended that dietary supplements containing stimulant laxatives carry a warning label listing adverse effects.

Green tea: Extracts from green tea, often sold in capsule form, are claimed to increase fat oxidation, thereby leading to weight loss. Studies of green tea extract combined with caffeine[36] or capsaicin[44] showed increased metabolic rate in users; but several clinical trials found little evidence of weight loss or maintenance of weight loss.[45]

How Can Fraudulent Products and Practices Be Avoided?

The information in this section has probably increased your awareness of the possible harmful effects of dietary supplements. Many of these products have been studied in limited, uncontrolled trials, and few studies have assessed the effects of taking products in combination. For people seeking help in losing weight, dietary supplements may raise hopes, waste money, and cause irreparable harm.

The Internet has increased consumer awareness of weight-loss dietary supplements and made it very easy to buy new products. Individuals who are reluctant or embarrassed to browse in a health food store for weight-loss remedies can browse privately on the World Wide Web. Most dietary supplement distributors have Web sites, and these often look very clinical and scientific. How can we help clients—and ourselves—to guard against fraudulent health claims?

Signs that a Product Is Not All that It Claims to Be

Wary consumers should be alert for the following signs, which are good indicators that a product may not do what it claims:

- Use of words like "miracle," "breakthrough," and "new discovery": If a real breakthrough in weight loss had occurred, it would be reported in scientific journals and newspapers and would be heralded in news releases by government health agencies.

- Intimations that the federal government or health organizations are covering up the effectiveness of the product for poorly explained reasons: The DSHEA created an Office of Dietary Supplements at the National Institutes of Health to ensure that information would be made available to consumers and scientists quickly.
- Suggestions that a product can be used to treat or cure a variety of disorders: It is unlikely that a single supplement could have widespread effects.
- Misuse of medical terms or overuse of pseudomedical terminology: Supplement manufacturers are particularly fond of the terms metabolic and thermogenic.
- Repeated reference to the fact that a product is "natural": Natural does not necessarily equal safe. Toxic mushrooms and berries are natural, but they can still make you very sick or kill you.
- Poorly referenced scientific data backing up the product: There should be some way to document claims made about a product. A lack of references or a listing of references that are impossible to obtain is a sign that no studies exist to back up claims made about a supplement.
- Failure to acknowledge side effects: Anything powerful enough to stimulate the effects claimed by some supplement manufacturers is sure to have side effects.

Identifying Quality in Nutritional Supplements

According to the FDA, most dietary supplement manufacturers are responsible and careful about what goes into their products. Still, manufacturing practices for supplements in the United States are not as standardized as they are in Europe. How can consumers who want to use supplements select quality products?

- Buy brands from nationally known manufacturers or distributors: Manufacturers who have been in business for many years are likely to have better experience with quality control simply because they market more products.
- Look for the USP Verified Dietary Supplement Mark (Figure 10-3) on the supplement container: This means that the manufacturer has met U.S. Pharmacopeia's

FIGURE 10-3 USP verified dietary supplement mark.

Source: Reprinted with permission of U.S. Pharmacopeial Convention, 12601 Twinbrook Parkway, Rockville, MD 20852.

stringent standards for strength, quality, purity, packaging, and labeling. USP also sets standards for botanicals (plants and herbs) used as dietary supplements. To learn more about standards development for botanicals, visit the Web site of U.S. Pharmacopeia at http://www.usp.org.

Visit the Web sites of the FDA (www.fda.gov) and the National Institutes of Health Office of Dietary Supplements at (ods.od.nih.gov) to read more about dietary supplements. The Office of Dietary Supplements, in cooperation with the Food and Nutrition Information Center of the U.S. Department of Agriculture, maintains an International Bibliographic Information on Dietary Supplements database, which contains more than 750,000 citations. The European Scientific Cooperative on Phytotherapy (www.escop.com) reports latest scientific information on the use of plant drugs (herbal remedies).

If you or your clients experience an adverse reaction from a dietary supplement, report it to the FDA's MedWatch program through the FDA Web site (www.fda.gov/Safety/MedWatch/default.htm) or at 1-800-332-1088.

Dietary Supplements: Summary

Dietary supplements include vitamins, minerals, amino acids, and herbs or botanicals. These products are estimated to be used by half of the adult U.S. population. Dietary supplements are regulated by the FDA under the 1994 DSHEA. Several supplements claim to promote weight loss, but there is no scientific evidence of their effectiveness either alone or in combination. Serious side effects may occur, even when supplements are used correctly. Consumers should be wary of products making claims about weight loss.

LOW-CALORIE DIETS

Dieting is the most common method used by Americans who want to lose weight. Half of adult women and one-quarter of adult men will go on diets this year, most on so-called fad diets, a smaller number in structured programs, and still fewer on medically supervised very-low-calorie diets. This section of the chapter focuses on very-low-calorie (fewer than 800 kcal/day) and low-calorie (800–1,800 kcal/day) diets for weight loss and weight maintenance, including fad diets and meal replacements.

What Are the Physiological Effects of Caloric Reduction?

Any time caloric intake is reduced below the level necessary to support normal metabolic functioning, physiological changes occur. For example, as discussed in Chapter 3, when daily caloric intake falls below approximately 1,200 kcal, resting metabolic rate decreases as the body adapts to insufficient calories. And as covered in Chapter 4, caloric restriction triggers dozens of hormonal and neurochemical changes that affect hunger, appetite, and satiety. The result is often failure to lose predicted amounts of weight and, most commonly, failure to keep off lost weight. The physiological effects of caloric

Energy Expenditure Decreases
- ↓ metabolic rate
- ↓ diet-induced thermogenesis
- ↓ physical activity

Hunger Increases
- Low circulating glucose stimulates ↑ orexin production in brain
- ↑ NPY secretion
- ↑ galanin secretion
- ↓ POMC secretion
- ↑ ghrelin secretion
- ↓ leptin secretion

Changes Occur in Fat Cells
- Enlarged fat cells shrink
- Hyperinsulinemia is normalized
- Gradual ↑ in lipoprotein lipase assures fat storage when eating resumes

FIGURE 10-4   Physiological effects of caloric restriction.

restriction are summarized in Figure 10-4. Other side effects of reduced caloric intake include the following, with more noticeable effects seen in those who consume the fewest calories:

- Weakness and fatigue
- Dizziness
- Constipation
- Menstrual irregularity
- Cold intolerance
- Edema

More severe effects are evident in people who consume less than 1,000 kcal/day:

- **Gallstones:** The risk of developing gallstones is increased when people lose weight rapidly. About 25% of people on very-low-calorie diets (600–800 kcal/day) have evidence of gallstones.[46]
- **Gout:** Gout occurs when excess uric acid is deposited in the joints, particularly the big toe. Individuals with a history of gout may have an acute episode when consuming a very-low-calorie diet.
- Sudden cardiac death: Loss of lean body mass and electrolyte imbalances that may occur during diets have been implicated in some cases of sudden cardiac death. Fortunately, this is very rare.

People who use low-calorie diets are often either desperate to lose weight or want a quick fix to what they perceive as a weight problem. Do the benefits of low-calorie diets outweigh the risks?

gallstones
Deposits made of cholesterol, bile, and calcium salts that form in the gallbladder and may block the duct between the gallbladder and the duodenum, causing pain.

gout *Deposition of excess uric acid in the joints, especially the big toe, causing swelling and pain.*

What Are Very-Low-Calorie Diets?

very-low-calorie diets (VLCDs)

Diets providing 800 kcal or less each day; usually the diet is in liquid form, rich in protein, and supplemented with essential nutrients.

Very-low-calorie diets (VLCDs) contain no more than 800 kcal/day or less than 12 kcal/kg body weight/day. Between 1988, when Oprah Winfrey very publicly lost almost 70 lb on a liquid diet, and 1990, when she revealed that she had regained all the lost weight, VLCD popularity peaked in the United States. More than 200,000 Americans are estimated to have followed a VLCD in 2004.[46]

Because of the severity of the caloric restriction, VLCDs in the United States are limited to people whose BMI is higher than 30, and they are always medically supervised. In Europe VLCDs are available without prescription and can be used by obese individuals who do not have comorbidities for up to 3 weeks without medical supervision.

Characteristics of VLCDs

VLCDs are both low in calories and high in protein (0.8–1.5 g of protein per kilogram body weight per day up to a maximum of 125 g protein/day). The protein supplied in a VLCD must be equivalent in quality to the protein found in milk and eggs, meaning that it is made up of amino acids in a proportion needed by the body for health. In the 1970s, some commercial VLCD preparations made from collagen, a low-quality protein, caused several deaths. Maintaining sufficient protein in the diet is also essential for preventing loss of excess amounts of lean body mass. Fat content should not exceed 30% of calories.

Calories are usually obtained from a liquid formulation that is prepared from a powder mixed with water. The formulation provides the recommended dietary allowance for vitamins, minerals, electrolytes, and essential fatty acids. Sometimes a food-based, high-protein diet is used, but in those cases vitamin and mineral supplements must be taken, because such a small quantity of food could never supply enough of the essential nutrients.

Most VLCDs last for 12 weeks. Health would be compromised if the diet were continued for a longer time. In the initial 12 weeks of very low caloric intake, coupled with lifestyle changes, the individual loses weight rapidly. This is followed by another 12–14 weeks of refeeding, where regular foods are gradually reintroduced. Nutrition education and encouragement to be physically active are important during this time. The cost of a 6-month program is about $3,500.

Short-Term Weight Loss and Health Outcomes of VLCDs

People who go on a VLCD always lose a large amount of weight in the early stages of the diet. Losses of 11 lb (5 kg) in the first 2 weeks are not unexpected. The success of these diets at inducing weight loss is the reason for their popularity. People who have never been able to lose large amounts of weight before finally find a way that works. The average person loses 15–25% of his baseline weight over 12 to 24 weeks.[46]

People who lose weight in a VLCD program generally experience improved health, often within weeks:

- Reductions in systolic and diastolic blood pressure
- Reduced blood cholesterol
- Better glycemic control

- Loss of abdominal fat
- Some reversal of sleep apnea and cardiac failure
- Reduced risk of surgery

Programs appear to vary considerably, but up to half of those who begin a VLCD drop out during the first 3 to 6 months.[46] Why people drop out has not been systematically studied, but it is unlikely to be due to failure to lose weight (everyone who complies with the diet will lose weight) or to hunger. Individuals on most VLCDs go into ketosis, so they tend to experience less hunger than people on low-calorie diets. Several likely explanations for dropping out are the desire for food or the pleasure of eating, inability to resist the temptations of social situations, and—most likely—lack of readiness for such a radical weight-loss program.

Long-Term Weight Maintenance Outcomes

A substantial amount of weight—up to half of weight lost—is typically regained in the year after the VLCD. Some people are able to maintain significant weight losses by exercising and selecting food carefully. A meta-analysis of VLCDs reported that 5 years after program completion, slightly over half of men and slightly under half of women maintain a 5% reduction in body weight, and 28% of men and 31% of women are able to keep off 10% of baseline weight.[47] However, in one study more than one-third of participants had regained more weight than they had lost.[48] What you can expect from VLCDs is summarized in Figure 10-5.

Why do so many people who participate in this rigorous and expensive procedure regain most of the weight they lost when the program ends? In addition to the physiological effects summarized in Figure 10-4, there are several explanations:

- Severe dietary restriction is known to precipitate binge eating, and people who have successfully completed a 12-week VLCD are not immune. The refeeding period is a

---

Weight Loss/Health
- Loss of 15–25% of initial weight in 12–16 weeks
- Regain of up to half of lost weight 1 year following program end
- Maintenance of 5% weight loss at 5 years by about half of users
- Improvements in blood pressure, blood lipids, glycemic control, sleep apnea with weight loss
- Reduced risk of surgery

Complications/Side Effects
- Weakness and fatigue
- Dizziness
- Constipation
- Menstrual irregularity
- Cold intolerance
- Edema
- Gallstones
- Gout
- Sudden cardiac death (very rare)

FIGURE 10-5 What to expect from Very Low Calorie Diets (VLCDs).

very risky time for episodes of bingeing, unless people have support and resistance skills.

- A highly prescriptive diet that includes no "real" foods does not prepare people for the realities of planning, preparing, and eating real meals, unless skills-based nutrition education is part of the VLCD program.
- People who undertake a VLCD are not necessarily in the action stage of change. Without readiness for behavior change, even a large initial weight loss cannot guarantee maintenance of habits and attitudes needed to keep weight off.
- Physical activity may not be adequately emphasized in the VLCD program. You have already learned that the strongest predictor of weight maintenance is participation in regular physical activity.

How Effective Are Low-Calorie Diets?

Low-calorie diets provide 800–1,800 kcal/day. These include numerous medically unsupervised weight-loss regimens—liquid, frozen, or shelf-stable meal replacements; reduced calorie diets with various macronutrient combinations (low-fat; low-carbohydrate; high-protein); commercial programs like Jenny Craig and Weight Watchers; and fad diets that use special foods or food combinations. These are widely promoted in popular books, in infomercials, and on the Internet. Most offer a simple solution to obesity—follow this program and you'll finally lose that weight!

Quite often low-calorie diets are developed by people with legitimate credentials (medical, doctoral, or master's degrees) who claim to be experts in weight management or nutrition. Or, diets may be written by a celebrity, usually one who has lost weight or is lean and trim and claims that the diet keeps him or her looking so good. Scientific references backing up the diet are often included. If these references were examined closely, then readers would notice that in many cases references are incomplete, difficult to obtain, or are only peripherally related to the diet being promoted.

Despite widespread publicity about the failure of popular diets to help people keep off lost weight, people continue to be attracted to them. The lure of promised weight loss is just too great, and everyone wants to believe that it will work for them. Participation in a VLCD program is expensive and limited to people with a BMI above 30, and medications require a prescription. Because of this, people with varying degrees of overweight/obesity who cannot afford a medically supervised regimen may be attracted to commercially available low-calorie diets.

Partial Meal Replacements

Commercially available meal replacements call for replacing one or more meals and snacks each day with flavored shakes, snack bars, frozen foods, and other calorie-reduced products fortified with vitamins and minerals. Beverages like SlimFast and Nestlé's Sweet Success are only loosely based on the liquid-diet formularies found in VLCDs. Each shake provides about 200 kcal, so total daily caloric intake depends on the caloric value of regular meal(s). Although shakes are fairly nutritionally balanced, they should not be used exclusively (like a VLCD).

A review of studies of meal replacements concluded that people who use them for a year can lose 7–8% of starting weight, exceeding weight lost by 5½ to 8 lb over other reduced calorie diets.[49] This is comparable to losses seen with orlistat and sibutramine.

Why do meal replacements work? There are a number of theories, supported by evidence:

- They help with portion—and, therefore, calorie—control.
- For the meals and snacks replaced, they eliminate the need to make decisions about "what to eat."
- Because they do not replace all the food a person eats, there is an opportunity to incorporate food preferences into the meal that isn't replaced.

An additional benefit of partial meal replacements is that they do not have complications that are as severe as VLCDs. For example, gall bladder disorders are less likely with 1,200 to 1,800 kcals than with 800 kcals.

Diets with Varying Macronutrient Content

Thousands of popular diet books are available, with approaches ranging from diets that claim to reduce waist and abdomen size to those that claim to balance hormones or metabolism to those that promote eating like a caveman or getting detoxified. Many diets are based upon reorganizing the energy nutrients that one consumes, so instead of a diet that follows general dietary recommendations (approximately 55% carbohydrate, 30% fat, and 15% protein), these diets may be low-carbohydrate (<20%), very-low-fat/high-carbohydrate (≤10% fat and >55% carbohydrate), or moderate-fat/high-carbohydrate (20–30% fat, >60% carbohydrate).

Low-Carbohydrate Diets A diet that is low in carbohydrate is by its nature high in fat and protein. The first commercially successful high-protein diets surfaced in the 1970s—The Complete Scarsdale Medical Diet and Dr. Atkins' Diet Revolution. These classics were followed by The Zone and Protein Power, all based on a similar premise: that carbohydrates prevent the body from burning fat, cause an insulin imbalance, and make you hungry, so dietary carbohydrates should be limited and replaced with fat and protein.

As low-carbohydrate diets became more widely used, the sale of low-carbohydrate foods similarly increased. The popularity of this approach is explained by the rapid weight loss that almost always happens. The loss of stored glycogen (and, with it, lots of water) causes an immediate and gratifying weight reduction. In addition, people who like meat, cheese, and butter feel indulgent, not deprived, on these diets (although The Zone recommends more plant sources of protein than animal). And the protein in the diet does help maintain blood glucose levels and prevent hunger.

The problems with high-protein diets are plentiful, however. Foremost is the failure of such diets to promote maintenance of lost weight. Although the type of foods being eaten seems indulgent, these are essentially low-calorie diets, poorly adhered to, and impossible to follow for a prolonged period. When compared with a conventional low-calorie diet, individuals on the Atkins diet lost more weight at 6 months but thereafter weight loss maintenance was similar between the two groups.[50] Very few individuals are able to continue to meet carbohydrate intake goals for the duration of the diet.[50–51]

In the meantime, increased consumption of saturated fat and cholesterol may have unfortunate effects on the risk of cardiovascular disease. LDL-C levels increase in people on low carbohydrate diets.[50] Some people even develop ketosis from insufficient carbohydrate, and there is always a risk of overburdening the kidneys, which must process so much protein. People who limit their intake of fruits and vegetables do not obtain sufficient dietary antioxidants, which are needed for the prevention of cardiovascular disease, cancer, eye diseases, and other health problems.

Very-Low-Fat Diets   The Ornish Program for Reversing Heart Disease and the Pritikin diet are examples of diets that limit fat intake to ≤10% of calories. Carbohydrate content of these diets is fairly high, generally exceeding 60% of calories. The Ornish diet is essentially a high-fiber vegetarian diet that allows users to eat as many fruits, vegetables, grains, and legumes as they wish, until they feel full. Meats, avocados, olives, nuts, sugar, and alcohol are prohibited.

Very-low-fat diets have demonstrated significant weight losses and reasonable adherence.[52,53] They seem to help avoid the sensation of hunger, because people can eat more due to the diets' low energy density.[54] Studies comparing diets with varying carbohydrate, fat, and protein content—the Atkins, Zone, Ornish, Weight Watchers, and LEARN diets—found similar weight losses with all the programs and similar reductions in cardiac risk factors.[51,55] In addition, all these diet programs had similar adherence rates—fairly low.

Diets that Use Special Foods or Food Combinations   The Cabbage Soup Diet is an example of a fad diet based around a particular food. The dieter is allowed to eat unlimited quantities of cabbage soup, which is prepared from cabbage, onions, green peppers, canned tomatoes, celery, soup mix, and water. Caffeinated beverages are also allowed. On particular days, small quantities of additional foods are permitted. The New Beverly Hills Diet does not include cabbage soup, but it does restrict foods to particular times. Fruits must be eaten alone, for example, and carbohydrates are prohibited until at least 2 hours after the last fruit is eaten. Carbohydrates may not be eaten with protein. Once protein is consumed almost all remaining foods eaten that day must be protein foods. The Grapefruit Diet combines grapefruit at every meal with lots of caffeinated beverages and small quantities of other foods.

The theory behind these kinds of diets is that certain foods, eaten alone or in combination, have fat-burning properties or are processed efficiently only in the presence (or absence) of other specific foods. Scientific evidence backs up none of these claims. Eliminating whole categories of foods from the diet and centering the diet on low-calorie foods (like cabbage soup or fruit) increases the likelihood that nutrient deficiencies will occur. Including lots of caffeine, a diuretic, increases loss of body water and may lead to dehydration. Diarrhea and gastric discomfort are other side effects of such diets. A more severe effect is loss of lean body mass caused by lack of dietary protein.

Fasting   For centuries, people have fasted for religious occasions or as part of meditation and other rituals. Today, people may still fast periodically for religious reasons, but fasting is also promoted as a part of health spa programs and for general "body cleansing." Usually water, juice, or broth is permitted during a fast.

Most healthy people can tolerate periodic fasting without medical complications, other than feeling weak or dizzy. Exercising on a fast day is inadvisable because electrolytes needed for normal heart and muscle contraction may be reduced, and energy level may be low. If sufficient liquids are not consumed, then dehydration results. Weight loss will certainly occur after several days of fasting, but water—not fat—is the component of weight lost. Periodic fasting does not accelerate fat loss during non-fasting periods. Once normal intake is resumed, weight is regained. In cases when fasting is prolonged, a decrease in resting metabolic rate might even make weight regain easier.

A modified fast having some success was recently reported in the scientific literature.[56] For 4 weeks, 16 obese women and men followed their regular diet (but with less than 30% of calories from fat) and on alternate days consumed a controlled diet providing 25% of their baseline energy needs (the modified fast). For another 4 weeks they continued the alternate-day modified fast, but on fast days were allowed to self-select foods with the help of a dietitian. Weight losses averaged 12 lb; fat losses averaged 3%; and all experienced reductions in total cholesterol, LDL-C, triglycerides, and blood pressure. HDL-C did not change. While this approach seems to have had good short-term results, whether weight losses can be sustained needs further research.

Commercial Low-Calorie Diet Programs Several commercial weight-loss programs are available to people who do not want or do not qualify for a VLCD and who are skeptical of diet books. Some of these programs, like Jenny Craig, offer diet plans as low as 1,000 kcal/day using prepackaged foods and liquid supplements that are purchased from the program. Others, like Weight Watchers and Nutri/System, generally start dieters at about 1,200–1,500 kcal/day. Other than Weight Watchers, which is the most flexible and realistic of the commercial programs, most are reliant on selling their own foods, drinks, and nutritional supplements. Some even offer diet drugs and herbal supplements.

Commercial weight-loss programs are technically "medically supervised" because they employ a medical director at the national level. Most participants have little or no contact with health professionals during the course of their diets, however. Although commercial programs have maintenance programs, none publish data on the effectiveness of the initial weight-loss phase, maintenance phase, or long-term outcomes, so it is impossible to say with certainty that they work. The expense of these programs can be considerable. Given the uncertainty of the outcome, a relatively inexpensive program that offers group support and flexibility in food choice, like Weight Watchers, is preferable to an expensive program that relies on prepackaged foods and supplements.

Recently an independent research group completed a 2-year study of over 400 overweight/obese men and women randomly assigned to either a self-help group or Weight Watchers. The self-help group lost about 3 lb in the first 12 months but gradually returned to their starting weight at 2 years. The Weight Watchers group lost 9–11 lb during the first year and maintained a 6–6½ lb weight loss at the end of the second year.[57] Almost three-quarters of participants stuck with the study for 2 years.

Outcomes of Medically Unsupervised Low-Calorie Diets

The most likely long-term outcome of any calorie restricted diet is dissatisfaction, unless weight loss goals are realistic. Approximately 70% of people who begin a commercial low-calorie diet program with either regular or prepackaged foods drop out within 12 weeks.[58] For people who complete these diets, very little information (other than testimonials) is available on their short- or long-term effectiveness in promoting weight loss. Low-carbohydrate diets are poorly adhered to beyond 2–3 months, and weight losses on most reduced calorie diets, whether low-carbohydrate or low-fat, are similar at 1 year and rarely exceed 5–10% of starting weight; beyond 1 year weight regain usually occurs.

This is understandable, given that few people will sustain diets that are so different from their typical eating behaviors and food preferences, while they are fighting against physiological forces driving them to eat. Only Weight Watchers has published data that shows maintenance of significant weight loss after a year, but even those individuals regained some weight. In Chapter 4 you learned that a group of researchers recently found evidence that there may be low-fat and low-carbohydrate genotypes that determine the success of different types of diets. Stanford University researchers analyzed DNA from overweight women for genes needed to metabolize fats and carbohydrates and classified women as having a low-carbohydrate diet responsive genotype, a low-fat diet responsive genotype, or a balanced diet responsive genotype. Women who followed a diet that matched their genotype lost 2–3 times more weight than women who followed a diet that did not match their genotype.[59] Being able to identify a person's genotype might allow us to someday suggest the specific low-calorie diet for that individual. Until then, the general outcomes of low-calorie diets are summarized in Figure 10-6.

With weight regain, dieters will doubtless blame themselves for not having enough willpower, for not following the diet correctly, or for picking the wrong diet book. The author of the diet may even suggest these failures to undermine the confidence of the dieter and to encourage additional dieting. To avoid self-blame and disappointment, individuals contemplating a medically unsupervised low-calorie diet should ask three questions: (1) Will this diet harm my health? (2) Is the diet scientifically valid? and (3) Can I realistically follow it? The criteria listed in Figure 10-7 should help your clients answer those questions.

Weight Loss/Health
- Loss of 5–10% of initial weight at 6 months to 1 year (meal replacements, Weight Watchers, low-carbohydrate, and low-fat diets)
- Poor adherence and high drop-out rates
- Regain of most or all lost weight at 1 year (exception: Weight Watchers)
- Maintenance of 5% weight loss at 5 years by about half of users
- Improvements in blood pressure, blood lipids, glycemic control with weight loss

Complications/Side Effects
- At lowest calorie levels:
  - Weakness
  - Dizziness
  - Constipation
  - Dehydration
  - Inability to maintain intake of essential nutrients without supplements
- Low-carbohydrate diets:
  - Possible increased LDL-C
  - Possible ketosis

FIGURE 10-6 What to expect from low calorie diets.

To rate a medically unsupervised diet, answer "yes" or "no" to each of the following questions. Use the scoring table at the bottom of the page to evaluate the validity, safety, and effectiveness of the diet.

Criteria	Yes	No
1. Is the diet monotonous and boring?	—	—
2. Does the diet focus on a few (but not all) of the food groups?	—	—
3. Do foods have to be eaten in a specific pattern or particular days?	—	—
4. Must the diet be followed for a set period of time?	—	—
5. Does the diet fail to emphasize physical activity?	—	—
6. Are personal food likes and dislikes ignored by the diet?	—	—
7. Will the diet provide less than 1,200 kcal/day?	—	—
8. Does the diet rely on dietary supplements?	—	—
9. Is snacking prohibited?	—	—
10. Will it be difficult to follow the diet in typical social situations (restaurants, parties)?	—	—
11. Does the diet author and/or diet endorser(s) lack credentials in human nutrition?	—	—
12. Is the diet described as miraculous or as some type of breakthrough?	—	—
13. Does the diet's basic premise seem to contradict what mainstream health organizations say?	—	—
14. Does the diet lack scientific references that could be obtained for verification?	—	—
15. Does the diet seem too good to be true?	—	—

If you have marked "yes" less than four times, you have found a diet that is somewhat restrictive and has some elements of faddism but that probably won't harm your health.

If you have marked "yes" more than seven times, you have a diet that is not only restrictive but also unlikely to be based on sound nutritional practices and is possibly harmful to your health.

If you answered "yes" to questions 1, 6, 9, and 10, your likelihood of following the diet for more than a few weeks is slim.

FIGURE 10-7 A fifteen-question form can help you to evaluate medically unsupervised diets.

How Can Consumers Protect Themselves from Fraudulent Low-Calorie Diet Programs?

The Federal Trade Commission (FTC) regulates the advertising of weight-loss products and programs, including dietary supplements for weight loss. The FTC and the Partnership for Healthy Weight Management developed, "Voluntary Guidelines for Providers of Weight Loss Products or Services."[60] It provides sound guidance to the general public for achieving and maintaining a healthy weight. The voluntary guidelines recommend that providers of weight-loss products or services voluntarily disclose the following to consumers:

- Qualifications of staff and central program components
- Risks associated with overweight and obesity and the benefits of modest weight loss
- Risks associated with the provider's product or program
- Program costs

- Information about the difficulty of maintaining weight loss and ways to increase the probability of success

If a company is reluctant to disclose this information, then consumers should take their business elsewhere. Although guidelines are voluntary, the FTC does mandate truth in advertising. Companies that wish to advertise "typical" weight losses for a particular program must base their estimates on a representative sample of all participants entering the program (excluding those who drop out in the first 2 weeks). Companies may not claim that weight loss is maintained over the long term unless there is evidence that participants have kept off the weight for at least 2 years from the end of the program or when the client terminated the program.

Low-Calorie Diets: Summary

Half of adult women and one-quarter of adult men will go on diets this year. Although most of them will lose weight, few low-calorie diets—even medically supervised VLCDs—promote long-term maintenance of lost weight. Any health improvements gained from weight loss are reversed when lost weight is regained. Consumers considering a medically unsupervised, low-calorie diet or a fad diet need to set realistic goals for weight loss and be wary of quick fixes that are based on little scientific evidence and are harmful to health.

Case Study. The Traveling Man, Part 2	John took Xenical for about 1 year. For the first 7 months, until November, he was losing weight fairly steadily. Dr. Parnell had stressed the importance of watching his caloric intake and maintaining physical activity, and John complied pretty well. When he knew there was a social event on a particular day, he ate just a piece of toast for breakfast and had an apple at his desk for lunch, so he could eat more at dinner. He was able to get some exercise at least twice a week. His weight dropped to 218 lb.

Starting in November, the firm received lots of invitation to social events, and John found himself dining out 4–5 times each week. He also started dating more. Some weeks he was too busy to go to the gym. When he started experiencing more severe gastrointestinal side effects, Dr. Parnell advised him to stop taking the Xenical. By January John had regained 30 lb. A woman at work told John that he could easily lose that weight by going on a diet that she had tried. She was eating 900 kcal/day with very few carbohydrates and was losing lots of weight. John thought that diet might work for him, but he felt that 1,000 kcal a day would be better due to his larger size.

- Design a 1,000 kcal/day low-carbohydrate (20% of kcals) diet for John that has at least 300 kcal from protein).
- What do you imagine would be the barriers to staying on such a diet?

WEIGHT-LOSS SURGERY

bariatric surgery
Surgical procedures for the treatment of obesity.

Bariatric, or weight-loss, surgery has increased exponentially as a treatment for severe obesity. Fewer than 15,000 surgeries were performed in 1998; 150,000 in 2005; and 200,000 in 2007.[61] Today the most common procedures are gastric bypass and laparoscopic adjustable gastric banding. These procedures, as well as other types of **bariatric surgery**, have helped very large people to lose excess weight quickly. They are not for everyone and are generally used only when other weight-loss methods fail. Liposuction, a type of cosmetic surgery that is used by less obese (and sometimes nonobese) people to remove unwanted fat from specific areas, is discussed in the next section of the chapter.

Who Should (and Who Should Not) Have Bariatric Surgery?

Only very obese people are candidates for gastric surgery. Surgery is usually restricted to people whose BMI is 40 or higher, although individuals with comorbidities (like sleep apnea and diabetes) whose BMI is 35 might be considered. Other weight-loss methods, such as behavioral programs and drug therapy, usually are tried before an extreme measure like surgery. Individuals in the action stage who are severely obese and have been unable to achieve weight loss through changes in diet and activity are good candidates for surgical intervention. Individuals who binge and demonstrate an inability to make dietary or activity modifications are generally poor candidates because their lack of adherence to the post-surgery dietary regimen could put them in danger of serious physical harm. Careful screening and education of surgical patients is essential if there is to be any hope of success.

Usually a team approach is used to screen and educate patients before surgery. The typical team includes a bariatric surgeon (a specialist in the surgical treatment of obesity), an internist, a psychologist, and a dietitian. Before surgery, the team will evaluate the patient's medical, psychological, and nutritional status and ensure that the individual receives information about diet and other restrictions that are absolutely essential to prevent post-surgery complications. The patient is given information about the potential side effects of surgery, management of postoperative diet and stress, and the necessity of follow-up. A person resistant to changing eating and activity behaviors who believes that surgery will "cure" his obesity is not psychologically ready for bariatric surgery. For this person, the risks of surgery will clearly outweigh the benefits.

What Are the Most Common Bariatric Procedures?

Bariatric surgery accomplishes one of three things: (1) gastric restriction, (2) significant intestinal malabsorption of nutrients, or (3) a combination of gastric restriction and some intestinal malabsorption.

Jejunoileal bypass surgery was introduced in the 1950s as a treatment for severe obesity. The jejunoileal bypass causes significant malabsorption of nutrients by shortening the small intestine by about 90%. Follow-up studies on people who had this surgery revealed a number of complications, such as severe diarrhea, vomiting, dehydration, electrolyte imbalances, and abdominal pain. Protein malnutrition, liver dysfunction, kidney stones, and gallstones were also reported. Because of these complications, several safer forms of gastric surgery have almost completely replaced jejunoileal bypass surgery.

Gastric Bypass

Gastric bypass (Roux-en-Y gastric bypass) is the most common surgical treatment for severe obesity, accounting for about 80% of bariatric procedures in the United States.[61] Gastric bypass partitions the stomach into a small pouch (15–60 ml) using several rows of surgical staples. The pouch is attached directly to the jejunum (the middle

gastric bypass (Roux-en-Y gastric bypass) *Surgery that partitions the stomach into a small pouch using surgical staples. The pouch empties into the jejunum, the middle part of the small intestine.*

part of the small intestine) by a narrow opening that is maintained by a ring or band to prevent stretching. This procedure thus involves both gastric restriction and some degree of intestinal malabsorption of nutrients, resulting in three effects:

1. Bypassing the stomach limits the amount of food that can be consumed in one sitting.
2. Bypassing the duodenum (the upper part of the small intestine) limits absorption of nutrients—and calories—from ingested food.
3. Creating a small opening between the stomach pouch and the jejunum reduces the speed with which food leaves the pouch.

Gastric bypass can be performed either by opening the abdomen or by **laparoscopy** (several small incisions). The procedure has a 30-day mortality rate of 0.5%.[62]

laparoscopy
Insertion of a small scope through the abdominal wall to examine the abdominal cavity or to perform surgery.

laparoscopic adjustable gastric banding
Also called lap-band, a surgical procedure in which the stomach is partitioned into two segments with a removable band, so that food empties into the small intestine more slowly.

Laparoscopic Adjustable Gastric Banding

Sometimes called lap-band surgery, adjustable gastric banding uses a silicon ring, or band, to divide the stomach into a small upper pouch and a larger pouch below the band. The tightness of the band determines how rapidly food from the upper pouch empties, which affects hunger. The ring is connected to an infusion port accessed just under the skin. After surgery, the band can be tightened by injecting saline into the port or loosened by removing saline.

This procedure involves only gastric restriction. Weight loss with gastric banding is not as dramatic or sustainable as gastric bypass surgery. However, the procedure is less invasive, reversible, does not involve stapling, and has a 30-day mortality rate of 0.1%.[62] Figure 10-8 shows the differences between gastric bypass and laparoscopic adjustable gastric banding.

Gastric banding

Gastric bypass

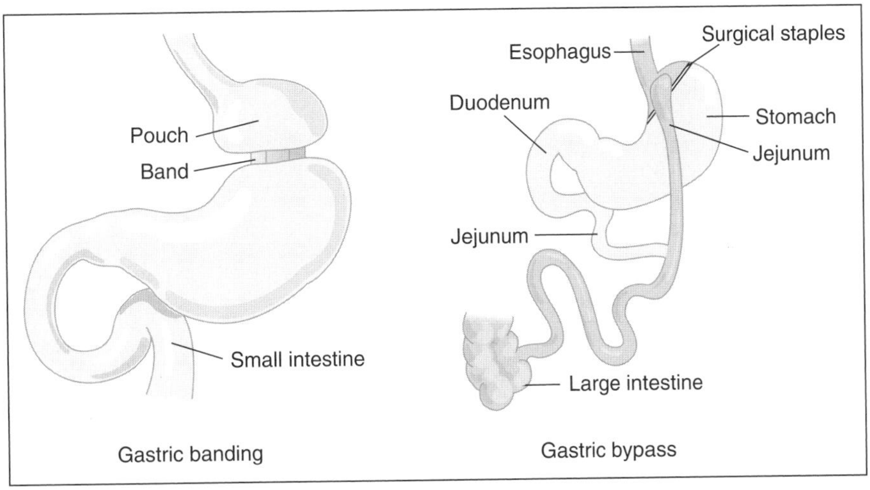

FIGURE 10-8 Surgical procedures used in the treatment of severe obesity. The dark gray areas highlight the movement of food through the digestive tract, while the light gray areas have been bypassed. Notice how most of the stomach, all of the duodenum, and some of the jejunum are bypassed in gastric bypass surgery.

Other Surgical Procedures

A drastic surgical procedure that is rarely performed in the United States is **biliopancreatic diversion** or **bypass**. This procedure involves stapling across two-thirds of the stomach and creating a very long Roux-en-Y reconstruction close to the valve between the large and small intestines. The result bypasses over half of the small intestine and diverts bile and pancreatic juice into the ileum (the lower segment of the small intestine), causing significant nutrient malabsorption. Weight loss is generally much greater than seen with conventional gastric bypass. The mortality rate at 30 days is about 1.1%.[62]

A new procedure called **sleeve gastrectomy** is sometimes performed on individuals at high risk who need to lose a lot of weight quickly. The stomach is stapled so it becomes a small vertical tube; connection to the small intestine is maintained. Individuals who have this surgery lose between 33% and 45% of body weight in a year. Once the person's health risk is reduced by weight loss, gastric bypass or biliopancreatic diversion may be performed as a follow-up. But if the individual loses enough weight with the sleeve gastrectomy, additional surgery may not be required.

How Effective Is Bariatric Surgery?

All of the obesity surgeries described in this section are considered effective treatments for severe obesity, producing average weight losses of more than 50% of excess body weight. The National Heart, Lung, and Blood Institute calls bariatric surgery the only treatment for severe obesity resulting in significant weight loss that is maintainable.[63] The results are caused by eating less or absorbing fewer nutrients, but may also be a product of altered hormone levels in the gastrointestinal tract, which reduces the desire to eat.[64]

The surgical procedure itself is not the principal determinant of success. Some severely obese individuals do not lose weight after surgery that was technically successful. The patient's readiness, willingness, and ability to change are critical to preventing inactivity, overeating, and poor food choices, all of which can sabotage a successful surgery.

Benefits of Surgery

The most obvious benefit of obesity surgery is weight loss, which is usually reported as a percentage of excess weight that is lost, where excess weight is defined as that which exceeds ideal body weight. Average losses of excess weight from bariatric surgery procedures are:

- 47% from gastric banding[62]
- 55% from sleeve gastrectomy[65]
- 62% from gastric bypass[62]
- 70% from biliopancreatic diversion[65]

In almost all cases, comorbidities like sleep apnea, type 2 diabetes mellitus, hypertension, and high blood lipids are improved with weight loss. Improvement of type 2 diabetes may occur within days, even before any weight loss is recorded.[62] The malabsorptive procedures have the greatest impact on comorbidities. Patients often report improved mood and quality of life, too. The 300-lb woman who loses 75 lb is still obese at 225 lb, but improvements in her ability to get around and to fit in smaller-sized clothing can make a big difference in her outlook. And, perhaps most significantly, life expectancy increases.[66]

Most weight loss takes place during the first year after surgery, and then plateaus at 18–24 months, with some regain expected. A long-term study of Swedish patients found that 10 years after surgery about one-quarter of gastric banding patients maintained 20% of weight lost, while 75% of gastric bypass patients did; one-quarter of gastric banding patients and 9% of those who had gastric bypass maintained less than 5% of lost weight.[67]

Why is surgery sometimes unsuccessful? Following bariatric surgery, regular foods are reintroduced over several months until pre-surgery dietary variety is restored. The quantity of food that can be consumed is drastically reduced. However, some patients have been reported to adapt after surgery by frequently consuming high-calorie liquids, which could prevent caloric reduction. Gastric bypass failure is often the result of eating junk foods like chips and popcorn that empty quickly from the pouch, instead of foods high in fiber, like many fruits and vegetables, which contribute few calories and take longer to empty from the pouch.

Complications of Surgery

Mortality rates for bariatric surgery are low, ranging from 0.1% for lap-band, 0.5% for gastric bypass, and 1.1% for biliopancreatic diversion. Post-surgical morbidity can be quite high. Insurance data reveal a complication rate of around 22% during initial hospitalization, and 30-day hospital readmission rates that range from 5–6%.[68] As with any medical procedure, the physician's expertise makes a difference. Choosing a surgeon who specializes in bariatric surgery and has considerable experience with the procedure can minimize the risks of surgery.

Gastric banding has fewer complications than gastric bypass. Among the reported complications of gastric bypass and biliopancreatic diversion are:

- Vomiting, which may occur once or twice a week for several months after surgery. This is generally the result of eating too much for the capacity of the pouch or not chewing food sufficiently.
- Dumping syndrome, characterized by a feeling of fullness, weakness, dizziness, and need to lie down after eating. Excess fluid in the small intestine is the culprit here. This results either from drinking too much liquid while eating or from eating foods high in sugar, which have high osmolarity and attract fluid into the small intestine.
- Dehydration, sometimes requiring hospitalization, occurs when individuals fail to sip liquids throughout the day, since they should not take liquids at the same time as food to avoid dumping syndrome.
- Gallstones happen in 71% of patients in the year after surgery.[69]
- Vitamin and mineral deficiencies are common. Vitamin B_{12} and calcium deficiencies occur in all gastric bypass patients, and menstruating women are at additional risk of iron deficiency anemia. To prevent vitamin and mineral deficiencies, supplements must exceed levels seen in typical multivitamin/mineral preparations.
- Surgical complications, including failure of the staple line and infections, may occur in 5–20% of cases
- Depression is reported in about one-quarter of cases. One medical center that followed more than 500 patients for 5 years after surgery reported that five to ten deaths were suicides.[70]

Typical outcomes from bariatric surgery are summarized in Figure 10-9. Nutrition education can prevent some of the side effects of bariatric surgery. Follow-up visits with

---

Weight Loss/Health
- Loss of excess weight as follows:
  - 47% from gastric banding
  - 55% from sleeve gastrectomy
  - 62% from gastric bypass
  - 70% from biliopancreatic diversion
- Maintenance of 20% weight loss at 10 years by about 75% of gastric bypass patients; 25% of gastric banding patients
- Almost immediate improvements in blood pressure, blood lipids, glycemic control, type 2 diabetes

Complications/Side Effects
- Complications are more common following gastric bypass, sleeve gastrectomy, and biliopancreatic diversion than gastric banding
- Surgical complications include failure of the staple line and infections; may be minimized by having surgery at a high-volume bariatric surgery center
- Other complications can be minimized by nutrition education:
  - Vomiting
  - Dumping syndrome
  - Dehydration
  - Vitamin and mineral deficiencies
- Gallstones happen in 71% of patients in the year after surgery

---

FIGURE 10-9   What to expect from bariatric surgery.

a nutrition professional may increase compliance and address nutritional concerns. Life-long medical surveillance is necessary after surgery.

Weight-Loss Surgery: Summary

The outcome of surgical intervention for obesity is much better than any other treatment presented in this text. About half of individuals maintain a weight reduction of 50% of excess weight or greater for 5 years after surgery. However, the risks of surgery can be considerable, and severely obese individuals who choose surgery must be prepared for a lifetime of compliance to a new pattern of eating and continued physical activity. Some regain of lost weight is inevitable.

LOCALIZED FAT REDUCTION

Many people would like to reduce stored fat from a particular area of their body. Not surprisingly, non-obese individuals are as concerned about localized fat deposits, including the rippled skin called cellulite, as obese people are. At the present time, localized fat reduction is not considered to be a treatment for obesity.

cellulite *Adipose tissue characterized by a dimpled appearance due to weak connective tissue that allows fat to penetrate into the dermis (true skin).*

What Is Cellulite?

The most hated fat on any woman's body has to be **cellulite**. Because of its characteristic dimpled appearance, cellulite is often thought of as a specialized type of tissue. In fact, cellulite is just fat, but it is fat with a twist. The fat that makes up cellulite has projected from the adipose tissue into the dermis, the layer of skin just under the epidermis, in which blood vessels, lymphatic vessels, nerves, hair follicles, and glands are found. In people with cellulite, the connective tissue border between the dermis and adipose tissue is irregular, not smooth and continuous. This occurs as a result of connective tissue weakening and stretching from clumps of fat cells pulling down on it.

Obesity does not cause cellulite. Obese men rarely have cellulite, whereas nearly all women—even lean women—have it. Cellulite makes its first appearance during adolescence, primarily in women, due to the effect of estrogen on the connective tissue matrix. It is particularly evident in the thighs and buttocks and may become more noticeable as fat deposits increase in size.

Cellulite Treatments

Cellulite is tricky to treat because it involves weak connective tissue as well as clumped fat cells. Liposuction, discussed later, sometimes worsens skin dimpling. Nonsurgical treatments for cellulite include:

- **Exercise:** Enhancing underlying muscle tone may improve the appearance of areas affected by cellulite. Because cellulite results not only from adipose but also from connective tissue, even being physically active will not eliminate it.
- **Lactic acid:** The alpha-hydroxy acids, including lactic acid, are known as humectants because they help retain moisture in the skin, which keeps it soft and flexible. These "anti-aging" compounds could theoretically minimize the appearance of cellulite by maintaining skin integrity. No published reports have shown this to be the case.
- **Xanthine creams:** Creams that contain aminophylline and theophylline claim to stimulate β-adrenergic receptors on the fat cells and increase lipid breakdown. These creams are also theorized to break bonds between the fat cells, allowing them to return to their original positions and reduce stress on the connective tissue. A frequently cited xanthine cream success story found average reductions in girth of 3.08 cm among women who applied 0.5% aminophylline cream to their thighs for 5 weeks.[71] Only twelve women participated in the study, and researchers were vague about how thigh measurements were taken. More recently, 134 women participated in a study that used a methylxanthine cream (7% caffeine) applied to the thighs and hips for 30 days. More than 80% of the 99 women who completed the study experienced significant reduction in thigh diameter, and about 70% saw reduced hip circumference.[72]
- **Herbal treatments:** Several botanical extracts contain xanthine derivatives (sweet clover, ivy, horse chestnut, algae, and barley), suggesting that they may be of use in breaking up cellulite. In one published study, 50 women applied only a botanical cream or a botanical and caffeine combination to the thigh once a day for 30 days. Changes in thigh circumference were obtained with girth measurements and ultrasound. The botanical-only group decreased thigh circumference by an average of 1.9 mm, while the combination group decreased thigh girth by an average of 2.8 mm.[73] Of the women studied, 72% reportedly were satisfied with the result, 5% were extremely pleased, and 23% were dissatisfied with cellulite reduction. No adverse effects were reported.

Should you or your clients invest in cellulite creams? At the present time, the expense of purchasing these creams is probably not worth the results. The results reported to date in published studies suggest that only small changes in thigh or hip diameter follow regular use of xanthine or some botanical creams.

Is Liposuction an Effective Treatment for Obesity?

Liposuction is the most common cosmetic procedure performed in the United States. It involves removing superficial fat tissue by means of a suction pump and can be performed in a hospital on an inpatient or outpatient basis or, more commonly, in a physician's office. Plastic surgeons are most likely to perform the procedure, but dermatologists, family practitioners, gynecologists, and even oral surgeons may do liposuction.

Types of Liposuction

Tumescent liposuction is the preferred procedure in use today. **Cannulae** of various sizes are attached to a vacuum pump. The physician infuses under the skin a solution containing saline, epinephrine, and lidocaine (a local anesthetic). Cannulae are then inserted into subcutaneous fat tissue and manipulated to suction out the desired quantity of fat. The tumescent infusion prevents loss of body fluids and excessive bleeding, although there is some removal of small blood vessels that run through adipose tissue. This technique is considered to have an excellent safety record when performed by a trained physician.[74,75]

Ultrasound-assisted liposuction was introduced in the United States in 1992. In this procedure, sound waves are transmitted from the end of the cannula into subcutaneous tissue, causing disruption and emulsification of fat cells. Suction is used either simultaneously or in a later step to remove liquefied fat. This procedure has had so many complications that its use has been largely stopped in the United States.

Laser-assisted lipolysis is a new liposuction technology. This procedure is similar to tumescent liposuction, but a laser breaks up fat cells and collagen and coagulates small blood vessels, and fat is removed with tiny cannulae. To date there is no evidence that this is more safe or effective than tumescent liposuction.[75]

Complications of Liposuction

Although liposuction is widely perceived as no more than a minor surgical procedure, it can have life-threatening complications, and patients occasionally die. Any deaths from elective surgery performed primarily for cosmetic reasons should be cause for scrutiny.

The first complication of liposuction was reported by the man who introduced the technique in the 1920s, Dr. Charles Dujarrier, who damaged femoral blood vessels to such an extent that the patient's affected leg had to be amputated.[76] Today, liposuction is generally considered to be a safe procedure with few complications when it is performed on appropriately screened patients by experienced physicians. Adverse effects reported as a result of liposuction include:[77]

- Pain, experienced by almost everyone
- Anemia in about 18% of patients
- **Seromas**, or pockets of fluid, in about 5% of patients. Fluid is typically aspirated with a needle
- Fibrosis, or scarring from excess connective tissue, about 2% of the time
- Burns may result from during laser-assisted lipolysis, although this is rare (fewer than 0.9% of cases.)[78]

More severe complications are very rare:

- **Pulmonary embolism** (where a breakaway blood clot obstructs blood flow in the lungs) and deep vein thrombosis (a blood clot forms in a deep vein causing

fat embolism
Blockage of an artery by an embolus of fat.

necrotizing fasciitis *Rare infection of the subcutaneous tissue, typically not involving muscle or skin.*

obstruction and possibly becoming a pulmonary embolism) happen in about .03% of cases.[77] When a piece of fat tissue lodges in an artery, it is called a **fat embolism**.

- Complications from anesthesia may occur
- Infection is possible. Severe infection can lead to **necrotizing fasciitis**, a rare disease in which subcutaneous tissue (not muscle or skin) is infected to such an extent that immediate and radical surgical debridement of the affected area is necessary to save the patient. Depending on how quickly treatment is started, mortality rates from necrotizing fasciitis range from 12% to 76%.[76] In one case, an apparently healthy 48-year-old woman died 25 days after liposuction, when even almost complete debridement of the anterior abdominal wall and large doses of antibiotics failed to contain massive infection. In another case, a healthy 31-year-old woman spent 41 days at a burn center, where necrotizing fasciitis involving the entire abdominal wall, the fronts and backs of both thighs, and the lower back was treated by debridement and skin grafts.

The mortality rate from liposuction is .01% and mainly due to pulmonary embolism. There are several strategies to minimize the risk of complications and death:

- Liposuction should not be combined with other surgical procedures. Tumescent liposuction in which a large volume of fluid is removed should never be performed at the same time as other cosmetic surgical procedures under general anesthesia.
- Physicians need to know about any medications or dietary supplements being taken. People taking dietary supplements or medications that act as anticoagulants are at particular risk of problems during and after surgery.
- Patients should stop smoking for 2 weeks before and during the healing process after surgery. This prevents cardiopulmonary complications and promotes healing.

Liposuction as a Treatment for Obesity

Large amounts of fat can be removed with liposuction, making it a tempting "treatment" for obesity. However, people need to be realistic in their expectations. Liposuction is expensive; it can have complications, particularly when large amounts of fat are removed; there will be pain and may be bruising and considerable drainage for several days after the procedure; body contour irregularities may result; and, without changes in diet and exercise behavior, excess fat will return.

Should you or your clients consider liposuction for fat reduction? There may be some benefits to liposuction for people willing to take the risk of undergoing a cosmetic procedure with potentially serious complications. For example, liposuction is extremely effective in the treatment of gynecomastia, the abnormal enlargement of breast tissue in men. People who have successfully adopted new eating and exercise behaviors and want to reduce the size of specific body areas resistant to change may get a psychological boost from cosmetic surgery that keeps them motivated. However, weight-management professionals need to help people understand what liposuction cannot do:

- Liposuction cannot remove enough fat to make a fat person skinny
- Liposuction is not a substitute for lifelong healthy eating and activity. Fat deposits do return.
- Liposuction cannot change the inner person. Taking 1,200 cc of fat out of a thigh will not improve an individual's relationship with herself or others.

Localized Fat Reduction: Summary

Cellulite occurs in both lean and obese women. Neither cellulite-reducing creams nor liposuction are treatments for obesity. There is not consistent evidence supporting the value of xanthine or herbal creams for reducing cellulite. Liposuction perhaps has its greatest value in reducing the size of specific body areas resistant to change in people who have successfully adopted new eating and exercise behaviors. However, liposuction is not a substitute for lifelong healthy eating and activity, and fat deposits do return.

CONCLUSION

The weight-loss methods presented in this chapter are different from the approaches described in other chapters of this book in one fundamental way: They create the illusion that they are not based on behavior change. On closer examination, however, it should be apparent that all of these non-behavioral approaches require behavior change for long-term effectiveness. The diet pill cannot be taken forever and works better when caloric intake is reduced and physical activity is increased; the VLCD lasts for only 12 to 16 weeks and then refeeding begins; the cabbage soup diet is safe for only 7 days. While bariatric surgery is the only consistently effective treatment for morbid obesity, post-surgery caloric intake has to be modified with a stomach pouch the size of a packet of gum.

The biggest surprise about these more extreme weight-control approaches is that, with the exception of gastric surgery, they result in about the same weight loss as behavioral programs. The bottom line: There is no simple solution for obese individuals who want to lose weight. Unless weight loss is needed to maintain health, a better approach to weight management is to focus on behaviors that promote health, good attitudes toward food, and prevention of further weight gain.

| **Case Study. The Traveling Man, Part 3** | Five years later, John has gained and lost about 500 lbs through a succession of diets. He even tried a VLCD, but a year later gained back all the weight he had so diligently lost. At 370 lb, John can no longer fly without using a seat-belt extender and recently he had to buy a second seat. He has arthritis in both knees, hypertension, and type 2 diabetes. Three weeks ago John broke a chair at a co-worker's housewarming party. Because of the embarrassment about his size, John rarely attends social events where food is served. However, he still loves to eat—he now does so mainly by himself.

Because of his health issues, Dr. Parnell suggested to John that he consider gastric surgery. They had a lengthy discussion about it, and Dr. Parnell referred John to an experienced bariatric surgeon.

- Is John a good candidate for bariatric surgery?
- Calculate the excess weight that John might lose from (a) gastric banding, (b) sleeve gastrectomy, (c) gastric bypass, or (d) biliopancreatic diversion.
- Pick one of the types of bariatric surgery and explain to John what he can expect postoperatively.
- What will John need to do to maximize the weight-loss potential of this type of bariatric surgery? |

REFERENCES

1. *Guidance for industry developing products for weight management.* (2007). Rockville, MD: Office of Training and Communications, Center for Drug Evaluation and Research, Food and Drug Administration.

2. Li, Z., Maglione, M., Tu, W., Mojica, W., Arterburn, D., Shugarman, L. R., et al. (2005). Meta-analysis: Pharmacologic treatment of obesity. *Annals of Internal Medicine, 142*(7), 532–546

3. Hansen, D. L., Toubro, S., Stock, M. J., Macdonald, I. A., & Astrup, A. (1998). Thermogenic effects of sibutramine in humans. *American Journal of Clinical Nutrition, 68*(6), 1180–1186.

4. Berg, F. M. (1998, March/April). Fen-phen tragedy triggers uproar. *Healthy Weight Journal, 12*(2), 17, 32.

5. Bray, G. A., & Greenway, F. L. (2007). Pharmacological treatment of the overweight patient. *Pharmacological Reviews, 59*(2), 151–184.

6. Wadden, T. A., Berkowitz, R. I., Womble, L. G., Sarwer, D. B., Phelan, S., Cato, R. K., et al. (2005). Randomized trial of lifestyle modification and pharmacotherapy for obesity. *New England Journal of Medicine, 353*(20), 2111–2120.

7. James, W. P. T., Caterson, I. D., Coutinho, W., Finer, N., Van Gaal, L. F. Maggioni, A. P., et al. (2010). Effect of sibutramine on cardiovascular outcomes in overweight and obese subjects. *New England Journal of Medicine, 363*(10), 905–917.

8. European Medicines Agency. (2008 October). *The European Medicines Agency recommends suspension of the marketing authorisation of Acomplia* [press release]. Available from http://www.associazionemediciendocrinologi.it/download/download_file_147342591.pdf. Accessed July 19, 2010.

9. Gadde, K. M., Franciscy, D. M., Wagner, H. R., 2nd, & Krishnan, K. R. (2003). Zonisamide for weight loss in obese adults: A randomized controlled trial. *JAMA, 289*(14), 1820–1825.

10. Cooke, D., & Bloom, S. (2006). The obesity pipeline: Current strategies in the development of anti-obesity drugs. *Nature Reviews Drug Discovery, 5*(11), 919–931.

11. Torgerson, J. S., Hauptman, J., Boldrin, M. N., & Sjostrom, L. (2004). XENical in the prevention of diabetes in obese subjects (XENDOS) study: A randomized study of orlistat as an adjunct to lifestyle changes for the prevention of type 2 diabetes in obese patients. *Diabetes Care, 27*(1), 155–161.

12. Brass, E. P. (2001). Changing the status of drugs from prescription to over-the-counter availability. *New England Journal of Medicine, 345*(11), 810–816.

13. Anderson, J. W., Schwartz, S. M., Hauptman, J., Boldrin, M., Rossi, M., Bansal, V., et al. (2006). Low-dose orlistat effects on body weight of mildly to moderately overweight individuals: A 16 week, double-blind, placebo-controlled trial. *Annals of Pharmacotherapy, 40*(10), 1717–1723.

14. Schwartz, S. M., Bansal, V. P., Hale, C., Rossi, M., & Engle, J. P. (2008). Compliance, behavior change, and weight loss with orlistat in an over-the-counter setting. *Obesity, 16*(3), 623–629.

15. Batterham, R. L., Cohen, M. A., Ellis, S. M., Le Roux, C. W., Withers, D. J., Frost, G. S., et al. (2003). Inhibition of food intake in obese subjects by peptide YY3–36. *New England Journal of Medicine, 349*(10), 941–948.

16. Wynne, K., Park, A. J., Small, C. J., Patterson, M., Ellis, S. M., Murphy, K. G., et al. (2005). Subcutaneous oxyntomodulin reduces body weight in overweight and obese subjects: A double-blind, randomized, controlled trial. *Diabetes, 54*(8), 2390–2395.

17. Wynne, K., Park, A. J., Small, C. J., Meeran, K., Ghatei, M. A., Frost, G. S., et al. (2006). Oxyntomodulin increases energy expenditure in addition to decreasing energy intake in overweight and obese humans: A randomised controlled trial. *International Journal of Obesity, 30*(12), 1729–1736.

18. Oral, E. A., Simha, V., Ruiz, E., Andewelt, A., Premkumar, A., Snell, P., et al. (2002). Leptin-replacement therapy for lipodystrophy. *New England Journal of Medicine, 346*(8), 570–578.

19. Chapman, I., Parker, B., Doran, S., Feinle-Bisset, C., Wishart, J., Strobel, S., et al. (2005). Effect of pramlintide on satiety and food intake in obese subjects and subjects with type 2 diabetes. *Diabetologia, 48*(5), 838–848.

20. Roth, J. D., Maier, H., Chen, S., & Roland, B. L. (2009). Implications of amylin receptor agonism: Integrated neurohormonal mechanisms and therapeutic applications. *Archives of Neurology, 66*(3), 306–310.

21. Walston, J., Silver, K., Bogardus, C., Knowler, W. C., Celi, F. S., Austin, S., et al. (1995). Time of onset of non-insulin-dependent diabetes mellitus and genetic variation in the beta 3-adrenergic-receptor gene. *New England Journal of Medicine, 333*(6), 343–347.

22. Widen, E., Lehto, M., Kanninen, T., Walston, J., Shuldiner, A. R., & Groop, L. C. (1995). Association of a polymorphism in the beta 3-adrenergic-receptor gene with features of the insulin resistance syndrome in Finns. *New England Journal of Medicine, 333*(6), 348–351.

23. Clement, K., Vaisse, C., Manning, B. S., Basdevant, A., Guy-Grand, B., Ruiz, J., et al. (1995). Genetic variation in the beta 3-adrenergic receptor and an increased capacity to gain weight in patients with morbid obesity. *New England Journal of Medicine, 333*(6), 352–354.

24. Ravussin, E., & Kozak, L. P. (2009). Have we entered the brown adipose tissue renaissance? *Obesity Reviews, 10*(3), 265–268.

25. Padwal, R. S., & Majumdar, S. R. (2007). Drug treatments for obesity: Orlistat, sibutramine, and rimonabant. *Lancet, 369*(9555), 71–77.

26. National Task Force on the Prevention and Treatment of Obesity. (1996). Long-term pharmacotherapy in the management of obesity. *JAMA, 276*(23), 1907–1915.

27. Radimer, K., Bindewald, B., Hughes, J., Ervin, B., Swanson, C., & Picciano, M. F. (2004). Dietary supplement use by US adults: Data from the National Health and Nutrition Examination Survey, 1999–2000. *American Journal of Epidemiology, 160*(4), 339–349.

28. *Dietary Supplement Information Bureau, reporting data on consumer spending on dietary supplements from the Natural Products Foundation.* Available from http://www.supplementinfo.org. Accessed July 17, 2010.

29. Proposed rules. dietary supplements containing ephedrine alkaloids. (1997, June 4). *Federal Register, 62*(107), 30677–30724.

30. Pittler, M. H., Abbot, N. C., Harkness, E. F., & Ernst, E. (1999). Randomized, double-blind trial of chitosan for body weight reduction. *European Journal of Clinical Nutrition, 53*(5), 379–381.

31. Saper, R. B., Eisenberg, D. M., & Phillips, R. S. (2004). Common dietary supplements for weight loss. *American Family Physician, 70*(9), 1731–1738.

32. Blanck, H. M., Serdula, M. K., Gillespie, C., Galuska, D. A., Sharpe, P. A., Conway, J. M., et al. (2007). Use of nonprescription dietary supplements for weight loss is common among Americans. *Journal of the American Dietetic Association, 107*(3), 441–447.

33. Shekelle, P. G., Hardy, M. L., Morton, S. C., Maglione, M., Mojica, W. A., Suttorp, M. J., et al. (2003). Efficacy and safety of ephedra and ephedrine for weight loss and athletic performance: A meta-analysis. *JAMA, 289*(12), 1537–1545.

34. Astrup, A., Breum, L., Toubro, S., Hein, P., & Quaade, F. (1992). The effect and safety of an ephedrine/caffeine compound compared to ephedrine, caffeine and placebo in obese subjects on an energy restricted diet. A double blind trial. *International Journal of Obesity and Related Metabolic Disorders, 16*(4), 269–277.

35. Heishman, S. J., & Henningfield, J. E. (1992). Stimulus functions of caffeine in humans: Relation to dependence potential. *Neuroscience and Biobehavioral Reviews, 16*(3), 273–287.

36. Dulloo, A. G., Geissler, C. A., Horton, T., Collins, A., & Miller, D. S. (1989). Normal caffeine consumption: Influence on thermogenesis and daily energy expenditure in lean and postobese human volunteers. *American Journal of Clinical Nutrition, 49*(1), 44–50.

37. Trumbo, P. R., & Ellwood, K. C. (2006). Chromium picolinate intake and risk of type 2 diabetes: An evidence-based review by the United States Food and Drug Administration. *Nutrition Reviews, 64*(8), 357–363.

38. Anderson, R. A. (1998). Effects of chromium on body composition and weight loss. *Nutrition Reviews, 56*(9), 266–270.

39. Wasser, W. G., Feldman, N. S., & D'Agati, V. D. (1997). Chronic renal failure after ingestion of over-the-counter chromium picolinate. *Annals of Internal Medicine, 126*(5), 410.

40. Cerulli, J., Grabe, D. W., Gauthier, I., Malone, M., & McGoldrick, M. D. (1998). Chromium picolinate toxicity. *Annals of Pharmacotherapy, 32*(4), 428–431.

41. Heymsfield, S. B., Allison, D. B., Vasselli, J. R., Pietrobelli, A., Greenfield, D., & Nunez, C. (1998). Garcinia cambogia (hydroxycitric acid) as a potential antiobesity agent: A randomized controlled trial. *JAMA, 280*(18), 1596–1600.

42. Kalman, D., Colker, C. M., Wilets, I., Roufs, J. B., & Antonio, J. (1999). The effects of pyruvate supplementation on body composition in overweight individuals. *Nutrition, 15*(5), 337–340.

43. Kurtzweil, P. (1997, July-August). Dieter's brews make tea time a dangerous affair. *FDA Consumer Magazine, 31*(5). Available from http://permanent.access.gpo.gov/lps1609/www.fda.gov/fdac/597_toc.html. Accessed July 17, 2010.

44. Belza, A., & Jessen, A. B. (2005). Bioactive food stimulants of sympathetic activity: Effect on 24–h energy expenditure and fat oxidation. *European Journal of Clinical Nutrition, 59*(6), 733–741.

45. Phung, O. J., Baker, W. L., Matthews, L. J., Lanosa, M., Thorne, A., & Coleman, C. (2010). Effect of

green tea catechins with or without caffeine on anthropometric measures: A systematic review and meta-analysis. *American Journal of Clinical Nutrition, 91*(1), 73–81.

46. Tsai, A. G., & Wadden, T. A. (2006). The evolution of very-low-calorie diets: An update and meta-analysis. *Obesity, 14*(8), 1283–1293.

47. Anderson, J. W., Konz, E. C., Frederich, R. C., & Wood, C. L. (2001). Long-term weight-loss maintenance: A meta-analysis of US studies. *American Journal of Clinical Nutrition, 74*(5), 579–584.

48. Wadden, T. A., Foster, G. D., Letizia, K. A., & Stunkard, A. J. (1992). A multicenter evaluation of a proprietary weight reduction program for the treatment of marked obesity. *Archives of Internal Medicine, 152*(5), 961–966.

49. Heymsfield, S. B., van Mierlo, C. A., van der Knaap, H. C., Heo, M., & Frier, H. I. (2003). Weight management using a meal replacement strategy: Meta and pooling analysis from six studies. *International Journal of Obesity and Related Metabolic Disorders, 27*(5), 537–549.

50. Foster, G. D., Wyatt, H. R., Hill, J. O., McGuckin, B. G., Brill, C., Mohammed, B. S., et al. (2003). A randomized trial of a low-carbohydrate diet for obesity. *New England Journal of Medicine, 348*(21), 2082–2090.

51. Gardner, C. D., Kiazand, A., Alhassan, S., Kim, S., Stafford, R. S., Balise, R. R., et al. (2007). Comparison of the Atkins, Zone, Ornish, and LEARN diets for change in weight and related risk factors among overweight premenopausal women: The A TO Z weight-loss study: A randomized trial. *JAMA, 297*(9), 969–977.

52. Franklin, T. L., Kolasa, K. M., Griffin, K., Mayo, C., & Badenhop, D. T. (1995). Adherence to very-low-fat diet by a group of cardiac rehabilitation patients in the rural southeastern United States. *Archives of Family Medicine, 4*(6), 551–554.

53. Astrup, A. (2005). The role of dietary fat in obesity. *Seminars in Vascular Medicine, 5*(1), 40–47.

54. Rolls, B. J. (2000). The role of energy density in the overconsumption of fat. *Journal of Nutrition, 130*(2S Suppl), 268S–271S.

55. Dansinger, M. L., Gleason, J. A., Griffith, J. L., Selker, H. P., & Schaefer, E. J. (2005). Comparison of the Atkins, Ornish, Weight Watchers, and Zone diets for weight loss and heart disease risk reduction: A randomized trial. *JAMA, 293*(1), 43–53.

56. Varady, K. A., Bhutani, S., Church, E. C., & Klempel, M. C. (2010), Short-term modified alternate-day fasting: A novel dietary strategy for weight loss and cardioprotection in obese adults. *American Journal of Clinical Nutrition, 90*(5), 1138–1143.

57. Heshka, S., Anderson, J. W., Atkinson, R. L., Greenway, F. L., Hill, J. O., Phinney, S. D., et al. (2003). Weight loss with self-help compared with a structured commercial program: A randomized trial. *JAMA, 289*(14), 1792–1798.

58. Wadden, T. A., & Bartlett, S. J. (1992). Very low calorie diets: An overview and appraisal. In T. A. Wadden, & T. B. vanItallie (Eds.), *Treatment of the seriously obese patient* (pp. 44–79). New York: Guilford Press.

59. Nelson, M. D., Prabhakar, P., Kondragunta, V., Kornman, K. S., & Gardner, C. (2010). *Genetic phenotypes predict weight loss success: The right diet does matter.* Abstract from the American Heart Association Joint Conference: Nutrition, Physical Activity and Metabolism and 50th Cardiovascular Disease Epidemiology and Prevention. Available from http://americanheart.org/downloadable/heart/1267139896704EPI_NPAM_FinalProg_ABSTRACTS.pdf. Accessed July 17, 2010.

60. Partnership for Healthy Weight Management. (1999). Voluntary guidelines for providers of weight loss products or services. Washington, D.C.: Federal Trade Commission. Available from www.ftc.gov/bcp/edu/pubs/business/adv/bus38.shtm. Accessed July 17, 2010.

61. Mechanick, J. I., Kushner, R. F., Sugerman, H. J., Gonzalez-Campoy, J. M., Collazo-Clavell, M. L., Guven, S., et al. (2008). American Association of Clinical Endocrinologists, the Obesity Society, and American Society for Metabolic & Bariatric Surgery medical guidelines for clinical practice for the perioperative nutritional, metabolic, and nonsurgical support of the bariatric surgery patient. *Surgery for Obesity and Related Diseases, 4*(5 Suppl), S109–84.

62. Buchwald, H., Avidor, Y., Braunwald, E., Jensen, M. D., Pories, W., Fahrbach, K., et al. (2004). Bariatric surgery: A systematic review and meta-analysis. *JAMA, 292*(14), 1724–1737.

63. National Heart, Lung, and Blood Institute. (2004). *Think tank on enhancing obesity research at the NHLBI.* Bethesda, MD: NIH.

64. Naslund, E., Melin, I., Gryback, P., Hagg, A., Hellstrom, P. M., Jacobsson, H., et al. (1997). Reduced food intake after jejunoileal bypass: A possible association with prolonged gastric emptying and altered gut hormone patterns. *American Journal of Clinical Nutrition, 66*(1), 26–32.

65. Clinical Issues Committee of the American Society for Metabolic and Bariatric Surgery. (2010). Updated position statement on sleeve gastrectomy as a bariatric

procedure. *Surgery for Obesity and Related Diseases*, 6(1), 1–5.

66. Sjostrom, L., Narbro, K., Sjostrom, C. D., Karason, K., Larsson, B., Wedel, H., et al. (2007). Effects of bariatric surgery on mortality in Swedish obese subjects. *New England Journal of Medicine*, 357(8), 741–752.

67. Sjostrom, L., Lindroos, A. K., Peltonen, M., Torgerson, J., Bouchard, C., Carlsson, B., et al. (2004). Lifestyle, diabetes, and cardiovascular risk factors 10 years after bariatric surgery. *New England Journal of Medicine*, 351(26), 2683–2693.

68. Encinosa, W. E., Bernard, D. M., Chen, C. C., & Steiner, C. A. (2006). Healthcare utilization and outcomes after bariatric surgery. *Medical Care*, 44(8), 706–712.

69. Fujioka, K. (2005). Follow-up of nutritional and metabolic problems after bariatric surgery. *Diabetes Care*, 28(2), 481–484.

70. Yale, C. E. (1989). Gastric surgery for morbid obesity complications and long-term weight control. *Archives of Surgery*, 124(8), 941–946.

71. Greenway, F. L., Bray, G. A., & Heber, D. (1995). Topical fat reduction. *Obesity Research*, 3, (Suppl 4), 561S–568S.

72. Lupi, O., Semenovitch, I. J., Treu, C., Bottino, D., & Bouskela, E. (2007). Evaluation of the effects of caffeine in the microcirculation and edema on thighs and buttocks using the orthogonal polarization spectral imaging and clinical parameters. *Journal of Cosmetic Dermatology*, 6(2), 102–107.

73. Bascaglia, D. A., Conte, E. T., McCain, W., & Frideman, S. (1996). The treatment of cellulite with methylxanthine and herbal extract based cream: An ultrasonographic analysis. *Cosmetic Dermatology*, 9, 30–40.

74. Lawrence, N., Clark, R. E., Flynn, T. C., & Coleman, W. P., 3rd. (2000). American Society for Dermatologic Surgery guidelines of care for liposuction. *Dermatologic Surgery*, 26(3), 265–269.

75. Ahern, R. W. (2009). The history of liposuction. *Seminars in Cutaneous Medicine and Surgery*, 28(4), 208–211.

76. Barillo, D. J., Cancio, L. C., Kim, S. H., Shirani, K. Z., & Goodwin, C. W. (1998). Fatal and near-fatal complications of liposuction. *Southern Medical Journal*, 91(5), 487–492.

77. Triana, L., Triana, C., Barbato, C., & Zambrano, M. (2009). Liposuction: 25 years of experience in 26,259 patients using different devices. *Aesthetic Surgery Journal*, 29(6), 509–512.

78. Katz, B., & McBean, J. (2008). Laser-assisted lipolysis: A report on complications. *Journal of Cosmetic and Laser Therapy*, 10(4), 231–233.

Eating Disorders

CHAPTER

11

R. Gino Santa Maria/Shutterstock.com

CHAPTER OUTLINE

EATING DISORDERS: ANOREXIA NERVOSA, BULIMIA NERVOSA, AND BINGE-EATING DISORDER
What Is "Normal" Eating?
What Is Anorexia Nervosa?
What Is Bulimia Nervosa?
What Is Binge-Eating Disorder?
What Other Types of Disordered Eating Are There?
Eating Disorders: Summary

PHYSIOLOGICAL EFFECTS OF EATING DISORDERS
What Are the Effects on the Cardiovascular System?
What Are the Effects on the Digestive Tract and Kidneys?
What Are the Effects on the Endocrine System?
What Are the Effects on the Skeletal System?
Physiological Effects of Eating Disorders: Summary

PREDISPOSING FACTORS FOR EATING DISORDERS
Do People with Eating Disorders Have a Common Psychological Profile?
Is There an "Eating Disorder" Personality?
Is There a Biologic Cause of Eating Disorders?
What Family Issues Are Risk Factors?
Do Cultural Factors Increase Risk?
Are Athletes More Susceptible to Eating Disorders?
Can Dieting Cause Eating Disorders?
What Is the Connection Between Diabetes and Eating Disorders?
Predisposing Factors for Eating Disorders: Summary

TREATMENT AND PREVENTION OF EATING DISORDERS
When Is Hospitalization Required?
What Does Nutritional Treatment Include?
Which Psychotherapeutic Approaches Are Effective?
What Other Treatments Are Used?
What Is the Prognosis for Recovery?
Can Eating Disorders be Prevented?
Treatment and Prevention of Eating Disorders: Summary

REFERENCES

cappi thompson/Shutterstock.com

Iorga Studio/Shutterstock.com

Between 1% and 19% of the adult and adolescent U.S. population has an eating disorder, and in certain groups many more have a partial or subclinical eating disorder.[1,2] Eating disorders are seen in men, minorities, and people of all ages, although they are most prevalent in young white females. Much evidence points to an increased prevalence of eating disorders since the 1980s. This is not surprising: As you have learned, our biological makeup encourages weight gain to ensure survival of the species. Brain chemicals and hormones discussed in Chapter 4 provide a powerful drive to eat. Eating feels good, which reinforces the behavior. Lots of delicious food is readily available in our society. Advances in agriculture, manufacturing, and technology have significantly reduced our daily caloric expenditure. Consequently, most Americans are overweight. In contrast, consider the socio-culturally promoted weight "ideal," particularly for women—thinness. Battle and Brownell eloquently articulate this disconnect: "It is difficult to envision an environment more effective than ours for producing nearly universal body dissatisfaction, preoccupation with eating and weight, clinical cases of eating disorders, and obesity."[3]

Much of this book has focused on overweight and obesity. While obesity and eating disorders may seem like opposite ends of a continuum, in fact they share several characteristics: dieting, fasting, skipping meals, and issues with food; lack of awareness of hunger and satiety cues; disruption of biological factors that turn eating on and off; family environment in which there may be a focus on the child's weight; anxiety about one's body arising from media portrayals, teasing, or other factors; and difficulties in treatment. Extreme weight control behaviors, such as use of diet pills, laxatives, and diuretics, as well as purging and fasting, are not just characteristic of young people with eating disorders but also prevalent among overweight adolescents.[4,5] So while this chapter focuses on eating disorders, the reader should not lose sight of the relationships with overweight and obesity (Figure 11-1).

EATING DISORDERS: ANOREXIA NERVOSA, BULIMIA NERVOSA, AND BINGE-EATING DISORDER

Eating disorders are complex, multifaceted conditions that stem from physiological factors, family dynamics, and psychological vulnerability, in conjunction with an obesity-promoting environment. The American Psychological Association classifies eating disorders as mental disorders, but they may also be thought of as nutritional disorders because of their effects on body weight, nutrient intake, and metabolism.

Many adolescents and young adults engage in unhealthy practices that could lead to an eating disorder. The 2009 Youth Risk Behavior Survey of American high school

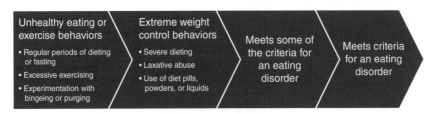

FIGURE 11-1   The continuum of disordered eating.

students found that 40% of students had made dietary adjustments (eaten smaller quantity of food, reduced caloric intake, or used low-fat foods) to lose weight. During the 30 days just before the survey, about 4% of students had taken laxatives or vomited, and about 5% had used diet pills for weight control. During the same time period, 11% of students had fasted for 24 hours or more and 62% had exercised to lose or maintain weight.[6] These behaviors were seen in all racial and ethnic groups. Although only a small number of people who diet, fast, purge, exercise excessively, or use laxatives and other drugs go on to develop a full-blown eating disorder, the frequency of these behaviors is alarming, the effect on health unpredictable, and the fact that they are gateway behaviors for more serious eating disorders unmistakable.

The most common eating disorders are anorexia nervosa, bulimia nervosa, and binge-eating disorder. Anorexia and bulimia are psychiatric disorders that have specific diagnostic criteria in the fourth edition of the American Psychiatric Association's **Diagnostic and Statistical Manual of Mental Disorders (DSM-IV)**.[7] Binge-eating disorder is included in the category called EDNOS, a catch-all grouping for eating disorders that do not yet have specific diagnostic criteria or for partial syndromes.

People with eating disorders spend an inordinate amount of time thinking about food, body weight, and appearance. They rarely eat in response to hunger, and they often continue eating despite satiety. They may engage in any or all of the following behaviors:

- **Restrained eating**: Caloric intake is severely reduced either all of the time or sporadically. Fasting is common.
- Binge eating: A **binge** is the consumption of large quantities of food over a limited period of time, usually less than 2 hours. A binge is typically followed by feelings of guilt, remorse, and self-loathing.
- Inappropriate **compensatory behaviors**: These behaviors include excessive exercise or abuse of laxatives, **diuretics**, and/or other drugs to compensate for caloric intake from eating or bingeing.

People with eating disorders come in all shapes and sizes. The stereotype is of an emaciated female adolescent, which certainly describes a girl with anorexia nervosa. However, individuals with bulimia nervosa may be of average weight; individuals with binge-eating disorder are typically overweight; and people who have not yet been diagnosed with an eating disorder but who have food obsessions, body image distortions, and unusual eating behaviors driven by anxiety may be underweight, overweight, or of average weight.

What Is "Normal" Eating?

Eating disorders are characterized by abnormal eating patterns. But what is "normal" eating? Ellyn Satter, a registered dietitian and psychiatric social worker, describes normal eating as follows:

Normal eating is being able to eat when you are hungry and continue eating until you are satisfied. It is being able to choose food you like and eat it and truly get enough of it—not just stop eating because you think you should. Normal eating is being able to use some moderate constraint in your food selection to get the right food, but not being so restrictive that you miss out on pleasurable foods. Normal

DSM-IV *Fourth edition of the American Psychiatric Association's Diagnostic and Statistical Manual of Mental Disorders (DSM-IV), a reference that classifies and describes psychiatric illnesses.*

restrained eating *Extreme reduction in caloric intake, including fasting.*

binge *Consumption of large quantities of food during a limited period of time, usually less than two hours.*

compensatory behavior *Behaviors used to compensate for caloric intake from eating or bingeing, such as excessive exercise or abuse of laxatives as well as diuretics and/or other drugs.*

diuretic *Drug that increases the excretion of urine.*

eating is giving yourself permission to eat sometimes because you are happy, sad or bored, or just because it feels good. Normal eating is three meals a day, most of the time, but it can also be choosing to munch along. It is leaving some cookies on the plate because you know you can have some again tomorrow, or it is eating more now because they taste so wonderful when they are fresh. Normal eating is over-eating at times: feeling stuffed and uncomfortable. It is also under-eating at times and wishing you had more. Normal eating is trusting your body to make up for your mistakes in eating. Normal eating takes up some of your time and attention, but keeps its place as only one important area of your life. (Reprinted from Satter, E. [1987]. *How to Get Your Kid to Eat...But Not Too Much.* Menlo Park, CA: Bull Publishing, pp. 69–70.)

So, many different eating patterns can constitute "normal" eating behavior. What is not normal is spending a great deal of time and energy thinking about food and avoiding food and food-related situations. People with eating disorders exhibit, among other things, an abnormal relationship with food.

What Is Anorexia Nervosa?

anorexia nervosa *Eating disorder characterized by significant weight loss, intense fear of becoming fat, distorted body image, and amenorrhea in women.*

amenorrhea *Absence of menstruation.*

People with **anorexia nervosa** experience significant weight loss, intense fear of becoming fat, and a distorted body image. Body image distortion may be so extreme that an emaciated person believes that she is overweight and continues to try to lose more weight.

Weight loss is the most obvious symptom of anorexia nervosa. Other indicators that might arouse suspicion are rigid dieting practices, strict avoidance of high-fat foods, excessive exercise, dry skin, thinning hair, and abdominal bloating. In men, sexual impotence may be present. In women, **amenorrhea**, the absence of menstruation for at least three consecutive cycles, is an additional criterion. Amenorrhea is problematic as a diagnostic criterion for several reasons: (1) it would not be noticeable in pre-puberty, (2) menstrual cycles may already be irregular in early puberty or in the pre-menopausal period, and (3) post-menopausal women do not menstruate. Thus, some cases may be classified as Eating Disorder Not Otherwise Specified (EDNOS). Because the psychological profiles of women with anorexia nervosa are the same, whether or not amenorrhea is present, some researchers are advocating that amenorrhea be removed as a criterion for anorexia when the DSM-V is published in 2012.[8] Table 11-1 summarizes the characteristics of anorexia nervosa.

TABLE 11-1 Characteristics of Anorexia Nervosa

- Occurs in about 1% of the adult population
- Occurs in up to 3% of the adolescent population
- More common in women than men
- Weight status: significant weight loss (BMI under 17.5 kg/m^2)
- Extreme fear of becoming fat
- Distorted body image
- Amenorrhea in women; sexual impotence in men

The *DSM-IV* provides two methods of determining whether an individual is significantly underweight: (1) the individual weighs less than 85% of normal weight, based on the Metropolitan Insurance Company's height–weight tables or based on pediatric growth tables, or (2) the body mass index (BMI) is 17.5 kg/m^2 or less.

Individuals who reach this weight by severe caloric restriction are said to have the restricting type of anorexia. Caloric restriction is more common than the other method of weight loss—purging by vomiting or abusing laxatives and/or diuretics. Individuals who engage in purging, whether or not they also periodically binge, are said to have the binge-eating/purging type of anorexia nervosa. Many individuals with anorexia also exercise compulsively and smoke cigarettes to further promote weight loss.

Anorexia occurs in under 1% of the general U.S. population, affecting more than 1.2 million Americans.[9] The typical onset occurs between ages 15 and 19 years, with the median age of onset being 17 years. About 20 times more women than men are diagnosed with anorexia nervosa, although there may be some underreporting, because males lack the very visible diagnostic criterion of amenorrhea. Gay or bisexual men are more likely than heterosexual men to have an eating disorder, including anorexia nervosa, perhaps because of an over-emphasis on physical appearance, which thereby increases susceptibility to many of the same socio-cultural forces that affect women.[10]

Anorexia nervosa is a very serious disorder. About one in ten people with anorexia die as a result of the disorder, most commonly from cardiac failure, ventricular arrhythmia, or suicide.[9]

What Is Bulimia Nervosa?

bulimia nervosa
Eating disorder characterized by repeated episodes of binge eating and inappropriate compensatory behaviors (vomiting, laxative abuse, excessive exercise) adopted to prevent weight gain.

Bulimia nervosa is characterized by repeated episodes of binge eating and inappropriate compensatory behaviors adopted to prevent weight gain. To meet the *DSM-IV* criteria for the diagnosis of bulimia, the individual must engage in these behaviors at least twice a week for 3 months. People with bulimia often begin this self-destructive cycle by binge eating in response to emotions that make them feel out of control—anger, guilt, depression, and stress. They then feel even guiltier, angrier, and out of control and use purging to not only restore a sense of control but to relieve themselves of excess calories consumed in the binge. The characteristics of bulimia nervosa are summarized in Table 11-2.

Symptoms of bulimia are very subtle, in part because there may be no obvious weight changes. Purging does not get rid of all calories consumed (as many as 1,200 calories may remain, depending on size of binge and timing of purge), so bulimics do not lose significant amounts of weight.[11] However, preoccupation with food is universal, and often there is a rigid classification of foods as "good" and "bad." Individuals who have been purging for some time may have gum disease or tooth decay. Occasionally, excessive

TABLE 11-2 Characteristics of Bulimia Nervosa

- Occurs in 1–3% of the adolescent and adult population
- More common in women than men
- Weight status: average or slightly overweight
- Use of binge eating and inappropriate compensatory behaviors
- Body dissatisfaction and low self-esteem

Tony Latham/Stone/Getty Images

Bulimic binges are often followed by self-induced vomiting and feelings of shame and disgust.

exercise is used as a compensatory behavior and may be the most visible sign to alert friends and family members to the presence of an eating disorder.

The compensatory behavior adopted—purging or nonpurging—is used to classify the type of bulimia. Vomiting is the most common purging compensatory behavior seen in bulimia, followed by abuse of laxatives, diuretics, and enemas. Approximately 80% of bulimics induce vomiting after a binge, and about 30% use laxatives. Smaller numbers of individuals have the nonpurging type of bulimia, where compensatory behaviors include severe food restriction or excessive exercise.

Bulimia occurs in 1–3% of the U.S. population, and about 90% of bulimics are female. The disorder is seen at about this same rate in all industrialized nations. Individuals with bulimia tend to be older than anorexics when they are diagnosed. This is partly due to the secretive nature of the disease—individuals with bulimia do their bingeing and purging privately—and also due to the average body weight of most people with bulimia. People with bulimia do not usually seek treatment for medical complications or because family or friends urged them to, but rather as a result of many years of personal pain.

What Is Binge-Eating Disorder?

binge-eating disorder *Eating disorder in which individuals consume large amounts of food in a short period of time, feel out of control when eating, and experience guilt, shame, and depression after eating; at one time, this disorder was known as compulsive eating.*

Individuals with **binge-eating disorder** consume large amounts of food in short periods of time; feel out of control when eating; and experience guilt, shame, and depression after eating. They engage in compensatory behaviors on an irregular basis or not at all. People with this disorder were once called compulsive eaters but, since the publication of the *DSM-IV* in 1994, the term binge-eating disorder has become more common. Binge-eating disorder and other nondistinct psychiatric disorders are categorized in the *DSM-IV* as EDNOS. Disorders in this category are considered to be worthy of additional study.

TABLE 11-3 Characteristics of Binge-Eating Disorder

- Occurs in 0.7–4% of adult population and up to half of those in weight-control programs
- Approximately equal numbers of women and men
- Weight status: overweight or obese
- Binge eating used, but compensatory behaviors rarely or never used

For diagnostic purposes, a person with binge-eating disorder must engage in binge eating at least 2 days a week for 6 months. Most binge-eaters develop the disorder in their late teens or twenties, but treatment may not be sought for many years.[12] The characteristics of binge-eating disorder are summarized in Table 11-3.

Binge-eating disorder occurs in 0.7–4% of the U.S. population, although as many as half of people in weight-control programs may have the disorder. In addition, up to half of people seeking gastric bypass surgery have been reported to have binge-eating disorder.[13] Approximately equal numbers of men and women have the condition.[14,15] Unlike anorexia and bulimia, binge-eating disorder is common in all racial and ethnic groups.[16] Affected individuals tend to be more obese than those with other eating disorders, and they often have long histories of dieting and weight cycling, becoming overweight earlier in life. They tend to eat more than other obese individuals, even when not bingeing.[14]

Binge eating may be triggered by depression, anxiety, or stress. Individuals with binge-eating disorder have been described as having a sensation of numbness when they are bingeing, and they are more likely to experience disgust with their weight than other obese individuals.[7]

What is a Binge?

After reading about bulimia nervosa and binge-eating disorder, you might be curious about what exactly defines a binge. Even the scientists who study eating disorders lack consensus on the criteria for a binge episode. What about people who snack throughout the day on small amounts of food? What about the person who loves to eat and seems to eat large quantities of food all the time? Is bingeing three times in one day the same as bingeing once on three separate days?

While *DSM-IV* provides some criteria for defining a binge, researchers continue to seek consensus on better defining binge behavior. Most commonly the following components help to describe a binge:

- Amount of food eaten—while difficult to precisely quantify, this would include eating more than most people would eat during a similar time period and under similar circumstances. This also might involve eating faster than usual, eating when not hungry, and eating even though feeling uncomfortably full.
- Time period in which eating occurred—consuming this large quantity of food during a 2-hour period has been suggested as the time frame to define the binge period. However, it can be difficult to determine when a binge actually begins (the individual might have started snacking on an appropriate number of cookies or chips and then escalated) and, if there are no compensatory behaviors, when it ends.
- Sense of loss of control—a key element of a binge, the individual may not feel that she or he can stop eating or control the quantity of food consumed. This is often

accompanied by feeling guilty, upset, or embarrassed, and may result in the person eating only when alone.

What Other Types of Disordered Eating Are There?

Anorexia nervosa, bulimia nervosa, and binge-eating disorder are the most well-defined and well-known eating disorders, but they are by no means the only ones. Many other atypical eating behaviors occur but have received less attention in the psychological or medical literature. These eating disorders may cause extreme disruption in affected people's lives and have severe medical consequences.

Variations of Anorexia and Bulimia

You now know that anorexia nervosa and bulimia nervosa are psychiatric disorders that have specific diagnostic criteria in the *DSM-IV*, and binge-eating disorder appears in the *DSM-IV* category EDNOS. EDNOS disorders have been studied and reported on far less comprehensively than anorexia or bulimia, but they affect a sizable number of people. In fact, there are probably many more individuals affected by subclinical or partial anorexia and bulimia, which includes:

- Females who continue to menstruate but otherwise meet all of the diagnostic criteria for anorexia nervosa
- Individuals who meet all of the diagnostic criteria for anorexia nervosa but have normal or near-normal body weight
- Individuals who meet all of the diagnostic criteria for bulimia nervosa but binge or purge less than two times a week or for less than 3 months
- Individuals who engage in inappropriate compensatory behaviors after eating only a small quantity of food
- Individuals who chew and spit out (but do not swallow) large amounts of food

Most adolescents seen in eating disorder treatment centers would be classified as EDNOS.[1] Yet these young people clearly have potentially serious issues with food, eating, or body image. Predisposing factors, physiological effects, and treatments for these eating disorders may be similar to those for anorexia and bulimia.

Pica

pica *Eating disorder characterized by intense craving for and consumption of nonfoods.*

The magpie is a bird known to eat almost anything. The Latin word for magpie is **pica**, which is also the term for an eating disorder characterized by intense craving for and chronic consumption of nonfoods for at least a month.[7] Common nonnutritive substances ingested by people with pica are dirt and clay (geophagia), laundry or cornstarch (amylophagia), ice (pagophagia), and hair (trichophagia). Other ingested substances include soap, dried paint, broken crockery, chalk, cigarette ashes, rust, foam rubber, charcoal, stones, coffee grounds, and clothing.

The prevalence of pica in the population is unknown but is believed to be increasing.[17] The disorder is most common during childhood and is classified in *DSM-IV* under Feeding and Eating Disorders of Infancy or Early Childhood. Among adults, women are more likely to have pica than men, and perhaps as many as one-fifth of pregnant women do.[17] Pica is common in some families and in some cultures. The *DSM-IV* specifically excludes culturally sanctioned consumption of nonfoods from the definition of pica.

The cause of pica is open to a great deal of speculation. One popular theory holds that people who are deficient in specific minerals consume nonfoods rich in the particular minerals that are missing from their diet. There is no scientific evidence to support this theory; however, in some cases, pica occurs as a symptom of iron-deficiency anemia and, once the anemia is successfully treated, pica stops.[17,18] In most cases, pica is thought to be a behavior that develops in susceptible people in response to stress; perhaps it is even in the family of obsessive-compulsive disorders.

Pica can be hazardous. The consumption of dirt, clay, or starch often leads to iron deficiency because these substances inhibit iron absorption. Some nonfoods contain high levels of environmental contaminants, such as the protozoans and hookworms found in soil. In pregnancy, these hazards affect not only the mother but the developing fetus. Hard substances like ice may cause tooth damage. Chronic abdominal pain commonly follows ingestion of certain substances. Cloth and hair consumption may promote the development of a solid mass known as a **bezoar** in the gastrointestinal tract. The November 1995 issue of the *Western Journal of Medicine* reports the case of a young woman who developed a bezoar and pancreatitis after chewing on and swallowing three to four athletic tube socks a week to "relax."[19] A recent report chronicles the removal of 275 coins from the gastrointestinal tract of a man with schizophrenia.[20] In both cases, the individuals were successfully treated.

bezoar *Mass of hair, cloth, or other material that accumulates in the gastrointestinal tracts of some people with pica who eat these substances.*

Nocturnal Eating

Some people eat or binge at night, either fully aware or in a sleeplike state. The most common of the night-eating disorders are night (nocturnal) eating syndrome and sleep-related eating disorder. The prevalence of these disorders in the general population is estimated at between 1.5% and 6%.[21] The prevalence increases with increasing BMI.

Sleep-related eating disorder (SRED) shares characteristics of both binge-eating disorder and sleep walking. The affected individual partially wakes after sleeping 2–3 hours, consumes a large quantity of food very rapidly, returns to bed, and retains no knowledge of the episode the next morning. Foods consumed are often similar to those eaten during a binge-eating episode—sweets, pie, and ice cream—and frequently include unusual foods or food combinations—cat food, raw bacon, uncooked spaghetti, butter on cigarettes. Some individuals may eat more than once per night, and eating may be out of control.

Prevalence data for this disorder have mainly been gathered from eating-disorder or sleep-disorder clinics. Up to one-third of these clinical populations have sleep-related eating disorder.[22] Most of those seeking treatment are overweight women who experienced their first episode of sleep-related eating during adolescence or early adulthood. Many have a personal and family history of sleep disorders. Psychiatric comorbidities are common, including depression and post-traumatic stress disorder. Almost none of those described in clinical case studies use compensatory behaviors. The similarities to binge-eating disorder suggest that there may be a common cause, although this has not yet been identified.[22]

sleep-related eating disorder *Type of eating disorder in which people engage in binge eating while in a state between wakefulness and sleep and do not remember eating when they are awake.*

night eating syndrome *Binge eating that occurs after retiring for the night and of which the individual is aware.*

Night eating syndrome (NES) is described by the International Classification of Sleep Disorders as "recurrent awakenings associated with an inability to return to sleep without eating or drinking."[23] Individuals with this disorder consume more than half of their daily intake after 7 PM, either before going to bed or during the night. They retain awareness and memory of their nocturnal eating behavior. Foods consumed during nocturnal binges are often very much like foods consumed in daytime binges—candy, sandwiches, cookies, and sugary cereals. There may be more than one episode per night, rarely in response to hunger. Individuals are considered to have NES if the behavior occurs for at least three months and binge-eating disorder or bulimia are absent.

Those with NES are typically obese women, and the presence of the disorder makes weight loss very difficult. The prevalence of the disorder is unknown, although it is suspected to occur with greater frequency than SRED. Stress and biological factors, some of which might be inherited, are theorized to be causal factors for NES. The hypothalamus, which regulates both sleep and appetite, is suspected to be involved. Overweight men and women with this disorder sometimes have abnormally low plasma **melatonin** levels at night.[24] Melatonin, a hormone secreted by the brain's pineal gland, is sometimes called the darkness hormone because its levels rise at night and fall during the daytime. It plays a role in biological rhythms and induces and maintains sleep. In addition, abnormal circadian rhythms of insulin and ghrelin, which you read about in Chapter 4, may be involved.[25]

Muscle dysmorphia is a form of body dysmorphic disorder, a relatively common syndrome characterized by obsessive attention to perceived defects in one's appearance. In muscle dysmorphia the individual is overly concerned with his or her muscularity. Sometimes this condition is referred to as "reverse anorexia," because the affected person seeks extreme muscle development on a lean physique. Body dissatisfaction is common, with the person feeling that optimal muscle size is never attained. Some researchers suggest that this might be related to other obsessive-compulsive disorders, because individuals become obsessed with working out, often to the detriment of job, family, and other obligations. There may also be obsessive dietary behaviors.

Muscle dysmorphia is more common in men than women, is particularly prevalent in bodybuilders, and is seen in both male and female athletes. Psychiatric comorbidities, including depression, anxiety disorders, and even other eating disorders, are typical.[26]

Cyclic vomiting syndrome is a rare disorder in which repeated episodes of nausea and vomiting occur that are not caused by any identifiable disease. It seems to be similar to migraine headaches and is sometimes accompanied by headaches, nausea, and abdominal pain, but a cause and an effective course of treatment have not been identified.

Based on published case reports, cyclic vomiting syndrome begins in preadolescence in most instances but does affect all ages. The prevalence is estimated to be 0.04–1.9% in children.[27] Vomiting episodes occur at characteristic times and for fairly predictable durations in each individual. Emotional stress—either positive (holidays, birthdays) or negative (parental conflict)—triggers episodes in most cases. Infections have also been reported to cause episodes. Most sufferers have migraines or a family history of migraines; anxiety and depression are also common. While the cause of cyclic vomiting syndrome is unknown, there may be mitochondrial dysfunction that increases motility in the gastrointestinal tract during times when increased energy is needed, such as stress and infection.[27]

Medical treatment of the disorder is important to prevent electrolyte depletion, weight loss, and damage to the esophagus. Sedation is one of the most useful treatment approaches. Anti-nausea drugs may also be used if the individual is aware that an episode is about to begin. Drugs used in the treatment of migraines may be helpful. Psychological treatments may also identify triggers and devise behavioral strategies for avoiding vomiting episodes.

Unlike the other eating disorders reviewed in this chapter, **gourmand syndrome** develops as a result of medical conditions in which the right frontal lobe of the brain is affected. Individuals with gourmand syndrome are obsessively preoccupied with buying,

melatonin
Hormone secreted by the brain's pineal gland that regulates biological rhythms and induces and maintains sleep.

muscle dysmorphia
Excessive concern with one's perceived muscularity, driving affected individuals to seek extreme muscle development.

cyclic vomiting syndrome
Disorder characterized by repeated episodes of nausea and vomiting not caused by any identifiable disease, although it may have some relationship to migraine headaches.

gourmand syndrome
Disorder in which people are obsessively preoccupied with buying, preparing, serving, and consuming gourmet food, typically resulting from brain injury or stroke.

preparing, serving, and consuming gourmet food. Typically as a result of a brain tumor, injury, or stroke, affected individuals (many of whom had little interest in food before) find themselves obsessed with fine food.

A study of 723 patients admitted to a neuropsychiatry unit in Switzerland reported that 36 patients met the criteria for gourmand syndrome:[28]

- Normal hunger and satiety mechanisms but persistent craving for good food and preoccupation with food and eating
- Cerebral lesion
- Food preoccupation onset coinciding with cerebral lesion
- No endocrine or metabolic disorders or medications that could account for eating behavior
- No history of eating disorders, neurological problems, or psychiatric illnesses

In addition to a passion for good food, many of the patients studied in Switzerland exhibited reduced impulse control in other areas, including excessive talking, aggression, and irritability. Why this occurs is not known; however, the fact that gourmet eating results from some cases of right frontal lobe lesion, mainly in the anterior corticolimbic region, may someday point to a specific cause. The corticolimbic area of the brain is the same region involved in other obsessive-compulsive disorders and, as you found in Chapter 4, is associated with appetite and satiety.

Eating Disorders: Summary

Anorexia nervosa, bulimia nervosa, and binge-eating disorder are characterized by distorted body image and extreme preoccupation with food, appearance, and weight. The prevalence of these disorders has been increasing since the 1980s. Up to 4% of the adult and adolescent U.S. population is now believed to have an eating disorder. Caucasian women are more likely than men or women of other races to develop anorexia nervosa or bulimia nervosa. Binge-eating disorder occurs about equally in women and men and in all racial and ethnic groups. Specific criteria have been established for the diagnosis of some eating disorders. In addition, many more individuals affected by subclinical or partial anorexia and bulimia, as well as other types of disordered eating.

Case Study. The Cross- Country Runner, Part I	Jean is an 18-year-old freshman cross-country runner at a large Division I college. She has always been lean. In high school she had a hearty appetite but was so active—running and playing sports—that she rarely gained weight. When her parents brought her to college in August she weighed 110 lbs (50 kg) on a 5′ 4″ (164 cm) frame.
	At the cross-country coach's first team meeting Coach Davis commented on the "freshman 15" and stressed that runners who weigh less run faster. One athlete asked him how she could gain some speed and he replied, "Don't get fat."
	Jean was recruited by several top universities, and she really wants to run well for this coach. Initial practices went well for Jean, and she happily put in 40 miles/week of training. Her coach frequently complimented her on her work ethic. Because of her class schedule, Jean grabbed a granola bar and cup of orange juice for breakfast and rarely at more than a candy bar for lunch. She almost

(Continued)

always had time for dinner, and enjoyed hanging out with her friends in the dining hall, typically having an entrée and salad. She and her friends also would occasionally go out for pizza in the evening.

After a 3-day weekend at home where she "ate everything in sight" and felt like she had gained 10 lbs, Jean returned to campus determined to stick to a strict diet. One of her teammates, who had been struggling on the team, invited Jean to join a group of athletes who were competing to lose weight and train harder. Soon, Jean was running 20 more miles/week and eating mainly salads and fruit. "The extra running really hurts," she confided to her teammate Maria, "but I don't think I can keep my weight down without it." "Do what I do," said Maria. "Eat your meal and then throw up."

Jean hated throwing up, but it was easier than running an extra 15–20 miles/week. She started to lose more weight, especially when she only ate candy. But at least once a week she was so hungry that she overate, and then she felt really guilty and restricted her intake even more.

As the season ended, Jean felt more tired than she had ever been. She missed the last meet due to a nagging pain in her left leg. Academic stress increased at the end of the semester, and sometimes Jean didn't leave her room to eat. She started swimming 2 hours a day to keep in shape.

At Christmas break, Jean's mother was alarmed at how thin she looked and made an appointment with the family physician. The doctor was concerned about her weight, which had dropped to 97 lbs (44 kg). Jean also told the doctor than she had not menstruated since October.

- What was Jean's BMI in August? What is her current BMI?
- Does Jean meet any of the *DSM-IV* criteria for an eating disorder?
- Is there any type of intervention that might be helpful to Jean at this point?

PHYSIOLOGICAL EFFECTS OF EATING DISORDERS

Anorexia nervosa, bulimia nervosa, and, to a lesser extent, binge-eating disorder cause harmful and often permanent effects to almost every body system. Because children and adolescents are still growing, they experience more severe medical complications than adults. Individuals with anorexia nervosa, who are essentially starving themselves, are especially adversely affected. The American Anorexia/Bulimia Association, Inc. states that approximately 1,000 females die each year from anorexia. Both morbidity and mortality are highest for those with the lowest body weights and those who purge. Physiological effects of eating disorders are summarized in Table 11-4.

What Are the Effects on the Cardiovascular System?

bradycardia
Resting heart rate slowed to less than 60 beats per minute that can occur both as a result of physical fitness and disease.

myopathy *Muscle weakness and wasting.*

Cardiovascular health may be impaired by behaviors associated with eating disorders, and cardiovascular complications are the most common causes of death among anorexics. The most often reported cardiovascular complications of eating disorders are:

- Slowed heart rate: **Bradycardia** is a heart rate slowed to 50 or fewer beats per minute (bpm) during the daytime and below 46 bpm at night.[2] Heart rates as low as 25 bpm have been reported.[9] Bradycardia affects circulatory capacity and may lead to abnormal heart rhythms in a small number of cases.
- Cardiac **myopathy**: Atrophy of the heart muscle occurs as a result of starvation and from chronic use of syrup of ipecac to induce vomiting. Ipecac is absorbed by muscle

TABLE 11-4 Physiological Effects of Eating Disorders

	Anorexia nervosa	Bulimia nervosa	Binge-eating disorder
Cardiovascular	Hypotension Irregular heart rate Bradycardia Cardiac myopathy Electrolyte abnormalities (if purging type) Refeeding syndrome	Arrhythmias Palpitations Cardiac myopathy Electrolyte abnormalities	Hypertension High blood lipids
Digestive system and kidneys	Bloating Constipation	Erosion of tooth enamel Chipped teeth Dental caries Esophageal irritation Bloating Constipation G-I bleeding	Dental caries Gum disease Gallbladder disease
Endocrine	Amenorrhea Menstrual irregularity Cold intolerance Fatigue Hypoglycemia	Menstrual irregularity	Type 2 diabetes mellitus
Musculoskeletal	Osteopenia Osteoporosis Stress fracture Stunted growth Muscle wasting	Muscle wasting	

hypokalemia
Reduced level of circulating potassium in the bloodstream.

hypochloremia
Reduced level of circulating chloride in the bloodstream.

hyponatremia
Reduced level of circulating sodium in the bloodstream.

leucopenia
Decreased white blood cell count.

tissue, including heart muscle, and interferes with muscle contraction. In the heart this produces less forceful heart contraction, fatigue, and bradycardia. Caffeine and exercise can be fatal to those with heart muscle wasting and slowed heart rate.

• Fluid and electrolyte imbalances: This occurs most often in people who purge. The most common electrolyte imbalances are low potassium (**hypokalemia**), low chloride (**hypochloremia**), and low sodium (**hyponatremia**). These conditions produce tiredness; muscle weakness; and, in the case of potassium, cardiac rhythm abnormalities.

• Blood pressure abnormalities: Abuse of diet pills that contain phenylpropanolamine (PPA) used to be a factor in some cases of high blood pressure and, in severe cases, stroke, until the Food and Drug Administration compelled manufacturers to remove PPA from their products. Unfortunately, some dietary supplements still contain ingredients that elevate blood pressure. Obese individuals with binge-eating disorder are at risk of hypertension. Conversely, both anorexia and bulimia may lead to low blood pressure when insufficient calories are consumed.

• Blood disorders: Iron-deficiency anemia and mild **leukopenia** (decreased white blood cell count) are most common.

• Dyslipidemia: High blood cholesterol and triglyceride levels are common in binge-eating disorder.

What Are the Effects on the Digestive Tract and Kidneys?

You are correct if you assumed that the digestive tract could be damaged by disorders characterized by the consumption of very little food, periodic food binges, abuse of laxatives, and/or frequent self-induced vomiting. Most commonly affected are:

parotid glands
Largest of the salivary glands.

edema *Fluid accumulation in the tissues, which may result from congestive heart failure, kidney disease, corticosteroid therapy, liver diseases, or other metabolic conditions.*

- Mouth: Permanent erosion of tooth enamel, particularly on the upper teeth, may occur as a result of the high acidity of vomited stomach contents. Tooth brushing immediately after vomiting aggravates the damage because teeth are more vulnerable when they have been exposed to gastric acid. Chipped front teeth and enlargement of the **parotid glands** often occur in people who vomit regularly. Frequent binges on high-sugar foods promote development of dental cavities and gum problems.
- Esophagus: Irritation and even rupture of the esophagus may occur from forced vomiting.
- Stomach: In rare circumstances, the stomach may dilate after bingeing, and the individual is unable to vomit. Death occurs if the stomach ruptures before medical attention is received.
- Intestines: Bloating and constipation are gastrointestinal complaints that follow water retention-induced dehydration due to starvation, vomiting, or laxative use. Once normal levels of food intake are restored, these tend to slowly resolve. Chronic laxative abuse can be more difficult to treat, as significant **edema** (10–20 pounds of water) can result from abrupt laxative withdrawal. Abuse of laxatives that contain phenolphthalein may irreversibly damage nerves in the colon and make restoration of normal bowel functions difficult.
- Kidneys: Individuals with eating disorders who lose large amounts of weight may be at risk for impaired kidney function and even renal failure.
- Gallbladder disease: Obese individuals with binge-eating disorder are at a high risk for gallbladder disease.

follicle-stimulating hormone
Hormone secreted by the anterior pituitary gland that is responsible for production of estrogen and ovarian follicles in females and sperm in males.

luteinizing hormone
Hormone secreted by the anterior pituitary gland that controls ovulation in women and sex hormone production in males and females.

What Are the Effects on the Endocrine System?

The endocrine system is particularly affected in people with anorexia nervosa. The overall effect of anorexia nervosa is to return the individual to a prepubertal state by interfering with hypothalamic and pituitary hormones. Pituitary secretion of **follicle-stimulating hormone** and **luteinizing hormone** are reduced, which results in low serum estrogen levels in women and low serum testosterone levels in men. These low levels cause several abnormalities, most of which can be reversed with eating and weight gain:

- Amenorrhea: The cessation of menstruation is a defining characteristic of anorexia nervosa in women who are post-puberty and pre-menopausal. Amenorrhea can even occur in bulimic women, probably due to stress, changes in weight, and nutrient deficiencies. Most women with amenorrhea will restore menstruation within 6 months of achieving 90% of ideal body weight.[9]
- Sexual dysfunction: Men and women with eating disorders often lose interest in sex.

- Thyroid hormone levels: The thyroid gland produces fewer hormones, which also occurs in starvation, as a way to reduce energy expenditure and conserve calories. This can cause constipation, dry skin, intolerance to cold, fatigue, and bradycardia.
- Glucose intolerance: Cortisol, a hormone that regulates nutrient metabolism and is involved in the body's stress response, is secreted in higher amounts in people with eating disorders, often leading to glucose intolerance. Individuals who have high BMIs are at greater risk of developing type 2 diabetes mellitus.
- Complications of diabetes: Type 1 diabetics who also have an eating disorder are far more likely to experience complications, such as diabetic ketoacidosis, severe **hypoglycemia**, and painful **neuropathy**.

hypoglycemia
Low blood glucose level.

neuropathy
Complication of diabetes in which nerves in the hands and the feet degenerate.

What Are the Effects on the Skeletal System?

During childhood and adolescence, bones grow longitudinally until the growth plates close. During these developmental stages, bones are continually remodeled through a process that breaks down existing bone tissue (a process known as resorption) and deposits new bone (bone formation). Adolescence is a critical period for building peak bone density—about half of bone mass is built during the teenage years. Bone health is compromised when resorption exceeds formation or when growth plates close prematurely. Eating disorders affect bone health in several ways:

osteopenia
Condition in which bone mineral density is reduced 1–2.5 standard deviations below average bone density for a healthy adult; this condition may lead to osteoporosis.

osteoporosis
Condition characterized by reduction in bone mineral density more than 2.5 standard deviations below average bone density for a healthy adult; the condition increases susceptibility to fractures.

- Reduced bone formation and increased resorption: The estrogen deficiency that occurs in women with amenorrhea affects calcium balance in bone.
- **Osteopenia**: Osteopenia is a loss of bone mineral. In eating disorders, reduction in bone mineral density happens in both types of bone tissue: trabecular (the spongy bone found in the interior of bone) and cortical (the harder bone that surrounds trabecular), even in young women. While weight-bearing exercise is recommended to prevent osteopenia, excessive exercise coupled with amenorrhea actually promotes it. About half of female anorectics have **osteoporosis**, a severe loss of bone mineral.[9] When peak bone mass is not attained in adolescence, it cannot be restored later in life.
- Stunted growth: Individuals with anorexia who have not yet attained peak height may experience stunted growth, which is irreversible.
- Skeletal myopathy: Skeletal muscles may waste away from both starvation and abuse of syrup of ipecac to induce vomiting.

Physiological Effects of Eating Disorders: Summary

Eating disorders cause unacceptably high morbidity and mortality rates. At highest risk of increased mortality and morbidity are severely underweight anorexics and individuals who purge. Cardiovascular complications are the leading causes of death and result in significant morbidity, including cardiac myopathy and heart rate abnormalities. Digestive system problems range from erosion of tooth enamel to severe constipation and bloating to stomach and esophageal rupture. The endocrine system is particularly affected in individuals with anorexia nervosa, who develop amenorrhea, sexual dysfunction, and impaired glucose tolerance. Reduced estrogen production often leads to osteopenia and, later in life, osteoporosis.

PREDISPOSING FACTORS FOR EATING DISORDERS

Although a single specific cause of eating disorders has not been identified, experts know that certain predisposing factors increase the risk of developing an eating disorder. These factors are summarized in Table 11-5. When one or more of these predisposing factors occurs, an eating disorder may develop in response to stressful life events, such as leaving home, a negative remark about one's body weight, a death in the family, or onset of puberty.

Psychological, personality, biological, family, and socio-cultural factors may increase the likelihood of developing an eating disorder. In addition, individuals who participate in athletics, who diet frequently, and who have diabetes are at increased risk. Health professionals who understand these risks may be able to prevent some cases of eating disorders. Because of the links between obesity and disordered eating, obesity prevention programs should also take these risk factors into consideration.

Do People with Eating Disorders Have a Common Psychological Profile?

As noted in Chapter 1, obese people are generally no more happy or unhappy than the nonobese population. This observation is not true for people with eating disorders. Depression, anxiety, and other psychiatric symptoms have a greater overall prevalence in people with eating disorders, even partial or subclinical conditions. In some cases, psychological disturbances occur after an eating disorder develops, so the disturbance could

TABLE 11-5 Predisposing Factors for the Development of Eating Disorders

Psychological
    Body image distortion
    Depression
    Anxiety disorders
    Impulse-control problems
Personality
Biological
    Neurochemicals
    Immunity
Familial
    Biology/genetics
    Poor communication
    Less nurturing family
    Parental issues about food and weight
Cultural
    Thinness ideal
Behavioral
    Athletics
    Dieting
    Diabetes

not have "caused" the disorder. In other cases, psychological disturbances were already present and were expressed as an eating disorder.

Body Image Distortion and Dissatisfaction

body image
Perceptions and feelings about one's body.

Body image is the mental picture that you have of your body and how you feel about it. You were not born with your present body image. It developed during infancy and early childhood from your interactions with others. Being touched affectionately, having your feelings acknowledged, and being allowed to express thoughts and views without repercussions are important for the development of a healthy body image. In contrast, being physically or sexually abused, deprived of touch, or required to suppress thoughts and feelings may result in body image disturbance, which can lead to profound body dissatisfaction.

Distorted body image is considered to be a major causal factor in the development of eating disorders. Many people are dissatisfied with their bodies and may perceive themselves as unattractive or too large. Those whose body image is profoundly distorted become overly preoccupied with imagined body flaws. Any negative comment about their appearance is long remembered, and any positive comment is quickly forgotten. All achievements and failures are attributed to how they look. This distorted perspective affects all areas of their lives. When body image distortion is combined with other predisposing factors, the risk of eating disorders increases dramatically.

Young women are far more susceptible than young men (except perhaps homosexual men) to media portrayals of a thinness ideal and cultural pressures to be thin. *Glamour* magazine recently surveyed 16,000 women and found that 40% were unhappy with their bodies, about the same number as in 1984.[29] Scientific reports echo this finding.[30] Obese individuals may resort to severe dieting to reach an unattainable body size, which can lead to binge eating or bulimia.

The National Women's Health Information Center, part of the U.S. Department of Health and Human Services, has an excellent Web site where women can get information about body image, eating disorders, and other related subjects (www.womenshealth.gov/bodyimage/).

Psychological Comorbidities

depression
Feelings of sadness and despair, usually accompanied by sleep problems, low energy, and inability to concentrate.

obsessive-compulsive disorder *Anxiety disorder in which the individual adopts compulsive, repetitive behaviors to cope with obsessive thoughts that increase anxiety.*

phobias *Reactions characterized by heightened anxiety in response to specific feared situations or things.*

Several psychological comorbidities may accompany eating disorders, most commonly **depression** and anxiety disorders. Depressive symptoms include sadness, low self-esteem, difficulty sleeping, and irritability. These symptoms sometimes occur before the onset of bulimia and binge-eating disorder, and overeating results from attempts to feel better through food. In other cases, depression appears to result from food deprivation, and treatment of the eating disorder relieves depressive symptoms.

Anxiety is a sense of apprehension or uneasiness that can arise in response to a variety of events or conditions. Chronic high levels of anxiety may lead to an anxiety disorder. The most frequently diagnosed anxiety disorders in individuals who have an eating disorder are **obsessive-compulsive disorder** and **phobias**, both of which are more common in people who have anorexia or bulimia than in the general population. Pica and muscle dysmorphia have also been linked to obsessive-compulsive disorder.

Obsessions are persistent thoughts that increase anxiety. For example, food obsessions often occur in people who are deprived of food or who believe certain foods are "bad." Compulsions are repetitive behaviors that serve to neutralize anxiety brought on by the obsession. Not surprisingly, many individuals with anorexia adopt compulsive behaviors that center around eating, such as calorie counting, recipe clipping, and cutting food into dozens of tiny, uniform pieces before eating. Some individuals have obsessions and compulsions unrelated to food, such as an obsession with cleanliness that is accompanied by compulsive hand washing.

Phobias are anxiety reactions to specific feared situations or things. Social phobias are common among people with eating disorders, especially women with anorexia but also obese binge eaters and lean people with muscle dysmorphia. Intense body dissatisfaction may lead them to believe that people are looking, or even laughing, at them, which increases anxiety and reinforces the desire to lose more weight or to lift more weights.

Impulse-Control Problems

Individuals with bulimia nervosa, binge-eating disorder, and the binge-eating/purging type of anorexia frequently have problems resulting from difficulty in controlling impulses. Shoplifting and substance abuse are examples of impulse-control problems. Various reviews have reported that substance abuse (including alcohol abuse) occurs in up to half of people with bulimia, up to 19% of people with anorexia, and 10–44% of individuals with binge-eating disorder.[31]

Is There an "Eating Disorder" Personality?

Psychologists have identified several personality traits shared by individuals who develop eating disorders. Among these traits are an overall sense of anxiety, difficulties in coping with stress, awkwardness with interpersonal relationships, lack of assertiveness, and an inability to express emotions. Many anorexics, particularly those with the restricting type, are rigid, inhibited, constrained, and overly concerned with being perfect in all aspects of their lives. They demonstrate a high degree of self-control. Because thinness is perceived in our society as the perfect body type, it is not surprising that an individual seeking perfection would want to control weight. Unfortunately, no matter how much control an anorexic exerts, a flawed self-image will never allow her to attain perfection in body weight. Even many years after treatment, this personality characteristic may persist.

Is There a Biologic Cause of Eating Disorders?

neuropeptide Y
Peptide found in the hypothalamus that stimulates food consumption.

In Chapter 4, you learned about some of the systems involved in regulating food intake. Many researchers have documented disturbances in these systems and their associated hormones and chemicals among individuals with eating disorders. Evidence continues to be gathered to determine whether these disturbances are a cause or an effect of eating disorders.

peptide YY
Peptide secreted in the digestive tract that suppresses appetite.

Neuropeptide Y and Peptide YY

Neuropeptide Y (NPY), a powerful appetite stimulant produced in the hypothalamus, and **peptide YY (PYY)**, an appetite suppressant secreted in the digestive tract, are among several brain chemicals being studied in individuals with eating disorders. Some researchers refer to the gut-brain axis to describe the phenomenon of neurotransmitters produced in the central nervous system having an impact on what happens in the digestive system. Among their findings:

- Elevated NPY levels have been measured in normal-weight bulimics who avoided bingeing for 30 days, setting the stage for a binge.
- PYY levels are typically low in obesity. Recently, significantly increased levels of PYY were discovered in adolescent girls with anorexia nervosa.[32]

Immune System Malfunction

Recently researchers discovered that autoantibodies, components of the immune system, have the capacity to work against the neuropeptides that control appetite and satiety, thereby affecting the ability to turn eating on and off. When autoantibodies acted against a melanocortin peptide called α-melanocortin stimulating hormone (α-MSH), either anorexia nervosa or bulimia nervosa occurred.[33,34] One theory is that a benign autoantibody might become harmful in response to stress, which would then allow suppression of neuropeptides than control eating and satiety.

Opioid Peptides

opioids
Neuropeptides, such as β-endorphin, that promote eating.

Opioid peptide levels have been found to vary among individuals with eating disorders. The **opioids** are narcotic-like substances produced in the brain that are known to play a role in food intake (they increase it), response to pain (they reduce it), and mood (they induce a sensation of calmness). "Runner's high" has been attributed to release of opioids in long-distance runners. Purging may also stimulate the release of opioids, which could result in "addiction" to vomiting and, ultimately, to the development of bulimia in some people. The opioid peptides may be involved in regulating pleasure responses to foods, especially sweets and high-fat foods. Not coincidentally, these are the kinds of foods most likely to be consumed during a binge.

Some women with bulimia produce lower levels of the opioid peptide β-endorphin.[35] Bingeing and vomiting stimulate secretion of several peptides from which β-endorphin is derived. If these peptides were oversecreted, they might become depleted between bingeing and vomiting episodes. The resulting lack of β-endorphin could alter mood, perception, and behavior, leading to depression, body image distortion, and compulsive behaviors. Abnormally low levels of β-endorphin could also cause bulimic behaviors. An individual might discover that binge eating and vomiting make her feel better (by stimulating release of the opioid peptides). Continued bingeing and purging are attempts to stimulate secretion of higher levels of opioids.[36]

The opioids may also play an important role in anorexia. Elevated opioid levels in anorexics have been postulated to make fasting tolerable; inhibit production of luteinizing hormone, which would cause amenorrhea; and induce an energy-conserving state, which could magnify the tendency to see one's body as too large.[36] Drugs called opiate antagonists work in opposition to the opioid peptides. When drugs like naltrexone and naloxone are given to individuals with bulimia or binge-eating disorder, binges diminish.[37] Naltrexone has also been used to prevent relapses in recovering alcoholics.

What Family Issues Are Risk Factors?

The family makes biological, environmental, and psychological contributions to an individual's predisposition for an eating disorder. Twin and family studies confirm that the risk of anorexia nervosa and bulimia nervosa is higher in people who have siblings or parents with an eating disorder.[38] Specific inherited factors have not been identified but might include neurochemical levels; susceptibility to obsessive-compulsive disorder or depression; personality type; and even tendency toward obesity, which could promote dieting.

In addition, family dynamics may contribute to risk. People with eating disorders often come from families in which communication is poor, conflict is avoided, affection is rare, and relationships between parents are strained. There may be less nurturing and more hostility or even physical, psychological, or sexual abuse in these families.[39] Parents may be over-involved and controlling. The combination of biological vulnerability, psychological factors, and family dynamics can lead to an eating disorder.

Parents who have their own unresolved issues about body weight and food may pass along these feelings and unusual food- and weight-related behaviors to their children. Parents may even encourage excessive dieting due to unwarranted fears that their child will be overweight. About 40% of women with bulimia and 31% of women with binge-eating disorder reported being overweight as children, compared with under 20% of those without an eating disorder.[40,41] Teasing and negative comments can make a child extremely self-conscious about weight and, when this is coupled with physical, psychological, or sexual abuse, may disturb the development of a healthy body image.

Do Cultural Factors Increase Risk?

From the bound feet of the twelfth-century Chinese to the bound waists of the nineteenth-century Victorians, a culturally determined aesthetic ideal has driven women to extreme behaviors. Today, the aesthetic ideal for women in western cultures is thinness, and this ideal is promoted in both subtle and not-so-subtle ways. Women's magazines, especially those aimed at teens and pre-teens, feature reed-like models and articles about dieting and weight loss; fashionable clothing is often unavailable to those who wear sizes larger than 12; most television and movie actresses are noticeably thin; and even Miss America Pageant contestants have gotten smaller. And men are not immune to these cultural pressures; everything from media images to action figures tell them that lean muscularity is the ideal.

At the same time, most developed nations, particularly the United States, have seen a surge in obesity. Such a large disparity between the real and the ideal creates a problem. For some young men and women, restrained eating or excessive exercise is the solution.

Eating disorders are rare in racial/ethnic groups in which the concept of attractiveness embraces larger people. For example, African American girls and women have been found to be more accepting of higher body weights, less concerned about weight-related discrimination, and less likely to express feeling social pressure to be thin.[16] In addition, African American women are more likely to equate large size with strength and stamina, and they have a more flexible definition of beauty than Caucasian women. Still, African American, Hispanic, and Native American adolescents trying to compete in predominantly Caucasian schools may experience the same risk of eating disorders as their Caucasian peers.[42]

Of concern in the Internet age is the proliferation of pro-ana (anorexia) and pro-mia (bulimia) Web sites that depict eating disorders as a lifestyle choice. People at these Web sites provide tips for hiding an eating disorder and portray thinness as an ideal. For the adolescent, this sense of acceptance and community can be very tempting and make identification and treatment of an eating disorder difficult for parents and health professionals.

Are Athletes More Susceptible to Eating Disorders?

Eating disorders are more prevalent among male and female athletes than in the general population. Prevalence estimates vary. The American Academy of Pediatrics Committee on Sports Medicine and Fitness reports that 10–15% of high school boys who participate in sports with a weight class either have an eating disorder or exhibit eating disordered behaviors.[43] Up to 8% of elite and college female athletes may have a diagnosed eating disorder, with up to 25% having a partial or subclinical eating disorder.[44] Detection of an eating disorder can be very difficult in athletes.

At particular risk are:

- High-performance athletes
- Athletes in sports where appearance is important and/or where low body weight enhances performance, such as diving, gymnastics, figure skating, horse racing, running, and ballet; or where muscularity is expected, such as football and weightlifting
- Athletes in sports in which participation is determined by weight classification, such as wrestling, boxing, judo, and rowing

The Female Athlete Triad

Surveys of female college athletes report that many are trying to lose weight, primarily through excessive exercise, but also through self-induced vomiting or abuse of laxatives, diuretics, or diet pills.[43,44] A meatless diet may also act as a smoke screen for restrictive eating behaviors. Many female athletes with eating disorders do not eat meat.[45] Vegetarianism, unless practiced with care, increases the risk of anemia and menstrual irregularities.

female athlete triad Set of three problems often seen in female athletes: disordered eating, premature osteoporosis, and amenorrhea.

So prevalent are eating disorders among female athletes that exercise physiologists have a term, the **female athlete triad**, to refer to the three common problems seen in this population—nutritional (disordered eating), endocrine (amenorrhea), and musculoskeletal (premature osteoporosis). Up to two-thirds of female athletes may have amenorrhea, compared with fewer than 5% of women in the general population.[46] The triad of problems is pictured in Figure 11-2. Each of these disorders is a health concern

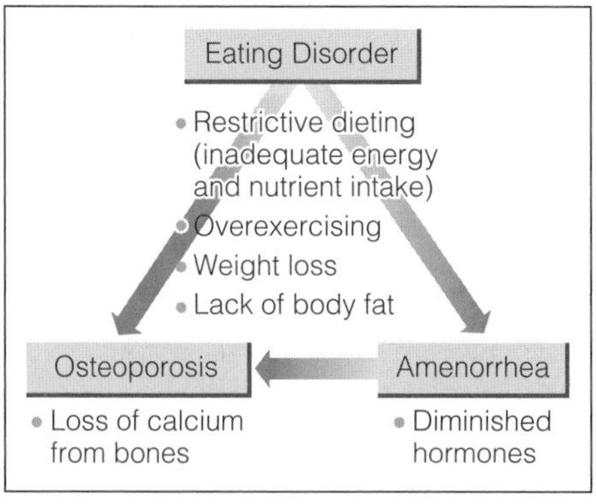

FIGURE 11-2 The female athlete triad.

by itself, but in combination they can be extremely harmful, particularly for adolescent athletes.

Making Weight

Male and female athletes in sports that require them to meet a weight classification (popularly called making weight) are also at risk of using dangerous eating and exercise behaviors. The purpose of weight classification in sports is to let skill, not one competitor's weight advantage, decide the outcome. Some athletes believe that they will gain an advantage by competing in the weight class one category lower than their preseason weight, even if this places them below their normal weight. Well over half of wrestlers, for example, lose weight during their season.[43] Unfortunately, the only way to attain and maintain an abnormally low weight is through dangerous weight-loss practices—food and water restriction, laxative and diuretic use, exercising in a sauna, and wearing rubber suits while exercising, for example. Some athletes who engage in these practices develop eating disorders, and some die.

In 1997, the first deaths associated with intentional weight loss in collegiate wrestling were widely reported. Three wrestlers in different states engaged in vigorous exercise in hot environments while wearing vapor-impermeable suits. Two hoped to compete in the 153-lb weight class and had preseason weights of 178 lbs and 180 lbs, respectively. The third had a preseason weight of 233 lbs and hoped to compete in the 195-lb weight class. All three died between November 7 and December 9, 1997, from complications of hyperthermia, cardiac stress, and rapid and extreme weight loss. As a result of these tragedies, the National Collegiate Athletic Association (NCAA) immediately made several rules changes, such as banning the use of saunas and rubber suits, increasing the weight allowance for each class, and moving the pre-match weigh-in time from 24 hours before a match to 2 hours, so that a wrestler who has lost dramatic amounts of weight for a weigh-in will not have time to hyperhydrate and recover before wrestling.

Can Dieting Cause Eating Disorders?

Dieting, especially severe dieting, is consistently cited as a predictor of both disordered eating and development of an eating disorder. In fact, one review concluded that dieting, weight and shape concerns, and body dissatisfaction were the strongest risk factors for the development of an eating disorder.[47] Dieting is dangerous because:

- Low-calorie diets can result in dramatic weight loss, but weight loss attained in that way is not permanent. The only way to keep off weight lost by severe dieting is to vigilantly exert superhuman control over eating (which occurs in anorexia) or to try to compensate for eating by purging, abusing laxatives, or compulsively exercising (as seen in bulimia).
- People who diet may become obsessed with the "bad" foods that they are restricting. This often leads to bingeing.
- An already negative self-image may be reinforced by failure to keep weight off or by depression associated with food deprivation.

The vicious cycle of dieting and bingeing is shown in Figure 11-3. Dieting almost always precedes binge eating, in people with bulimia nervosa. In contrast, in binge-eating

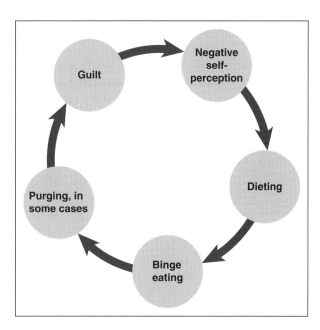

FIGURE 11-3 The vicious cycle of dieting and binge eating.

disorder, the onset of bingeing often occurs before any attempts to diet.[14] For those who have binge-eating disorder, dieting may be a result—trying to control their eating—rather than a cause of disordered eating.

What Is the Connection Between Diabetes and Eating Disorders?

type 1 diabetes mellitus *Form of diabetes in which the pancreas produces insufficient insulin, and insulin injections are required for normal glucose metabolism; also known as insulin-dependent diabetes mellitus (IDDM).*

diabetic ketoacidosis *Consequence of uncontrolled type 1 diabetes, wherein electrolyte imbalance, high blood glucose level, and abnormal fat metabolism increase blood acidity and may lead to fatal coma.*

Type 1 diabetes mellitus is the form of diabetes in which insulin is no longer produced by the pancreas, and insulin injections are needed to remove glucose from the bloodstream. Approximately 5–10% of diabetics have this form of the disease. People with type 1 diabetes need to monitor their diets carefully, paying particular attention to the carbohydrate content of food and the timing of meals so they can regulate insulin requirements. Type 1 diabetics lose weight when they do not take enough insulin.

Several characteristics shared by people with eating disorders and diabetes may increase the risk of eating disorders in the diabetic population. Both groups classify foods as good and bad; the diabetic's classification is based on carbohydrates and sugars, whereas the anorexic's or bulimic's is based on fat. Both carefully monitor the timing and content of meals. Both have control issues—the eating disordered person with weight and the diabetic with blood sugar.

Some individuals with diabetes deliberately avoid injecting insulin as a form of weight control. One estimate suggests that up to one-third of young women with diabetes lose weight by intentionally omitting insulin injections or not taking enough insulin.[48] Besides the risk of developing an eating disorder to compensate for the weight gain that occurs when these diabetics use insulin, they develop serious health problems—permanent damage to the retinas, the kidneys, and the nerves, and **diabetic ketoacidosis**.

Predisposing Factors for Eating Disorders: Summary

Several factors are known to increase the risk of developing an eating disorder. Physiological factors are increasingly implicated. Psychological disturbances, such as depression, anxiety, and impulse-control problems, are more common in the eating-disordered population than the general population. Personality traits, such as perfectionism and rigidity, may be an underlying factor. There may be imbalances in neuropeptides that influence hunger, satiety, and mood. Family dynamics, including poor communication, lack of nurturance, and physical or psychological abuse, play a role in some cases. Western culture itself, which promotes the thinness ideal, is a causal factor for many people, especially young women. Certain subgroups in the population, including type 1 diabetics and athletes, have a high prevalence of eating disorders. By understanding these predisposing factors, we may be able to identify individuals at increased risk and better target prevention efforts.

Case Study. The Cross-Country Runner, Part 2	During the holiday break Jean made a valiant attempt to eat more, at the urging of her mother. She was reluctant to eat too much, because she knew that after missing the final meet of the year her track coach was expecting her to come back strong for next season. She was also considering playing spring lacrosse and wanted to be lean for speed. When she had to eat a big meal in front of her parents, Jean would throw up afterwards. The leg pain had become an increasing concern, so she returned to the doctor, who took an x-ray and diagnosed a stress fracture. He discouraged her from playing lacrosse, because he wanted the stress fracture to have time to heal; he also wanted her to regain some of the lost weight.

Jean agreed to a referral to a sports nutritionist.

Kay, a sports nutritionist who works with Jean's physician, quickly learned that Jean was terrified about gaining weight. In addition, Jean confided that there were foods that she would absolutely NOT eat—meat, pizza, milk, cheese, bread—because she was convinced that these foods caused her to gain weight and feel sluggish. With her workouts limited due to the stress fracture, Jean was eating less than 1,000 kcal/day and still felt like that was too much to maintain her running weight. However, Jean was interested in working with Kay to identify low-calorie, high nutrition foods. Jean did not tell Kay that she sometimes made herself throw up after a big meal or a binge.

• How many kcals does a female competitive runner need to eat every day to support 3 hours of training and racing? (You will find some help for this in Appendix B-3)
• What are some food options that you think Kay might suggest for Jean?
• In addition to the stress fracture, what physiological issues are of concern with Jean's low caloric intake and vomiting? |

TREATMENT AND PREVENTION OF EATING DISORDERS

You might assume that the focus of treatment for an eating disorder would be restoring normal eating. Although it is true that a priority for individuals with eating disorders is to reestablish healthy eating patterns, treatment also must attend to psychological issues, or the nutritional intervention will fail. Therefore, the most effective approach is one in which a treatment team—a registered dietitian, a mental health professional, a physician, a nurse, and other clinical specialists—considers medical, nutritional, and psychological issues. However, when a team is not available, a primary care physician or nurse practitioner and registered dietitian, who are experienced in working with individuals with eating disorders and are able to consult with a mental health professional, can provide effective treatment.[49] Treatment may take place in inpatient, partial inpatient, or outpatient settings.

The goals of treatment for eating disorders are (1) treating medical complications; (2) correcting nutrient deficiencies, reestablishing good nutrition, and, for adolescents, promoting normal growth; (3) correcting disordered eating behaviors and attitudes toward food; (4) reestablishing normal responses to feelings of hunger and satiety; and (5) treating underlying psychological issues driving the behavior.

Treatment of eating disorders may include nutritional, psychotherapeutic, and pharmacologic approaches. The process is often lengthy, punctuated by periods of relapse. Even more difficult than restoring physical health is psychological healing. Denial that a problem exists and resistance to treatment, particularly for individuals with anorexia nervosa, are common barriers to treatment.

When Is Hospitalization Required?

Individuals with bulimia nervosa are rarely hospitalized for treatment because their medical complications tend to be less severe than those of people with anorexia nervosa, who sometimes require hospitalization. Inpatient treatment is recommended under these circumstances:

- Severe depression and/or suicide risk
- Dangerous medical complications, such as abnormal heart rhythms, gastric dilation, severe anemia, edema, electrolyte abnormalities, and, in children and adolescents, arrested growth and development
- Severe malnutrition, defined as a BMI under 13 kg/m² or a body weight that is at or below 75% of expected weight for age, gender, and height
- Refusal to eat
- No outpatient treatment available or outpatient treatment that was ineffective
- Family circumstances that necessitate removing the individual from the household
- Drug withdrawal
- Anorexia and type 1 diabetes, requiring close monitoring of blood glucose and insulin injections

When hospitalization is recommended, psychiatrists and nutritionists agree that tube feedings or intravenous therapy should be used only if the individual is in medical danger. The notion of forced feeding could irreparably harm the recovery process.

Recently inpatient treatment duration for eating disorders has become dangerously curtailed. The average length of stay for eating disorder patients decreased from 149.5 days in 1984 to 23.7 days in 1998. While discharge BMI of patients with anorexia nervosa was 19.3 in 1988, it had decreased to 17.7 in 1998. During the same time period, readmissions increased from 0% to 27%.[50]

What Does Nutritional Treatment Include?

A registered dietitian is a key member of the treatment team for the nutritional treatment of eating disorders. Registered dietitians educate patients and their families, assess nutritional status, provide meal-planning assistance, and work closely with others on the team. Dietitians may be involved in treatment in an educational or therapeutic role, according to the American Dietetic Association:[51]

- Brief consultation with the client to develop a food plan and answer questions
- Longer series of meetings with the client to discuss nutritional aspects of eating disorders
- Continuous or intermittent contact with the client to work on behavior change and be directly involved in counseling

Education Component of Nutritional Treatment for Eating Disorders

The main purpose of the educational aspect of treatment is to develop a plan for changing food and weight behaviors, not to address underlying psychological issues. Nutrition education for eating disorders typically begins with a nutritional assessment, including taking a history of weight changes, patterns of eating, and compensatory behaviors. Data are used to set weight goals and develop all aspects of the treatment plan.

Individuals with eating disorders need accurate information about the causes and physiological effects of their condition, what constitutes a healthy body weight, caloric intake needed to achieve good health, ineffectiveness of diets for the long-term management of body weight, and how hunger and appetite are regulated. In addition, they need information about planning nutritious meals and snacks. The family should be informed about the plan for meals, snacks, and appropriate caloric and nutrient intake. This knowledge allows the family to offer support but removes the pressure of meal planning and monitoring from them. As part of family education, the dietitian can correct the family's misconceptions about food, eating, and eating disorders.

The typical course of recovery must be discussed so the individual knows what to expect. It is helpful to present information about caloric intake and eating patterns typical of those who have recovered from the specific eating disorder. In addition, the dietitian should prepare the person for situations that might precede relapse, such as illness and stress.

Behavioral Component of Nutritional Treatment for Eating Disorders

Experienced dietitians may also remain involved while the plan for changing food- and weight-related behaviors is carried out. This aspect of treatment is coordinated with the individual's mental health professional because psychological issues surrounding food are likely to surface as new eating behaviors are implemented. Various psychotherapeutic techniques may be used to gradually change food behaviors, recognize hunger and satiety, and confront feelings that have been acted on through food- and weight-related behaviors.

Goals for Nutritional Therapy

Starvation and impaired nutritional status cause depression, obsession with food, and inability to concentrate. Therefore, an immediate goal of treatment is to reestablish good nutrition.

- Goals for individuals with anorexia nervosa center on weight gain to attain a healthy body weight (for adults and adolescents, a BMI of at least 20; for children, use pediatric growth tables). This can be extremely difficult, and significant resistance may be encountered. Usually a meal and snack plan starts at around 1,500 kcal per day and is gradually increased. An absolute goal for weight gain should be set, usually 1–2 lb/week (or 0.3—0.4 lb/day at the beginning of treatment, if malnutrition was present).[49] Reliance on nutritional supplements, liquid diets, low-fat foods, and artificial sweeteners is discouraged. As the individual gains weight, some exercise may be recommended to relieve constipation and bloating and improve mood.

- Goals for individuals with bulimia nervosa focus on regaining control over eating by eliminating fasting and restrictive dieting. A structured meal plan is created to teach the individual to eat normal meals, recognize hunger, and know what constitutes a normal quantity of food. The dietitian should provide information and assurances to counter the notion that the client will "get fat" by resuming normal eating. Misperceptions about so-called good and bad foods must be corrected.
- Goals for individuals with binge-eating disorder involve regaining control over eating, but not with restrictive dieting. With binge-eating disorder, an additional conflict facing practitioners is whether first to treat binge eating or to treat obesity. Weight loss may be a goal for obese binge-eaters, but it is not realistic until bingeing is under control. Modest weight loss goals may be set later in treatment. Physical activity is an important component of treatment.

Refeeding Syndrome

When calorically dense nutrients, especially those containing a lot of glucose, are ingested either orally or parenterally (intravenously) during the early stages of treatment, cardiac collapse may result. This most commonly occurs during the first week of reestablishing food intake, probably due to a starved and weakened cardiac muscle being confronted with a sudden infusion of phosphate, stimulated by insulin, as cardiac tissue rebuilds itself. The sudden drop in serum phosphate, as more phosphate enters the cells to fuel the increasing cardiac mass, can cause arrhythmia, reduced cardiac output, or even seizures.[52] Thus, restoring nutrition in severely malnourished individuals should be medically supervised.

Which Psychotherapeutic Approaches Are Effective?

Psychotherapeutic approaches used in the treatment of eating disorders include various types of behavioral therapy, interpersonal therapy, and family therapy. These approaches are more effective for bulimia and binge-eating disorder, where effects may be observed within 3–5 months. In anorexia nervosa, where treatment goes against the very core of the disease, periods of up to 2 years may be required to see progress.

Cognitive Behavior Therapy

Bulimia nervosa, binge-eating disorder, and adult cases of anorexia nervosa have been successfully treated with various techniques of cognitive behavior therapy. In fact it is the most common treatment used for bulimia and binge-eating disorder. Cognitive behavior therapy focuses on the present, not the past. Its purpose is to help people control eating by changing the thought processes that maintain the disorder. This is accomplished by:

- Self-monitoring intake, compensatory behaviors, and feelings that accompany eating and purging
- Monitoring body weight

- Replacing bingeing with regular meals, including previously avoided foods
- Cognitive restructuring
- Relapse prevention

Cognitive restructuring is an important component of cognitive behavior therapy. As you learned in Chapter 9, it helps individuals "think about thinking." Automatic thoughts and beliefs, such as "Chocolate chip cookies are fattening," are gradually replaced with reasoned thoughts, such as "One chocolate chip cookie contains 10 g of fat, and I need some fat every day." The counselor works with the client to examine and critically appraise beliefs. Behavioral exercises are used to change beliefs.

Interpersonal Psychotherapy

Unlike cognitive behavior therapy, interpersonal psychotherapy does not focus on changing attitudes about food, weight, and shape. Rather, it involves identifying and addressing problems with interpersonal relationships that have made it difficult for the person to function in a family or society without having an eating disorder. It may identify for the individual the disconnect between the effects of the eating disorder and their life goals. Interpersonal therapy helps people understand how emotions can trigger eating-disordered behavior. In anorexia, it can help the person accept the importance of eating. It can also improve self-esteem and mood. When used in bulimia and binge-eating disorder the course of treatment is fairly short, but in anorexia typically lasts for several years.

Family Therapy

Because numerous family factors contribute to the development of eating disorders, involving the family in the treatment process is recommended. Family therapy has been most commonly used in the treatment of adolescents who have anorexia nervosa and is considered very effective with this population. In fact, it is often the first treatment tried. Family therapy is less common for the treatment of bulimia nervosa because affected individuals are usually older when they enter treatment. Studies with young bulimics suggest promise for this form of treatment. Sometimes therapy that involves the family is conjoint, when the individual and family members (usually parents) meet together with a therapist. In other situations, family counseling takes place separately from individual counseling.

The Maudsley approach, which originated in England, is an intensive family-based outpatient treatment for adolescent anorexia that has had success rates of 50–75%. Rather than blame parents for the disease, the treatment regimen looks to parents as important resources for treatment. Generally there are 15–20 treatment sessions over one year. First, the goal is to restore lost weight by having parents monitor meals and snacks and restrict activity. Second, once 90% of ideal body weight has been achieved, control over eating is returned to the child. Parents are taught behavioral techniques for restoring healthy weight and eating. Therapeutic sessions also focus on the impact of the eating disorder on healthy adolescent identity.[53] The approach is most successful when duration of anorexia nervosa is under three years.

What Other Treatments Are Used?

Other approaches may be used in conjunction with nutritional and psychotherapeutic treatments. Two such treatments are drug therapy and exercise.

Pharmacotherapy

Because many individuals with eating disorders have accompanying psychological disorders that are treated with medication, you might wonder whether drug treatments for eating disorders could hasten recovery. In fact, drug therapy is neither the first line of treatment nor the sole form of treatment for eating disorders.

Several drugs, including antidepressants and antipsychotics, have been studied in a small number of clinical trials with anorexics, but these have not improved weight gain or long-term prognosis. To date, no pharmacological treatment for anorexia has been found effective.[51]

Bulimia and binge-eating disorder, in contrast, do seem to respond well to antidepressant therapy. Drugs effective in the treatment of bulimia include tricyclic antidepressants; specific serotonin-reuptake inhibitors, such as fluoxetine (Prozac); monoamine oxidase (MAO) inhibitors; bupropion; and trazodone. Pharmacotherapy can be an effective addition to other forms of therapy.

Exercise

Fitness programs may have real value for people who are overly concerned about weight and body image. Overweight individuals who have an eating disorder can be encouraged to become more physically active as a method of weight loss or weight maintenance. In addition, the antidepressant effects of exercise may have an immediate effect on mood. Because exercise can be used as a compensatory behavior, limits must be placed on the type and the duration of exercise permitted for some individuals. Whereas strength training is valuable for building muscle mass, strengthening bone, and increasing weight, strenuous aerobic activities may cause too much weight to be lost and are not recommended in the treatment of anorexia nervosa.

What Is the Prognosis for Recovery?

Full recovery from an eating disorder varies greatly and depends not only on the type of eating disorder but also characteristics of the individual. About half of individuals with anorexia or bulimia are successfully treated.

In one study of people with anorexia nervosa, at 10-year follow-up, about 24% had recovered, 25% had normal weight and menstrual cycles but continued to battle with abnormal eating and disturbed body perceptions, 43% had some diagnostic characteristic of anorexia (low body weight, amenorrhea, and/or body image disturbance), and 7% were dead.[54] A more recent study estimated that about 50% recover from anorexia, 26% have poor outcomes, and 21% have intermediate outcomes.[55] Slightly better outcomes have been reported by clinicians using the Maudsley method. Between 50–75% of adolescents with anorexia regain expected weight by the end of treatment, and only 10–15% are seriously ill at 5-year follow-up.[53]

In anorexia nervosa, the prognosis is more favorable when the individual has the purging type rather than the restricting type. Other factors that improve prognosis include:

- Shorter duration of the disorder
- Regain of a higher percentage of ideal body weight at discharge, if hospitalized
- Fewer psychiatric comorbidities
- Onset later in adolescence

The average time to full recovery is about 6 years. Still, half of individuals with anorexia may develop bulimia.[2]

Full recovery from bulimia happens in about half of affected individuals within about 2 years.[2] Cognitive behavior therapy has a long-term effectiveness of about 60%.[56] Unfortunately, relapse is common. Six years after treatment 20–46% may still be symptomatic, and rates of substance abuse are high. A favorable prognosis for recovery is more common when:

- The individual is in a satisfactory relationship
- There is less frequent purging behavior
- Behavior change is initiated early in the treatment process

Treatment of binge-eating disorder is more effective when the focus is on binge eating rather than weight control. Six years after treatment, 80–90% of those with binge-eating disorder no longer meet criteria for the condition; however, most are still obese.[14] It is certainly easier to treat binge eating than to treat excess weight. Interestingly a substantial number of people with binge-eating disorder improve without treatment.[14]

Treatment and the prognosis for recovery are frequently hindered by health insurance restrictions. Although evidence supports a team approach to treatment, medical and psychiatric benefits are usually regarded as separate approaches by insurers. A seriously malnourished person with anorexia may initially qualify for medical hospitalization, but when weight gain begins, insurance regulations may require discharge. If the initial hospitalization is classified as psychiatric, medical and nutritional services could have minimal coverage.

Outpatient treatment is often similarly compromised. Many people have no mental health benefits, or they are restricted to a small number of outpatient visits with a cap on lifetime coverage. Co-pays are often high, which limits ability to pay. Problems are compounded when an adolescent reaches the age under which he is no longer covered under his parents plan, or if he has to leave college to obtain treatment and he subsequently loses insurance coverage. Financial ruin for the family may be the result.

Recently families have begun to fight back. There have been several successful lawsuits against health insurance companies. In addition, health care reform may improve coverage for some. Information about health insurance issues in eating disorders is available at the Web site of the National Association of Anorexia Nervosa and Associated Eating Disorders (www.anad.org).

Can Eating Disorders be Prevented?

primary prevention *Prevention efforts directed toward reducing the number of new cases of a disease or a condition.*

incidence *Number of new cases of a disease or a condition.*

This section of the chapter has shown that treatment of an eating disorder has many challenges, and the prognosis for recovery is dependent on severity of the eating disorder and access to care. The links between disordered eating and/or exercise behavior and obesity adds to the urgency of prevention.

Public health specialists refer to primary, secondary, and tertiary prevention to reduce the incidence and the prevalence of disease. These levels of prevention are defined as follows:

- **Primary prevention**—efforts are directed toward reducing the number of new cases (the **incidence**) of a disease. Primary prevention of eating disorders involves

TABLE 11-6 Primary, Secondary, and Tertiary Prevention Interventions in Eating Disorders

Primary prevention interventions	Secondary prevention interventions	Tertiary prevention interventions
Comprehensive school health education that includes nutrition education with a focus on prevention of obesity AND eating disorders	Familiarizing school health personnel and coaches with warning signs of eating disorders	Referral to appropriate treatment program for eating disorders and for nutritional and medical support
Information about eating disorders in diabetes education programs	Screening athletes for disorders of the female athlete triad	
Coaching education to prevent dangerous weight-loss practices	Developing a system of referral for individuals who need help	

secondary prevention
Prevention directed at reducing the rate of established cases of a disease or a condition.

prevalence
Number of established cases of a disease or a condition.

tertiary prevention
Prevention directed at reducing chronic conditions or disabilities associated with a particular disease or condition.

interventions that take place *before* people engage in severe dieting, excessive exercise, or other behaviors. From a public health standpoint, this is the most effective strategy for lowering the number of cases of a disease.

- **Secondary prevention**—includes interventions that aim to reduce the rate of established cases (the **prevalence**) of a disease. In the prevention of eating disorders, this involves identifying individuals at risk and intervening before behaviors get more serious or become entrenched.
- **Tertiary prevention**—the focus shifts from disease prevention to reducing other chronic conditions or disabilities associated with the disease. In eating disorders, this means not only treating behaviors but addressing osteopenia, osteoporosis, malnutrition, and other disease complications.

Table 11-6 summarizes this approach to the prevention of eating disorders.

Primary Prevention

Overweight individuals, particular adolescent girls, often engage in unhealthy behaviors to lose weight. A study of almost 5,000 adolescents in St. Paul and Minneapolis public schools uncovered an alarming rate of extreme weight control behaviors among overweight girls—18% used diet pills, laxatives, diuretics, or vomiting, and dieting and binge eating were also common.[5] Earlier in this chapter you also read results of the Centers for Disease Control and Prevention's (CDC's) Youth Risk Behavior Survey, which reported that 40% of girls and boys had dieted to lose weight or to keep from gaining weight; 11% had fasted for at least 24 hours; and 5% had used unprescribed diet pills, powders, or liquids.[6] The media—fashion magazines, television programs and movies, even video games—increases exposure to messages about the "ideal" body, which can increase anxiety about size and appearance.

As a result, primary prevention efforts that focus on adopting healthy lifestyle behaviors and preventing unhealthy behaviors are essential. For young people, this can take place in health or physical education classes. Planet Health, an obesity-prevention program described in Chapter 12, had positive effects on unhealthy weight control practices in girls. Fewer than 3% of middle school girls in intervention schools reported using diet pills or purging for weight control, whereas 6% of girls in non-intervention schools were using those practices.[4]

This chapter has established that athletes are at a high risk of eating disorders. Athletes may not know how to attain a healthy weight, and coaches may pressure student athletes to maintain a particular weight but provide no guidance on how to do this. Sports programs for children and youth provide an excellent venue for the primary prevention of eating disorders. Educated coaches and other athletic personnel are key for:

- Clarifying acceptable and unacceptable weight loss practices
- Modeling appropriate attitudes by avoiding group weigh-ins and public commentary on body size, weight, and shape
- Prohibiting excessive training techniques primarily to burn calories

Secondary Prevention

In secondary prevention, we aim to identify individuals at risk of an eating disorder and intervene early. Only professionals trained in the treatment of eating disorders should attempt to treat these conditions. However, fitness, health, and nutrition professionals can help teach others to recognize warning signs and locate sources of referral and assistance.

You can help someone with a suspected eating disorder by:

- Raising the issue of eating disorders when one is suspected. A gentle way to introduce this topic is to inquire whether a person has used diet aids, laxatives, or vomiting in the past; characterizes some foods as "bad" and avoids them; considers weight to be an issue; or has amenorrhea.
- Requiring that pre-participation physical examinations for athletes include screening for disorders of the female athlete triad (eating behaviors; amenorrhea; and repetitive musculoskeletal injuries, which might be associated with loss of bone density).
- Being supportive, caring, and helpful without nagging, criticizing, or expressing hostility.
- Encouraging individuals (and their families) to seek treatment, while realizing that you cannot force the person to get help, nor are you responsible for their failure to do so.

Treatment and Prevention of Eating Disorders: Summary

About half of people with anorexia nervosa and bulimia nervosa make a satisfactory recovery, and 80–90% of those with binge-eating disorder do so, depending on the treatment regimen and severity of the disease. Some medications are helpful in treating the psychological comorbidities that accompany eating disorders. Most effective are multidisciplinary treatment approaches that include education, nutritional therapy, and cognitive behavior therapy or psychotherapy, with involvement of the family as appropriate. Without attention to the psychological aspects of these disorders, treatment is likely to fail. Prevention of eating disorders is the best way to reduce incidence and prevalence of these conditions.

Case Study. The Cross-Country Runner, Part 3	Jean returned to college weighing 97 lb and with her leg feeling a bit better. Kay had helped Jean identify some acceptable foods to add to her diet, although Jean was still reluctant to eat too much. She attended an organizational lacrosse meeting and immediately liked Coach Martin, who talked to the young women about the importance of eating well for strength and energy. After the meeting Jean told Coach Martin that she was recovering from a stress fracture and was seeing a sports nutritionist to try to normalize her eating. Coach Martin encouraged Jean to continue seeing the nutritionist at home but to also talk with the athletic trainer.

As the spring semester progressed, Jean began falling back into her habits of eating very little, training a lot, and binging and vomiting several times a week. One night at a team party, where Jean gorged on pizza and ice cream, her lacrosse teammate Casey caught her throwing up in the bathroom. "This is crazy, Jean," said Casey. "I'm taking you to see the trainer tomorrow so you can get straightened out." The next day, when Jean met with the trainer and Coach Martin, she was exhausted and wanted help.

- What approach might Jean's track coach have taken to support the health of his runners?
- What treatment approaches might help Jean recover?

REFERENCES

1. Golden, N. H., Katzman, D. K., Kreipe, R. E., Stevens, S. L., Sawyer, S. M., Rees, J., et al. (2003). Eating disorders in adolescents: Position paper of the Society for Adolescent Medicine. *Journal of Adolescent Health, 33*(6), 496–503.
2. Rome, E. S., & Ammerman, S. (2003). Medical complications of eating disorders: An update. *Journal of Adolescent Health, 33*(6), 418–426.
3. Battle, E. K., & Brownell, K. D. (1996). Confronting a rising tide of eating disorders and obesity: Treatment vs. prevention and policy. *Addictive Behaviors, 21*(6), 755–765.
4. Austin, S. B., Field, A. E., Wiecha, J., Peterson, K. E., & Gortmaker, S. L. (2005). The impact of a school-based obesity prevention trial on disordered weight-control behaviors in early adolescent girls. *Archives of Pediatrics & Adolescent Medicine, 159*(3), 225–230.
5. Neumark-Sztainer, D., Story, M., Hannan, P. J., Perry, C. L., & Irving, L. M. (2002). Weight-related concerns and behaviors among overweight and nonoverweight adolescents: Implications for preventing weight-related disorders. *Archives of Pediatrics & Adolescent Medicine, 156*(2), 171–178.
6. Eaton, D. K., Kann, L., Kinchen, S., Shanklin, S., Ross, J., Hawkins, J., et al. (2010). Youth risk behavior surveillance – United States, 2009. *Surveillance Summaries, MMWR, 59*(SS–5), 1–148.
7. American Psychiatric Association. (1994). *Diagnostic and statistical manual of mental disorders (DSM-IV)* (4th ed.). Washington, D.C.: American Psychiatric Association.
8. Attia, E., & Roberto, C. A. (2009). Should amenorrhea be a diagnostic criterion for anorexia nervosa? *International Journal of Eating Disorders, 42*(7), 581–589.
9. Mehler, P. S., & Krantz, M. (2003). Anorexia nervosa medical issues. *Journal of Women's Health, 12*(4), 331–340.
10. Feldman, M. B., & Meyer, I. H. (2007). Eating disorders in diverse lesbian, gay, and bisexual populations. *International Journal of Eating Disorders, 40*(3), 218–226.
11. Kaye, W. H., Weltzin, T. E., Hsu, L. K., McConaha, C. W., & Bolton, B. (1993). Amount of calories retained after binge eating and vomiting. *American Journal of Psychiatry, 150*(6), 969–971.
12. Bruce, B., & Wilfley, D. (1996). Binge eating among the overweight population: A serious and prevalent problem. *Journal of the American Dietetic Association, 96*(1), 58–61.
13. de Zwaan, M., Mitchell, J. E., Howell, L. M., Monson, N., Swan-Kremeier, L., Crosby, R. D., et al. (2003). Characteristics of morbidly obese patients before gastric bypass surgery. *Comprehensive Psychiatry, 44*(5), 428–434.

14. Dingemans, A. E., Bruna, M. J., & van Furth, E. F. (2002). Binge eating disorder: A review. *International Journal of Obesity and Related Metabolic Disorders, 26*(3), 299–307.

15. Spitzer, R. L., Stunkard, A., Yanovski, S., Marcus, M. D., Wadden, T., Wing, R., et al. (1993). Binge eating disorder should be included in *DSM-IV*: A reply to Fairburn et al.'s "The classification of recurrent overeating: The binge eating disorder proposal". *International Journal of Eating Disorders, 13*(2), 161–169.

16. National Heart, Lung, and Blood Institute. (1998). *Clinical guidelines on the identification, evaluation, and treatment for overweight and obesity in adults: The evidence report.* Bethesda, MD: National Institutes of Health.

17. Mills, M. E. (2007). Craving more than food: The implications of pica in pregnancy. *Nursing for Women's Health, 11*(3), 266–273.

18. Gupta, A., Rajput, S., Maduabuchi, G., & Kumar, P. (2007). Sponge eating: Is it an obsessive compulsive disorder or an unusual form of pica? *Acta Paediatrica, 96*(12), 1853–1854.

19. Adler, A. I., & Olscamp, A. (1995). Toxic 'sock' syndrome bezoar formation and pancreatitis associated with iron deficiency and pica. *Western Journal of Medicine, 163*(5), 480–482.

20. Pawa, S., Khalifa, A. J., Ehrinpreis, M. N., Schiffer, C. A., & Siddiqui, F. A. (2008). Zinc toxicity from massive and prolonged coin ingestion in an adult. *American Journal of the Medical Sciences, 336*(5), 430–433.

21. Stunkard, A. J., Allison, K. C., Geliebter, A., Lundgren, J. D., Gluck, M. E., & O'Reardon, J. P. (2009). Development of criteria for a diagnosis: Lessons from the night eating syndrome. *Comprehensive Psychiatry, 50*(5), 391–399.

22. Winkelman, J. W. (1998). Clinical and polysomnographic features of sleep-related eating disorder. *Journal of Clinical Psychiatry, 59*(1), 14–19.

23. American Academy of Sleep Medicine. (2001). *The international classification of sleep disorders, revised. Diagnostic and coding manual.* Westchester, IL: American Academy of Sleep Medicine. Available from esst.org/adds/ICSD.pdf. Accessed June 30, 2010.

24. Birketvedt, G. S., Florholmen, J., Sundsfjord, J., Osterud, B., Dinges, D., Bilker, W., et al. (1999). Behavioral and neuroendocrine characteristics of the night-eating syndrome. *JAMA, 282*(7), 657–663.

25. Goel, N., Stunkard, A. J., Rogers, N. L., Van Dongen, H. P., Allison, K. C., O'Reardon, J. P., et al. (2009). Circadian rhythm profiles in women with night eating syndrome. *Journal of Biological Rhythms, 24*(1), 85–94.

26. Olivardia, R., Pope, H. G., Jr., & Hudson, J. I. (2000). Muscle dysmorphia in male weightlifters: A case-control study. *American Journal of Psychiatry, 157*(8), 1291–1296.

27. Abell, T. L., Adams, K. A., Boles, R. G., Bousvaros, A., Chong, S. K., Fleisher, D. R., et al. (2008). Cyclic vomiting syndrome in adults. *Neurogastroenterology and Motility, 20*(4), 269–284.

28. Regard, M., & Landis, T. (1997). "Gourmand syndrome": Eating passion associated with right anterior lesions. *Neurology, 48*(5), 1185–1190.

29. Dreisbach, S. (2009, March 23). Exclusive body-image survey: 16,000 women tell their body confidence secrets. *Glamour Magazine.* Available from http://www.glamour.com/health-fitness/2009/03/women-tell-their-body-confidence-secrets. Accessed June 30, 2010.

30. Cash, T. F., & Henry, P. E. (1995). Women's body images: The results of a national survey in the USA. *Sex Roles, 33*(1/2), 19–28.

31. Varner, L. M. (1995). Dual diagnosis: Patients with eating and substance-related disorders. *Journal of the American Dietetic Association, 95*(2), 224–225.

32. Misra, M., Miller, K. K., Tsai, P., Gallagher, K., Lin, A., Lee, N., et al. (2006). Elevated peptide YY levels in adolescent girls with anorexia nervosa. *Journal of Clinical Endocrinology Metabolism, 91*(3), 1027–1033.

33. Fetissov, S. O., Hamze Sinno, M., Coquerel, Q., Do Rego, J. C., Coeffier, M., Gilbert, D., et al. (2008). Emerging role of autoantibodies against appetite-regulating neuropeptides in eating disorders. *Nutrition, 24*(9), 854–859.

34. Fetissov, S. O., Harro, J., Jaanisk, M., Järv, A., Podar, I., Allik, J., et al. (2005). Autoantibodies against neuropeptides are associated with psychological traits in eating disorders. *Proceedings of the National Academy of Sciences of the United States of America, 102*(41), 14865–14870.

35. Brewerton, T. D., Lydiard, R. B., Laraia, M. T., Shook, J. E., & Ballenger, J. C. (1992). CSF beta-endorphin and dynorphin in bulimia nervosa. *American Journal of Psychiatry, 149*(8), 1086–1090.

36. Gillman, M. A., & Lichtigfeld, F. J. (1986). The opioids, dopamine, cholecystokinin, and eating disorders. *Clinical Neuropharmacology, 9*(1), 91–97.

37. Neumeister, A., Winkler, A., & Wober-Bingol, C. (1999). Addition of naltrexone to fluoxetine in the treatment of binge eating disorder. *American Journal of Psychiatry, 156*(5), 797.

38. Wilmshurst, L. (2004). *Child and adolescent psychopathology: A casebook.* Thousand Oaks, CA: Sage Publications.

39. Kinzl, J. F., Traweger, C., Guenther, V., & Biebl, W. (1994). Family background and sexual abuse associated with eating disorders. *American Journal of Psychiatry, 151*(8), 1127–1131.

40. Fairburn, C. G., Welch, S. L., Doll, H. A., Davies, B. A., & O'Connor, M. E. (1997). Risk factors for bulimia nervosa. A community-based case-control study. *Archives of General Psychiatry, 54*(6), 509–517.

41. Fairburn, C. G., Doll, H. A., Welch, S. L., Hay, P. J., Davies, B. A., & O'Connor, M. E. (1998). Risk factors for binge eating disorder: A community-based, case-control study. *Archives of General Psychiatry, 55*(5), 425–432.

42. Pate, J. E., Pumariega, A. J., Hester, C., & Garner, D. M. (1992). Cross-cultural patterns in eating disorders: A review. *Journal of the American Academy of Child and Adolescent Psychiatry, 31*(5), 802–809.

43. American Academy of Pediatrics. (2005). Promotion of healthy weight-control practices in young athletes. *Pediatrics, 116*(6), 1557–1564.

44. Greenleaf, C., Petrie, T. A., Carter, J., & Reel, J. J. (2009). Female collegiate athletes: Prevalence of eating disorders and disordered eating behaviors. *Journal of American College Health, 57*(5), 489–495.

45. Loosli, A. R., & Ruud, J. S. (1998). Meatless diets in female athletes: A red flag. *Physician and Sportsmedicine, 26*(11), 45–48, 55.

46. Sanborn, C. F., Horea, M., Siemers, B. J., & Dieringer, K. I. (2000). Disordered eating and the female athlete triad. *Clinics in Sports Medicine, 19*(2), 199–213.

47. Jacobi, C., Hayward, C., de Zwaan, M., Kraemer, H. C., & Agras, W. S. (2004). Coming to terms with risk factors for eating disorders: Application of risk terminology and suggestions for a general taxonomy. *Psychological Bulletin, 130*(1), 19–65.

48. Rydall, A. C., Rodin, G. M., Olmsted, M. P., Devenyi, R. G., & Daneman, D. (1997). Disordered eating behavior and microvascular complications in young women with insulin-dependent diabetes mellitus. *New England Journal of Medicine, 336*(26), 1849–1854.

49. Rome, E. S., Ammerman, S., Rosen, D. S., Keller, R. J., Lock, J., Mammel, K. A., et al. (2003). Children and adolescents with eating disorders: The state of the art. *Pediatrics, 111*(1), e98–e108.

50. Wiseman, C. V., Sunday, S. R., Klapper, F., Harris, W. A., & Halmi, K. A. (2001). Changing patterns of hospitalization in eating disorder patients. *International Journal of Eating Disorders, 30*(1), 69–74.

51. ADA. (2006). Position of the American Dietetic Association: Nutrition intervention in the treatment of anorexia nervosa, bulimia nervosa, and other eating disorders. *Journal of the American Dietetic Association, 106*, 2073–2082.

52. Kohn, M. R., Golden, N. H., & Shenker, I. R. (1998). Cardiac arrest and delirium: Presentations of the refeeding syndrome in severely malnourished adolescents with anorexia nervosa. *Journal of Adolescent Health, 22*(3), 239–243.

53. leGrange, D., & Eisler, I. (2008). Family interventions in adolescents with anorexia nervosa. *Child and Adolescent Psychiatric Clinics of North America, 18*, 159–173.

54. Eckert, E. D., Halmi, K. A., Marchi, P., Grove, W., & Crosby, R. (1995). Ten-year follow-up of anorexia nervosa: Clinical course and outcome. *Psychological Medicine, 25*(1), 143–156.

55. Zipfel, S., Lowe, B., Reas, D. L., Deter, H. C., & Herzog, W. (2000). Long–term prognosis in anorexia nervosa: Lessons from a 21-year follow-up study. *Lancet, 355*(9205), 721–722.

56. Fairburn, C. G., Norman, P. A., Welch, S. L., O'Connor, M. E., Doll, H. A., & Peveler, R. C. (1995). A prospective study of outcome in bulimia nervosa and the long-term effects of three psychological treatments. *Archives of General Psychiatry, 52*(4), 304–312.

Prevention of Childhood Obesity

R. Gino Santa Maria/Shutterstock.com

CHAPTER OUTLINE

cappi thompson/Shutterstock.com

Iorga Studio/Shutterstock.com

Y ou are now aware that obesity is becoming increasingly prevalent in the U.S. population at all ages, that obesity and several chronic health problems are linked, and that treating obesity is challenging. Genetic factors may promote the development of obesity by affecting the drive to eat, satiety, energy-conserving mechanisms, and fat storage. Social, cultural, and environmental forces encourage the consumption of high-calorie foods and discourage physical activity. Children are not immune to these forces and factors.

THE EXTENT OF CHILDHOOD OVERWEIGHT AND OBESITY

Childhood obesity has become a matter of grave concern in both the United States and many countries around the world. In the United States, 20% of 6- to 11-year-old children and 18% of 12- to 19-year old adolescents are obese (having a BMI-for-age at or above the 95th percentile).[1] Refer back to Table 1-4 and you will note that these rates increased almost 3% for children since 2003 but have begun to stabilize for adolescents. Nevertheless, since 1976, obesity rates have tripled in children (see Figure 12-1).

A relatively new phenomenon is the presence of obesity at much earlier ages. Almost 10% of U.S. children up to age 2 years are in the 95th percentile for weight-to-recumbent-length.[1] About 18% of 4-year-old children in the United States are overweight.[2] Racial, ethnic, and socioeconomic differences in obesity prevalence are apparent even in preschoolers.[1] The Pediatric Nutrition Surveillance Survey found that 12.4% of 2- to 5-year-old low-income children were obese in 1998, 14.5% in 2003, and 14.6% in 2008.[2] Table 1-4 in Chapter 1 illustrates the higher prevalence of obesity among African American, Mexican American, and Hispanic children and adolescents at all ages.

This is not exclusively a U.S. issue. Globally, as many as 10% of school-age children are overweight or obese.[3]

Overweight in childhood is considered to be a good predictor of overweight in adolescence and adulthood[4] and accounts for about 20% of adult overweight and obesity.[5] The BMI at even 6 to 8 years of age may predict BMI later in life.[6,7] There are also some indications that obesity beginning at earlier ages, and continuing into adulthood, results in higher adult BMIs than in adult-onset obesity.[8]

Healthy-weight children can become obese adults, too. The Bogalusa Heart Study, a long-term epidemiologic study of children who attended school in Bogalusa, La., between 1973 and 1991, gives us a good picture of this. Over one-third of study participants who were healthy weight in childhood became overweight in young adulthood, and the greatest shift from healthy weight to overweight from childhood to young adulthood was among African Americans.[9]

Increased rates of childhood overweight and obesity are poised to have a profound negative effect on health and longevity, possibly lowering life expectancy for the first time in decades.[10] Prevention and early intervention are clearly an imperative.

PREVENTION CONCEPTS

You learned in Chapter 11 about primary, secondary, and tertiary prevention to reduce the incidence and the prevalence of eating disorders. These same public health concepts can be applied to identifying risk factors and designing prevention interventions for obesity.

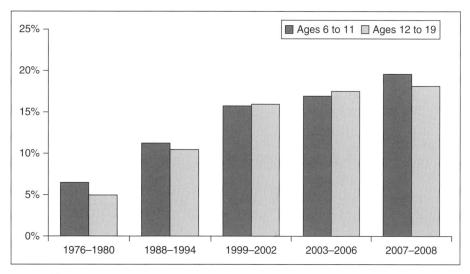

FIGURE 12-1 Prevalence of obesity among U.S. children and adolescents, 1976–2008.

How Does the Public Health Model of Prevention Apply to Obesity?

The three levels of prevention can guide obesity prevention:

- **Primary prevention** of obesity involves interventions that take place before children become overweight or obese. From a public health standpoint, this is the most effective strategy for lowering the number of cases of obesity in the U.S. population.
- **Secondary prevention** involves identifying overweight or obese children and intervening before BMI increases further.
- **Tertiary prevention** of obesity not only aims at slowing excess weight gain but also preventing complications of overweight, such as heart disease, type 2 diabetes, metabolic syndrome, and others.

Table 12-1 summarizes these concepts with applications to childhood obesity.

What Are the Risk Factors for Childhood Obesity?

Before looking at interventions that might be effective in obesity prevention, consider the factors that have contributed to the epidemic of childhood obesity. Four are examined here: parental influences—particularly those that are behavioral, environmental, and cultural; inactivity; television, computers, and other electronic media; and diet.

Fat Children and Fat Parents

The risk of both adult and childhood obesity is greatest in children with overweight parents.[11] A longitudinal study in Great Britain reported that the risk of obesity at age

TABLE 12-1 Primary, Secondary, and Tertiary Preventions in Obesity

Primary prevention interventions	Secondary prevention interventions	Tertiary prevention interventions
Limited "screen" time	Family-based weight management programs that include nutrition, activity, and behavior	Family-based weight management programs aimed at weight maintenance so children can grow into their weight
Quality, daily physical education in all schools		
School food services that reinforce nutrition education messages		· DASH diet
No food advertising on school premises	School-based programs that promote healthy diet and physical activity for all children	Pharmacological treatments for abnormal blood lipids, type 2 diabetes, hypertension, and obesity
Accessible community programs that encourage physical activity		
Walk/bike to school programs		Bariatric surgery
Sidewalks and bikeways		
Limits on food marketing to children		
Restaurant menu labeling		

7 years was increased when a child had one overweight parent, but was highest when both parents were obese.[12] This is partly due to genetic factors and partly due to the environment and behaviors shared by children and their parents.

Chapter 4 discussed how obese parents may pass along genes that promote obesity when the environment is favorable, such as when parents model obesity-promoting eating behaviors and inactivity or allow ready access to high-fat foods. Families also share cultural values that shape dietary choices and acceptance of weight; attitudes and beliefs about food and physical activity; and inappropriate feeding practices that start at an early age. Family income is a factor, as lower income families living in **food deserts** might not have access to supermarkets or fresh fruits and vegetables, or be able to afford healthier foods. The prenatal environment may also play a role. Children born to mothers who were obese early in their pregnancy have twice the risk of obesity at age 2 to 4 years.[13] This may be due to higher circulating insulin levels or even to early proliferation of fat cells.

food deserts

Areas, often in low-income communities and neighborhoods, where there is limited access to healthy foods.

Low Energy Expenditure

In 1953, Jean Mayer was already cautioning that physical inactivity is a factor in the development of obesity.[14] Many experts continue to believe that lack of physical activity is more important than a high-energy diet in explaining increased rates of obesity in the United States. While we think of children as being very active—and most are—a substantial proportion of them have limited activity, both inside and outside of school.

According to latest School Health Policies and Programs Study (SHPPS), 21% of the nation's schools do not require physical education (31% of elementary schools, 16% of middle schools, and 5% of high schools).[15] Of those schools that do require physical education, only 14–15% of elementary and middle schools and 3% of high schools provide it at least 3 days per week for the entire school year for students in all grades. About 20% of high schools allow students to substitute other activities, including band and chorus, for physical education.[15]

Engagement in out-of-school physical activity may also be limited. A 2002 Centers for Disease Control and Prevention (CDC) survey found that only 38% of 9- to 13-year-olds participated in organized physical activity during their nonschool hours, and that 23% did not engage in any free-time physical activity.[16]

The time of greatest risk for developing a pattern of inactivity is adolescence. The CDC's 2009 Youth Risk Behavior Surveillance System (YRBSS) determined that males are generally more active than females but that activity levels of all young people decrease during adolescence.[17] Here are some of the CDC's other findings about activity levels of ninth to twelfth graders:

- About 37% of students met recommended levels of physical activity by engaging in moderate to vigorous activity for at least 60 minutes per day, 5 days per week. More boys (46%) than girls (28%) met the recommendation.
- Almost one-quarter of students did not participate in an hour or more of physical activity on any day during the week before the survey.
- White students were more likely to have met recommended levels of physical activity than black and Hispanic students.
- A quarter of students used computers for something other than schoolwork 3 or more hours per day on an average weekday, and this was more common among boys than girls.
- 33% of students watched television 3 or more hours per day on an average school day.
- About 72% of ninth graders attended physical education classes, while 44% of twelfth graders did. One-third of ninth to twelfth graders had daily physical education, up from 25% in the 1995 YRBSS.
- Just over half of students had played on a school or community sports team, 64% of boys and 52% of girls.

Young people are inactive for many of the same reasons that adults are. Modern conveniences have reduced the necessity to expend energy on everyday tasks, more people have cars, and many neighborhoods are pedestrian unfriendly. In households where both parents or caregivers work, children may have fewer opportunities to participate in physical activities after school because they lack transportation or have safety concerns. Economic factors may keep some from activities that charge a fee or require equipment. Available opportunities for sports participation typically decline as children get older.

This trend should concern health professionals. Late childhood and early adolescence are pivotal periods for developing habits that may prevent chronic diseases later in life. For example, arterial lesions that mark the beginning of coronary artery disease make an appearance during this time. Bone density increases at this point in life, stimulated not only by diet but by activity, giving protection against osteoporosis in adulthood. Physically active children tend to be leaner than their inactive peers. Physical activity can also improve physical fitness and enhance psychological well-being. According to findings from the National Longitudinal Study of Adolescent Health, physical activity and—even more significantly—physical inactivity, track into young adulthood.[18] Since physical activity levels have already declined by the time most children reach adolescence, prevention efforts in this area must begin early.

Television, Computers, and Video Games

Electronic devices play a role—and a complicated one—in childhood obesity. In its study of the role of media in the lives of children and adolescents, a Kaiser Family Foundation report observed, "The term 'media rich' does not fully capture this environment."[19] The report describes the typical 8- to18-year-old U.S. child in 2004 as living in a home with three television sets, three VCRs, three radios, three CD/tape players, two video game consoles, and a personal computer. Five years later, we could add to that DVDs, portable video games, cell phones, and netbooks. This typical child watches screen media (television and VCR/DVD) 4¼ hours per day and spends just over an hour daily in recreational computer time.

Use of screen media is not limited to older children. A study of preschoolers reported daily television viewing by 63% of those ages 6 months to 2 years (82% of 3- to 4-year-olds, and 78% of 5- to 6-year-olds).[20] On a typical day, the average preschool child watches television for 1 hour and 19 minutes. Twenty percent of 3- to 4-year-olds and 27% of 5- to 6-year-olds use the computer on a typical day, for an average of 50 minutes, and 71% of 5- to 6-year-olds can use a mouse. Up to one-third of the preschoolers in this study had televisions in their bedrooms.

Overweight children watch more television than their lean peers.[21,22] Television viewing is linked to increased risk of obesity.[13] Skinfold thicknesses of children followed from preschool into early adolescence were greater among those who watched 3 or more hours of television each day than those who watched less than 1¾ hours per day.[23] Similarly, the BMI growth trajectory of children from kindergarten to grade 5 is directly linked to hours of television watched each day.[24] The child who watches 4 hours, as opposed to 1 hour per day, could gain enough BMI units to propel him or her up to or over the 85th percentile by age 10–11 years.

While logic suggests that increased screen time results in less physical activity, thereby promoting weight gain, the relationship between television, physical activity, and obesity is likely more complicated than that. A small study of nonoverweight 12- to 16-year-olds found that when sedentary behaviors were targeted and decreased (mainly by reductions in television and computer time), energy intake decreased by 17% and total fat intakes decreased by 31%.[25]

Food is the most highly advertised product on television, and high-fat, high-sugar foods are advertised the most. More than half of televised advertising during children's shows is food related, and about 80% is for convenience foods, fast foods, and sweets.[26] An Expert Committee of the Institute of Medicine found that food marketing through television influences children's food preferences, purchase requests, and diets.[27]

On a positive note, there are now some active electronic games on the market that are quite popular. Children expend considerably more energy when playing active electronic games than seated games.[28] It remains to be seen if children can use these games to meet physical activity recommendations.

Diet and Weight Gain

Like physical inactivity, dietary habits that promote weight gain start early in life, and dietary trends for children and adolescents are not good. Data from the U.S. Department of Agriculture's Continuing Surveys of Food Intakes by Individuals, 1989–1998, and the NHANES (National Health and Nutrition Examination Survey) 2005–2006 indicate that 6- to 19-year olds obtain about one-third of their daily caloric intake from fat, and the typical child or adolescent exceeds current recommendations for saturated fat (10% of total calories).[29,30,31] Intake of sugar-sweetened beverages has increased dramatically over the past 20 years, accounting for as much as 10–15% of children's and adolescent's daily caloric intake.[32] As soda consumption rises, milk consumption has declined; intake of grain-based snack foods, such as chips, popcorn, and pretzels, has increased, while whole grain consumption has dropped; and caloric intake has increased for both children (6–11 years old) and adolescents (12–19 years old).

Several behavioral and environmental factors contribute to these dietary trends and may be promoting increased rates of childhood obesity:

- Fast food: Children and adolescents are big consumers of fast-food products, many of which are high in fat and calories and low in price. Fried chicken, chicken nuggets, burgers, French fries, macaroni and cheese, and soda predominate on chain restaurants' children's meals. According to a Center for Science in the Public Interest (CSPI) study 93% of children's meals in chain restaurants exceed 430 calories, 86% are high in sodium, and 45% are high in saturated fat.[33] Perhaps not surprisingly, when children and adolescents eat at restaurants, they consume significantly more energy from fat and saturated fat than when they eat at other places.[34] There is some evidence that overweight adolescents not only consume more fast food but are less likely than their lean peers to cut back on food consumption during the rest of the day to compensate.[35]

- Sugar-sweetened beverages: Sodas and fruit drinks provide calories but virtually no nutrients. People who drink these beverages generally do not compensate for the excess calories by eating less at subsequent meals. One study estimated that children's risk of becoming obese increased 1.6 times for every sugar-sweetened beverage consumed in a day.[36]

- Food costs: A study in Seattle supermarkets revealed that while the price of low energy dense foods (mainly fruits and vegetables) increased 20% over two years, the cost of energy dense foods (those highest in fat and sugar) remained unchanged.[37] Individuals from lowest socioeconomic strata eat foods that cost less, and many of these are energy dense and of low nutritional value.[38]

- Eating while watching television: Compared with children who watch little television, children and adolescents who watch more television are also more likely to have unhealthy eating habits and unhealthy beliefs about food. They tend to consume more fat.[39]

- Microwaveable meals and snacks: The popularity of microwave ovens among busy American families has greatly expanded the range of quickly prepared meals and snacks. Because microwave ovens are safe and easy to use, children can take a more active role in preparing food. Foods that once were reserved for mealtimes—like pizza, macaroni and cheese, and hot sandwiches—can now be prepared as snacks, even by young children.

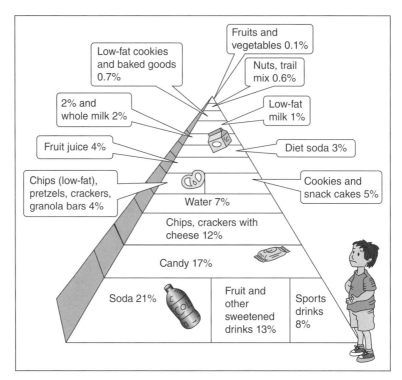

FIGURE 12-2 School vending machine pyramid.

Source: Center for Science in the Public Interest.

competitive foods *Foods sold in schools outside the oversight of the National School Lunch and School Breakfast Programs, such as vending machine, snack bars, à la carte cafeteria lines, and fund-raisers.*

- Food served in schools: Federal regulations govern the quality of school meals provided under the National School Lunch Program (NSLP) and the School Breakfast Program (SBP). However, foods are available at school from many additional sources—cafeteria à la carte items; vending machines; school stores, canteens, or snack bars; fund-raising items. These foods, known as **competitive foods**, are largely exempt from regulations. Not unexpectedly, such foods tend to be high in fat, sodium, and added sugars as depicted in Figure 12-2. During the 2004–2005 school year, almost 90% of schools offered competitive foods.[40] Most high schools, more than half of middle schools, and almost one-quarter of elementary schools have at least one vending machine available to students.[41] Although nearly one-third of the states require that schools prohibit students from using vending machines for part of the school day, Table 12-2 illustrates that a considerable proportion of students continue to have access to vending machines, school stores, canteens, and snack bars.

Prevention Concepts: Summary

Principles of primary, secondary, and tertiary prevention can be applied to preventing childhood obesity and the health complications of overweight. Primary prevention efforts should start as early in life as possible, particularly in families where one or both parents are overweight. Inactivity, excessive use of television and computers, and poor diet (especially sugar-sweetened beverages and foods high in fat) are key factors promoting high rates of overweight in children and adolescents.

TABLE 12-2   Sources of Competitive Foods Available to U.S. School Children

	Elementary schools	Middle schools	High schools
	(percentage)		
Have one or more vending machines	21.1	62.4	85.8
Have school store, canteen, or snack bar	16.7	33.0	50.1
Allow students to purchase foods and beverages high in fat, sodium, or added sugars from vending, school store, canteen, or snack bar ...			
... during lunch	11.9	25.4	48.0
... before classes begin	7.3	27.7	64.0
... during school hours when meals not being served	6.6	20.6	51.3

Source: Data from O'Toole, T. P., Anderson, S., Miller, C., & Guthrie, J. (2007). Nutrition services and foods and beverages available at school: Results from the School Health Policies and Programs Study 2006. *The Journal of School Health, 77*(8), 500–521.

Case Study. Penny and Pablo, Part 1

Penny is a 4-year-old Caucasian girl who lives with her parents in single-family home in a middle class suburb of a major metropolitan area. Penny's father is an accountant, who commutes into the city by car, and her mother works part-time as an editor. Penny is dropped off by car at a nearby preschool while her mother is at work. There are four television sets in the home, including one in Penny's bedroom, and there is generally at least one television set turned on whenever someone is at home. Penny's parents are both slightly overweight, and the family eats out several times a week. There is lots of pre-prepared and frozen food in the house, but neither adult pays much attention to meal planning or nutritional guidelines. Penny's father leaves most household decisions to his wife. Penny has eaten solid food since she was four months old and is allowed to have unlimited amounts of 100% fruit juice but no soda. Penny's father works long hours but belongs to a health club in the city and sometimes stays even later after work to exercise. Her mother diets constantly and occasionally walks in the neighborhood. Penny weighs 44 lbs and is 41 inches tall.

Pablo lives with his parents and older brother in a 2-bedroom apartment in the city where Penny's father works. He is also 4 years old. His parents are both Hispanic, having come from Guatemala six years ago. Pablo's father speaks some English and works seasonally in construction; his mother's English is slightly better, and she is a full-time home health aide for two elderly residents of an assisted living facility in the city. The family does not have a car but lives near public transportation. The household has one television set in the living/dining room, and the set is on all the time. When both of his parents are working Pablo is looked after by his brother or a neighbor. There is a city park within walking distance, but Pablo is not allowed to go there without his

(Continued)

Case Study.
(Continued)

parents. The family tries to go to the park on Saturdays when the weather is good. Pablo's father is very lean and his mother is slightly overweight. Pablo's mother cooks most of their meals, although it is sometimes late before she gets home; she also tries to cook staple foods ahead on weekends. The family eats inexpensive traditional foods that can be purchased from small Guatemalan markets or convenience stores in the city, and when Pablo's father is working, they occasionally eat at a nearby fast food restaurant. Pablo weighs 46 lbs and is 42 inches tall.

1. Calculate Penny's and Pablo's BMI and use the tables in Appendix A to determine their BMI percentile.
2. What kinds of foods make up a traditional Guatemalan diet?
3. For both Penny and Pablo, what risk factors for obesity seem particularly apparent?
4. What obesity-preventing opportunities exist in their family situations?

PRIMARY PREVENTION OF CHILDHOOD OBESITY: CHILDHOOD PHYSICAL ACTIVITY AND NUTRITION

The most effective way to reduce the number of new cases of obesity is to start prevention interventions early in life—when health habits, beliefs, and attitudes are being established. This section of the chapter explores prevention interventions in the areas of physical activity and nutrition on three levels: (1) the individual level, which focuses on the home and family; (2) the environmental level, which aims at settings outside the home where children spend a great deal of time, most notably the school and community; and (3) the policy level, which involves governmental regulations that target such things as the food industry and the built environment.

Level two (the environment) and level three (policy) approaches to obesity prevention are sometimes called "upstream" approaches. These tend to reach the most people, including disadvantaged groups, and are the most cost-effective. As interventions become a part of the structure of a community, behavior change by individuals is more likely. Individual interventions are called "downstream" approaches. These can be quite effective with children and adolescents, but they are sometimes more costly than upstream approaches, they don't reach as many people, and they are less likely to change the behavior of a population. Figure 12-3 illustrates these concepts.

How Much Physical Activity Do Children Need?

Physical activity can have a powerful impact on preventing obesity. In our modern obesogenic society, where a child or adult would be challenged to avoid excess caloric intake, failure to engage in some sort of physical activity is almost certain to lead to weight gain.

Children should be physically active every day. In addition to preventing obesity, physical activity promotes many aspects of physical and mental health, as you have seen in Chapters 7 and 8. The 2005 U.S. Dietary Guidelines for Americans recommend at least 60 minutes of moderate-to-vigorous physical activity on most, if not every, day for children, and the 2010 Guidelines will most certainly continue activity recommendations.

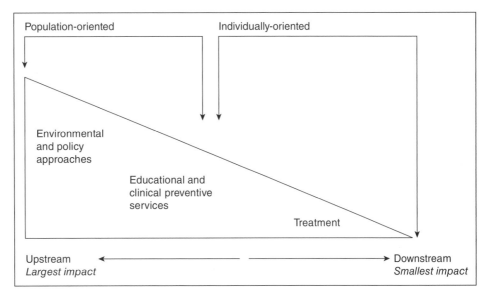

FIGURE 12-3 The prevention-to-treatment continuum.

Source: Reprinted with Permission Circulation. 2005; 111: 1999–2012 © 2005 American Heart Association, Inc.

The Role of Parents in Assuring Adequate Physical Activity

As already discussed in this chapter, obesity prevention needs to begin early in life, before children start school. While school and community settings ultimately reach more children, during the preschool years the home is central to this effort. Here are some ways that parents can help:

- Limit screen time. Parents who spend more time playing with and reading to their children and who restrict the time that children spend watching television and playing electronic games not only help prevent weight gain but also promote language development and socialization. The American Academy of Pediatrics (AAP) recommends that parents of children under the age of 2 years discourage them from watching television and that older children watch television no more than 2 hours a day. The AAP also strongly recommends against children having television sets in their bedrooms (Table 12-3).[42]
- Physical activity. Parental activity influences children's participation in physical activity.[43] Parents can help by finding time every day to play actively with their children, by providing opportunities for activity, and—perhaps most importantly–encouraging their children to be active.

The Role of Schools in Providing Physical Activity

Schools are logical places for interventions aimed at the primary prevention of obesity. Schools employ educators and other personnel who have the skills—or who can be trained—to enhance knowledge and change behavior. With approximately two-thirds of adults either mostly or completely sedentary, many children will not get enough opportunity or encouragement to be active at home.

TABLE 12-3 Summary of Physical Activity Recommendations for Children

Physical activity recommendations for children
• At least 60 minutes of moderate-to-vigorous physical activity on most, if not every day (*Source: Dietary Guidelines for Americans*)
• Little or no television for children under 2 years old (*Source: American Academy of Pediatrics*)
• No more than 2 hours of television each day for older children (*Source: American Academy of Pediatrics*)
• No television sets in children's bedrooms (*Source: American Academy of Pediatrics*)
• At least 30 minutes of activity available during the school day (*Source: Institute of Medicine*)
• Daily physical education for all children, grades K–12, for the entire school year (*Source: Various groups*)

The Institute of Medicine recommends that at least 30 minutes of physical activity be available during the school day.[44] This can take many forms in a school setting: physical education, recess, exercise breaks, walk/bike to school, and before- and after-school programs. Anyone who has observed a school playground knows that children are not necessarily active during recess, and classroom exercise breaks offer limited time and space for movement. Ideally, school-age children will meet most recommendations for physical activity through quality, daily physical education in school.

The Task Force on Community Preventive Services, an independent group of volunteer public health and prevention experts, has identified school-based physical education as a recommended intervention for increasing children's and adolescent's physical activity.[45] In addition, the American Dietetic Association's panel of experts has found fair evidence that increasing physical activity opportunities, including physical education, at schools, can reduce body fat and help children achieve healthy weight.[46] By one estimate, offering just one hour a week of additional physical education instruction to children in kindergarten and first grade could reduce by 10% the number of overweight 5- and 6-year-old girls.[47] The National Association of Sport and Physical Education (NASPE), the American Heart Association (AHA), the American Academy of Pediatrics, the National Association of State Boards of Education, and the American College of Sports Medicine support federal recommendations for daily physical education for all children, from kindergarten through grade 12 for the entire school year.

State data are beginning to show that fitter children have higher academic achievement. New York City recently published findings that children in all weight categories who had higher physical fitness ratings also scored higher on standardized achievement tests.[48] So far it is impossible to say if fitness results in better test scores or if children who score better on achievement tests are already more fit. But these findings may help build a case for physical education as an adjunct to learning.

The Role of Community-Based Physical Activity Programs

Without adequate amounts of physical education, many children cannot meet most of their daily physical activity requirements during school. The availability of community-based programs and the encouragement of physical activity by parents is critical.

Accessible community physical activity programs complement school-based programs and, in communities where children do not take physical education, may be the child's primary source of activity. If we want children to be active, then we must provide them with access to programs, facilities, and spaces. Examples of community efforts that can increase physical activity include:

- Affordable, accessible intramural programs sponsored by community organizations
- Community parks, recreation centers, fitness trails, bike paths, and sidewalks
- Noncompetitive alternatives, such as walking clubs, inline skating, jump-rope teams, and water aerobics
- Arrangements with schools to use facilities before and after school hours and during school vacations
- Community fitness events

The Trial of Activity for Adolescent Girls (TAAG), a multiple site, community-based program designed to increase physical activity among middle school girls, has had interesting results. With funding from NIH's National Heart, Lung, and Blood Institute, schools in six geographic regions of the United States were linked with community partners (the YMCA or YWCA, local health clubs, and community recreation centers). Middle school girls in those areas had modified school health education, physical education in which they engaged in moderate to vigorous physical activity for at least half of class time, and opportunities for weekend, lunchtime, and before- and after-school physical activity. After three years (but not two years), the programs increased the time spent in moderate-to-vigorous physical activity by about 2 minutes per day, or 80 calories a week, and reduced sedentary behavior by 8 minutes a day.[49] Weekday afternoon activities (between 2–5 p.m.) were most effective. If girls sustained this level of activity, they might be prevented from becoming overweight as teenagers or adults.

Others have also found success with after-school programs to supplement in-school physical activity. Eleven different after-school physical activity interventions were evaluated in a recent meta-analysis, which concluded that physical activity levels, physical fitness, body composition, and blood lipid profiles can all be improved by these programs.[50] Better program attendance results in better program outcomes and transportation assistance may be needed to promote attendance.

See Table 12-4 for ways to promote physical activity in your community.

TABLE 12-4 What YOU Can Do to Promote Physical Activity in Your Community

- Advocate for quality community physical activity programs and daily physical education
- Be a physically active role model: use public transportation, walk and bike to work, and organize lunchtime walks
- Participate in community coalitions that work to provide access to equipment, facilities, and spaces that promote physical activity
- Advocate for sidewalks and pedestrian-friendly communities

Physical Activity Policies for the Prevention of Obesity

Public policy at the local, state, and national level can have a great impact on individual behavior. Legislation/regulations in the following areas might help prevent obesity.

- **States—**
  - make mandatory daily physical education that meets national standards (150 minutes per week for grades K to 8 and 225 minutes per week for grades 9 to 12) and that lasts for the entire school year. Moderate-to-vigorous physical activity should occupy at least half of class time.
  - require that physical education is taught by certified physical education teachers at all levels.
- **Schools—**
  - work with communities to provide a variety of competitive and noncompetitive physical activity opportunities, including clubs, classes, and sports, before- and after-school and on weekends.
  - create a "walking school bus" by instituting walk/bike-to-school programs.
- **Communities—**
  - form local coalitions to promote physical activity and reduce barriers to participation.
  - provide funding to make communities safer and improve access to physical; activity facilities and programs.
  - provide sidewalks. A study of 11 countries, including the United States, found the presence of sidewalks to be the most important neighborhood characteristic for the promotion of physical activity.[51]
- **Federal government—**allocate highway money to the development of bikeways.

What Should Children Eat to Prevent Obesity?

The 2005 Dietary Guidelines for Americans, currently under revision, form the basis for most nutrition education programs and all USDA-supported meal programs. Table 12-5 provides guidelines for children's intake of calories, fat, fiber, and various food groups. Serving sizes are smaller for children than adults. Portion size is an important consideration in preventing childhood obesity. The USDA also provides a MyPyramid for children (www.mypyramid.gov). MyPyramid includes information for parents of preschoolers, as well as lesson plans and worksheets for the elementary school classroom.

The Role of Parents in Assuring Healthy Diets

Parents and caregivers who are aware of factors that promote childhood obesity, model appropriate eating and activity behaviors, provide healthy food and opportunities for activity, and set household rules that promote healthy eating and activity are the best defenders against childhood obesity. Unfortunately, with two-thirds of the adult population overweight or obese, many parents are poorly equipped for this effort. Education of parents thus becomes a high priority for childhood obesity prevention.

Breastfeeding. Women who breastfeed their babies for up to 6 months may be conveying protection from obesity for their children.[52] This is thought to be related to the breastfed child's better self-regulation of intake, compared with the formula-

TABLE 12-5 Daily Estimated Calories and Recommended Fat, Fiber, and Daily Amount of Food from Fruits, Vegetables, Grains, Lean Meat, and Milk/Dairy, by Age and Gender

	1 yr	2–3 yrs	4–8 yrs	9–13 yrs	14–18 yrs
Calories[1]					
Females	900[2]	1,000	1,200	1,600	1,800
Males	900[2]	1,000	1,400	1,800	2,200
Fat[2], % cals	30–40	30–35	25–35	25–35	25–35
Fiber[3] (g)					
Females	19	19	25	26	26
Males	19	19	25	31	38
Fruits[4] (cups)					
Females	1[2]	1	1	1.5	1.5
Males	1[2]	1	1.5	1.5	2
Vegetables[5] (cups)					
Females	¾[2]	1	1.5	2	2.5
Males	¾[2]	1	1.5	2.5	3
Grains[6] (oz.)					
Females	2[2]	3	4	5	6
Males	2[2]	3	5	6	7
Lean meat and beans[7] (oz.)					
Females	1.5[2]	2	3	5	5
Males	1.5[2]	2	4	5	6
Milk/dairy[8] (cups)					
Females	2[2]	2	2	3	3
Males	2[2]	2	2	3	3

Source: (Unless otherwise noted in footnotes): U.S. Department of Health and Human Services and U.S. Department of Agriculture. (2005). *Dietary guidelines for Americans, 2005, 6th edition, Appendix A–2.* Washington, D.C.: U.S. Government Printing Office.
[1]Calories based on estimated daily caloric needs for sedentary children. Add about 600 kcals for active lifestyle (800 kcals for 9- to 13-year-old boys and 1,000 kcals for 14- to 18-year-old boys. From USDA *MyPyramid* (www.mypyramid.gov)
[2]Servings for 1-year-old children from Daniels, S. R., Greer, F. R., & Committee on Nutrition. (2008). Lipid screening and cardiovascular health in childhood. *Pediatrics, 122(1),* 198–208.
[3]From Otten, J. J., Hellwig, J. P., & Meyers, L. D. (Eds.) (2006). *Dietary reference intakes: The essential guide to nutrient requirements.* Washington, D.C.: The National Academies Press. (Based upon 14 grams of fiber/1,000 kcals)
[4]Fruits—All fresh, frozen, canned, and dried fruits and fruit juices. In developing the food patterns, only fruits and juices with no added sugars or fats were used. Added sugars or fats need to be counted as discretionary calories.
[5]Vegetables—Includes all fresh, frozen, and canned dark green, orange, and starchy vegetables and cooked and dry beans, peas, and soybean products. (See the note under the Lean meat and beans group about counting legumes in that group.) In developing the food patterns, only vegetables with no added sugars or fats were used. Added sugars or fats need to be counted as discretionary calories.
[6]Grains—Includes all whole-grain products, refined grain products, and whole grains and refined grains used as ingredients. In developing the food patterns, only grains in low-fat and low-sugar forms were used. Added sugars or fats need to be counted as discretionary calories.
[7]Lean meats and beans—The following count as 1-ounce equivalents: 1 ounce of lean meat, poultry, or fish; 1 egg; ¼ cup of cooked dry beans or tofu; 1 Tbsp peanut butter; ½ ounce nuts or seeds. Dry beans and soybean products may be counted in this group, but if counted here should not be counted in the vegetable group.
[8]Milk/dairy—All milks, yogurts, frozen yogurts, dairy desserts, cheeses (except cream cheese), including lactose-free and lactose-reduced products. Most choices should be fat-free or low-fat. In developing the food patterns, only fat-free milk was used. Added sugars or fats need to be counted as discretionary calories.

fed child, who is generally fed until the bottle is empty. In addition, breastfed babies seem to have more tolerance for the taste of healthy foods after weaning. The AAP recommends that, when possible, breastfeeding continue for at least 12 months and that foods rich in iron be gradually introduced at around 6 months.[53] Over three-quarters of mothers breastfeed their babies initially, but fewer than 20% continue for 4–6 months.[52] In fact, by the time they reach 6 months of age, 66% of infants have received grain products, around 40% have eaten fruits and vegetables, and 14% have tried meat. By the time they reach 9–11 months, almost all infants have received grain products, three-quarters have eaten fruits, vegetables, and meat, and 11% have had a sweetened beverage. Juice should not be given before the age of 6 months.[54] Of additional concern is the use of non-infant cereals before age 1 year, as these may not contain adequate levels of vitamins and minerals and may increase children's preference for sweet foods.

Toddlers and Young Children. Up to the age of 7 years, children's caloric needs rise, and they should consume whole grains, fruits and vegetables, beans, fish, lean meat, and low fat and nonfat dairy products (under age 2 years children should consume 2% or whole milk). Parents who perceive that fruit drinks and 100% fruit juice offer equivalent nutritional benefit are giving their children unnecessary non-nutritive calories. Overconsumption of 100% fruit juice is also a problem, particularly for those households where fruit juice is provided to offset thirst. The AAP recommends limiting 100% fruit juice to 4–6 ounces per day and only given in a cup as part of a meal or snack.[54] Up to 10% of 2- to 5-year-olds actually drink more than twice that much,[54] and one-third of 19- to 24-month-old children consume no fruit.[52]

An additional peril for young children is introduction of fast foods, low-nutrition snack foods, whole milk, and sugar-sweetened beverages. The 2002 Feeding Infants and Toddlers Study found that almost half of 7- to 8-month-olds had eaten a dessert, sweet, or sweetened beverage. Among 15- to 18-month-olds, French fries were the most common vegetable; many young children eat no vegetables.[55] For a child who requires only 1,000 calories per day, a serving of French fries, 4 pieces of chicken nuggets, and an 12-ounce soda could account for half of daily caloric intake.

Parents need to provide a variety of nutrient-dense foods at regular times and preferably in a family meal setting, providing child-sized portions. Children should never be forced to "clean their plates," because their caloric intake tends to vary from day to day; some days they will simply be hungrier than other days. Parents should limit snacks and, for children over age 2 years, limit the use of whole-fat dairy products. Dietary practices for children that emphasize regular mealtimes and snack times can help children learn to eat normally.

Parents should be taught to recognize at-risk eating behaviors in children: eating in response to emotional state; lack of awareness of feeling of fullness; and over-responsiveness to food, such as overconsumption of almost anything offered. Children who exhibit these behaviors may be especially prone to gaining excess weight.

Older Children. By 10 years, caloric requirements increase to about 1,800 kcal/day. Dietary Guidelines recommend balancing food intake with physical activity. Parents who have their own issues with food and weight may be tempted to put their children on excessively restrictive diets. Even children who are inactive and/or overweight must not be deprived of food. Caloric restriction is dangerous at any age, but in childhood the consequences are not only physiological but possibly psychological as well. It is far better to encourage physical activity than to discourage eating. As children grow older, they have more opportunity to consume food away from home, which challenges even the most vigilant parents, and they have more money to purchase foods of their choosing. Among older children, breakfast consumption declines and percentage of calories from snacks rises; more fried and nutrient-poor foods are consumed; dairy, fruit, and vegetable consumption decrease (except French fries); and sodium intake and portion sizes increase.[52] According to the CDC's YRBSS, fewer than one-fourth of ninth to twelfth graders meet recommended fruit and vegetable intake, and only 14.5% meet recommendations for dairy.[17]

The Role of School-Based Nutrition Programs in Obesity Prevention

Most young people attend school. More than half eat at least one meal there, and 1 in 6 eat both breakfast and lunch at school.[41] Schools that serve poor, minority children—a high-risk group for the development of obesity—often provide two meals a day to *every* child in the school. School food services certainly influence children's intake, but they also affect childrens' attitudes toward food.

School Food Services   School food services must reinforce nutrition education messages that children hear in their health and physical education classes. Children are better able to adopt healthier eating behaviors if foods available at school cafeterias, in vending machines, and during special events are consistent with classroom messages. Ideally, schools should increase the availability of healthy foods (fruits, vegetables, and whole grains), present a variety of foods, and limit the availability of foods high in fat, sodium, and added sugars.

> **nutrition education** *Broad range of learning experiences, conducted both in the classroom and the school cafeteria that promote healthy eating and other nutrition-related behaviors.*

The Food and Nutrition Service of the U.S. Department of Agriculture (USDA) oversees the NSLP, established in 1946, and the SBP, established in 1975. Both programs aim to assure that school children have access to nutritious, low-cost meals at school. NSLP and SBP meals follow the Dietary Guidelines for Americans, although critics contend that they are still too high in fat and saturated fat in many schools. The USDA has yet to offer guidelines for sodium, *trans* fat, and whole grain content of school meals and probably will not do so until the new Dietary Guidelines are released in 2010.

Recently Congress directed the CDC to work with the Institute of Medicine to devise recommendations for foods in schools, including competitive foods. The result was *Nutrition Standards for Foods in Schools: Leading the Way toward Healthier Youth.* Some of their recommendations are summarized in Table 12-6.

> **social-cognitive theory** *Theory that explains the adoption of new behaviors as resulting from interactions between factors within a person and the person's environment. New behaviors that are reinforced and promote self-efficacy are most likely to be continued.*

Classroom Nutrition Education Programs   While the American Dietetic Association says that there is only "limited evidence" to support that changing the school food environment can reduce body fatness of elementary or secondary school students, the ADA concludes that there is "fair evidence" that nutrition education can play a larger role in changing the way children eat.[46] **Nutrition education** might help change body fatness in elementary and, even more so, secondary students.

Most successful school-based nutrition education programs are grounded in **social-cognitive theory**. Recall from Chapter 9 that people adopt new behaviors when they feel confident and when the new behaviors are reinforced. The interaction between factors within the person and the person's environment influences behavior change. Using

TABLE 12-6   Recommended Nutritional Standards for Competitive Foods in Schools

Foods	Beverages
Tier 1 foods and beverages for all students	
≤ 200 calories per portion as packaged ≤ 35% of calories from fat < 10% of calories from saturated fats Zero trans fat (≤ 0.5 g per serving) ≤ 35% of calories from total sugars, except for yogurt with no more than 30 g of total sugars per 8-oz portion as packaged ≤ 200 mg sodium per portion as packaged à la carte entrée items must meet above fat and sugar limits, be NSLP menu items, and have ≤ 480 mg sodium	Water without flavoring, additives, or carbonation 1% and nonfat milk in 8-oz portions (includes lactose-free and soy milks) Flavored milk with no more than 22 g total sugars per 8 oz portion 100% fruit juice in 4-oz portion as packaged for elementary/middle school and 8 oz (two portions) for high school Caffeine-free, except naturally-occurring trace amounts
Tier 2 foods and beverages for high school students after school	
≤ 200 calories per portion as packaged ≤ 35% of calories from fat < 10% of calories from saturated fats Zero trans fat (≤ 0.5 g per serving) ≤ 35% of calories from total sugars ≤ 200 mg sodium per portion as packaged	Non-caffeinated, non-fortified beverages with less than 5 calories per portion as packaged (with or without nonnutritive sweeteners, carbonation, or flavoring)

Source: IOM (Institute of Medicine). (2007). *Nutrition standards for foods in schools: Leading the way toward healthier youth.* By the National Academy of Sciences. Reprinted with permission from the National Academies Press, Washington, D.C.

social-cognitive theory as a guide, effective nutrition education programs have the following elements:

- Role models who demonstrate healthy eating (teachers, parents, celebrities, or cartoon characters, for example)
- Reinforcement of healthy eating behaviors (stickers and verbal praise are examples)
- Development of practical skills for planning meals, purchasing food, and preparing food
- Opportunities to examine sociocultural influences on eating and weight (such as television advertising and programming) and to develop skills to resist social pressures
- Analysis of one's own eating patterns, goal setting for behavior change, and reanalysis to monitor progress

developmentally appropriate

Activity that is suitable for a child's level of cognition and development.

Nutrition education must be **developmentally appropriate**. At the elementary level, nutrition education should offer active experiences with food. Younger children benefit from being exposed to many different kinds of food, and they respond to cues for healthy eating, like posters displayed in the classroom and the cafeteria. The educational program should be built around skills activities, such as measuring quantities, assigning foods to food groups, creating healthy meals and snacks, and even growing food. Because elementary-age children depend on their parents and caregivers for meals, activities that involve the entire family are particularly effective. Homework packets, newsletters, and family health fairs are examples of activities that can involve the whole family.

At the middle school and high school levels, students are able to make the connection between diet and health. They benefit from using computer software to analyze their own diets. Older students are capable of setting goals and strategizing ways to improve their eating habits. Activities that allow older students to look critically at social pressures to purchase particular foods or eat at certain types of restaurants are helpful in promoting decision-making and resistance skills. Older students should also be given information to dispel myths about dieting and body weight.

Nutrition Policies for the Prevention of Obesity The Healthy Eating Active Communities (HEAC) program, established in 2005 by the California Endowment, was the first program of its kind to use policy and environmental change strategies as the primary approach to reversing the childhood obesity epidemic. HEAC focuses on the environment—schools, after-school hours, neighborhoods, health care, and marketing and advertising—not the individual (see www.healthyeatingactivecommunities.org for more information). Table 12-7 suggests nutrition policy approaches that might impact childhood obesity. Population-based approaches are the key. For example, a nutrition education class will touch all students in the classroom, perhaps influencing half of them to reduce consumption of soda. A tax on sweetened beverages could affect intake of thousands of people. Health advocates in New York estimate that a 1¢ per ounce tax on soda and fruit drinks would reduce consumption by 13%, with a 2 lb weight loss per year for the average person.[32]

TABLE 12-7 Nutrition Policies for Primary Prevention of Obesity

Schools and communities	States	Federal
• Prohibit food advertising on school premises • Prohibit brand-name fast foods in schools • Eliminate school meals that contain trans fats or >10% saturated fat • Require that fruits and vegetables be available during school functions • Provide free drinking water • Adopt the IOM recommendations for foods in schools	• Join the 20 states that have more stringent school nutrition guidelines than the USDA • Prohibit elementary students' access to vending machines • Restrict beverage sales to students from any source on school grounds other than low- and non-fat milk, non-dairy milks, 100% fruit and vegetable juices, and water • Tax soda and snack foods at a higher rate, and eliminate the tax on healthy foods to make them more affordable	• Restrict food advertising and marketing to young children • Pass legislation for restaurant menu labeling • Address strategies for reducing health disparities among socioeconomic, racial, and ethnic groups

Primary Prevention of Obesity: Summary

Primary prevention efforts aimed at increasing physical activity and adopting healthy eating patterns begin in the home but can be supported by school and community efforts. Coordinated national, state, and local activities and policies must commit adequate resources to primary prevention efforts. In its review of individual, family, school, and community interventions for prevention of overweight in children, the American Dietetic Association concluded that many school-based interventions have been quite successful in changing diet and activity behaviors that not only reduce body fat but improve health.[46]

Case Study. Penny and Pablo, Part 2	Penny and Pablo are now 8 years old and attending elementary school.

Pablo is a third grader at the elementary school eight blocks from his home. He is a good student and loves going to school, in part because he has had some wonderful teachers and also because he gets plenty of food at school. His family's income qualifies him for the National School Lunch Program and the School Breakfast Program, and he eats both breakfast and lunch at school during the academic year. Pablo frequently trades food with other kids who bring their lunch, and his brother, who meets him after school, is almost always good for a soda on the way home. After school, Pablo is on his own to make a snack and, sometimes, to make his own dinner or re-heat leftovers in the family's new microwave oven. At Pablo's school there is no playground, and recess typically takes place on a concrete pad. His school does not employ a physical education (PE) teacher, but his classroom teacher provides PE twice a week for 20 minutes, which Pablo enjoys. Pablo is now 52 inches tall and weighs 75 lbs.

Penny attends third grade at an elementary school about 1 mile from her home, and she is ferried to school by car by her mother, who now works 30 hours a week and picks her up by car after school. Penny has an infant sister who goes to day care. No one in the family has time to prepare breakfast, so Penny generally eats a Pop Tart and has a juice box in the car on the way to school. She purchases her lunch from the school cafeteria. Sometimes after school she gets a soda from the vending machine in the cafeteria and drinks it while she waits for her mother. Dinner is generally pre-prepared food that can be picked up at the supermarket, or fast food delivery. Penny's school has a physical education teacher, and every class gets PE three times a week for about 25 minutes. Penny developed asthma this year, and she never cared much for PE anyway, so she frequently sits on the sidelines during class. Penny is now 52 inches tall and weighs 80 lbs. Her mother is concerned that Penny seems heavy, and the pediatrician mentioned that they should limit sweets, so her mother no longer keeps regular soda in the house and does not allow Penny to have dessert or candy.

1. What BMI percentile are Pablo and Penny in now?
2. Use MyPyramid and determine what their caloric requirements are.
3. With the dietary recommendations from MyPyramid, and considering Pablo's and Penny's circumstances, develop a diet plan for each of them for one day.
4. What suggestions for increasing physical activity can you make?

SECONDARY PREVENTION: TREATING OVERWEIGHT AND OBESITY IN YOUNG PEOPLE

The purpose of secondary prevention is to reduce the prevalence (number of new cases) of a condition. This section of the chapter looks at interventions specifically designed for overweight children and adolescents.

What Is the Definition of Overweight/Obesity in Children?

Chapter 2 presented information about assessment methods that can help target populations at risk of weight gain or at risk of health problems associated with excess weight. Unlike adults (considered to be overweight at or above a BMI of 25 and obese at a BMI of 30), children are a bit of a moving target when it comes to BMI, and boys and girls grow at different rates. So any cut-off point to define overweight or obesity in children has to be age- and gender-specific.

In the United States the CDC BMI-for-age charts are now widely used to define overweight and obesity in children ages 2 to 20 years (for children under age 2 years, weight-for-length curves are used). These may be found in Appendix A, Tables 8 and 9. Healthy weight for children and adolescents is between the 5th and 84th CDC BMI percentile; overweight is the 85th–94th BMI percentile; and obesity is at or above the 95th BMI percentile. An Expert Committee convened in 2007 by the AAP proposed that severe obesity be defined as at or above the 99th BMI percentile, which would be a BMI of 30–32 kg/m^2 for 10- to 12-year-olds and a BMI of > 34 kg/m^2 for 14- to 16-year-olds.[56] There is general agreement that for children under age 2 years who have a weight-for-length $\geq$ 95th percentile, the term overweight (not obese) should be used.[57]

Percent fat can also be estimated in children over age 6 years using skinfold measures. The risk of health problems increases in boys whose fat percentage is at or above 25% and in girls whose fat percentage is at or above 30%. Table 2-5 in Chapter 2 provides an interpretation of percent fat in children.

Recall from Chapter 2 that waist circumference is used with adults in addition to the BMI to assess disease risk. While health risks over the long-term among children with large waists have not been widely researched, there have been several studies indicating that children with larger waist circumferences have a greater risk of cardiovascular disease (CVD), hypertension, and high blood lipids.[58] Racial and ethnic differences in optimal waist circumference have been developed.[59] Generally children with waist circumference values at or above the 75th percentile may be at risk of obesity-related comorbidities. Table 12-8 provides 75th and 90th waist circumference percentiles for European American, African American, and Mexican American children.

Development of Overweight in Children

A normally developing child should experience fat gain during the first year of life, which should then slow as fat cells shrink in size, until an **adiposity rebound** begins at between 3–6 years of age. The rebound is a typical period of increased body fat accumulation, as fat cells increase in number, in the early preschool years. This is reflected in the

adiposity rebound *Increased accumulation of body fat that typically occurs at around age 6 years after fat storage slows during the preschool years.*

TABLE 12-8 75th and 90th Waist Circumference Percentiles for Boys and Girls, by Racial/Ethnic Group (Waist circumference in cm)

	European-American			
	Percentile for boys		Percentile for girls	
AGE (yrs)	75th	90th	75th	90th
2	48.6	50.6	49.6	52.5
3	51.2	54.0	51.9	55.4
4	53.8	57.4	54.2	58.2
5	56.5	60.8	56.5	61.1
6	59.1	64.2	58.8	64.0
7	61.7	67.6	61.1	66.8
8	64.3	71.0	63.4	69.7
9	67.0	74.3	65.7	72.6
10	69.6	77.7	68.0	75.5
11	72.2	81.1	70.3	78.3
12	74.9	84.5	72.6	81.2
13	77.5	87.9	74.9	84.1
14	80.1	91.3	77.2	86.9
15	82.8	94.7	79.5	89.8
16	85.4	98.1	81.8	92.7
17	88.0	101.5	84.1	95.5
18	90.6	104.9	86.4	98.4

	African-American			
	Percentile for boys		Percentile for girls	
AGE (yrs)	75th	90th	75th	90th
2	48.5	50.0	47.7	50.1
3	50.7	53.2	50.6	53.8
4	52.9	56.4	53.4	57.5
5	55.1	59.6	56.2	61.1
6	57.3	62.8	59.0	64.8
7	59.5	66.1	61.8	68.5
8	61.7	69.3	64.7	72.2
9	63.9	72.5	67.5	75.8
10	66.1	75.7	70.3	79.5
11	68.3	78.9	73.1	83.2
12	70.5	82.1	75.9	86.9

13	72.7	85.3	78.8	90.5
14	74.9	88.5	81.6	94.2
15	77.1	91.7	84.4	97.9
16	79.3	94.9	87.2	101.6
17	81.5	98.2	90.0	105.2
18	83.7	101.4	92.9	108.9

Mexican-American				
	Percentile for boys		Percentile for girls	
AGE (yrs)	75th	90th	75th	90th
2	49.8	53.2	50.0	53.5
3	52.5	56.7	52.6	56.7
4	55.3	60.2	55.2	59.9
5	58.0	63.6	57.8	63.0
6	60.7	67.1	60.4	66.2
7	63.4	70.6	63.0	69.4
8	66.2	74.1	65.6	72.6
9	68.9	77.6	68.2	75.8
10	71.6	81.0	70.8	78.9
11	74.4	84.5	73.4	82.1
12	77.1	88.0	76.0	85.3
13	79.8	91.5	78.6	88.5
14	82.6	95.0	81.2	91.7
15	85.3	98.4	83.8	94.8
16	88.0	101.9	86.4	98.0
17	90.7	105.4	89.0	101.2
18	93.5	108.9	91.6	104.4

Source: Reprinted from Journal of Pediatrics, 145, Fernández, J.R., Redden, D.T., Pietrobelli, A., & Allison, D.B. (2004). Waist circumference percentiles in nationally representative samples of African-American, European-American, and Mexican-American Children and Adolescents, 439–444, with permission from Elsevier.

BMI. Average BMI levels increase rapidly up to age 1 year, then decrease and reach their lowest point between ages 4–8 years, after which they begin to rise again. When the adiposity rebound begins before age 6, there is increased risk that accumulation of an excess number of fat cells will promote obesity that persists into adulthood, perhaps because there is a longer time for fat to accumulate.[8,11,60] In addition to fluctuations in the accumulation of fat, children experience changes in the composition of lean body mass. These changes affect body density and make adult equations for calculating percent fat extremely inaccurate.

At greater risk for becoming obese adults are (1) children from ages 3 to 9 who are overweight and have at least one overweight or obese parent, (2) any child who is severely overweight, and (3) older children who are overweight. These children may benefit from interventions.

What Types of Programs Are Effective for Overweight Children?

Many weight-management strategies used by adults are inappropriate for children and adolescents. Caloric restriction can impair growth and health and may lead to eating disorders. Weight-management approaches that focus primarily on weight loss, rather than on healthy eating and activity patterns, are no more appropriate for young people than they are for most adults. Weight-loss drugs and bariatric surgery should only be used in severely obese children who have **comorbidities**; these are discussed in the section on Tertiary Prevention.

comorbidities

Diseases or conditions that often exist in conjunction with another disease or condition. For example, sleep apnea is a comorbidity that often accompanies severe obesity.

Some weight-management strategies used with obese adults are very effective with overweight children. In fact, behavioral interventions that address eating and activity have been found to work better with children than adults.

The Role of Parents and Families

Family eating patterns and family dynamics affect all young people. For the following reasons, treatment programs for overweight children are much more effective when there is family involvement in the program:

- Parents and caregivers model dietary and activity behaviors that contribute to the child's behaviors.
- Frequently, obese children are in families where one or both parents are also obese or overweight. Others in the family may, therefore, benefit from dietary and activity interventions aimed at the child.
- Positive reinforcement is helpful in changing the child's eating and activity behaviors, and the family provides most reinforcement.

Skills-based programs that teach children and their parents or caregivers new eating and activity behaviors, like those already discussed in Chapter 9, are particularly effective. Behavior change may not have to be extreme. Younger children have fewer fat cells than adolescents or adults, so, in most cases, interventions need only shrink a normal number of cells. Because children are still growing, the goal is not weight loss but weight maintenance—thereby giving time for height to catch up with weight.

Overweight children should never "diet," which is not the same as saying that their eating habits will not need to be adjusted. Some parents who think that their children are overweight or obese may actually start to restrict food. This practice is in itself considered to increase the risk for obesity, because it may lead to binge eating.[43] Far better to focus on common problem areas indicated by dietary assessment of overweight children, such as eating too fast, selecting too many foods high in fat and added sugars, and eating excessively large portions. These problems can be corrected without using a calorie-restricted diet. The child—and the parent—may need to develop skills in menu planning, label reading, food selection, and food preparation methods that use less fat, sugar, and/or salt. Foods that the child likes—even biscuits, ice cream, and French fries—should still be offered, but not as dietary staples and not in unlimited quantities. If these foods happen to be dietary staples in the household, then the family will have to change its eating habits, too.

One of the easiest ways to eliminate empty calories from the diet is to modify beverage choices. Because soda and juice drinks are inexpensive and widely available, even in school cafeterias, many children and adolescents get a substantial number of calories from them. Consider that 12 oz of most soda contains 150 kcal. Drinking three sodas a day would provide 450 kcal. Tap water, club soda, and mineral water should be the beverages of choice to quench thirst, and soda and juice drinks should be saved for occasional use. This is not a matter of "counting calories," but sensibly limiting sugar intake.

Including adequate amounts of fiber in the diet is also important in treating obesity and preventing the development of cancer, CVD, and type 2 diabetes mellitus. The daily recommended intake for fiber may be found in Table 12-5.

Mealtimes and snack times may need to be better regulated so that children know when they are going to eat and are not forced to forage for food when they are ravenous. The television should be turned off at mealtimes, and snacking should be done at the dining room or kitchen table. Time for physical activity must be worked into the daily schedule. Obviously, the support of parents or caregivers and other family members is crucial to these efforts.

What Are Some Guidelines for Physical Activity for Overweight Children?

Expending more energy has two components: (1) Spending less time in sedentary pursuits; and (2) engaging in some type of physical activity, exercise, or sports program. For many children, reinforcing *decreased* time spent in sedentary pursuits is more effective than reinforcing *more* time spent in physical activity.[61] This strategy gives children a greater sense of control over which behaviors will replace their sedentary activities.

The family plays an important role in helping children to be less sedentary and more active. The first hurdle is limiting television and computer time for all family members, not just the overweight child and having ready substitutes for television/computer time:

- Walking with a family member or a friend to the library, a store, or just around the neighborhood
- Helping to do yard work, such as planting flowers or trimming hedges
- Doing household chores, such as drying dishes and emptying wastebaskets
- Playing a game, doing a craft project, or flying a kite

Parents can also provide opportunities for more formal physical activities. Fortunately, several childhood obesity researchers have demonstrated that it is easier for children to increase their activity level than adults, and children usually respond well to opportunities to participate in formal sports and exercise programs. About an hour of physical activity every day will improve fitness and health. This can be at a low to moderate intensity, depending on the child's physical capabilities.

To avoid orthopedic injuries during exercise, overweight children should engage in low-impact activities at first. Swimming, bicycling, and walking are easier on the joints than running or most sports. Although the risk of a cardiac event in children is low, severely overweight children and adolescents should see their pediatrician before beginning a physical activity program.

The whole family should limit passive television viewing; restrict snacking during television time; and make an effort to find active, engaging leisure pursuits. Bicycling, brisk walking, ice skating, hiking, skateboarding, swimming, dancing, and slow rope-skipping are appropriate activities for children and adolescents. Weight training is also a challenging activity for overweight children, who tend to score well on strength tests. As you read in Chapter 7, prepubertal children can engage in supervised light to moderate weight training with little risk of injury, although heavy weight lifting must be avoided to prevent damage to the growth plates of the bones. Most health clubs will not let children under age 18 years use weight-training equipment, but some community recreation departments offer classes for children. Parents might also invest in light weights for the home.

Parents can also encourage participation in extramural and community sports, the only requirement being the child's interest, moderate level of fitness, and good motor skills. For example, overweight children who are strong often excel in sports like baseball and softball. If the child is not embarrassed by wearing a bathing suit, swim team, water polo, and synchronized swimming are aquatic sports that build fitness and confidence.

School-Based Programs

With increasing rates of overweight among U.S. children and adolescents, schools have the potential to reach the majority of this population. However, programs designed solely for overweight students may unnecessarily stigmatize them and perhaps add to teasing, embarrassment, and social isolation. Framing secondary prevention (interventions specifically for overweight children) within the context of primary prevention (activity and nutrition programs for *all* students) will be beneficial to everyone.

Several multi-component programs—those that include physical activity, nutrition, and family involvement—have reported success in increasing children's consumption of fruits and vegetables, reducing intake of fat and sweetened beverages, increasing levels of moderate to vigorous physical activity, and reducing waist circumference, although most have not found significant reductions in children's body fat.[62–64] In its comprehensive review of interventions for pediatric overweight, the American Dietetic Association stated that "limited evidence supports using a multi-component school-based secondary prevention program to decrease overweight in elementary or secondary school students."[46]

In fact, studies of school programs have had mixed results. Some have promoted weight loss in participating children, but others have produced no change. Even in the absence of weight loss, however, many of these programs have increased children's activity levels and improved nutrition, which can have long-term health implications. Here are some effective programs:

- An innovative school-based intervention for sixth to eighth grades called Planet Health has had encouraging results. Ethnically diverse children at five schools received health and physical education classes that emphasized less television viewing, more activity, fewer high-fat foods, and increased intake of fruits and vegetables. When compared with children at five control schools after 2 years, Planet Health girls (but not boys) had a lower prevalence of obesity and a greater remission of obesity. Both boys and girls in Planet Health schools watched less television and ate more fruits and vegetables.[21]

- Sports, Play, and Active Recreation for Kids (SPARK PE) was developed in the late 1980s by a group of researchers and educators from San Diego State University with funding from the National Heart, Lung, and Blood Institute of the National Institutes of Health (NHLBI/NIH). The goal of the program was to reduce the risk of CVD by increasing elementary-age children's physical activity levels during school-based physical education classes as well as outside of school. Students in SPARK PE intervention schools got much more in-school physical activity than students in control schools and expended more energy in physical education class, although there were no reductions in body fat over the 2 years of the program.[51] Four years after the program's initiation, most SPARK schools continued to implement the program.[65]

- The Child and Adolescent Trial for Cardiovascular Health (CATCH), which was also funded by NHLBI/NIH, is the largest, most comprehensive, and most rigorously evaluated school health program in the United States. More than 5,000 ethnically diverse children from twelve school districts in four states were followed for 3 years (1991–1994) to determine the effectiveness of the program in promoting cardiovascular health and preventing cardiovascular risk factors. The CATCH program was developed for students in third to fifth grade. It includes a heart-health curriculum, family activities, a school food service program, and physical education. Evaluation of the CATCH program found that teachers liked CATCH and followed the health and physical education curricula; CATCH cafeterias were able to modify their school lunch offerings to lower fat, saturated fat, and cholesterol content; and CATCH PE resulted in significant increases in activity time during class, and physical activity outside of school increased.[66] Five years later, student energy expenditure and amount of moderate to vigorous physical activity in physical education classes was maintained.[64]

BMI Reporting in Schools

Although the idea of having children's BMIs measured in school and then reported to their parents in the form of a health report card is still controversial, it has been tried in some states with modest success. In 2004 baseline data from Arkansas, the first state to begin reporting children's BMI, showed that 20.9% of public school students were overweight and 17.1% were at-risk-of-overweight. Three years later, 20.4% of students were overweight and 17.1% were at-risk-of-overweight.[67]

Still, the CDC does not recommend that schools conduct BMI screening, mainly because there is a lack of research on the impact of this practice. If screening is conducted, schools need to decide:

- Who will do the screening? The school nurse would be an obvious choice, if available, or health or physical education teachers. Schools will have to identify a standard protocol and provide training for all personnel involved in screening. This may not be the best use of limited personnel resources in the schools.

- What will be done with the information? Certainly it needs to be provided to parents, but in the form of a health report card or some other format? Collecting information, putting it in a format for parents, and transmitting the information to parents will require personnel resources.

- What kinds of additional or follow-up information will be provided to parents? Schools would want to use the report card as an educational tool and avoid causing punishments or restrictive feeding.

Overweight Children and Adolescents Who Want to Diet

What if an overweight adolescent wants to diet to lose weight? Data from the YRBS indicate that more than half of girls and 28% of boys had dieted to lose weight or to keep from gaining weight; 11% had fasted for at least 24 hours; and 6% had used unprescribed diet pills, powders, or liquids.[17] Caloric reduction that occurs as part of an overall plan to improve the quality of one's diet is not inherently dangerous, but caloric reduction for the sole purpose of dropping weight quickly is inadvisable. Particularly during adolescence, caloric restriction can increase food obsessions, leading to eating disorders. At any age, weight loss attained through severe caloric restriction is poorly maintained.

Adolescents who want to try to lose weight through dieting should be offered guidance for approaching it in a healthy way. Important points to convey when counseling adolescents about weight loss include:

- Maintaining an intake of at least 1,500 kcal/day
- Eating three meals and two snacks every day (spread calories out through the day)
- Making changes in eating behaviors that can be maintained, including finding ways to eat in restaurants and at social occasions without deprivation or bingeing
- Avoiding fad diets, such as those described in Chapter 10
- Incorporating physical activity into the weight-loss plan

Outcomes of Treatments for Overweight Children

Clinical programs for the treatment of childhood obesity have been modestly effective. For example, an evaluation of one family-based treatment program found that almost one-third of 158 obese children who were 6–12 years old during the program were no longer obese at 10-year follow-up.[68] In another intervention, half of 4 children who lost weight through an individual or group behavioral program had maintained a lower body weight 5 years after the program ended.[69] An intensive family-based program at the Yale Pediatric Obesity Clinic reported improvements in body composition and insulin resistance among 8- to16-year-old participants.[70] Overweight children have greater success at weight maintenance than their obese parents, for as long as 10 years after completion of a family-based program.[71]

Initial weight loss is about the same for overweight children treated in family-based programs versus other types of programs, but those in family programs tend to regain significantly less weight over the ensuing years.[72] When parents are taught to restructure the eating and exercise environment, use problem-solving skills, and employ positive reinforcement, dramatic and lasting outcomes may result for the child.

Secondary Prevention: Summary

Secondary prevention interventions are aimed at reducing the prevalence of a health problem. For obesity, behavioral interventions that focus on eating and activity behaviors can prevent further weight gain, allowing overweight children to "grow into" their weight and avoid obesity later in life.

Case Study. **Part 3**	Penny and Pablo are now 11 years old and in middle school in their communities.

Penny and Pablo are now 11 years old and in middle school in their communities.

Penny has had a difficult transition from elementary to middle school. She is acutely aware of her weight and is one of the heaviest girls in her class. Her school recently started measuring heights and weights and reporting BMI percentiles to parents in the form of a Health Report Card, and, according to Penny, her mother "freaked out" when she got Penny's result. So Penny's mother now has her on a diet. As a result of the diet, Penny is no longer allowed to buy lunch at school, and her mother packs a lunch of carrot and celery sticks, yogurt, and low-fat cheese and crackers. Because she is frequently hungry and continues to skip breakfast, Penny uses her allowance to buy à la carte items in the cafeteria. The family continues to eat convenience foods at dinner, although Penny's portions are regulated by her mother. Penny's middle school provides physical education taught by a certified teacher 3 days a week for about 35 minutes. While Penny's asthma has actually improved as a result of getting allergy shots, she frequently asks for and receives notes from her mother to exempt her from PE. Neither she nor her mother typically exercises, although her father continues to use his health club regularly and has lost about 10 lbs over the past year. Penny now weighs 110 lbs and is 58 inches tall; she has a 28 inch waist.

Pablo is also 11-years-old and is allowed to ride the city bus by himself to the junior high school, which is not near his apartment. One of the tallest boys in his class, Pablo is popular with his classmates and teachers. He excels in many sports and is always one of the first called when they choose up sides in physical education class. Pablo gets PE 3 times a week for the first half of the school year, and he really likes his teacher, who is a junior varsity football coach not certified in physical education. Pablo participates in an after-school sports program. He no longer eats breakfast at school, preferring to sleep a bit later before catching the bus, so he grabs whatever is available at home; he does get a subsidized school lunch. Pablo still loves soda and a snack after school, which he usually buys at a convenience store near the bus stop. Pablo's mother works long hours, and even his father has been working more, so he frequently makes dinner or re-heats whatever dishes his mother has made over the weekend and eats in front of the television. The family continues to enjoy traditional Guatemalan dishes and fast food. Pablo is 60 inches tall and weighs 105 lbs.

1. Calculate Pablo's and Penny's BMIs and determine their BMI percentiles. What is the significance, if any, of Penny's waist measurement?
2. Assuming that Penny grows only three inches over the next year, in order for her to be in the 85th BMI percentile, what should she weigh at age 12? What will happen to Pablo's BMI percentile if his weight increases by 5 lbs and he gains 5 inches in height?
3. What behavioral strategies could Pablo's and Penny's families help them to use to make changes in their eating and activity habits?

TERTIARY PREVENTION: REDUCING THE RISK OF CHRONIC DISEASE

Tertiary prevention is not actually prevention in the true sense of the word because children who are candidates for tertiary prevention interventions are already overweight or obese. Some experts prefer the term *maintenance intervention* over tertiary prevention, thinking that it better reflects the reality of a condition not being prevented but rather maintained in a way that reduces the risk of developing comorbidities.[73] This discussion focuses on several complications associated with obesity, which until recently were seen mainly in adults but are now seen in overweight adolescents and children as well—type 2 diabetes mellitus, cardio-vascular changes, and the metabolic syndrome. This section also discusses treatments formerly reserved for adults—pharmacological agents and bariatric surgery.

What Are Health Problems of Overweight and Obese Children?

Much of what is known about the effect of excess fat on children's health comes from the Bogalusa Heart Study, a community-based study started in 1973 and that, for more than 25 years, assessed almost every 5- to 17-year-old child living in Ward 4 of Washington Parish, Louisiana, for cardiovascular risk factors. Bogalusa Heart Study researchers have found overweight to be associated with high low-density lipoprotein cholesterol (LDL-C) levels, high triglyceride levels, low high-density lipoprotein cholesterol (HDL-C) levels, high systolic blood pressure, and hyperinsulinemia.[74] Even children as young as 7 years who are overweight can have multiple cardiovascular risk factors.

Compared with healthy-weight children, overweight children are more likely to have higher levels of total cholesterol (16% vs. 7%), LDL cholesterol (11% vs. 8%), triglycerides (7% vs. 2%), fasting glucose (3% vs. 0%), and systolic blood pressure (9% vs. 2%); and lower HDL cholesterol levels (15.5% vs. 3%).[75] They also suffer in unacceptably high numbers from sleep apnea and nonalcoholic fatty liver disease. When these health problems persist into adulthood or worsen, morbidity and mortality are increased later in life. In addition, social problems and low self-esteem may make childhood an unhappy time for an obese child, and feelings of social isolation may persist into adulthood.

Cardiovascular Risk Factors in Obese and Overweight Children

Cardiovascular disease (CVD) is the leading cause of death in the United States. At the root of most CVD is atherosclerosis, and the risk of atherosclerosis is increased by two risk factors seen in about 60% of obese 5- to 10-year-olds: hypertension and abnormal blood lipids.[76] Scientists know that atherosclerosis is a slow, progressive process that begins with fatty streaks in the arteries, beginning in childhood, and culminates in the formation of plaque. Recall that plaque formation initiates in the innermost layer of endothelial cells that line an artery (the intima) and migrates into the smooth muscle layer just above it (the media), where smooth muscle cells, lipids, and fibroblasts (scar-tissue forming cells) grow into the opening of the artery.

Intima-media thickness (IMT) of the common carotid artery, which is located in the neck, is considered to be a good predictor of cardiac atherosclerosis in adults. A study of pre-pubertal obese (BMI ≥ 97th percentile) and nonobese children found the obese children to have greater IMT, as well as other CVD risk factors (hypertension, abnormal blood lipids, and impaired glucose metabolism).[77]

Abnormal Blood Lipids Overweight is the strongest predictor of abnormal blood lipid levels (dyslipidemia) in children, other than family history, and dyslipidemia is a significant risk factor for CVD.[78] Blood lipid levels tend to track from childhood to adulthood. A U.S. Preventive Services Task Force Evidence review did not find sufficient evidence that routinely screening children for blood lipid abnormalities delays the onset

TABLE 12-9   Classification of Blood Lipid Levels in Children and Adolescents[1,2]

Category	Total cholesterol (mg/dL)	LDL cholesterol (mg/dL)	HDL cholesterol (mg/dL)	Triglycerides (mg/dL)
Acceptable	< 170	< 110	--	--
Borderline	170–199	110–129	--	--
High risk	≥ 200	≥ 130	< 35	> 200

Source: Created by the author from : NCEP Expert Panel on Blood Cholesterol Levels in Children and Adolescents (1992). National cholesterol education program (NCEP): Highlights of the report of the expert panel on blood cholesterol levels in children and adolescents. *Pediatrics, 89*(3), 495–501; and Williams, C. L., Hayman, L. L., Daniels, S. R., Robinson, T. N., Steinberger, J., Paridon, S., et al. (2002). Cardiovascular health in childhood: A statement for health professionals from the committee on atherosclerosis, hypertension, and obesity in the young (AHOY) of the council on cardiovascular disease in the young, American Heart Association. *Circulation, 106*(1), 143–160.

of CVD or reduces mortality from heart disease.[78] In addition, there is no consensus on what normal blood lipid values are for children. Still, the AHA recommends monitoring blood lipids in children with a family history of CVD and endorses the National Cholesterol Education Program of the Expert Panel on Blood Cholesterol Levels in Children and Adolescents, whose findings are summarized in Table 12-9. AHA's Committee on Atherosclerosis, Hypertension, and Obesity in the Young recommends that all children and adolescents with BMI above the 95th percentile have an assessment of blood lipids, blood pressure, and fasting insulin and glucose.[79]

Hypertension   Children and adolescents whose BMI is above the 95th percentile are at greater risk of hypertension, with half of obese[80] and 13% of overweight[56] children and adolescents having high blood pressure. Data from the NHANES indicates that both systolic and diastolic blood pressures have increased in 8- to 17-year-old children since 1988, and high blood pressure in childhood predicts high blood pressure in adulthood.[79,81,82] Normal blood pressure for 1- to 17-year-olds is based upon age, gender, and height. Children whose systolic or diastolic blood pressure is below the 90th percentile are considered to have normal blood pressure. When the systolic or diastolic blood pressure, taken on a least three different occasions, is at or above the 95th percentile, stage 1 hypertension is diagnosed; when it is between the 90th and < 95th percentile, or when blood pressure exceeds 120/80 mm Hg, the child is considered prehypertensive.

Table 12-10 provides 90th and 95th blood pressure percentiles for children. Like high blood lipids, high blood pressure has been shown to predict increased carotid IMT, as well as increased left ventricular thickness.[82]

Sleep Disordered Breathing   Nighttime snoring, breathing difficulties, or excessive tiredness during the day are all possible signs of obstructive sleep apnea, in which the child experiences blockage of the upper airways while sleeping. Sleep apnea is a potentially serious risk factor for the development of CVD. The condition is more common in obese than nonobese children.[83]

TABLE 12-10 Blood Pressure Values for Children, by Gender, Height Percentile, and Age

Age (Yr)	BP 'Tile	Boys										Girls									
		Systolic blood pressure					Diastolic blood pressure					Systolic blood pressure					Diastolic blood pressure				
		← Percentile of height →					← Percentile of height →					← Percentile of height →					← Percentile of height →				
		5	25	50	90	95	5	25	50	90	95	5	25	50	90	95	5	25	50	90	95
1	90th	94	97	99	102	103	49	51	52	53	54	97	98	100	102	103	52	53	54	55	56
	95th	98	101	103	106	106	54	55	56	58	58	100	102	104	106	107	56	57	58	59	60
2	90th	97	100	102	105	106	54	56	57	58	59	98	100	101	104	105	57	58	59	61	61
	95th	101	104	106	109	110	59	60	61	63	63	102	104	105	108	109	61	62	63	65	65
3	90th	100	103	105	108	109	59	60	61	63	63	100	102	103	106	106	61	62	63	64	65
	95th	104	107	109	112	113	63	64	65	67	67	104	105	107	109	110	65	66	67	68	69
4	90th	102	105	107	110	111	62	64	65	66	67	101	103	104	107	108	64	65	66	67	68
	95th	106	109	111	114	115	66	68	69	71	71	105	107	108	111	112	68	69	70	71	72
5	90th	104	106	108	111	112	65	67	68	69	70	103	105	106	109	109	66	67	68	69	70
	95th	108	110	112	115	116	69	71	72	74	74	107	108	110	112	113	70	71	72	73	74
6	90th	105	108	110	113	113	68	69	70	72	72	104	106	108	110	111	68	69	70	71	72
	95th	109	112	114	117	117	72	73	74	76	76	108	110	111	114	115	72	73	74	75	76
7	90th	106	109	111	114	115	70	71	72	74	74	106	108	109	112	113	69	70	71	72	73
	95th	110	113	115	118	119	74	75	76	78	78	110	112	113	116	116	73	74	75	76	77
8	90th	107	110	112	115	116	71	72	73	75	76	108	110	111	114	114	71	71	72	74	74

Age (Year)	BP Percentile	Systolic BP (mm Hg) — Percentile of Height							Diastolic BP (mm Hg) — Percentile of Height						
		5th	10th	25th	50th	75th	90th	95th	5th	10th	25th	50th	75th	90th	95th
9	90th	107	109	110	112	114	115	116	70	70	71	72	73	73	74
9	95th	111	112	114	116	118	119	120	74	74	75	76	77	78	78
10	90th	109	110	112	114	115	117	118	72	73	73	74	75	76	76
10	95th	113	114	116	118	119	121	121	76	77	77	78	79	80	80
11	90th	111	112	114	116	118	119	120	74	74	75	76	77	78	78
11	95th	115	116	118	120	121	123	123	78	78	79	80	81	82	82
12	90th	113	114	116	118	120	121	122	75	75	76	77	78	79	79
12	95th	117	118	120	122	123	125	125	79	79	80	81	82	83	83
13	90th	115	116	118	120	122	123	124	75	76	76	77	78	79	79
13	95th	119	120	122	124	126	127	128	79	80	81	82	83	83	83
14	90th	117	118	120	122	124	125	126	76	76	77	78	79	80	80
14	95th	121	122	124	126	128	129	130	80	80	81	82	83	84	84
15	90th	120	121	123	125	126	128	128	77	77	78	79	80	81	81
15	95th	124	125	127	129	131	132	133	81	81	82	83	84	85	85
16	90th	122	124	125	127	129	130	131	78	78	79	80	81	82	82
16	95th	126	127	129	131	133	134	135	82	82	83	84	85	86	86
17	90th	125	126	128	130	131	133	134	80	80	81	82	83	84	84
17	95th	129	130	132	134	135	137	137	84	85	86	87	87	88	89

Source: U.S. Department of Health and Human Services. (2005). *The fourth report on the diagnosis, evaluation, and treatment of high blood pressure in children and adolescents.* Bethesda, MD: National Institutes of Health, National Heart Lung, and Blood Institute. Available from: http://www.nhlbi.nih.gov/health/prof/heart/hbp/hbp_ped.pdf.
(NOTE: The source provides complete blood pressure tables for the 50th through the 99th blood pressure percentiles and for the 5th, 10th, 25th, 50th, 75th, 90th, and 95th height percentiles.)

Type 2 Diabetes Mellitus and the Obese Child

Approximately 16 million Americans have diabetes mellitus, and 90–95% are type 2 diabetics. As you learned in Chapter 1, obesity is the biggest risk factor for type 2 diabetes. As many as one-fourth of obese children and adolescents may have impaired glucose tolerance, which may lead to diabetes.[84] Glucose intolerance can cause insulin resistance in children, like adults. Almost one-third of diabetes diagnosed in children today is type 2, and researchers estimate that the rate of type 2 diabetes is 10 times greater than it was 25 years ago.[85] The morbidity and mortality associated with type 2 diabetes can be significantly reduced by keeping blood glucose under control.

Because of the links between insulin resistance, overweight, dyslipidemia, and hypertension in children and adolescents, the rise in type 2 diabetes mellitus portends potentially serious health issues in adulthood. One group of researchers estimates that diabetes has the potential to shorten life by 13 years.[10]

Screening can help identify children with impaired glucose tolerance. A fasting plasma glucose test is recommended for overweight children with a family history of type 2 diabetes, particularly when they are members of a racial/ethnic group with high rates of diabetes (American Indian, Hispanic, African American, or Asian/Pacific Islander).[79] Impaired glucose tolerance in children and adolescents would be reflected by a fasting blood glucose of greater than 110 mg/dL; and an oral glucose tolerance test of greater than 140 mg/dL.

Metabolic Syndrome in Children and Adolescents

In adults, a cluster of risk factors associated with CVD and type 2 diabetes mellitus—obesity, insulin resistance, abnormal blood lipids, and hypertension—has been called the metabolic syndrome. Adults with metabolic syndrome are more likely to develop early CVD and type 2 diabetes.

Chapter 1 provides the International Diabetes Federation (IDF) criteria for metabolic syndrome in adults. A definition of the metabolic syndrome in children has been more difficult to develop. But recently the IDF published criteria for children and adolescents ages 6–16 years, which are presented in Table 12-11.

Based upon this definition and the population studied, 0.4–29% of obese adolescents may have the metabolic syndrome.[86–88] Scientists do not yet know how metabolic syndrome in childhood will predict morbidity and mortality in adulthood. There is some evidence that it predicts CVD in adulthood.[89] Certainly it has adverse health consequences during childhood and adolescence. Overweight and obese children with the metabolic syndrome may be more likely to have nonalcoholic fatty liver disease (NAFLD), discussed below, and children with NAFLD are more likely than their age- and sex-matched peers to have significant CVD risks.[90]

Nonalcoholic Fatty Liver Disease (NAFLD)

The most common cause of liver disease in children, NAFLD is a cluster of abnormalities in which triglycerides accumulate in the liver and, untreated, may result in fibrosis, cirrhosis, and liver failure. In addition to its association with obesity in general, NAFLD is associated with central fat deposits, high blood lipids, hypertension, and impaired fasting glucose.[90] Rates of NAFLD in children have increased as obesity prevalence has risen. A retrospective study of autopsy results from children and adolescents reported 38% of obese children and 10% of nonobese children to have NAFLD.[91] NAFLD is thought to be related to insulin resistance and the metabolic syndrome.

TABLE 12-11 Criteria for Metabolic Syndrome in Children

IDF criteria*	
Ages 6 years to < 10 years	Obesity (defined as > 90th percentile of waist circumference)
Ages 10 years to < 16 years	Obesity (defined as > 90th percentile of waist circumference) Two of these criteria:
	• Triglycerides ≥ 150 mg/dL
	• HDL-C < 40 mg/dL
	• Systolic blood pressure ≥ 130 mm Hg or Diastolic blood pressure ≥ 85 mm Hg
	• Fasting plasma glucose ≥ 100 mg/dL or previously diagnosed type 2 diabetes mellitus
AAP criteria**	
Children and Adolescents	BMI > 97th percentile
	Waist circumference > 90th percentile in boys; ≥ 90th percentile in girls
	Triglycerides > 110 mg/dL
	HDL-C < 40 mg/dL
	Systolic/diastolic blood pressure > 90th percentile for age, gender, and height
	Fasting plasma glucose > 110 mg/dL; oral glucose tolerance test > 140 mg/dL

*Source: Ford, E. S., Li, C., Zhao, G., Pearson, W. S., & Mokdad, A. H. (2008). Prevalence of the metabolic syndrome among U.S. adolescents using the definition from the international diabetes federation. *Diabetes Care, 31*(3), 587–589.

**Source: Promoting healthy weight. In Hagan, J. F., Shaw, J. S., & Duncan, P. (2008). *Bright Futures: Guidelines for Health Supervision of Infants, Children, and Adolescents*, 3rd edition. Elk Grove, IL: American Academy of Pediatrics.

Treatments for the Comorbidities of Obesity in Children

Tertiary prevention of childhood obesity focuses on preventing or treating complications associated with obesity and preventing further weight gain. Computer modeling projects that the prevalence of CVD will increase among young adults, as a result of higher rates of adolescent overweight, with a possible 100,000 additional cases by 2035.[92] Tertiary prevention is essential to avert this projection.

Lifestyle Changes

Nonpharmacologic methods involving the whole family in lifestyle change—diet, activity, and behavior change—are generally preferred in children. The Dietary Guidelines for Americans provide an excellent starting point for dietary intervention. Obese children over age 4 should consume no more than 30% of calories from fat, and saturated fat, cholesterol, and sodium intake should be limited. The DASH diet, described in Chapter 5, is considered appropriate for children, as long as adequate protein and calorie intake are assured.[52]

The importance of physical activity cannot be overstated. Physical activity has been shown to improve insulin sensitivity, lower LDL cholesterol, and prevent age-related increases in blood pressure.[93] Seriously obese children may benefit from a supervised exercise program (to prevent injury as they gain fitness), but it is critical that activity be fun.

Changes in diet and exercise can reverse CVD risk factors in obese children. In one study of obese pre-pubertal children, a 1-year family-based program of nutrition (focus on lowering dietary fat), exercise, and behavioral therapy resulted in 24 of 32 children substantially lowering weight. For those who lost weight, but not those who did not, carotid artery IMT decreased.[77]

When BMI exceeds the 95th percentile and comorbidities are present, more than lifestyle interventions may be needed. Pharmacological treatment and, as a last resort, bariatric surgery are sometimes recommended for severely obese children and adolescents. Lack of health insurance reimbursement could present a barrier to accessing these services for some children.

Pharmacological Treatment

Several medications are available to treat the comorbidities associated with childhood obesity, as well as drugs developed specifically to target excess weight. There are currently no guidelines for which children would benefit the most from obesity-reducing drugs.

Medications for High Blood Lipids Statins are widely prescribed for adults who have high blood cholesterol. Statins up-regulate LDL-C receptors, thereby reducing blood LDL-C levels by 25–45%. The FDA has approved four statins for use in children: simuvastin (Zocor), lovastatin (Mevacor), and atorvastatin (Lipitor) for children 10 years and older; and pravastatin (Pravachol) for children 8 years and older.[94]

In July 2008 the AAP published new recommendations for management of high blood cholesterol in children, lowering from 10 years to 8 years the age that drug therapy can start and adding statins to the list of approved medications.[95] The AAP recommends pharmacological therapy under these conditions:[96]

familial hypercholesterolemia Disorder in which approximately 1:1,000 individuals inherit a gene increasing the risk of high cholesterol beginning at a very early age.

- For children under 8 years, only if LDL-C > 500 mg/dL (this would likely only occur in **familial hypercholesterolemia**);
- For children 10 years and older, only if LDL-C remains above 160 mg/dL after dietary changes and there are CVD risk factors (obesity, hypertension, family history);
- For diabetic children 10 years and older, if LDL-C is persistently at or above 130 mg/dL.

Statins are not routinely prescribed, because they can have side effects, such as elevated liver enzymes, gastrointestinal issues, and decreased serum levels of vitamins and minerals. Although short-term studies suggest that statins are safe for children, there have been no long-term follow-up studies.

Medications for Hypertension Unless blood pressure exceeds the 99th percentile, lifestyle interventions will be used first to reduce blood pressure. Several antihypertensive medications prescribed for adults, such as diuretics and beta-blockers, have been successfully used by children without serious medical complications.[97] Antihypertensive drug therapy would be indicated if: other comorbidities are present; high blood pressure persists despite lifestyle change; or there is hypertensive target-organ damage.

Medications for Type 2 Diabetes Mellitus Metformin is an insulin sensitizer that is tolerated well by adults and adolescents. It has been used with obese adolescents who have hyperinsulinemia as an adjunct to a program of diet, exercise, and behavioral therapy. The FDA has approved several drugs that contain metformin for children ages 9 years and older. In addition to improving insulin levels and insulin sensitivity, metformin also produces small weight losses[98,99] and positive changes in body composition.[100] The FDA recently issued an advisory about increased bone fracture risk in adult women taking some of the metformin-containing drugs used to treat type 2 diabetes mellitus (reported on the May 2007 FDA Patient Safety News show. Available from: www.access-data.fda.gov/psn/printer-full.cfm?id = 67; Accessed July 16, 2010).

Medications for Obesity Treatment Until recently, two drugs were approved by the FDA for the treatment of childhood obesity: sibutramine (for ages 16 years and older) and orlistat (for ages 12 years and older). Sibutramine suppresses appetite by inhibiting reuptake of norepinephrine and serotonin in the brain; and orlistat inhibits fat absorption.

Larger weight losses have been reported with sibutramine than with orlistat. In a year-long behavioral treatment program for severely obese 12- to 16-year-olds, the group taking sibutramine lost significantly more weight (BMI decreased by 3.1 kg/m^2) than the group taking a placebo.[101] At the end of the year 17% of the sibutramine group had lost enough weight to fall below the 95th BMI percentile, in addition to improving waist circumference, triglycerides, HDL-C, and insulin levels. However, the sibutramine group reported significantly more events of **tachycardia** and slightly more occurrences of dry mouth, constipation, dizziness, and insomnia. Recall from Chapter 10 that sibutramine's adverse health effects prompted the FDA to request its withdrawal from the United States market.

A year-long study of orlistat resulted in less dramatic BMI changes. Adolescents (12 to 16 years) participated in a program of behavior modification, exercise, and calorie reduction while taking orlistat or a placebo. The orlistat group significantly decreased BMI (by 0.86 kg/m^2), waist circumference, and body fat.[102] The primary adverse effects were gastrointestinal tract issues. However, seven children in this study developed gall bladder disease, and one had to have her gall bladder removed.

No obesity drug study with children and adolescents has lasted more than 1 year. Several have identified side effects of concern. Whether medications provide to be safe and effective over the long term remains to be seen.

tachycardia
Resting heart rate exceeding 100 beats per minute.

Bariatric Surgery

Recently bariatric surgery has emerged as a potential treatment for severely obese adolescents. This should really be a treatment of last resort—and, at least for now—it is. Although adolescent cases account for only about 0.7% of bariatric surgery nationwide, between 2000–2003 the number of bariatric surgeries performed on adolescent patients more than tripled.[103]

Criteria for bariatric surgery include:[56,104]

- Age: at least 13 years old for females and 15 for males
- BMI: > 50; or > 40 with serious comorbidity
- Maturity: Adolescents need to be not only physically mature but also emotionally and cognitively ready for the surgery and its aftermath
- Failure of other approaches: lifestyle modification and pharmacotherapy should have been tried and been ineffective
- No pregnancy for at least one year

Both the Roux-en-Y gastric bypass and adjustable gastric banding have been used with adolescents and are effective for weight loss.[103,105] Adjustable gastric banding is not approved for use in people under age 18 years, and very few insurance plans cover it. Roux-en-Y gastric bypass results in greater weight losses than gastric banding but has more post-operative complications. Both techniques have the potential to improve comorbidities like type 2 diabetes mellitus, obstructive sleep apnea, hypertension, abnormal blood lipids, and NFLD.

Bariatric surgery requires medical monitoring for a lifetime. Adolescents who have this surgery need to adhere to specific nutritional requirements—vitamin and mineral supplementation, very low carbohydrate intake, and very low calorie intake—as well as engage in regular physical activity. Compliance with these requirements is low among adolescents. One six-year follow-up found that fewer than one-fourth of post-bariatric surgery adolescent patients exercised weekly, most continued to eat high-fat snacks, and only 13% supplemented as directed with vitamins and minerals.[106]

Tertiary Prevention: Summary

Tertiary prevention (or maintenance intervention) involves preventing the development of comorbidities associated with obesity, most notably type 2 diabetes mellitus, metabolic syndrome, and CVD. Obese children and adolescents who engage in regular physical activity and make prudent dietary modifications may prevent, delay the development of, or reduce the complications of these comorbidities. For severely obese adolescents, pharmacological or surgical treatments may be indicated.

Case Study. Penny and Pablo, Part 4

Pablo and Penny are now 15 years old and in high school.

Pablo has grown into a powerfully built young man who does well in school and is hoping to get an athletic scholarship to go to college. This year he has experienced some health problems. His knees sometimes hurt when he runs, and he has been so tired that he actually fell asleep in class twice. He plays defensive tackle on the football team, and his coach frequently tells him to stay bulked up if he wants that college scholarship. When Pablo had his blood pressure taken by the team physician, it was 134/85, so the doctor recommended that Pablo see his family doctor for follow-up. Pablo's family does not have a doctor or even medical insurance, so his mother took him to a church health fair, where his blood pressure was still about the same, his total cholesterol was 200 mg/dL, and his glucose level was in normal ranges. At the health fair his mother observed a

Case Study. (Continued)

cooking demonstration about lowering the fat and sodium content of traditional Hispanic foods, which interested her, although she did not think her husband would want her to adjust any of her recipes. Pablo now weighs 230 lbs and is 71 inches tall.

Penny is having a difficult time in high school. She is a good student but struggles with the social aspects of school. She sings in the chorus so is allowed to opt out of physical education. She recently visited her grandmother in Florida and swam daily in the retirement community pool. She would like to keep swimming but balks at the thought of her classmates seeing her in a bathing suit. Penny's mother believes that she should have gastric bypass surgery, since she is still heavy and dieting doesn't seem to be working. At a recent doctor's visit, Penny's blood pressure was 125/82, HDL-C = 30 mg/dL, and fasting glucose = 120 mg/dL. She weighs 189 lbs and is 64 inches tall, with a waist measurement of 36 inches.

1. Calculate Pablo's and Penny's BMIs and determine their BMI percentiles.
2. Evaluate Pablo's and Penny's blood pressure, total cholesterol, and blood glucose results.
3. Summarize Pablo's and Penny's health risks.
4. Is Penny a good candidate for bariatric surgery?
5. Modify a diet plan for Penny and Pablo using the DASH diet.
6. What "upstream" approaches would have helped Penny and Pablo to achieve a healthy weight?

REFERENCES

1. Ogden, C. L., Carroll, M. D., Curtin, L. R., Lamb, M. M., & Flegal, K. M. (2010). Prevalence of high body mass index in US children and adolescents, 2007–2008. *JAMA, 303*(3), 244–245.

2. Centers for Disease Control and Prevention. (2009). Obesity prevalence among low-income, preschool-aged children—United States, 1998–2008. *Morbidity and Mortality Weekly Reports, 58*(28), 769–773.

3. Maziak, W., Ward, K. D., & Stockton, M. B. (2008). Childhood obesity: Are we missing the big picture? *Obesity Reviews, 9*(1), 35–42.

4. Guo, S. S., Huang, C., Maynard, L. M., Demerath, E., Towne, B., Chumlea, W. C., et al. (2000). Body mass index during childhood, adolescence and young adulthood in relation to adult overweight and adiposity: The Fels longitudinal study. *International Journal of Obesity and Related Metabolic Disorders, 24*(12), 1628–1635.

5. Freedman, D. S., Khan, L. K., Dietz, W. H., Srinivasan, S. R., & Berenson, G. S. (2001). Relationship of childhood obesity to coronary heart disease risk factors in adulthood: The Bogalusa Heart Study. *Pediatrics, 108*(3), 712–718.

6. Daniels, S. R., Arnett, D. K., Eckel, R. H., Gidding, S. S., Hayman, L. L., Kumanyika, S., et al. (2005). Overweight in children and adolescents: Pathophysiology, consequences, prevention, and treatment. *Circulation, 111*(15), 1999–2012.

7. Magarey, A. M., Daniels, L. A., Boulton, T. J., & Cockington, R. A. (2003). Predicting obesity in early adulthood from childhood and parental obesity. *International Journal of Obesity and Related Metabolic Disorders, 27*(4), 505–513.

8. Freedman, D. S., Kettel Khan, L., Serdula, M. K., Srinivasan, S. R., & Berenson, G. S. (2001). BMI rebound, childhood height and obesity among adults: The Bogalusa Heart Study. *International Journal of Obesity and Related Metabolic Disorders, 25*(4), 543–549.

9. Deshmukh-Taskar, P., Nicklas, T. A., Morales, M., Yang, S. J., Zakeri, I., & Berenson, G. S. (2006). Tracking of overweight status from childhood to young adulthood: The Bogalusa Heart Study. *European Journal of Clinical Nutrition, 60*(1), 48–57.

10. Olshansky, S. J., Passaro, D. J., Hershow, R. C., Layden, J., Carnes, B. A., Brody, J., et al. (2005). A

potential decline in life expectancy in the United States in the 21st century. *New England Journal of Medicine, 352*(11), 1138–1145.

11. Lagstrom, H., Hakanen, M., Niinikoski, H., Viikari, J., Ronnemaa, T., Saarinen, M., et al. (2008). Growth patterns and obesity development in overweight or normal-weight 13-year-old adolescents: The STRIP study. *Pediatrics, 122*(4), e876–883.

12. Reilly, J. J., Armstrong, J., Dorosty, A. R., Emmett, P. M., Ness, A., Rogers, I., et al. (2005). Early life risk factors for obesity in childhood: Cohort study. *BMJ (Clinical Research Ed.), 330*(7504), 1357.

13. Whitaker, R. C. (2004). Predicting preschooler obesity at birth: The role of maternal obesity in early pregnancy. *Pediatrics, 114*(1), e29–e36.

14. Mayer, J. (1953). Genetic, traumatic, and environmental factors in the etiology of obesity. *Physiological Reviews, 33*(4), 472–508.

15. Lee, S. M., Burgeson, C. R., Fulton, J. E., & Spain, C. G. (2007). Physical education and physical activity: Results from the School Health Policies and Programs Study 2006. *Journal of School Health, 77*(8), 435–463.

16. Physical activity levels among children aged 9–13 years—United States, 2002. *Morbidity and Mortality Weekly Reports, 52*(33), 785–788.

17. Eaton, D. K., Kann, L., Kinchen, S., Shanklin, S., Ross, J., Hawkins, J., et al. (2010). Youth risk behavior surveillance – United States, 2009. *Surveillance Summaries, MMWR, 59*(SS–5), 1–148.

18. Gordon-Larsen, P., Nelson, M. C., & Popkin, B. M. (2004). Longitudinal physical activity and sedentary behavior trends: Adolescence to adulthood. *American Journal of Preventive Medicine, 27*(4), 277–283.

19. Kaiser Family Foundation. (2005) *Generation M: Media in the lives of 8–18 year olds.* Menlo Park, CA: Kaiser Family Foundation, p. 10.

20. Vandewater, E. A., Rideout, V. J., Wartella, E. A., Huang, X., Lee, J. H., & Shim, M. S. (2007). Digital childhood: Electronic media and technology use among infants, toddlers, and preschoolers. *Pediatrics, 119*(5), e1006–e1015. doi:10.1542/peds.2006–1804.

21. Gortmaker, S. L., Peterson, K., Wiecha, J., Sobol, A. M., Dixit, S., Fox, M. K., et al. (1999). Reducing obesity via a school-based interdisciplinary intervention among youth: Planet Health. *Archives of Pediatrics & Adolescent Medicine, 153*(4), 409–418.

22. O'Brien, M., Nader, P. R., Houts, R. M., Bradley, R., Friedman, S. L., Belsky, J., et al. (2007). The ecology of childhood overweight: A 12-year longitudinal analysis. *International Journal of Obesity, 31*(9), 1469–1478.

23. Proctor, M. H., Moore, L. L., Gao, D., Cupples, L. A., Bradlee, M. L., Hood, M. Y., et al. (2003). Television viewing and change in body fat from preschool to early adolescence: The Framingham Children's Study. *International Journal of Obesity and Related Metabolic Disorders, 27*(7), 827–833.

24. Danner, F. W. (2008). A national longitudinal study of the association between hours of TV viewing and the trajectory of BMI growth among US children. *Journal of Pediatric Psychology, 33*(10), 1100–1107.

25. Epstein, L. H., Roemmich, J. N., Paluch, R. A., & Raynor, H. A. (2005). Influence of changes in sedentary behavior on energy and macronutrient intake in youth. *American Journal of Clinical Nutrition, 81*(2), 361–366.

26. Harrison K., & Marske, A. L. (2005). Nutritional content of foods advertised during the television programs children watch most. *American Journal of Public Health, 95*(9), 1568–1574.

27. McGinnis, J. M., Gootman, J. A., & Kraak, V. I. (Eds.). (2006). *Food marketing to children and youth: Threat or opportunity.* Washington, D.C.: National Academies Press.

28. Mellecker, R. R., & McManus, A. M. (2008). Energy expenditure and cardiovascular responses to seated and active gaming in children. *Archives of Pediatrics Adolescent Medicine, 162*(9), 886–891.

29. Enns, C. W., Mickle, S. J., & Goldman, J. D. (2002). Trends in food and nutrient intakes by children in the United States. *Family Economics & Nutrition Review, 14*(2), 56.

30. Enns, C. W., Mickle, S. J., & Goldman, J. D. (2003). Trends in food and nutrient intakes by adolescents in the United States. *Family Economics & Nutrition Review, 15*(2), 15–27.

31. U.S. Department of Agriculture, Agricultural Research Service. (2008). *Nutrient intakes from food: Mean amounts and percentages of calories from protein, carbohydrate, fat, and alcohol, one day, 2005–2006.* Available from http://www.ars.usda.gov/ba/bhnrc/fsrg. Accessed December 31, 2010.

32. Brownell, K. D., & Frieden, T. R. (2009). Ounces of prevention – the public policy case for taxes on sugared beverages. *New England Journal of Medicine, 360*(18), 1805–1808.

33. Wootan, M. G., Batada, A., & Marchlewicz, E. (2008). *Kids' meals: Obesity on the menu.* Washington, D.C.: Center for Science in the Public Interest.

34. Zoumas-Morse, C., Rock, C. L., Sobo, E. J., & Neuhouser, M. L. (2001). Children's patterns of macronutrient intake and associations with restaurant and home eating. *Journal of the American Dietetic Association, 101*(8), 923–925.

35. Ebbeling, C. B., Sinclair, K. B., Pereira, M. A., Garcia-Lago, E., Feldman, H. A., & Ludwig, D. S. (2004). Compensation for energy intake from fast food among overweight and lean adolescents. *JAMA, 291*(23), 2828–2833.

36. Ludwig, D. S., Peterson, K. E., & Gortmaker, S. L. (2001). Relation between consumption of sugar-sweetened drinks and childhood obesity: A prospective, observational analysis. *Lancet, 357,* 505–508.

37. Monsivais, P. & Drewnowski, A. (2007). The rising cost of low-energy dense foods. *Journal of the American Dietetic Association, 107*(12), 2071–2076.

38. Darmon, N., & Drewnowski, A. (2008). Does social class predict diet quality? *American Journal of Clinical Nutrition, 87*(5), 1107–1117.

39. Centers for Disease Control and Prevention. (1996). Guidelines for school health programs to promote lifelong healthy eating. *Morbidity and Mortality Weekly Reports, 45*(RR–9), 1–41.

40. GAO (U.S. Government Printing Office). (2005). *School meal programs: Competitive foods are widely available and generate substantial revenues for school.* No. GAO–05–563). Washington, D.C.: GAO.

41. O'Toole, T. P., Anderson, S., Miller, C., & Guthrie, J. (2007). Nutrition services and foods and beverages available at school: Results from the School Health Policies and Programs Study 2006. *Journal of School Health, 77*(8), 500–521.

42. American Academy of Pediatrics. Committee on Public Education. (2001). American Academy of Pediatrics: Children, adolescents, and television. *Pediatrics, 107*(2), 423–426.

43. Golan, M., & Crow, S. (2004). Parents are key players in the prevention and treatment of weight-related problems. *Nutrition Reviews, 62*(1), 39–50.

44. Koplan, J. P., Liverman, C. T., & Kraak, V. A. (Eds). (2005). *Preventing childhood obesity: Health in the balance.* Washington, D.C.: The National Academies Press.

45. Kahn, E. B., Ramsey, L. T., Brownson, R. C., Heath, G. W., Howze, E. H., Powell, K. E., et al. (2002). The effectiveness of interventions to increase physical activity. A systematic review. *American Journal of Preventive Medicine, 22*(4 Suppl), 73–107.

46. American Dietetic Association (ADA). (2006). Position of the American Dietetic Association: Individual-, family-, school-, and community-based interventions for pediatric overweight. *Journal of the American Dietetic Association, 106*(6), 925–945.

47. Datar, A., & Sturm, R. (2004). Physical education in elementary school and body mass index: Evidence from the early childhood longitudinal study. *American Journal of Public Health, 94*(9), 1501–1506.

48. Eggar, J. R., Bartley, K. F., Benson, L., Bellino, D., & Kerker, B. (2009). Childhood obesity is a serious concern in New York City: Higher levels of fitness associated with better academic performance. *NYC Vital Signs, 8*(1), 1–4. Available from http://www.nyc.gov/html/doh/downloads/pdf/survey/survey-2009fitnessgram.pdf. Accessed December 31, 2010.

49. Webber, L. S., Catellier, D. J., Lytle, L. A., Murray, D. M., Pratt, C. A., Young, D. R., et al. (2008). Promoting physical activity in middle school girls: Trial of activity for adolescent girls. *American Journal of Preventive Medicine, 34*(3), 173–184.

50. Beets, M. W., Beighle, A., Erwin, H. E., & Huberty, J. L. (2009). After-school program impact on physical activity and fitness a meta-analysis. *American Journal of Preventive Medicine, 36*(6), 527–537.

51. Sallis, J. F., McKenzie, T. L., Alcaraz, J. E., Kolody, B., Faucette, N., & Hovell, M. F. (1997). The effects of a 2-year physical education program (SPARK) on physical activity and fitness in elementary school students. sports, play and active recreation for kids. *American Journal of Public Health, 87*(8), 1328–1334.

52. Gidding, S. S. (2007). Special article: Physical activity, physical fitness, and cardiovascular risk factors in childhood. *American Journal of Lifestyle Medicine, 1*(6), 499–505.

53. American Academy of Pediatrics, Section on Breastfeeding. (2005). Breastfeeding and the use of human milk. *Pediatrics, 115*(2), 496–506.

54. American Academy of Pediatrics, Committee on Nutrition. (2001). The use and misuse of fruit juice in pediatrics. *Pediatrics, 107*(5), 1210–1213.

55. Fox, M. K., Pac, S., Devaney, B., & Jankowski, L. (2004). Feeding infants and toddlers study: What foods are infants and toddlers eating? *Journal of the American Dietetic Association, 104*(1 Suppl 1), s22–s30.

56. Barlow, S. E., & and the Expert Committee, (2007). Expert committee recommendations regarding the

prevention, assessment, and treatment of child and adolescent overweight and obesity: Summary report. *Pediatrics*, *120*(Supplement 4), S164–S192.

57. Krebs, N. F., Himes, J. H., Jacobson, D., Nicklas, T. A., Guilday, P., & Styne, D. (2007). Assessment of child and adolescent overweight and obesity. *Pediatrics*, *120*(Supplement 4), S193–228.

58. Katzmarzyk, P. T.; Srinivasan, S. R.; Chen, W.; Malina, R. M.; Bouchard, C.; Berenson, G. S. (2004) Body mass index, waist circumference, and clustering of cardiovascular disease risk factors in a biracial sample of children and adolescents. *Pediatrics*, *114*(2), 198–205.

59. Fernandez, J. R., Redden, D. T., Pietrobelli, A., & Allison, D. B. (2004). Waist circumference percentiles in nationally representative samples of African-American, European-American, and Mexican-American children and adolescents. *Journal of Pediatrics*, *145*, 439–444.

60. Rolland-Cachera, M. F., Deheeger, M., & Guilloud-Bataille, M. (1987). Tracking the development of adiposity from one month of age to adulthood. *Annals of Human Biology*, *14*, 219–229.

61. Epstein, L. H., Myers, M. D., Raynor, H. A., & Saelens, B. E. (1998). Treatment of pediatric obesity. *Pediatrics*, *101*(3 Pt 2), 554–570.

62. Angelopoulos, P. D., Milionis, H. J., Grammatikaki, E., Moschonis, G., & Manios, Y. (2009). Changes in BMI and blood pressure after a school-based intervention: The CHILDREN study. *European Journal of Public Health*, *19*(3), 319–325.

63. Taylor, R. W., McAuley, K. A., Barbezat, W., Strong, A., Williams, S. M., & Mann, J. I. (2007). APPLE project: 2-y findings of a community-based obesity prevention program in primary school–age children. *American Journal of Clinical Nutrition*, *86*(3), 735–742.

64. McKenzie, T. L., Li, D., Derby, C. A., Webber, L. S., Luepker, R. V., & Cribb, P. (2003). Maintenance of effects of the CATCH physical education program: Results from the CATCH-ON study. *Health Education & Behavior*, *30*(4), 447–462.

65. Dowda, M., James, F., Sallis, J. F., McKenzie, T. L., Rosengard, P., & Kohl, H. W., 3rd. (2005). Evaluating the sustainability of SPARK physical education: A case study of translating research into practice. *Research Quarterly for Exercise and Sport*, *76*(1), 11–19.

66. Luepker, R. V., Perry, C. L., McKinlay, S. M., Nader, P. R., Parcel, G. S., Stone, E. J., et al. (1996). Outcomes of a field trial to improve children's dietary patterns and physical activity the child and adolescent trial for cardiovascular health. CATCH collaborative group. *JAMA*, *275*(10), 768–776.

67. Plaza, C. I., & Henze, C. (2006). *Balance: A report on state action to promote nutrition, increase physical activity, and prevent obesity.* (Issue 3). Princeton, N.J.: Robert Wood Johnson Foundation.

68. Epstein, L. H., Valoski, A., Wing, R. R., & McCurley, J. (1994). Ten-year outcomes of behavioral family-based treatment for childhood obesity. *Health Psychology*, *13*(5), 373–383.

69. Nuutinen, O., & Knip, M. (1992). Long-term weight control in obese children: Persistence of treatment outcome and metabolic changes. *International Journal of Obesity and Related Metabolic Disorders*, *16*(4), 279–287.

70. Savoye, M., Shaw, M., Dziura, J., Tamborlane, W. V., Rose, P., Guandalini, C., et al. (2007). Effects of a weight management program on body composition and metabolic parameters in overweight children: A randomized controlled trial. *JAMA*, *297*(24), 2697–2704.

71. Epstein, L. H., Valoski, A. M., Kalarchian, M. A., & McCurley, J. (1995). Do children lose and maintain weight easier than adults: A comparison of child and parent weight changes from six months to ten years. *Obesity Research*, *3*(5), 411–417.

72. Epstein, L. H., & Wing, R. R. (1987). Behavioral treatment of childhood obesity. *Psychological Bulletin*, *101*(3), 331–342.

73. Thomas, P. R. (Ed.). (1995). *Weighing the options: Criteria for evaluating weight-management programs.* Washington, D.C.: National Academy Press.

74. Freedman, D. S., Serdula, M. K., Srinivasan, S. R., & Berenson, G. S. (1999). Relation of circumferences and skinfold thicknesses to lipid and insulin concentrations in children and adolescents: The Bogalusa Heart Study. *American Journal of Clinical Nutrition*, *69*, 308–317.

75. Skinner, A. C., Mayer, M. L., Flower, K., & Weinberger, M. (2008). Health status and health care expenditures in a nationally representative sample: How do overweight and healthy-weight children compare? *Pediatrics*, *121*(2), e269–e277.

76. Freedman, D. S., Dietz, W. H., Srinivasan, S. R., & Berenson, G. S. (1999). The relation of overweight to cardiovascular risk factors among children and adolescents: The Bogalusa Heart Study. *Pediatrics*, *103*(6 Pt 1), 1175–1182.

77. Wunsch, R., de Sousa, G., Toschke, A. M., & Reinehr, T. (2006). Intima-media thickness in

obese children before and after weight loss. *Pediatrics, 118*(6), 2334–2340.

78. Haney, E. M., Huffman, L. H., Bougatsos, C., Freeman, M., Steiner, R. D., & Nelson, H. D. (2007). Screening and treatment for lipid disorders in children and adolescents: Systematic evidence review for the US preventive services task force. *Pediatrics, 120*(1), e189–214.

79. Williams, C. L., Hayman, L. L., Daniels, S. R., Robinson, T. N., Steinberger, J., Paridon, S., et al. (2002). Cardiovascular health in childhood: A statement for health professionals from the Committee on Atherosclerosis, Hypertension, and Obesity in the Young (AHOY) of the Council on Cardiovascular Disease in the Young, American Heart Association. *Circulation, 106*(1), 143–160.

80. Maggio, A. B., Aggoun, Y., Marchand, L. M., Martin, X. E., Herrmann, F., Beghetti, M., et al. (2008). Associations among obesity, blood pressure, and left ventricular mass. *The Journal of Pediatrics, 152*(4), 489–493.

81. Muntner, P., He, J., Cutler, J. A., Wildman, R. P., & Whelton, P. K. (2004). Trends in blood pressure among children and adolescents. *JAMA, 291*(17), 2107–2113.

82. Cole, T. J., Bellizzi, M. C., Flegal, K. M., & Dietz, W. H. (2000). Establishing a standard definition for child overweight and obesity worldwide: International survey. *BMJ, 320*(7244), 1240–1243.

83. Verhulst, S. L., Schrauwen, N., Haentjens, D., Suys, B., Rooman, R. P., Van Gaal, L., et al. (2007). Sleep-disordered breathing in overweight and obese children and adolescents: Prevalence, characteristics and the role of fat distribution. *Archives of Disease in Childhood, 92*(3), 205–208.

84. Sinha, R., Fisch, G., Teague, B., Tamborlane, W. V., Banyas, B., Allen, K., et al. (2002). Prevalence of impaired glucose tolerance among children and adolescents with marked obesity. *New England Journal of Medicine, 346*(11), 802–810.

85. Isganaitis, E., & Lustig, R. H. (2005). Fast food, central nervous system insulin resistance, and obesity. *Arteriosclerosis, Thrombosis, and Vascular Biology, 25*(12), 2451–2462.

86. Johnson, W. D., Kroon, J. J. M., Greenway, F. L., Bouchard, C., Ryan, D., & Katzmarzyk, P. T. (2009). Prevalence of risk factors for metabolic syndrome in adolescents: National Health and Nutrition Examination Survey (NHANES), 2001–2006. *Archives of Pediatrics Adolescent Medicine, 163*(4), 371–377.

87. Cook, S., Weitzman, M., Auinger, P., Nguyen, M., & Dietz, W. H. (2003). Prevalence of a metabolic syndrome phenotype in adolescents: Findings from the third National Health and Nutrition Examination Survey, 1988–1994. *Archives of Pediatrics Adolescent Medicine, 157*(8), 821–827.

88. Chi, C. H., Wang, Y., Wilson, D. M., & Robinson, T. N. (2006). Definition of metabolic syndrome in preadolescent girls. *Journal of Pediatrics, 148*(6), 788–792.

89. Morrison, J. A., Friedman, L. A., & Gray-McGuire, C. (2007). Metabolic syndrome in childhood predicts adult cardiovascular disease 25 years later: The Princeton lipid research clinics follow-up study. *Pediatrics, 120*(2), 340–345.

90. Schwimmer, J. B., Pardee, P. E., Lavine, J. E., Blumkin, A. K., & Cook, S. (2008). Cardiovascular risk factors and the metabolic syndrome in pediatric nonalcoholic fatty liver disease. *Circulation, 118*(3), 277–283.

91. Schwimmer, J. B., Deutsch, R., Kahen, T., Lavine, J. E., Stanley, C., & Behling, C. (2006). Prevalence of fatty liver in children and adolescents. *Pediatrics, 118*(4), 1388–1393.

92. Bibbins-Domingo, K., Coxson, P., Pletcher, M. J., Lightwood, J., & Goldman, L. (2007). Adolescent overweight and future adult coronary heart disease. *New England Journal of Medicine, 357*(23), 2371–2379.

93. Gidding, S. S., Dennison, B. A., Birch, L. L., Daniels, S. R., Gillman, M. W., Lichtenstein, A. H., et al. (2005). Dietary recommendations for children and adolescents: A guide for practitioners: Consensus statement from the American Heart Association. *Circulation, 112*(13), 2061–2075.

94. Belay, B., Belamarich, P. F., & Tom-Revzon, C. (2007). The use of statins in pediatrics: Knowledge base, limitations, and future directions. *Pediatrics, 119*(2), 370–380.

95. de Ferranti, S., & Ludwig, D. S. (2008). Storm over statins—the controversy surrounding pharmacologic treatment of children. *New England Journal of Medicine, 359*(13), 1309–1312.

96. Daniels, S. R., Greer, F. R., & Committee on Nutrition. (2008). Lipid screening and cardiovascular health in childhood. *Pediatrics, 122*(1), 198–208.

97. U.S. Department of Health and Human Services. (2005). *The fourth report on the diagnosis, evaluation, and treatment of high blood pressure in children and adolescents.* Bethesda, MD: National Institutes of Health, National Heart Lung, and Blood Institute.

468   **PART VI** Identification, Prevention, and Treatment of Eating Disorders and Childhood Obesity

98. Caprio, S., Daniels, S. R., Drewnowski, A., Kaufman, F. R., Palinkas, L. A., Rosenbloom, A. L., et al. (2008). Influence of race, ethnicity, and culture on childhood obesity: Implications for prevention and treatment. *Diabetes Care, 31*(11), 2211–2221.

99. Atabek, M. E., & Pirgon, O. (2008). Use of metformin in obese adolescents with hyperinsulinemia: A 6-month, randomized, double-blind, placebo-controlled clinical trial. *Journal of Pediatric Endocrinology & Metabolism, 21*(4), 339–348.

100. Burgert, T. S., Duran, E. J., Goldberg-Gell, R., Dziura, J., Yeckel, C. W., Katz, S., et al. (2008). Short-term metabolic and cardiovascular effects of metformin in markedly obese adolescents with normal glucose tolerance. *Pediatric Diabetes, 9*(6), 567–576.

101. Berkowitz, R. I., Fujioka, K., Daniels, S. R., Hoppin, A. G., Owen, S., Perry, A. C., et al. (2006). Effects of sibutramine treatment in obese adolescents: A randomized trial. *Annals of Internal Medicine, 145*(2), 81–90.

102. Chanoine, J., Hampl, S., Jensen, C., Boldrin, M., & Hauptman, J. (2005). Effect of orlistat on weight and body composition in obese adolescents: A randomized controlled trial. *JAMA, 293*(23), 2873–2883.

103. Tsai, W. S., Inge, T. H., & Burd, R. S. (2007). Bariatric surgery in adolescents: Recent national trends in use and in-hospital outcome. *Archives of Pediatrics Adolescent Medicine, 161*(3), 217–221.

104. Inge, T. H., Krebs, N. F., Garcia, V. F., Skelton, J. A., Guice, K. S., Strauss, R. S., et al. (2004). Bariatric surgery for severely overweight adolescents: Concerns and recommendations. *Pediatrics, 114*(1), 217–223.

105. Strauss, R. S., Bradley, L. J., & Brolin, R. E. (2001). Gastric bypass surgery in adolescents with morbid obesity. *Journal of Pediatrics, 138*(4), 499–504.

106. Hinds, H. L. (2005). Pediatric obesity: Ethical dilemmas in treatment and prevention. *The Journal of Law, Medicine & Ethics, 33*(3), 599–602.

Appendix A

Body Composition Assessment Tools

TABLE A-1 Body Mass Index Calculator

Body weight (pounds)

Height	18	19	20	21	22	23	24	25	26	27	28	29	30	31	32	33	34	35	36	37	38	39	40
4'10"	86	91	96	100	105	110	115	119	124	129	134	138	143	148	153	158	162	167	172	177	181	186	191
4'11"	89	94	99	104	109	114	119	124	128	133	138	143	148	153	158	163	168	173	178	183	188	193	198
5'0"	92	97	102	107	112	118	123	128	133	138	143	148	153	158	163	168	174	179	184	189	194	199	204
5'1"	95	100	106	111	116	122	127	132	137	142	147	153	158	164	169	174	180	185	190	195	201	206	211
5'2"	98	104	109	115	120	126	131	136	142	147	153	158	164	169	175	180	186	191	196	202	207	213	218
5'3"	102	107	113	118	124	130	135	141	146	152	158	163	169	175	180	186	191	197	203	208	214	220	225
5'4"	105	110	116	122	128	134	140	145	151	157	163	169	174	180	186	192	197	204	209	215	221	227	232
5'5"	108	114	120	126	132	138	144	150	156	162	168	174	180	186	192	196	204	210	216	222	228	234	240
5'6"	112	118	124	130	136	142	148	155	161	167	173	179	186	192	198	204	210	216	223	229	235	241	247
5'7"	115	121	127	134	140	146	153	159	166	172	178	185	191	198	204	211	217	223	230	236	242	249	255
5'8"	118	125	131	138	144	151	158	164	171	177	184	190	197	203	210	216	223	230	236	243	249	256	262
5'9"	122	128	135	142	149	155	162	169	176	182	189	196	203	209	216	223	230	236	243	250	257	263	270
5'10"	126	132	139	146	153	160	167	174	181	188	195	202	209	216	222	229	236	243	250	257	264	271	278
5'11"	129	136	143	150	157	165	172	179	186	193	200	208	215	222	229	236	243	250	257	265	272	279	286
6'0"	132	140	147	154	162	169	177	184	191	199	206	213	221	228	235	242	250	258	265	272	279	287	294
6'1"	136	144	151	159	166	174	182	189	197	204	212	219	227	235	242	250	257	265	272	280	288	295	302
6'2"	141	148	155	163	171	179	186	194	202	210	218	225	233	241	249	256	264	272	280	287	295	303	311
6'3"	144	152	160	168	176	184	192	200	208	216	224	232	240	248	256	264	272	279	287	295	303	311	319
6'4"	148	156	164	172	180	189	197	205	213	221	230	238	246	254	263	271	279	287	295	304	312	320	328
6'5"	151	160	168	176	185	193	202	210	218	227	235	244	252	261	269	277	286	294	303	311	319	328	336
6'6"	155	164	172	181	190	198	207	216	224	233	241	250	259	267	276	284	293	302	310	319	328	336	345

Under-weight (<18.5)   Healthy weight (18.5–24.9)   Overweight (25–29.9)   Obese (≥30)

Source: Sizer, F., & Whitney, E. (2008). *Nutrition: Concepts and Controversies* (11th edition).

TABLE A-2 Metropolitan Desirable Weights for Men and Women, 1959 (according to height and frame, ages 25 and over)

Height (in shoes)	Weight (in lb in indoor clothing)		
	Small frame	Medium frame	Large frame
Men			
5'2"	112–120	118–129	126–141
5'3"	115–123	121–133	129–144
5'4"	118–126	124–136	132–148
5'5"	121–129	127–139	135–152
5'6"	124–133	130–143	138–156
5'7"	128–137	134–147	142–161
5'8"	132–141	138–152	147–166
5'9"	136–145	142–156	151–170
5'10"	140–150	146–160	155–174
5'11"	144–154	150–165	159–179
6'0"	148–158	154–170	164–184
6'1"	152–162	158–175	168–189
6'2"	156–167	162–180	173–194
6'3"	160–171	167–185	178–199
6'4"	164–175	172–190	182–204
Women			
4'10"	92–98	96–107	104–119
4'11"	94–101	98–110	106–122
5'0"	96–104	101–113	109–125
5'1"	99–107	104–116	112–128
5'2"	102–110	107–119	115–131
5'3"	103–113	110–122	118–134
5'4"	108–116	113–126	121–138
5'5"	111–119	116–130	125–142
5'6"	114–123	120–135	129–146
5'7"	118–127	124–139	133–150
5'8"	122–131	128–143	137–154
5'9"	126–136	132–147	141–158
5'10"	130–140	136–151	145–163
5'11"	134–144	140–155	149–168
6'0"	138–148	144–159	153–173

Source: From Metropolitan Life Insurance Company. (1959, November–December). New weight standards for men and women. *Statistical Bulletin*, p. 3; reprinted by permission.

TABLE A-3 Metropolitan Height and Weight Tables for Men and Women, 1983
Weights at ages 25–59 based on lowest mortality. Weights in pounds according to frame (in indoor clothing weighing 5 pounds for men or 3 pounds for women), shoes with 1-inch heels.

Men Height ft	inches	Small frame	Medium frame	Large frame
5	2	128–134	131–141	138–150
5	3	130–136	133–143	140–153
5	4	132–138	135–145	142–156
5	5	134–140	137–148	144–160
5	6	136–142	139–151	146–164
5	7	138–145	142–154	149–168
5	8	140–148	145–157	152–172
5	9	142–151	148–160	155–176
5	10	144–154	151–163	158–180
5	11	146–157	154–166	161–184
6	0	149–160	157–170	164–188
6	1	152–164	160–174	169–192
6	2	155–168	164–178	172–197
6	3	159–162	167–182	176–202
6	4	162–176	171–187	181–207

Women Height ft	inches	Small frame	Medium frame	Large frame
4	10	102–111	109–121	118–131
4	11	103–113	111–123	120–134
5	0	104–115	113–126	122–137
5	1	106–118	115–129	125–140
5	2	108–121	118–132	128–143
5	3	111–124	121–135	131–147
5	4	114–127	124–138	134–151
5	5	117–130	127–141	137–155
5	6	120–133	130–144	140–159
5	7	123–136	133–147	143–163
5	8	126–139	136–150	146–167
5	9	129–142	139–153	149–170
5	10	132–145	142–156	152–173
5	11	135–148	145–159	155–176
6	0	138–151	148–162	158–179

Source: Reproduced with permission of Metropolitan Life Insurance Company; Source of basic data; 1979 Build Study; Society of Actuaries and Association of Life Insurance Medical Directors of America, 1980.

TABLE A-4   Determination of Frame Size from Elbow Breadth

Height (in cm)[a]	Frame size from elbow breadth (in cm [and inches])		
	Small	Medium	Large
	Men		
158–161	<6.4 [2.5]	6.4–7.2 [2.5–2.8]	>7.2 [2.8]
162–171	<6.7 [2.6]	6.7–7.4 [2.6–2.9]	>7.4 [2.9]
172–181	<6.9 [2.7]	6.9–7.6 [2.7–3.0]	>7.6 [3.0]
182–191	<7.1 [2.8]	7.1–7.8 [2.8–3.0]	>7.8 [3.0]
192–193	<7.4 [2.9]	7.4–8.1 [2.9–3.1]	>8.1 [3.1]
	Women		
143–151	<5.6 [2.2]	5.6–6.4 [2.2–2.5]	>6.4 [2.5]
152–161	<5.8 [2.3]	5.8–6.5 [2.3–2.5]	>6.5 [2.5]
162–171	<5.9 [2.3]	5.9–6.6 [2.3–2.6]	>6.6 [2.6]
172–181	<6.1 [2.4]	6.1–6.8 [2.4–2.7]	>6.8 [2.7]
182–183	<6.2 [2.5]	6.2–6.9 [2.5–2.7]	>6.9 [2.7]

[a]Height includes 2.5 cm heels.

Source: From Metropolitan Life Insurance Company (1983, January–June). 1983 Metropolitan height and weight tables for men and women. *Statistical Bulletin;* adapted by permission.

TABLE A-5 Prediction Equations from Skinfold Measures

Note: In all of the following equations, ΣSKF means the sum of skinfolds. *Db(g/cc)* means body density in grams per cubic centimeter.

American Indian women, ages 18–60 years[1]
   Skinfold sites: triceps. Midaxillary, suprailiac ($\Sigma 3SKF$)
   Equation: Db (g/cc) = 1.061983 − 0.000385($\Sigma 3SKF$) − 0.00204(age)
   % fat = [(4.8l/Db) − 4.34] × 100

African American women, ages 18–55 years[2]
   Skinfold sites: Chest, abdomen, thigh, subscapular, triceps, suprailiac, midaxillary ($\Sigma 7SKF$)
   Equation: Db (g/cc) = 1.0970 − 0.00046971($\Sigma 7SKF$)
                          + 0.00000056($\Sigma 7SKF$)^2 − 0.00012828(age)
   % fat = [(4.85/Db) − 4.39] × 100

African American men, ages 18–61 years[3]
   Skinfold sites: chest, abdomen, thigh, subscapular, triceps, suprailiac, midaxillary ($\Sigma 7SKF$)
   Equation: Db (g/cc) = 1.1120 − 0.00043499($\Sigma 7SKF$)
                          + 0.00000055($\Sigma 7SKF$)^2 − 0.00028826(age)
   % fat = [(4.37/Db) − 3.93] × 100

Hispanic women, ages 20–40 years[2]
   Skinfold sites: chest, abdomen, thigh, subscapular, triceps, suprailiac, midaxillary ($\Sigma 7SKF$)
   Equation: Db (g/cc) = 1.0970 − 0.00046971($\Sigma 7SKF$)
                          + 0.00000056($\Sigma 7SKF$)^2 − 0.00012828(age)
   % fat = [(4.87/Db) − 4.4l] × 100

Caucasian women, ages 18–55 years[2]
   Skinfold sites: Triceps, thigh, suprailiac($\Sigma 3SKF$)
   Equation: Db (g/cc) = 1.0994921 − 0.0009929($\Sigma 3SKF$)
                          + 0.0000023($\Sigma 3SKF$)^2 − 0.0001392(age)
   % fat = [(5.01/Db) − 4.57] × 100

Caucasian men, ages 18–61 years[3]
   Skinfold sites: Chest, abdomen, thigh ($\Sigma 3SKF$)
   Equation: Db (g/cc) = 1.09380 − 0.0008267($\Sigma 3SKF$)
                          + 0.0000016($\Sigma 3SKF$)^2 − 0.0002574(age)
   % fat = [(4.95/Db) − 4.50] × 100

Anorexic women, ages 18–55 years[2]
   Skinfold sites: Triceps, thigh, suprailiac($\Sigma 3SKF$)
   Equation: Db (g/cc) = 1.0994921 − 0.0009929($\Sigma 3SKF$)
                          + 0.0000023($\Sigma 3SKF$)^2 − 0.0001392(age)
   % fat = [(5.01/Db) − 4.57] × 100

Female athletes, ages 18–29 years[2]
   Skinfold sites: Triceps, suprailiac, abdomen, thigh ($\Sigma 4SKF$)
   Equation: Db (g/cc) = 1.096095 − 0.0006952($\Sigma 4SKF$)
                          + 0.0000011($\Sigma 4SKF$)^2 − 0.0000714(age)
   % fat = [(5.01/Db) − 4.57] × 100

Male athletes, ages 18–29 years[3]
   Skinfold sites: chest, abdomen, thigh, subscapular, triceps, suprailiac, midaxillary ($\Sigma 7SKF$)
   Equation: Db (g/cc) = 1.112 − 0.00043499($\Sigma 7SKF$)
                          + 0.00000055($\Sigma 7SKF$)^2 − 0.00028826(age)
   % fat = [(4.95/Db) − 4.50] × 100

TABLE A-5 Prediction Equations from Skinfold Measures (*Continued*)

Children, ages 8–18 years[4]

Note: All equations use the triceps (abbreviated T) and subscapular (abbreviated S) skinfolds.

Prepubcscent Caucasian males

$$\% \text{ fat} = 1.21(T + S) - 0.008(T + S)^2 - 1.7$$

Prepubcscent African American males

$$\% \text{ fat} = 1.21(T + S) - 0.008(T + S)^2 - 3.2$$

Pubescent Caucasian males

$$\% \text{ fat} = 1.21(T + S) - 0.008(T + S)^2 - 3.4$$

Pubescent African American, males

$$\% \text{ fat} = 1.21(T + S) - 0.008(T + S)^2 - 5.2$$

Postpubescent Caucasian males

$$\% \text{ fat} = 1.21(T + S) - 0.008(T + S)^2 - 5.5$$

Postpubescent African American males

$$\% \text{ fat} = 1.21(T + S) - 0.008(T + S)^2 - 6.8$$

All females

$$\% \text{ fat} = 1.33(T + S) - 0.013(T + S)^2 - 2.5$$

When triceps + subscapular is greater than 35 mm, use the following equation:

All males: $\% \text{ fat} = 0.783 (T + S) + 1.6$

All females: $\% \text{ fat} = 0.546 (T + S) + 9.7$

Skinfold sites

- Abdominal—A horizontal fold 2 cm to the right of the umbilicus. (*Note:* In the generalized skinfold equations, this is a vertical fold.[2,3])
- Biceps—A vertical fold directly over the belly of the biceps muscle on the front of the arm; this fold is at approximately the same level as the triceps skinfold.
- Calf—A vertical fold on the medial (inner) calf at the point of maximum circumference.
- Chest—A diagonal fold between the anterior axillary line and the nipple. (*Note:* In the generalized skinfold equations, in women the measurement is taken one-third of the distance between the axillary line and the nipple; in men, half of the distance.[2,3])
- Midaxillary—A horizontal fold on the midaxillary line level with the xyphoid process (the bottom of the sternum). (*Note:* In the generalized skinfold equations, this is a vertical fold.[2,3])
- Subscapular—A diagonal fold following the skin's natural lines just below the lower edge of the scapula. (*Note:* In the generalized skinfold equations, a diagonal fold 1–2 cm below the lower edge of the scapula.[2,3])
- Suprailiac—A diagonal fold above the iliac crest but along the midaxillary line. (*Note:* In the generalized skinfold equations, the measurement is taken just above the iliac crest, anterior to the midaxillary line, and following the angle of the iliac crest.[2,3])
- Thigh—A vertical fold on the front of the thigh, midway between the top of the patella and the hip.
- Triceps—A vertical fold over the triceps muscle on the back of the arm, midway between the olecranon process (elbow) and the acromion process (tip of shoulder blade).

---

[1]Hicks, V., Heyward, V., Flores, A., Stolarczyk, L., Koppy, P., & Wotruba, E. (1993). Validation of near-infrared interactance (NIR) and skinfold (SKF) methods for estimating body composition of American Indian women. *Medicine and Science in Sports and Exercise, 25,* S152.

[2]Jackson, A. S., Pollock, M. L., & Ward, A. (1980). Generalized equations for predicting body density of women. *Medicine and Science in Sports and Exercise, 12,* 175–182.

[3]Jackson, A. S., & Pollock, M. L. (1978). Generalized equations for predicting body density of men. *British Journal of Nutrition, 40,* 497–504.

[4]Slaughter, M. H., Lohman, T. G., Boileau, R. A., Horswill, C. A., Stillman, R. J., Van Loan, M. D., & Bemben, D. A. (1988). Skinfold equations for estimation of body fatness in children and youth. *Human Biology, 60(5),* 709–723.

TABLE A-6   Prediction Equations from Bioelectrical Impedance Analysis

Note: FFM refers to fat-free mass; HT refers to height in cm; BW refers to body weight in kg; C refers to circumference in cm; R refers to resistance in ohms; and X_c refers to reactance in ohms.

Women, ages 50–70 years[1]

   FFM (kg) = 0.474 (HT^2/R) + 0.180 (BW) + 7.3

Women, ages 65–94 years[2]

   FFM (kg) = 0.28 (HT^2/R) + 0.27 (BW) + 0.31 (thigh C) − 1.732

Men, ages 50–70 years[1]

   FFM (kg) = 0.600 (HT^2/R) + 0.186 (BW) + 0.226 (X_c) − 10.9

Men, ages 65–94 years[1]

   FFM (kg) = 0.28 (HT^2/R) + 0.27 (BW) + 0.31 (thigh C) − 2.768

American Indian women, ages 18–60 years[3]

   FFM (kg) = 0.1555 (BW) + 0.001254 (HT^2) − 0.04904 (R) + 0.1417 (X_c) − 0.833 (age) + 20.05

American Indian men, ages 18–29 years[1]

   FFM (kg) = 0.485 (HT^2/R) + 0.338 (BW) + 5.32

American Indian men, ages 30–49 years[1]

   FFM (kg) = 0.549 (HT^2/R) + 0.163 (BW) + 0.092 (X_c) + 4.51

American Indian men, ages 50–70 years[1]

   FFM (kg) = 0.60 (HT^2/R) + 0.186 (BW) + 0.226 (X_c) − 10.9

Hispanic women, ages 20–40 years[4]

   FFM (kg) = 0.00151 (HT^2) − 0.0344 (R) + 0.140 (BW) − 0.158 (age) + 20.387

Hispanic men, ages 19–59 years[5]

   FFM (kg) = 13.74 + 0.34 (HT^2/R) + 0.33 (BW) − 0.14 (age) + 6.18

Caucasian women, ages 18–29 years[1]

   FFM (kg) = 0.476 (HT^2/R) + 0.295 (BW) + 5.49

Caucasian women, ages 30–49 years[1]

   FFM (kg) = 0.493 (HT^2/R) + 0.141 (BW) + 11.59

Caucasian women, ages 50–70 years[1]

   FFM (kg) = 0.494 (HT^2/R) + 0.180 (BW) + 7.3

Caucasian men, ages 18–29 years[1]

   FFM (kg) = 0.485 (HT^2/R) + 0.338 (BW) + 5.32

Caucasian men, ages 17–62 years (<20% fat)[6]

   FFM (kg) = 0.00066360 (HT^2) − 0.02117 (R) + 0.62854 (BW) + 0.12380 (age) + 9.33285

Obese women[4]

   FFM (kg) = 0.00151 (HT^2) − 0.0344 (R) + 0.140 (BW) − 0.158 (age) + 20.387

TABLE A-6 Prediction Equations from Bioelectrical Impedance Analysis (*Continued*)

Obese men[4]

$\quad$ FFM (kg) = 0.00139 (HT^2) − 0.0801 (R) + 0.187 (BW) + 39.830

Female college athletes[7]

$\quad$ FFM (kg) = 0.73 (HT^2/R) + 0.116 (BW) + 0.096 (X$_c$) − 4.03

Male college athletes[7]

$\quad$ FFM (kg) = 0.734 (HT^2/R) + 0.116 (BW) + 0.096 (X$_c$) − 3.152

[1]Lohman, T. G. (1992). *Advances in body composition assessment.* Champaign, IL: Human Kinetics.
[2]Baumgartner, R. N., Heymsfield, S. B., Lichtman, S., Wang, J., & Pierson, R. N. (1991). Body composition in elderly people: Effect of criterion estimates on predictive equations. *American Journal of Clinical Nutrition, 53,* 1–9.
[3]Stolarczyk, L. M., Heyward, V. H., Hicks, V. L., & Baumgartner, R. N. (1994). Predictive accuracy of bioelectrical impedance in estimating body composition of Native American women. *American Journal of Clinical Nutrition, 59,* 964–970.
[4]Gray, D. S., Bray, G. A., Gemayel, N., & Kaplan, K. (1989). Effect of obesity on bioelectrical impedance. *American Journal of Clinical Nutrition, 50,* 255–260.
[5]Rising, R., Swinburn, B., Larson, K., & Ravussin, E. (1991). Body composition in Pima Indians: Validation of bioelectrical resistance. *American Journal of Clinical Nutrition, 53,* 594–598.
[6]Segal, K. R., Van Loan, M., Fitzgerald, P. I., Hodgdon, J. A., & Van Itallie, T. B. (1988). Lean body mass estimation by bioelectrical impedance analysis: A four-site cross-validation study. *American Journal of Clinical Nutrition, 47,* 7–14.
[7]Lukaski, H. C., & Bolonchuk, W. W. (1987). Theory and validation of the tetrapolar bioelectrical impedance method to assess human body composition. In K. J. Ellis, S. Yasamura, & W. D. Morgan (Eds.), *In vivo body composition studies* (pp. 410–414). London: Institute of Physical Sciences in Medicine.

TABLE A-7 Prediction Equations from Circumference Measures

Note: All measurements are recorded in cm. The mean abdomen consists of the average of two abdominal measurements: (1) laterally, midway between the lowest lateral portion of the rib cage and the iliac crest, and anteriorly, midway between the xyphoid process of the sternum and the umbilicus; and (2) laterally, at the iliac crest, and anteriorly, at the umbilicus.

Caucasian women, ages 15–79 years[1]

$\quad$ Body density = 1.168297 − (0.002824 × mean abdomen) + (0.0000122098 × mean abdomen^2)
$\qquad$ − (0.000733128 × hips) + (0.000510477 × height) − (0.000216161 × age)

Obese women[2]

$\quad$ % fat = (0.11077 × mean abdomen) − (0.17666 × height)
$\qquad$ + (0.14354 × weight in kg) + 51.03301

Obese men[3]

$\quad$ % fat = (0.31457 × mean abdomen) − (0.10969 × weight in kg) + 10.8336

[1]Tran, Z. V., & Weltman, A. (1989). Generalized equations for predicting body density of women from girth measurements. *Medicine and Science in Sports and Exercise, 21,* 101–104.
[2]Weltman, A., Levine, S., Seip, R. L., & Tran, Z. V. (1988). Accurate assessment of body composition in obese females. *American Journal of Clinical Nutrition, 48,* 1179–1183.
[3]Weltman, A., Seip, R. L., & Tran, Z. V. (1987). Practical assessment of body composition in adult obese males. *Human Biology, 59(3),* 523–535.

TABLE A-8   Body Mass Index-for-Age Percentiles, Ages 2–20 Years: Boys

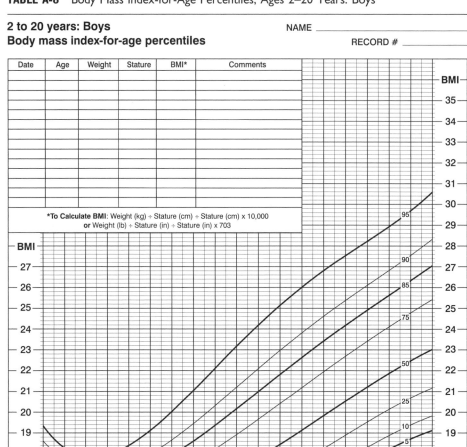

2 to 20 years: Boys
Body mass index-for-age percentiles

NAME _____

RECORD # _____

Date	Age	Weight	Stature	BMI*	Comments

*To Calculate BMI: Weight (kg) ÷ Stature (cm) ÷ Stature (cm) x 10,000
or Weight (lb) ÷ Stature (in) ÷ Stature (in) x 703

AGE (YEARS)

Published May 30, 2000 (modified 10/16/00).
SOURCE: Developed by the National Center for Health Statistics in collaboration with
the National Center for Chronic Disease Prevention and Health Promotion (2000).
http://www.cdc.gov/growthcharts

SAFER · HEALTHIER · PEOPLE™

TABLE A-9 Body Mass Index-for-Age Percentiles, Ages 2–20 Years: Girls

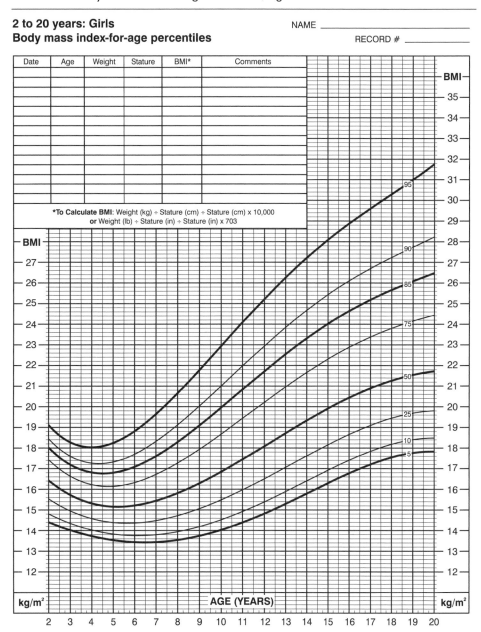

2 to 20 years: Girls
Body mass index-for-age percentiles

NAME _____

RECORD # _____

*To Calculate BMI: Weight (kg) ÷ Stature (cm) ÷ Stature (cm) x 10,000
or Weight (lb) ÷ Stature (in) ÷ Stature (in) x 703

Published May 30, 2000 (modified 10/16/00).
SOURCE: Developed by the National Center for Health Statistics in collaboration with
the National Center for Chronic Disease Prevention and Health Promotion (2000).
http://www.cdc.gov/growthcharts

SAFER·HEALTHIER·PEOPLE™

Nutrition and Physical Activity Assessment Tools

TABLE B-1   Food Record Form

Instructions:
Record everything you eat and drink for 2 week days and 1 weekend day, using the format below. Use a new form for each day that records are kept.

- In column 1, indicate time of day the food/beverage was eaten (indicate A.M. or P.M.).
- In column 2, identify the meal at which the food/beverage was eaten (indicate breakfast, lunch, brunch, dinner, or snack).
- In column 3, record the place where the food/beverage was eaten (car, dining room table, living room sofa)
- In column 4, record all foods and beverages, including water, alcohol, coffee/tea, candy, and chewing gum. Indicate cooking method used and brand names when available. For "combination" foods like salads and sandwiches, or for recipes that you prepare yourself, indicate individual ingredients.
- In column 5, estimate quantity of food/beverage that was actually consumed. Use household measures or portion size from a food label.
- Leave column 6 blank for the person who will code this form.

Time of day	Meal	Place eaten	Food/beverage	Quantity	Code

List any dietary supplements that you took today (vitamins, minerals, amino acids, etc). Give brand name and quantity:

_____

_____

TABLE B-2   Physical Activity Record Form

Instructions:

Use the Activity Record Form (b) to record all of your activities, including sleeping, eating, job activities, and exercise, for a 24-hour period.

Calculating personal MET values

The metabolic equivalent (MET) is a tool for estimating caloric expenditure. Once initial calculations have been performed, this is quicker and easier than calculating the caloric value of each daily activity. The MET system is based on oxygen consumption. For every liter of oxygen taken in and used by the body, 4.825 kcal are expended. At rest, a person uses 3.5 ml oxygen per minute for each kilogram of body weight.

1 MET = 3.5 ml oxygen consumption/kg body weight/minute, or

1 MET = 1 kcal/kg/hour

Convert your weight from pounds to kilograms:

Your body weight in pounds × 2.2 = your weight in kilograms = ——— kg

Your weight in kilograms is approximately equal to the number of calories you expend per hour when you are at 1 MET level of activity.

Now, fill out the chart on page 482 to calculate the caloric value of 0.9–9 METs and to determine the caloric value per 15 minutes of activity. You need to know the 15-minute caloric value because you will be recording your activities in 15-minute segments.

On the Activity Record Form, the four columns titled "Fifteen-minute intervals" represent the four 15-minute segments of each hour. In each space, make a brief notation that indicates what you were doing in each time interval. The eight columns under "MET levels" are for you to indicate the MET level of each of your 15-minute segments for each hour. Indicate in the appropriate MET levels column how many 15-minute segments were at that level during the hour. (*Note:* This can never be greater than 4.) For activities at or above 6 METs, indicate the MET level and the number of 15-minute segments at that MET level in the column marked > 6.

When you have completed the form, total each MET levels column. For the column marked > 6, indicate how many time segments were at each MET level at or above 6 METs.

Put the 15-minute caloric value for each MET level in the appropriate boxes (you calculated these on the Calculating Personal MET Values Form).

Multiply the 15-minute caloric value for each MET level by the number of times you were at that MET level, and record these numbers for subtotal caloric output.

Add all the subtotals to get an estimate of total energy expenditure for the 24-hour period.

(Continued)

TABLE B-2 Physical Activity Record Form (*Continued*)

| MET level | Caloric value | | Activity* |
	Per hour	Per 15 minutes	
0.90			Sleeping
1			Awake, resting quietly; any quiet sitting activity
1.5			Active sitting activities
2			Walking slowly; grooming; light housework (2.5 METs)
3			Jobs that require mostly standing; walking 2.5 mph; heavy cleaning
4			Water aerobics; bicycling < 10 mph; ping pong
5			Low-impact aerobics
6			Higher-intensity recreational activities, such as dancing, badminton, swimming, tennis doubles, cycling,
7			Jogging or running above 5.5 mph; many competitive sports
8			Higher-intensity running, swimming; tennis singles
9			
10			

*See Table B-3, MET Values of Common Activities, also in this appendix, for a more comprehensive list of activities at each MET level.

TABLE B-2 Physical Activity Record Form (*Continued*)

Clock time	Fifteen-minute intervals				MET levels							
	:00-:15	:16-:30	:31-:45	:46-:60	0.9	1	1.5	2	3	4	5	>6
Midnight												
1:00 A.M.												
2:00												
3:00												
4:00												
5:00												
6:00												
7:00												
8:00												
9:00												
10:00												
11:00												
Noon												
1:00 P.M.												
2:00												
3:00												
4:00												
5:00												
6:00												
7:00												
8:00												
9:00												
10:00												
11:00												
Total METS (add down in each level)												
15-Minute caloric value for each level												
Subtotal caloric output (multiply total METS times total caloric value)												
Total caloric output (add subtotals)												

TABLE B-3 MET Values of Common Activities

METs	Activity
0.9	Sleeping
1.0	Any quiet sitting or reclining activity (watching television or movie, listening to music or a lecture, riding in a car or bus)
1.2	Standing quietly (such as standing in line)
1.3	Sitting, reading a book or newspaper
1.5	Any active sitting activity (sewing, desk work, taking notes, light office work, attending a meeting, playing board games, talking on telephone, eating)
1.8	Typing, computer work, taking notes in class
2.0	Driving a car
2.0	Playing most musical instruments
2.0	Grooming (showering, dressing, shaving)
2.3	Grocery shopping
2.5	Household chores, light effort (sweeping, dusting, light cleaning, taking out trash, changing bed linens)
2.5	Cooking
2.5	Wash dishes, clear table, serve food
2.0	Walking very slowly (< 2 mph), strolling
2.5	Walking 2 mph
2.5	Stretching, hatha yoga
2.5	Billiards, croquet
Activities at 3–6 METs	
3.0	Heavy cleaning, vigorous effort
3.5	Vacuuming
4.5	Cultivating garden, planting seedlings, and shrubs
5.5	Mowing lawn with power mower
3.0	Walking, 2.5 mph; walking the dog
3.3	Walking, 3 mph
3.8	Walking briskly (3.5 mph)
6.0	Hiking
3.0	Frisbee playing
3.0	Weight lifting, light to moderate effort
6.0	Weightlifting, vigorous
3.0	Bowling
3.0	Volleyball, noncompetitive recreational activities
3.5	Calisthenics, home exercises, light to moderate effort
3.5	Stationary rowing, light effort (100 watts)
4.0	Bicycling < 10 mph

TABLE B-3   MET Values of Common Activities (*Continued*)

5.5	Stationary bicycle, light effort
4.0	Horseback riding
4.0	Table tennis, ping pong
4.5	Golf
5.0	Aerobics, low-impact
6.0	Tennis, doubles
4.0	Water aerobics
6.0	Swimming leisurely
Activities > 6 METs	
6	Heavy manual labor (chop wood, carry heavy loads, move furniture)
6	Walking uphill, 3.5 mph
7	Jogging
8	Running, 12-minute mile pace (5 mph)
10	Running, 10-minute mile pace
7	Skiing, cross-country, slow or light effort (2.5 mph)
7	Skiing, downhill
8	Skiing, cross-country, moderate effort (4–5 mph)
8	Snow shoeing
7	Racquetball
8	Tennis singles
7	Stationary bike, moderate effort (150 watts)
7	Stationary rowing, moderate effort (100 watts)
8	Bicycling, moderate effort (12–14 mph)
8	Swimming laps, freestyle, slow to moderate effort
10	Swimming laps, vigorous
10	Rope jumping, moderate

Source: The entire list of MET values of activities can be found in: Ainsworth, B.E., Haskell, W.L., Whitt, M.C., Irwin, M.L., Swartz, A.M., Strath, S.J., et al. (2000). Compendium of physical activities: An update of activity codes and MET intensities. *Medicine and Science in Sports and Exercise, 32*(9 Suppl), S498–S516.

Exchange Lists for Weight Management

Choose Your Foods: Exchange Lists for Weight Management (formerly called the U.S. Food Exchange System) was developed to help people lose or maintain weight by focusing on nutrient balance, moderation, and portion control without calorie counting. This is a complement to Choose Your Foods: Diabetes which helps people with diabetes to control their blood glucose levels and reduce some of the complications of diabetes, including elevated blood lipids.

EXCHANGE LISTS

The exchange lists classify foods by their carbohydrate, fat, and protein content. Foods are organized in four major lists—carbohydrate, meat and meat substitutes, fat, and alcohol—based on the similarity of their energy nutrient make-up. Within the carbohydrate and meat lists, there are sublists. One serving of a food on a particular exchange list contains the same amount of carbohydrate, fat, and protein and, therefore, the same number of kilocalories, as one serving of any other food on the same list. This system allows any food on a list to be "exchanged" for another food on the same list.

Table C-1 illustrates the make-up of one serving of each food on the exchange lists. Notice that the carbohydrate group includes five lists: starch; fruit; milk; sweets, desserts, and other carbohydrates; and nonstarchy vegetables. The meat and meat substitutes group contains four lists: lean, medium fat, high fat, and plant-based proteins. The fat list includes foods that should be consumed in limited quantities, particularly by individuals trying to attain a healthy blood lipid profile or trying to prevent excessive fat storage. And the alcohol list has various alcoholic beverages.

Here are some important points about the food lists:

- A serving of food from the starch, fruit, or sweets, desserts, and other carbohydrates list provides 15 g of carbohydrate. Individuals limited to a specific quantity of carbohydrates each day may exchange foods from these three lists; however, there are foods on the starch and sweets, desserts, and other carbohydrates lists that may provide substantially more fat. These are indicated with the symbol.

- All of the foods on the meat and meat substitutes list contain the same quantity of protein, but a high-fat meat may contain up to eight times the fat of a lean meat.

- Some foods provide more than 3 grams of dietary fiber per serving. These are indicated with the symbol.

- Foods that provide more than 480 mg sodium (or, in the case of main dish means on the combination foods and fast foods lists that provide more than 600 mg sodium) have the symbol.

TABLE C-1   Exchange Lists

List	Carbohydrate (g)	Protein (g)	Fat (g)	Energy (kcal/serving)
Carbohydrates				
Starch: breads, cereals, and grains; starchy vegetables; crackers and snacks; and beans, peas, and lentils	15	0–3	0–1	80
Fruit	15	—	—	60
Milk				
Fat-free, low fat, 1%	12	8	0–3	100
Reduced-fat, 2%	12	8	5	120
Whole	12	8	8	150
Sweets, desserts, and other carbohydrates	15	Varies	Varies	Varies
Nonstarchy vegetables	5	2	—	25
Meat and meat substitutes				
Lean	—	7	0–3	45
Medium-fat	—	7	4–7	75
High-fat	—	7	8+	100
Plant-based proteins	—	7	Varies	Varies
Fats	—	—	5	45
Alcohol	Varies	—	—	100

FOODS ON EACH LIST

Because the foods on the exchange lists are organized by their energy and nutrient content, you may not find familiar foods where you would expect:

- Corn, peas, and potatoes are on the starch list, not the vegetable list, because of their high carbohydrate content.
- Frozen yogurt and ice cream are on the sweets, desserts, and other carbohydrates list, not the milk list, due to their higher fat and carbohydrate content.
- Cheese is on the meat and meat substitute list because its fat and protein content make it more like meat than milk.
- On the fat list are olives, avocado, nuts, seeds, and bacon, in addition to the fats that you would expect to see there (such as butter, oil, and sour cream).

A WORD ABOUT PORTION SIZES

In the exchange lists that follow, you will notice that a serving, or portion, size is designated for each food. This information ensures that a portion of one food on a list is equivalent in energy and nutrient content to a portion of other foods on the same list. For example, on the starch list, ¼ of a large bagel can be exchanged for one 5-inch waffle, or 1½ cup of puffed rice cereal, or ¼ cup of low-fat granola. Although each portion size is different, all contain 15 g carbohydrate, 3 g protein, 0–1 g fat, and about 80 kcal. The person who eats a large bagel for breakfast has consumed four starch exchanges and about 320 kcal.

Part of the process of learning to use the exchange system is regularly referring to the exchange lists for information about serving sizes and measuring portion sizes, at least initially. A food scale can be very handy. For example, the ¼ bagel that counts as one starch exchange should weigh about 1 oz. A large bagel might weigh 4 oz. and would therefore count as four starch exchanges.

MEAL PLANNING WITH EXCHANGE LISTS

You are probably beginning to realize how a person could use the exchange system to plan a well-balanced diet with moderate caloric content. Because we know the caloric value of one serving from each exchange list, a diet that includes a specified number of exchanges from each group would provide a predictable quantity of fat and calories.

For example, a daily intake of seven servings from the starch list, five from lean meat and meat substitute, four vegetables, three fruit, two fat-free milk, and five fat would provide about 1,500 kcal and 40–53 g fat (between 24% and 31% of calories from fat, depending on food choice).

7 starch = 560 kcal and 0–7 grams fat

5 lean meat = 225 kcal and 15 g fat

4 nonstarchy vegetable = 100 kcal and 0 g fat

3 fruit = 180 kcal and 0 g fat

2 fat-free milk = 200 kcal and 0–3 g fat

5 fat = 225 kcal and 25 g fat

Table C-2 provides diet patterns for different energy intakes.

CHOOSE YOUR FOODS: EXCHANGE LISTS FOR WEIGHT MANAGEMENT

Twenty-three tables and subtables are included in this section, covering the following major areas: starch; fruits; milk; sweets, desserts, and other carbohydrates; nonstarchy vegetables; meat and meat substitutes; fats; free foods; combination foods; fast foods; and alcohol. The exchange lists are the basis of a meal planning system designed by a committee of the American Diabetes Association and The American Dietetic Association. While designed originally for people with diabetes and others who must follow special diets, the exchange lists are based on principles of good nutrition that apply to everyone. © 2008: American Diabetes Association & American Dietetic Association. *Choose your foods: Exchange lists for weight management.* Alexandria, VA: American Diabetes Association. (978-0-88091-379-9)

TABLE C-2A Starch: Bread

	Food	Serving size
	Bagel, large (4 oz)	¼
⚠	Biscuit, 2½ inches across	1
	Bread	
☺	reduced-calorie	2 slices
	white, whole-grain, pumpernickel, rye, unfrosted raisin	1 slice
	Chapatti, small, 6 inches across	1
⚠	Cornbread, 1¾-inch cube	1
	English muffin	½
	Hot dog bun or hamburger bun	½
	Naan, 8 inches by 2 inches	¼
	Pancake, 4 inches across, ¼-inch thick	1
	Pita, 6 inches across	½
	Roll, plain, small	1
⚠	Stuffing, bread	⅓ cup
⚠	Taco shell, 5 inches across	2
	Tortilla, corn, 6 inches across	1
	Tortilla, flour, 6 inches across	1
	Tortilla, flour, 10 inches across	⅓
⚠	Waffle, 4-inch square	1

TABLE C-2B Starch: Cereals and Grains

	Food	Serving size
	Barley, cooked	⅓ cup
	Bran, dry	
☺	oat	¼ cup
☺	wheat	½ cup
☺	Bulgur, cooked	½ cup
	Cereals	
☺	bran	½ cup
	cooked (oats, oatmeal)	½ cup
	puffed	1½ cup
	shredded wheat, plain	½ cup
	sugar-coated	½ cup
	unsweetened, ready-to-eat	¾ cup

(Continued)

TABLE C-2B Starch: Cereals and Grains (*Continued*)

Food	Serving size
Couscous	⅓ cup
Granola	
low-fat	¼ cup
regular	¼ cup
Grits, cooked	½ cup
Kasha	½ cup
Millet, cooked	⅓ cup
Muesli	¼ cup
Pasta, cooked	⅓ cup
Polenta, cooked	⅓ cup
Quinoa, cooked	⅓ cup
Rice, white or brown, cooked	⅓ cup
Tabbouleh, prepared	½ cup
Wheat germ, dry	3 Tbsp
Wild rice, cooked	½ cup

TABLE C-2C Starch: Starchy Vegetables

Food	Serving size
Cassava	⅓ cup
Corn	½ cup
on cob, large	½ cob
Hominy, canned	¾ cup
Mixed vegetables w/corn, peas, or pasta	1 cup
Parsnips	½ cup
Peas, green	½ cup
Plantain, ripe	⅓ cup
Potato	
baked with skin	¼ large
boiled, all kinds	½ cup of ½ med.
mashed, with milk and fat	½ cup
French fried, oven-baked	1 cup

TABLE C-2C   Starch: Starchy Vegetables (*Continued*)

	Food	Serving size
☺	Pumpkin, canned, no sugar added	1 cup
	Spaghetti/pasta sauce	½ cup
☺	Squash, winter (acorn, butternut)	1 cup
☺	Succotash	½ cup
	Yam, sweet potato, plain	½ cup

TABLE C-2D   Starch: Crackers and Snacks

	Food	Serving size
	Animal crackers	8
	Crackers	
▽!	round, butter-type	6
	saltine-type	6
▽!	sandwich-style, cheese or peanut butter filling	3
▽!	whole-wheat, regular	2–5 (¾ oz)
☺	whole-wheat lower fat or crispbreads	2–5 (¾ oz)
	Graham cracker, 2½-inch square	3
	Matzoh	¾ oz
	Melba toast, 2 inches by 4 inches	4
	Oyster crackers	20
	Popcorn	
▽! ☺	with butter	3 cups
☺	no fat added	3 cups
☺	lower-fat	3 cups
	Pretzels	¾ oz
	Rice cakes, 4 inches across	2
	Snack chips	
	fat-free or baked	15–20 (¾ oz)
▽!	regular	9–13 (¾ oz)

TABLE C-2E Starch: Beans, Peas, and Lentils
These count as 1 starch + 1 lean meat.

	Food	Serving size
😊	Baked beans	⅓ cup
😊	Beans, cooked (black, garbanzo, kidney, lima, navy, pinto, white)	½ cup
😊	Lentils, cooked (brown, green, and yellow)	½ cup
😊	Peas, cooked (black-eyed, split)	½ cup
😊 🗋	Refried beans, canned	½ cup

TABLE C-3 Fruits

	Food	Serving size
	Apple, unpeeled, small	1 (4 oz)
	Apples, dried	4 rings
	Applesauce, unsweetened	½ cup
	Apricots	
	canned	½ cup
	dried	8 halves
😊	fresh	4 whole
	Bananas, extra small	1 (4 oz)
😊	Blackberries	¾ cup
	Blueberries	¾ cup
	Cantaloupe, small	⅓ melon or 1 c cubed
	Cherries	
	sweet, canned	½ cup
	sweet, fresh	12
	Dates	3
	Dried fruits (blueberries, cherries, cranberries, mixed fruit, raisins)	2 Tbsp
	Figs	
	dried	1½
😊	fresh	1½ lg or 2 med
	Fruit cocktail	½ cup
	Grapefruit	
	large	½
	sections, canned	¾ cup

TABLE C-3 Fruits *(Continued)*

Food	Serving size
Grapes, small	17
Honeydew melon	1 slice or 1 c cubed
Kiwi	1
Mandarin oranges, canned	¾ cup
Mango, small	½ fruit or ½ cup
Nectarine, small	1
Orange, small	1
Papaya	½ fruit or 1 cup cubed
Peaches	
canned	½ cup
fresh, medium	1
Pears	
canned	½ cup
fresh, large	½
Pineapple	
canned	½ cup
fresh	¾ cup
Plums	
canned	½ cup
dried	3
small	2
Raspberries	1 cup
Strawberries	1¼ cup whole
Tangerines, small	2
Watermelon	1 slice or 1¼ cup cubed
Fruit juice	
Apple juice/cider	½ cup
Fruit juice blends, 100% juice	⅓ cup
Grape juice	⅓ cup
Grapefruit juice	½ cup
Orange juice	½ cup
Pineapple juice	½ cup
Prune juice	⅓ cup

TABLE C-4 Milk

Food	Serving size	Count as
Milk and yogurts		
Fat-free (skim) or low-fat (1%)		
Milk, buttermilk, acidophilus milk, Lactaid	1 cup	1 fat-free milk
Evaporated milk	½ cup	1 fat-free milk
Yogurt, plain or flavored with artificial sweetener	⅔ cup (6 oz)	1 fat-free milk
Reduced-fat (2%)		
Milk, acidophilus milk, kefir, Lactaid	1 cup	1 reduced-fat milk
Yogurt, plain	⅔ cup (6 oz)	1 reduced-fat milk
Whole		
Milk, buttermilk, goat's milk	1 cup	1 whole milk
Evaporated milk	½ cup	1 whole milk
Yogurt, plain	1 cup (8 oz)	1 whole milk
Dairy-like foods		
Chocolate milk		
fat-free	1 cup	1 fat-free milk + 1 carbohydrate
whole	1 cup	1 whole milk + 1 carbohydrate
Eggnog, whole milk	½ cup	1 carbohydrate + 2 fats
Rice drink		
flavored, low-fat	1 cup	2 carbohydrates
plain, fat-free	1 cup	1 carbohydrate
Smoothies, flavored regular	10 oz	1 fat-free milk + 1½ carbohydrate
Soy milk		
light	1 cup	1 carbohydrate + ½ fat
regular, plain	1 cup	1 carbohydrate + 1 fat
Yogurt		
and juice blends	1 cup	1 fat-free milk + 1 carbohydrate
low-carbohydrate (< 6 g carbohydrate per serving)	⅔ cup (6 oz)	½ fat-free milk
with fruit, low-fat	⅔ cup (6 oz)	1 fat-free milk + 1 carbohydrate

TABLE C-5 Sweets, Desserts, and Other Carbohydrates

Food	Serving size	Count as
Beverages, soda, and energy/sports drinks		
Cranberry juice cocktail	½ cup	1 carbohydrate
Energy drink	1 can (8.3 oz)	2 carbohydrates
Fruit drink or lemonade	1 cup (8 oz)	2 carbohydrates
Hot chocolate		
regular	1 envelope + 8 oz water	1 carbohydrate + 1 fat
sugar-free or light	1 envelope + 8 oz water	1 carbohydrate
Soft drink (soda), regular	1 can (12 oz)	2½ carbohydrates
Sports drink	1 cup (8 oz)	1 carbohydrate
Brownies, cake, cookies, gelatin, pie, and pudding		
Brownie, small, unfrosted	1¼-inch square	1 carbohydrate + 1 fat
Cake		
angel food, unfrosted	1/12 of cake	2 carbohydrates
frosted	2-inch square	2 carbohydrates + 1 fat
unfrosted	2-inch square	1 carbohydrate + 1 fat
Cookies		
chocolate chip	2 (2¼ inches across)	1 carbohydrate + 2 fats
gingersnap	3	1 carbohydrate
sandwich with crème filling	2 small (⅔)	1 carbohydrate + 1 fat
sugar-free	3 small or 1 large	1 carbohydrate + 1–2 fats
vanilla wafer	5	1 carbohydrate + 1 fat
Cupcake, frosted	1 small	2 carbohydrates + 1–1½ fats
Fruit cobbler	½ cup	3 carbohydrates + 1 fat
Gelatin, regular	½ cup	1 carbohydrate
Pie		
commercially prepared fruit, two crusts	⅙ of 8-inch pie	3 carbohydrates + 2 fats
pumpkin or custard	⅛ of 8-inch pie	1½ carbohydrates + 1½ fats
Pudding		
regular (made with reduced-fat milk)	½ cup	2 carbohydrates
sugar-free or sugar- and fat-free (made with fat-free milk)	½ cup	1 carbohydrate
Candy, spreads, sweets, sweeteners, syrups, and toppings		
Candy bar, chocolate/peanut	2 "fun size" bars	1½ carbohydrates + 1½ fats
Candy, hard	3	1 carbohydrate
Chocolate "kisses"	5	1 carbohydrates + 1 fats
Coffee creamer		
dry, flavored	4 tsp	½ carbohydrate + ½ fat
liquid, flavored	2 Tbsp	1 carbohydrate
Fruit snacks, chewy (pureed fruit concentrate)	1 roll	1 carbohydrate
Fruit spreads, 100% fruit	1½ Tbsp	1 carbohydrate

(Continued)

TABLE C-5   Sweets, Desserts, and Other Carbohydrates (*Continued*)

Food	Serving size	Count as
Honey	1 Tbsp	1 carbohydrate
Jam or jelly, regular	1 Tbsp	1 carbohydrate
Sugar	1 Tbsp	1 carbohydrate
Syrup		
chocolate	2 Tbsp	2 carbohydrates
light (pancake type)	2 Tbsp	1 carbohydrate
regular (pancake type)	1 Tbsp	1 carbohydrate
Condiments and sauces		
Barbeque sauce	1 Tbsp	1 carbohydrate
Cranberry sauce, jellied	¼ cup	1½ carbohydrates
Gravy, canned or bottled	½ cup	½ carbohydrate + ½ fat
Salad dressing, fat-free, low-fat, cream-based	3 Tbsp	1 carbohydrate
Sweet and sour sauce	3 Tbsp	1 carbohydrate
Doughnuts, muffins, pastries, and sweet breads		
Banana nut bread	1-inch slice	2 carbohydrates + 1 fat
Doughnut		
cake, plain	1 medium	1½ carbohydrates + 2 fats
yeast type, glazed	3¾ inches across	2 carbohydrates + 2 fats
Muffin (4 oz)	¼ muffin (1 oz)	1 carbohydrate + ½ fat
Sweet roll or Danish	1	2½ carbohydrates + 2 fats
Frozen bars, frozen desserts, frozen yogurt, and ice cream		
Frozen pops	1	½ carbohydrate
Fruit juice bars, frozen, 100% juice	1	1 carbohydrate
Ice cream		
fat-free	½ cup	1½ carbohydrates
light	½ cup	1 carbohydrate + 1 fat
no sugar added	½ cup	1 carbohydrate + 1 fat
regular	½ cup	1 carbohydrate + 2 fats
Sherbert, sorbet	½ cup	2 carbohydrates
Yogurt, frozen		
fat-free	⅓ cup	1 carbohydrate
regular	½ cup	1 carbohydrate + 0–1 fat
Granola bars, meal replacement bars/shakes, and trail mix		
Granola or snack bar, regular or low-fat	1 (1 oz)	1½ carbohydrates
Meal replacement bar	1 (1⅓ oz)	1½ carbohydrates + 0–1 fat
Meal replacement bar	1 (2 oz)	2 carbohydrates + 1 fat
Meal replacement shake, reduced calorie	1 can (10–11 oz)	1½ carbohydrates + 0–1 fat
Trail mix		
candy/nut-based	1 oz	1 carbohydrate + 2 fats
dried fruit-based	1 oz	1 carbohydrate + 1 fat

TABLE C-6 Nonstarchy Vegetables

½ cup cooked or 1 cup raw of:

Food	Food
Amaranthe or Chinese spinach	Jicama
Artichoke	Kohlrabi
Artichoke hearts	Leeks
Asparagus	Mixed vegetables (without corn, peas, or pasta)
Baby corn	Mung bean sprouts
Bamboo shoots	Mushrooms, all kinds, fresh
Bean sprouts	Okra
Beans (green, wax, Italian)	Onions
Beets	Pea pods
Borscht	Peppers, all varieties
Broccoli	Radishes
Brussels sprouts	Rutabaga
Cabbage (green, bok choy, Chinese)	Sauerkraut
Carrots	Soybean sprouts
Cauliflower	Spinach
Celery	Squash (summer, crookneck, zucchini)
Chayote	Sugar snap peas
Coleslaw, packaged, no dressing	Swiss chard
Cucumber	Tomato
Daikon	Tomatoes, canned
Eggplant	Tomato sauce
Gourds (bitter, bottle, luffa, bitter melon)	Tomato/vegetable juice
Green onions or scallions	Turnips
Greens (collard, kale, mustard, and turnip)	Water chestnuts
Hearts of palm	Yard-long beans

TABLE C-7A Meat and Meat Substitutes: Lean

Food	Serving size
Beef: select or choice grades trimmed of fat: ground round, roast (chuck, rib, rump), round, sirloin, steak (cubed, flank, porterhouse, T-bone), tenderloin	1 oz
Beef jerky	½ oz
Cheeses with 3 grams of fat or less per oz	1 oz
Cottage cheese	¼ cup
Egg substitutes, plain	¼ cup
Egg whites	2

(Continued)

TABLE C-7A Meat and Meat Substitutes: Lean (*Continued*)

Food	Serving size
Fish, fresh or frozen, plain: catfish, cod, flounder, haddock, halibut, orange roughy, salmon, tilapia, trout, tuna	l oz
Fish, smoked: herring or salmon (lox)	l oz
Game: buffalo, ostrich, rabbit, venison	l oz
Hot dog with 3 grams of fat or less per oz (8 dogs per 14 oz package). NOTE: *May be high in carbohydrate*	l
Lamb: chop, leg, roast	l oz
Organ meats: heart, kidney, liver NOTE: *May be high in cholesterol*	l oz
Oysters, fresh or frozen	6 medium
Pork, lean	
Canadian bacon	l oz
rib or loin chop/roast, ham, tenderloin	l oz
Poultry, without skin: chicken, Cornish hen, domestic duck or goose (well-drained of fat), turkey	l oz
Processed sandwich meats with 3 grams of fat or less per oz: chipped beef, deli-thin sliced meats, turkey ham, turkey kielbasa, turkey pastrami	l oz
Salmon, canned	l oz
Sardines, canned	2 small
Sausage with 3 grams of fat or less per oz	l oz
Shellfish: clams, crab, imitation shellfish, lobster, scallops, shrimp	l oz
Tuna, canned in water or oil, drained	l oz
Veal: loin, chop, roast	l oz

TABLE C-7B Meat and Meat Substitutes: Medium-Fat

Food	Serving size
Beef: corned beef, ground beef, meatloaf, Prime grades trimmed of fat (prime rib), short ribs, tongue	l oz
Cheeses with 4–7 grams of fat per oz: feta, mozzarella, pasteurized processed cheese spread, reduced-fat cheeses, string	l oz
Egg *High in cholesterol, limit to 3 per week*	l
Fish, any fried type	l oz
Lamb: ground, rib roast	l oz
Pork: cutlet, shoulder roast	l oz
Poultry: chicken with skin; dove, pheasant, wild duck, or goose; fried chicken; ground turkey	l oz
Ricotta cheese	2 oz (¼ cup)
Sausage with 4–7 grams of fat per oz	l oz
Veal, cutlet (no breading)	l oz

TABLE C-7C Meat and Meat Substitutes: High-Fat

	Food	Serving size
	Bacon	
	pork	2 slices (16 slices per lb or 1 oz each before cooking)
	turkey	3 slices (1/2 oz each before cooking)
	Cheese, regular: American, bleu, brie, cheddar, hard goat, Monterey jack, queso, Swiss	1 oz
	Hot dog: beef, pork, or combination (10 per 1-lb-sized package)	1
	Pork: ground, sausage, spareribs	1 oz
	Processed sandwich meats with 8 grams of fat or more per oz: bologna, hard salami, pastrami	1 oz
	Sausage with 8 grams of fat or more per oz: bratwurst, chorizo, Italian, knockwurst, Polish, smoked, summer	1 oz

TABLE C-7D Meat and Meat Substitutes: Plant-Based Proteins

	Food	Serving size	Count as
	"Bacon" strips, soy-based	3 strips	1 medium-fat meat
	Baked beans	⅓ cup	1 starch + 1 lean meat
	Beans, cooked (black, garbanzo, kidney, lima, navy, pinto, white)	½ cup	1 starch + 1 lean meat
	"Beef" or "sausage" crumbles, soy-based	2 oz	½ carbohydrate + 1 lean meat
	"Chicken" nuggets, soy-based	2 nuggets (1½ oz)	½ carbohydrate + 1 medium-fat meat
	Edamame	½ cup	½ carbohydrate + 1 lean meat
	Falafel (spiced chickpea and wheat patties)	3 patties (about 2 inches across)	1 carbohydrate + 1 high-fat meat
	Hot dog, soy-based	1 (1½ oz)	½ carbohydrate + 1 lean meat
	Hummus	⅓ cup	1 carbohydrate + 1 high-fat meat
	Lentils, brown, green, or yellow	½ cup	1 carbohydrate + 1 lean meat
	Meatless burger, soy-based	3 oz	½ carbohydrate + 2 lean meats
	Meatless burger, vegetable- and starch-based	1 patty (about 2½ oz)	1 carbohydrate + 2 lean meats
	Nut spreads: almond butter, cashew butter, peanut butter, soy nut butter	1 Tbsp	1 high-fat meat
	Peas, cooked: black-eyed and split peas	½ cup	1 starch + 1 lean meat
	Refried beans, canned	½ cup	1 starch + 1 lean meat
	"Sausage" patties, soy-based	1 (1½ oz)	1 medium-fat meat

(Continued)

TABLE C-7D   Meat and Meat Substitutes: Plant-Based Proteins (*Continued*)

Food	Serving size	Count as
Soy nuts, unsalted	¾ oz	½ carbohydrate + I medium-fat meat
Tempeh	¼ cup	I medium-fat meat
Tofu	4 oz (½ cup)	I medium-fat meat
Tofu, light	4 oz (½ cup)	I lean meat

TABLE C-8A   Fats: Unsaturated Fats—Monounsaturated Fats

Food	Serving size
Avocado, medium	2 Tbsp (I oz)
Nut butters (*trans* fat-free): almond butter, cashew butter, peanut butter (smooth or crunchy)	I½ Tbsp
Nuts	
almonds	6 nuts
Brazil	2 nuts
cashews	6 nuts
filberts (hazelnuts)	5 nuts
macadamia	3 nuts
mixed (50% peanuts)	6 nuts
peanuts	10 nuts
pecans	4 halves
pistachios	16 nuts
Oil: canola, olive, peanut	I tsp
Olives	
black (ripe)	8 large
green, stuffed	10 large

TABLE C-8B   Fats: Unsaturated Fats—Polyunsaturated Fats

Food	Serving size
Margarine: lower-fat spread (30%–50% vegetable oil, *trans* fat-free)	I Tbsp
Margarine: stick, tub (trans fat-free), or squeeze (*trans* fat-free)	I tsp
Mayonnaise	
reduced-fat	I Tbsp
regular	I tsp
Mayonnaise-style salad dressing	
reduced-fat	I Tbsp
regular	2 tsp

TABLE C-8B Fats: Unsaturated Fats—Polyunsaturated Fats (*Continued*)

Food	Serving size
Nuts	
Pignolia (pine nuts)	1 Tbsp
walnuts, English	4 halves
Oil: com, cottonseed, flaxseed, grape seed, safflower, soybean, sunflower	1 tsp
Oil: made from soybean and canola oil—Enova	1 tsp
Plant stanol esters	
light	1 Tbsp
regular	2 tsp
Salad dressing	
reduced-fat (*may be high in carbohydrate*)	2 Tbsp
regular	1 Tbsp
Seeds	
flaxseed, whole	1 Tbsp
pumpkin, sunflower	1 Tbsp
sesame seeds	1 Tbsp
Tahini or sesame paste	2 tsp

TABLE C-8C Fats: Unsaturated Fats—Saturated Fats

Food	Serving size
Bacon, cooked, regular or turkey	1 slice
Butter	
reduced-fat	1 Tbsp
stick	1 tsp
whipped	2 tsp
Butter blends made with oil	
reduced-fat	1 Tbsp
regular	1½ tsp
Chitterlings, boiled	2 Tbsp (½ oz)
Coconut, sweetened, shredded	2 Tbsp
Coconut milk	
light	⅓ cup
regular	1½ Tbsp
Cream	
half and half	2 Tbsp
heavy	1 Tbsp

(*Continued*)

TABLE C-8C Fats: Unsaturated Fats—Saturated Fats (*Continued*)

Food	Serving size
light	1½ Tbsp
whipped	2 Tbsp
whipped, pressurized	¼ cup
Cream cheese	
reduced-fat	1½ Tbsp
regular	1 Tbsp
Lard	1 tsp
Oil: coconut, palm, palm kernel	1 tsp
Salt pork	¼ oz
Shortening, solid	1 tsp
Sour cream	
reduced-fat or light	3 Tbsp
regular	2 Tbsp

TABLE C-9 Free Foods

Free foods have less than 20 calories and 5 grams of carbohydrate per serving. Because they do contain calories, no more than 3 servings per day should be consumed. If no serving size is listed, these can be eaten whenever you like.

Food	Serving size
Low-carbohydrate foods	
Cabbage, raw	½ cup
Candy, hard (regular or sugar-free)	1
Carrots, cauliflower, or green beans, cooked	¼ cup
Cranberries, sweetened with sugar-substitute	½ cup
Cucumber, sliced	½ cup
Gelatin	
dessert, sugar-free	
unflavored	
Gum	
Jam or jelly, light or no sugar added	2 tsp
Rhubarb, sweetened with sugar substitute	½ cup
Salad greens	
Sugar substitutes (artificial sweeteners)	
Syrup, sugar-free	2 Tbsp

TABLE C-9 Free Foods (*Continued*)

Food	Serving size
Modified-fat foods with carbohydrate	
Cream cheese, fat-free	1 Tbsp (½ oz)
Creamers	
nondairy, liquid	1 Tbsp
nondairy, powdered	2 tsp
Margarine spread	
fat-free	1 Tbsp
reduced-fat	1 tsp
Mayonnaise	
fat-free	1 Tbsp
reduced-fat	1 tsp
Mayonnaise-style salad dressing	
fat-free	1 Tbsp
reduced-fat	1 tsp
Salad dressing	
fat-free or low-fat	1 Tbsp
Fat-free, Italian	2 Tbsp
Sour cream, fat-free or reduced-fat	1 Tbsp
Whipped topping	
light or fat-free	2 Tbsp
regular	1 Tbsp
Condiments	
Barbeque sauce	2 tsp
Catsup (ketchup)	1 Tbsp
Honey mustard	1 Tbsp
Horseradish	
Lemon juice	
Miso	1½ tsp
Mustard	
Parmesan cheese, freshly grated	1 Tbsp
Pickle relish	1 Tbsp
Pickles	
dill	1½ medium
sweet, bread and butter	2 slices
sweet, gherkin	¾ oz
Salsa	¼ cup
Soy sauce, light or regular	1 Tbsp

(*Continued*)

TABLE C-9 Free Foods (*Continued*)

Food	Serving size
Sweet chili sauce	2 tsp
Taco sauce	1 Tbsp
Vinegar	
Yogurt, any type	2 Tbsp
Free snacks	
Baby carrots and celery sticks	5
Blueberries	½ c
Cheese, fat-free, sliced	½ oz
Goldfish-style crackers	10
Saltine-type crackers	2
Frozen cream pop, sugar-free	1
Lean meat	½ oz
Light popcorn	1 cup
Vanilla wafer	1
Drinks/mixes	
Bouillon broth, consomme	
Bouillon or broth, low-sodium	
Carbonated or mineral water	
Club soda	
Cocoa powder, unsweetened	1 Tbsp
Coffee, unsweetened or with sugar substitute	
Diet soft drinks, sugar-free	
Drink mixes, sugar-free	
Tea, unsweetened, or with sugar substitute	
Tonic water, diet	
Water	
Water, flavored, carbohydrate free	
Seasonings	
Flavoring extracts (vanilla, almond, peppermint)	
Garlic	
Herbs, fresh or dried	
Nonstick cooking spray	
Pimiento	
Spices	
Hot pepper sauce	
Wine, used in cooking	
Worcestershire sauce	

TABLE C-10   Combination Foods

	Food	Serving size	Count as
	Entrees		
	Casserole type (tuna noodle, lasagna, spaghetti with meatballs, chili with beans, macaroni and cheese)	1 cup (8 oz)	2 carbohydrates + 2 medium-fat meats
	Stews (beef/other meats and vegetables)	1 cup (8 oz)	1 carbohydrate + 1 medium-fat meat + 0–3 fats
	Tuna salad or chicken salad	½ cup (3½ oz)	½ carbohydrate + 2 lean meats + 1 fat
	Frozen meals/entrees		
	Burrito (beef and bean)	1 (5 oz)	3 carbohydrates + 1 lean meat + 2 fats
	Dinner-type meal	Generally 14–17 oz	3 carbohydrates + 3 medium-fat meats + 3 fats
	Entrée or meal with less than 340 calories	About 8–11 oz	2–3 carbohydrates + 1–2 lean meats
	Pizza		
	cheese/vegetarian, thin crust	¼ of a 12-inch (4½–5 oz)	2 carbohydrates + 2 medium-fat meats
	meat topping, thin crust	¼ of a 12-inch (5 oz)	2 carbohydrates + 2 medium-fat meats + 1½ fats
	Pocket sandwich	1 (4½ oz)	3 carbohydrates + 1 lean meat + 1–2 fats
	Pot pie	1 (7 oz)	2½ carbohydrates + 1 medium-fat meat + 3 fats
	Salads (deli-style)		
	Coleslaw	½ cup	1 carbohydrate + 1½ fats
	Macaroni/pasta salad	½ cup	2 carbohydrates + 3 fats
	Potato salad	½ cup	1½–2 carbohydrates + 1–2 fats
	Soups		
	Bean, lentil, or split pea	1 cup	1 carbohydrate + 1 lean meat
	Chowder (made with milk)	1 cup (8 oz)	1 carbohydrate + 1 lean meat + 1½ fats
	Cream (made with water)	1 cup (8 oz)	1 carbohydrate + 1 fat
	Instant	6 oz prepared	1 carbohydrate
	with beans or lentils	8 oz prepared	2½ carbohydrates + 1 lean meat
	Miso soup	1 cup	½ carbohydrate + 1 fat
	Ramen noodle	1 cup	2 carbohydrates + 2 fats
	Rice (congee)	1 cup	1 carbohydrate
	Tomato (made with water)	1 cup (8 oz)	1 carbohydrate
	Vegetable beef, chicken noodle, or other broth type	1 cup (8 oz)	1 carbohydrate

TABLE C-11A Fast Foods: Breakfast Sandwiches and Main Dishes

	Food	Serving size	Counts as
	Breakfast sandwiches		
	Egg, cheese, meat, English muffin	1	2 carbohydrates + 2 medium-fat meats
	Sausage biscuit sandwich	1	2 carbohydrates + 2 high-fat meats + 3½ fats
	Main dishes/entrees		
	Burrito (beef and beans)	1 (about 8 oz)	3 carbohydrates + 3 medium-fat meats + 3 fats
	Chicken breast, breaded and fried	1 (about 5 oz)	1 carbohydrate + 4 medium-fat meats
	Chicken drumstick, breaded and fried	1 (about 2 oz)	2 medium-fat meats
	Chicken nuggets	6 (about 3 ½ oz)	1 carbohydrate + 2 medium-fat meats + 1 fat
	Chicken thigh, breaded and fried	1 (about 4 oz)	½ carbohydrate + 3 medium-fat meats + 1½ fats
	Chicken wings, hot	6 (5 oz)	5 medium-fat meats + 1½ fats

TABLE C-11B Fast Foods: Asian and Pizza

	Food	Serving size	Counts as
	Asian		
	Beef/chicken/shrimp with vegetables in sauce	1 cup (about 5 oz)	1 carbohydrate + 1 lean meat + 1 fat
	Egg roll, meat	1 (about 3 oz)	1 carbohydrate + 1 lean meat + 1 fat
	Fried rice, meatless	½ cup	1½ carbohydrate + 1½ fats
	Meat and sweet sauce (orange chicken)	1 cup	3 carbohydrates + 3 medium-fat meats + 2 fats
	Noodles and vegetables in sauce (chow mein, lo mein)	1 cup	2 carbohydrates + 1 fat
	Pizza		
	Pizza		
	cheese, pepperoni, regular crust	⅛ of a 14-inch (about 4 oz)	2½ carbohydrates + 1 medium-fat meat + 1½ fats
	cheese/vegetarian, thin crust	¼ of a 12-inch (about 6 oz)	2½ carbohydrates + 2 medium-fat meats + 1½ fats

TABLE C-11C Fast Foods: Sandwiches and Salads

	Food	Serving size	Count as
	Chicken sandwich, grilled	1	3 carbohydrates + 4 lean meats
	Chicken sandwich, crispy	1	3½ carbohydrates + 3 medium-fat meats + 1 fat
	Fish sandwich with tartar sauce	1	2½ carbohydrates + 2 medium-fat meats + 2 fats

TABLE C-11C Fast Foods: Sandwiches and Salads (*Continued*)

	Food	Serving size	Count as
	Hamburger		
	large with cheese	1	2½ carbohydrates + 4 medium-fat meats + 1 fat
	regular	1	2 carbohydrates + 1 medium-fat meat + 1 fat
	Hot dog with bun	1	1 carbohydrates + 1 high-fat meat + 1 fat
	Submarine sandwich		
	less than 6 grams of fat	6-inch sub	3 carbohydrates + 2 lean meats
	regular	6-inch sub	3½ carbohydrates + 2 medium-fat meats + 1 fat
	Taco, hard or soft shell (meat and cheese)	1 small	1 carbohydrate + 1 medium-fat meat + 1½ fats
	Salads		
	Salad, main dish (grilled chicken type, no fressing or croutons)	1	1 carbohydrate + 4 lean meats
	Salad, side (no dressing or cheese)	Small	1 vegetable

TABLE C-11D Fast Foods: Sides/Appetizers and Desserts

	Food	Serving size	Count as
	Sides/appetizers		
	French fries, restaurant style	Small	3 carbohydrates + 3 fats
		Medium	4 carbohydrates + 4 fats
		Large	5 carbohydrates + 6 fats
	Nachos with cheese	Small (about 4½ oz)	2½ carbohydrates + 4 fats
	Onion rings	1 serving (about 3 oz)	2½ carbohydrates + 3 fats
	Desserts		
	Milkshake, any flavor	12 oz	6 carbohydrates + 2 fats
	Soft-serve ice cream cone	1 small	2½ carbohydrates + 1 fat

TABLE C-12 Alcohol
> *One alcohol equivalent (1½ oz of absolute alcohol) has about 100 calories.*

Alcoholic Beverage	Serving size	Count as
Beer		
light (4.2%)	12 fl oz	1 alcohol equivalent + ½ carbohydrate
regular (4.9%)	12 fl oz	1 alcohol equivalent + 1 carbohydrate
Distilled spirits: vodka, run, gin, whiskey 80 or 86 proof	1½ fl oz	1 alcohol equivalent
Liquer, coffee (53 proof)	1 fl oz	½ alcohol equivalent + 1 carbohydrate
Sake	1 fl oz	½ alcohol equivalent
Wine		
dessert (sherry)	3½ fl oz	1 alcohol equivalent + 1 carbohydrate
dry, red or white (10%)	5 fl oz	1 alcohol equivalent

Dietary Reference Intakes (DRI)

The Dietary Reference Intakes (DRI) include two sets of values that serve as goals for nutrient intake—Recommended Dietary Allowances (RDA) and Adequate Intakes (AI). The RDA reflect the average daily amount of nutrient considered adequate to meet the needs of most healthy people. If there is insufficient evidence to determine an RDA, an AI is set. AI are more tentative than RDA, but both may be used as goals for nutrient intakes. (Chapter 6 provides more details.)

In addition to the values that serve as goals for nutrient intakes (present in the tables on pages 509–514), the DRI include a set of values called Tolerable Upper Intake Levels (UL). The UL represent the maximum amount of a nutrient that appears safe for most healthy people to consume on a regular basis. Turn the page for a listing of the UL for selected vitamins and minerals.

Estimated Energy Requirements (EER), Recommend Dietary Allowances (RDA), and Adequate Intakes (AI) for Water, Energy, and the Energy Nutrients

Age (yr)	Reference BMI (kg/m²)	Reference Height cm (in)	Reference Weight kg (lb)	Water[a] AI (L/day)	Energy EER[b] (Kcal/day)	Carbohydrate RDA (g/day)	Total Fiber AI (g/day)	Total Fat AI (g/day)	Linoleic Acid AI (g/day)	Linolenic Acid[c] AI (g/day)	Protein RDA (g/day)	Protein RDA[d] (g/kg/day)
Males												
0–0.5	—	62(24)	6(13)	0.7[e]	570	60	—	31	4.4	0.5	9.1	1.52
0.5–1	—	71(28)	9(20)	0.8[f]	743	95	—	30	4.6	0.5	11	1.20
1–3[g]	—	86(34)	12(27)	1.3	1046	130	19	—	7	0.7	13	1.05
4–8[g]	15.3	115(45)	20(44)	1.7	1742	130	25	—	10	0.9	19	0.95
9–13	17.2	144(57)	36(79)	2.4	2279	130	31	—	12	1.2	34	0.95
14–18	2.5	174(68)	61(34)	3.3	3152	130	38	—	16	1.6	52	0.85
19–30	20.5	177(70)	70(154)	3.7	3067[h]	130	38	—	17	1.6	56	0.80
31–50	22.5[i]	177(70)[i]	70(154)[i]	3.7	3067[h]	130	38	—	17	1.6	56	0.80
>50	22.5[i]	177(70)[i]	70(154)[i]	3.7	3067[h]	130	30	—	14	1.6	56	0.80
Females												
0–0.5	—	62(24)	6(13)	0.7[e]	520	60	—	31	4.4	0.5	9.1	1.52
0.5–1	—	71(28)	9(20)	0.8[f]	676	95	—	30	4.6	0.5	11	1.20
1–3[g]	—	86(34)	12(27)	1.3	992	130	19	—	7	0.7	13	1.05
4–8[g]	15.3	115(45)	20(44)	1.7	1642	130	25	—	10	0.9	19	0.95
9–13	17.4	144(57)	37(81)	2.1	2071	130	26	—	10	1.0	34	0.95
14–18	20.4	163(64)	54(119)	2.3	2368	130	26	—	11	1.1	46	0.85
19–30	21.5	163(64)	57(126)	2.7	2403[j]	130	25	—	12	1.1	46	0.80
31–50	21.5[i]	163(64)[i]	57(126)[i]	2.7	2403[j]	130	25	—	12	1.1	46	0.80
>50	21.5[i]	163(64)[i]	57(126)[i]	2.7	2403[j]	130	21	—	11	1.1	46	0.80

(Continued)

Estimated Energy Requirements (EER), Recommend Dietary Allowances (RDA), and Adequate Intakes (AI) for Water, Energy, and the Energy Nutrients (continued)

Age (yr)	Reference BMI (kg/m²)	Reference Height cm (in)	Reference Weight kg (lb)	Water[a] AI (L/day)	Energy EER[b] (Kcal/day)	Carbohydrate RDA (g/day)	Total Fiber AI (g/day)	Total Fat AI (g/day)	Linoleic Acid AI (g/day)	Linolenic Acid[c] AI (g/day)	Protein RDA (g/day)[d]	Protein RDA (g/kg/day)
Pregnancy												
1st trimester				3.0	+0	175	28	—	13	1.4	71	0.80
2nd trimester				3.0	+340	175	28	—	13	1.4	71	1.10
3rd trimester				3.0	+452	175	28	—	13	1.4	71	1.10
Lactation												
1st 6 months				3.8	+330	210	29	—	13	1.4	71	1.30
2nd 6 months				3.8	+400	210	29	—	13	1.3	71	1.30

Note: For all nutrients, values for infants are AI. Dashes indicate that values have not been determined.

[a]The water AI includes drinking water, water in beverages, and water in foods; in general, drinking water and other beverages contribute about 70 to 80 percent, and foods, the remainder. Conversion factors: 1 L = 33.80 fluid oz; 1 L = 1.06 qt; 1 cup = 8 fluid oz.

[b]The Estimated Energy Requirement (EER) represents the average dietary energy intake that will maintain energy balance in a healthy person of a given gender, age, weight, height, and physical activity level. The values listed are based on an "active" person at the reference height and weight and at the midpoint ages for each group until age 19.

[c]The linolenic acid referred to in this table and text is the omega-3 fatty acid known as alpha-linolenic acid.

[d]The values listed are based on reference body weights.

[e]Assumed to be from human milk.

[f]Assumed to be from human milk and complementary foods and beverages. This includes approximately 0.6 L (~2½cups) as total fluid including formula, juices, and drinking water.

[g]For energy, the age groups for young children are 1–2 years and 3–8 years.

[h]For males, subtract 10 kcalories per day for each year of age above 19.

[i]Because weight need not change as adults age if activity is maintained, reference weights for adults 19 through 30 years are applied to all adult age groups.

[j]For females, subtract 7 kcalories per day for each year of age above 19.

Source: Adapted from the *Dietary Reference Intakes* series, National Academies Press. Copyright 1997, 1998, 2000, 2001, 2002, 2004, 2005, 2011 by the National Academies of Sciences.

Recommended Dietary Allowances (RDA) and Adequate Intakes (AI) for Vitamins

Age (yr)	Thiamin RDA (mg/day)	Riboflavin RDA (mg/day)	Niacin RDA (mg/day)[a]	Biotin AI (μg/day)	Pantothenic acid AI (mg/day)	Vitamin B6 RDA (mg/day)	Folate RDA (μg/day)[b]	Vitamin B12 RDA (μg/day)	Choline AI (mg/day)	Vitamin C RDA (mg/day)	Vitamin A RDA (μg/day)[c]	Vitamin D RDA (IU/day)[d]	Vitamin E RDA (mg/day)[e]	Vitamin K AI (μg/day)
Infants														
0–0.5	0.2	0.3	2	5	1.7	0.1	65	0.4	125	40	400	400 (10 μg)	4	2.0
0.5–1	0.3	0.4	4	6	1.8	0.3	80	0.5	150	50	500	400 (10 μg)	5	2.5
Children														
1–3	0.5	0.5	6	8	2	0.5	150	0.9	200	15	300	600 (15 μg)	6	30
4–8	0.6	0.6	8	12	3	0.6	200	1.2	250	25	400	600 (15 μg)	7	55
Males														
9–13	0.9	0.9	12	20	4	1.0	300	1.8	375	45	600	600 (15 μg)	11	60
14–18	1.2	1.3	16	25	5	1.3	400	2.4	550	75	900	600 (15 μg)	15	75
19–30	1.2	1.3	16	30	5	1.3	400	2.4	550	90	900	600 (15 μg)	15	120
31–50	1.2	1.3	16	30	5	1.3	400	2.4	550	90	900	600 (15 μg)	15	120
51–70	1.2	1.3	16	30	5	1.7	400	2.4	550	90	900	600 (15 μg)	15	120
>70	1.2	1.3	16	30	5	1.7	400	2.4	550	90	900	800 (20 μg)	15	120
Females														
9–13	0.9	0.9	12	20	4	1.0	300	1.8	375	45	600	600 (15 μg)	11	60
14–18	1.0	1.0	14	25	5	1.2	400	2.4	400	65	700	600 (15 μg)	15	75
19–30	1.1	1.1	14	30	5	1.3	400	2.4	425	75	700	600 (15 μg)	15	90
31–50	1.1	1.1	14	30	5	1.3	400	2.4	425	75	700	600 (15 μg)	15	90
51–70	1.1	1.1	14	30	5	1.5	400	2.4	425	75	700	600 (15 μg)	15	90
>70	1.1	1.1	14	30	5	1.5	400	2.4	425	75	700	800 (20 μg)	15	90
Pregnancy														
≤18	1.4	1.4	18	30	6	1.9	600	2.6	450	80	750	600 (15 μg)	15	75
19–30	1.4	1.4	18	30	6	1.9	600	2.6	450	85	770	600 (15 μg)	15	90
31–50	1.4	1.4	18	30	6	1.9	600	2.6	450	85	770	600 (15 μg)	15	90
Lactation														
≤18	1.4	1.6	17	35	7	2.0	500	2.8	550	115	1200	600 (15 μg)	19	75
19–30	1.4	1.6	17	35	7	2.0	500	2.8	550	120	1300	600 (15 μg)	19	90
31–50	1.4	1.6	17	35	7	2.0	500	2.8	550	120	1300	600 (15 μg)	19	90

Note: For all nutrients, values for infants are AI.

[a] Niacin recommendations are expressed as niacin equivalents (NE), except for recommendations for infants younger than 6 months, which are expressed as preformed niacin.

[b] Folate recommendations are expressed as dietary folate equivalents (DFE).

[c] Vitamin A recommendations are expressed as retinol activity equivalents (RAE).

[d] Vitamin D recommendations are expressed as cholecalciferol and assume an absence of adequate exposure to sunlight.

[e] Vitamin E recommendations are expressed as α-tocopherol.

Recommended Dietary Allowances (RDA) and Adequate Intakes (AI) for Minerals

Age (yr)	Sodium AI (mg/day)	Chloride AI (mg/day)	Potassium AI (mg/day)	Calcium RDA (mg/day)	Phosphorus RDA (mg/day)	Magnesium RDA (mg/day)	Iron RDA (mg/day)	Zinc RDA (mg/day)	Iodine RDA (μg/day)	Selenium RDA (μg/day)	Copper RDA (μg/day)	Manganese AI (mg/day)	Fluoride AI (mg/day)	Chromium AI (μg/day)	Molybdenum RDA (μg/day)
Infants															
0–0.5	120	180	400	200	100	30	0.27	2	110	15	200	0.003	0.01	0.2	2
0.5–1	370	570	700	260	275	75	11	3	130	20	220	0.6	0.5	5.5	3
Children															
1–3	1000	1500	3000	700	460	80	7	3	90	20	340	1.2	0.7	11	17
4–8	1200	1900	3800	1000	500	130	10	5	90	30	440	1.5	1.0	15	22
Males															
9–13	1500	2300	4500	1300	1250	240	8	8	120	40	700	1.9	2	25	34
14–18	1500	2300	4700	1300	1250	410	11	11	150	55	890	2.2	3	35	43
19–30	1500	2300	4700	1000	700	400	8	11	150	55	900	2.3	4	35	45
31–50	1500	2300	4700	1000	700	420	8	11	150	55	900	2.3	4	35	45
51–70	1300	2000	4700	1000	700	420	8	11	150	55	900	2.3	4	30	45
>70	1200	1800	4700	1200	700	420	8	11	150	55	900	2.3	4	30	45
Females															
9–13	1500	2300	4500	1300	1250	240	8	8	120	40	700	1.6	2	21	34
14–18	1500	2300	4700	1300	1250	360	15	9	150	55	890	1.6	3	24	43
19–30	1500	2300	4700	1000	700	310	18	8	150	55	900	1.8	3	25	45
31–50	1500	2300	4700	1000	700	320	18	8	150	55	900	1.8	3	25	45
51–70	1300	2000	4700	1200	700	320	8	8	150	55	900	1.8	3	20	45
>70	1200	1800	4700	1200	700	320	8	8	150	55	900	1.8	3	20	45
Pregnancy															
≤18	1500	2300	4700	1300	1250	400	27	12	220	60	1000	2.0	3	29	50
19–30	1500	2300	4700	1000	700	350	27	11	220	60	1000	2.0	3	30	50
31–50	1500	2300	4700	1000	700	360	27	11	220	60	1000	2.0	3	30	50
Lactation															
≤18	1500	2300	5100	1300	1250	360	10	13	290	70	1300	2.6	3	44	50
19–30	1500	2300	5100	1000	700	310	9	12	290	70	1300	2.6	3	45	50
31–50	1500	2300	5100	1000	700	320	9	12	290	70	1300	2.6	3	45	50

Note: For all nutrients, values for infants are AI.

Tolerable Upper Intake Levels (UL) for Vitamins

Age(yr)	Niacin (mg/day)[a]	Vitamin B$_6$ (mg/day)	Folate (µg/day)[a]	Choline (mg/day)	Vitamin C (mg/day)	Vitamin A (µg/day)[b]	Vitamin D (IU/day)	Vitamin E (mg/day)[c]
Infants								
0–0.5	—	—	—	—	—	600	1000 (25 µg)	—
0.5–1	—	—	—	—	—	600	1500 (38 µg)	—
Children								
1–3	10	30	300	1000	400	600	2500 (63 µg)	200
4–8	15	40	400	1000	650	900	3000 (75 µg)	300
9–13	20	60	600	2000	1200	1700	4000 (100 µg)	600
Adolescents								
14–18	30	80	800	3000	1800	2800	4000 (100 µg)	800
Adults								
19–70	35	100	1000	3500	2000	3000	4000 (100 µg)	1000
>70	35	100	1000	3500	2000	3000	4000 (100 µg)	1000
Pregnancy								
≤18	30	80	800	3000	1800	2800	4000 (100 µg)	800
19–50	35	100	1000	3500	2000	3000	4000 (100 µg)	1000
Lactation								
≤18	30	80	800	3000	1800	2800	4000 (100 µg)	800
19–50	35	100	1000	3500	2000	3000	4000 (100 µg)	1000

[a]The UL for niacin and folate apply to synthetic forms obtained from supplements, fortified foods, or a combination of the two.
[b]The UL for vitamin A applies to the preformed vitamin only.
[c]The UL for vitamin E applies to any form of supplemental α-tocopherol, fortified foods, or a combination of the two.

Tolerable Upper Intake Levels (UL) for Minerals

Age (yr)	Sodium (mg/day)	Chloride (mg/day)	Calcium (mg/day)	Phosphorus (mg/day)	Magnesium (mg/day)[d]	Iron (mg/day)	Zinc (mg/day)	Iodine (µg/day)	Selenium (µg/day)	Copper (µg/day)	Manganese (mg/day)	Fluoride (mg/day)	Molybdenum (µg/day)	Boron (mg/day)	Nickel (mg/day)	Vanadium (mg/day)
Infants																
0–0.5	—	—	1000	—	—	40	4	—	45	—	—	0.7	—	—	—	—
0.5–1	—	—	1500	—	—	40	5	—	60	—	—	0.9	—	—	—	—
Children																
1–3	1500	2300	2500	3000	65	40	7	200	90	1000	2	1.3	300	3	0.2	—
4–8	1900	2900	2500	3000	110	40	12	300	150	3000	3	2.2	600	6	0.3	—
9–13	2200	3400	3000	4000	350	40	23	600	280	5000	6	10	1100	11	0.6	—
Adolescents																
14–18	2300	3600	3000	4000	350	45	34	900	400	8000	9	10	1700	17	1.0	—
Adults																
19–50	2300	3600	2500	4000	350	45	40	1100	400	10,000	11	10	2000	20	1.0	1.8
51–70	2300	3600	2000	4000	350	45	40	1100	400	10,000	11	10	2000	20	1.0	1.8
>70	2300	3600	2000	3000	350	45	40	1100	400	10,000	11	10	2000	20	1.0	1.8
Pregnancy																
≤18	2300	3600	3000	3500	350	45	34	900	400	8000	9	10	1700	17	1.0	—
19–50	2300	3600	2500	3500	350	45	40	1100	400	10,000	11	10	2000	20	1.0	—
Lactation																
≤18	2300	3600	3000	4000	350	45	34	900	400	8000	9	10	1700	17	1.0	—
19–50	2300	3600	2500	4000	350	45	40	1100	400	10,000	11	10	2000	20	1.0	—

[d]The UL for magnesium applies to synthetic forms obtained from supplements or drugs only.

NOTE: An Upper Limit was not established for vitamins and minerals not listed and for those age groups listed with a dash (—) because of a lack of data, not because these nutrients are safe to consume at any level of intake. All nutrients can have adverse effects when intakes are excessive.

Source: Adapted with permission from the *Dietary Reference Intakes* series, National Academies Press. Copyright 1997, 1998, 2000, 2001, 2002, 2005, 2011 by the National Academies of Sciences.

Index

Homocysteine, 211
Honolulu Heart Study, 250
Hospitalization, for eating disorders, 413
HRR. *See* Heart rate reserve
Human Fitness Gene Map, 146
Human Genome Project, 143
Human Obesity Gene Map, 143
Hunger
    defined, 121
    digestive system and, 130
    insulin and, 131
Hydration, 296
Hydroxycitric Acid/Garcinia cambogia, 362
Hyperinsulinemia, 22, 131
Hyperplasia, 61, 105, 133
Hypertension, 21–22, 455, 460–461
Hypertrophic cardiomyopathy, 289
Hypertrophy, 133, 247
Hypochloremia, 401
Hypoglycemia, 403
Hypokalemia, 222, 401
Hyponatremia, 401
Hypothalamus, 122, 131

I
IBW. *See* Ideal body weight
Ideal body weight, 4–5
Identification, Evaluation, and Treatment of Overweight and Obesity in Adults, 73
IDF. *See* International Diabetes Federation
Imagery, 323
Inactivity
    body weight and, 13–15
    excess calories and, 15
    factor in development of obesity, 428
    weight and, 148
Incidence, 418
Incomplete protein, 181
Indirect calorimetry, 98
Inflammation, adipose tissue and, 138
Injuries
    muscle strain, 295
    musculoskeletal, 295
    reducing risk of musculoskeletal, 296
    warm up and warm down exercise to prevent, 295–296
Inositol, 214
Insoluble fiber, 158
Institute for Aerobics Research, 250
Institute of Medicine, 222
Insulin, 126, 131
Insulin resistance, 22
Internal abdominal visceral fat, 5–6
International Bibliographic Information on Dietary Supplements, 365
International Diabetes Federation (IDF), 24, 458
International Nutrient Databank Directory, 67
Interpersonal psychotherapy, 416
Interval training, 277
Intima-media thickness (IMT), 454
Intrinsic factors, 210
Iron, 223–226
    food sources of, 224–225
    functions of, 223–224
    recommended dietary allowances, 225

Iron supplements, 225–226
Isoflavones, 182
Isokinetic contractions, 244

J
JAMA (Journal of the American Medical Association), 362
Joints, 241
Journal of the American Dietetic Association, 67

K
Kaiser Family Foundation, 430
Ketosis, 90
Kilocalories (kcals), 12, 83
Knee height, 45–46
Krebs cycle. *See* Citric acid cycle

L
Lactic acid, 381
Lacto-ovo vegetarians, 181
Lap-band surgery. *See* Laparoscopic adjustable gastric banding
Laparoscopic adjustable gastric banding, 377
Laparoscopy, 377
Lapse, 326
Laser-assisted lipolysis, 382
LDLs. *See* Low-density lipoprotein
Legumes, 181–182
Leptin, 136–138, 354–355
Leukopenia, 401
Ligaments, 241
Limbic system, 125
Lipid profile, 64–65
Lipids
    abnormal blood, 454–455
    cholesterol, 168–171
    classification in children and adolescents, 455
    desirable values in adults, 65
    dietary fats, 171–176
    in energy metabolism, 87–90
    fat replacers, 176–178
    fatty acids, 167–168
    medications for high blood, 460
    overview of, 170–171
    role of carbohydrates in, 89–90
    trans fatty acids, 171
Lipogenesis, 134
Lipoic acid, 214
Lipolysis, 134
Lipoprotein lipase, 135
Lipoproteins, 169–170. *See also* Cholesterol; Lipids
Liposuction, 382–383
Localized fat reduction, 380–384
Low body weight. *See* Underweight
Low-calorie diets, 365–375
    effectiveness of, 369–374
    fraud and, 374–375
    medically unsupervised outcomes, 372–374
Low-carbohydrate diets, 370–371
Low-density lipoproteins (LDLs), 64–65, 169
Low-fat diets
    prevention of chronic disease and, 173–174
    weight loss and, 173

Lower-body fat distribution patterns, 140
Lungs, 240, 246
Lutein, 198
Luteinizing hormone, 402
Lycopene, 198

M
Ma huang. *See* Ephedra/Ma huang
Macular degeneration, 199
Magnesium, 219–221
    food sources of, 220
    functions of, 219
    recommended dietary allowances, 220–221
Magnesium supplements, 221
Maintenance stage, 325–326
Making weight, 410
Maltitol, 165
Mannitol, 165
Maximal oxygen consumption (VO₂max), 246–247
    estimation for men and women of cardiorespiratory fitness from, 261
    measurement of, 259–260
Meal planning. *See* U.S. Food Exchange System
Medical health history. *See* Health history
Medicine and Science in Sports and Exercise, 69
Meditation, 323
Mediterranean diet, 189
    pyramid, 190
    using exchange system to create, 191
Megaloblastic anemia, 210
Melanin-concentrating hormone antagonists, 355
Melatonin, 398
Men
    cardiorespiratory fitness from VO₂max (ml/kg/min) for, 261
    hamstring flexibility from sit and reach test for, 267
    muscular endurance tests for, 265–266
    muscular strength test from bench press for, 263
Mental health, improved, 250
Mercury. *See* Methylmercury
Meridia. *See* Sibutramine
MET. *See* Metabolic equivalent
Metabolic equivalent (MET), 69
Metabolic equivalents (METs), 280–281
Metabolic rate, 97–103
    excercise and, 101–102
    factors affecting resting metabolic rate, 98–101
    measuring resting energy expenditure, 97–98
    relationship between resting metabolic rate and obesity, 102–103
Metabolic syndrome, 24–26, 458
    criteria for children, 459
Methylmercury, 183
Metropolitan desirable weights for men and women, 49
Metropolitan height and weight tables for men and women, 49
Metropolitan Life Insurance Company, 49

extent of childhood obesity and, 426
gender and racial differences in, 9–10
global perspectives on the prevalence of, 11–12
health history, 61–62
hypertension and, 21–22
osteoarthritis and, 26–27
sleep apnea and, 22
Overweight children
and adolescents who want to diet, 452
defined, 445
development of, 445–448
guidelines for physical activity, 449–450
health problems and, 454–459
outcomes of treatments for, 452
role parents and families, 448–449
school-based programs for, 450–451
weight-management strategies for, 448–452
Oxidative phosphorylation, 87
Oxygen, caloric value of, 92, 93

P

Pacific Islanders, obesity and, 148–149
Pantothenic acid, 208–210
PAR-Q. *See* Physical Activity Readiness Questionnaire
Paraaminobenzoic acid (PABA), 214
Parents
and adequate physical activity for children, 435
fat children and fat, 427–428
and healthy diets for children, 438–441
Parotid glands, 402
Partial meal replacements, 369–370
Pauling, Linus, 212
Pediatric Nutrition Surveillance Survey, 426
Pedometers, 71–72
interpretation of, 72
limitations of, 72
Pennington Biomedical Research Center, 143
Peptide YY (PYY), 354
Peptides, 123, 124
Pernicious anemia, 210
PET/CT. *See* Positron emission tomography/computed tomography
PFK. *See* Phosphofructokinase
Pharmacotherapy, 417
Pharmalogical treatments, 460–461
Phenotype, 143
Phentermine, 349
Phenylpropanolamine (PPA), 353–354
Phosphofructokinase (PFK), 86, 161
Phosphorus, 219–221
food sources of, 220
functions of, 219
recommended dietary allowances, 220–221
Phosphorus supplements, 221
Physical activity
appetite and, 274–275
barriers to involving older adults in, 255
benefits, 245–256
body fat distribution and, 249
body systems involved in, 236–244

cardiorespiratory improvements from, 245–247
children and, 434–438
community-based programs, 436–437
culture and, 339
development of bone density and, 241, 248
in energy balance equation, 111–112
energy balance equation and, 111–112
exercise and, 235–270
guidelines for overweight children, 449–450
inactivity and, 13–15
incorporation into daily life, 285–287
indicators of exercise intensity and, 278–282
motivation, 297–303
obesity and, 285
policies for prevention of obesity, 438
prevention of obesity and, 428–430
recommendations for children, 426
resting metabolic rate and, 101
role of parents in assuring adequate, 435
role of schools in providing, 435–436
safe, 288–297
screening for participation in, 289–294
skill building for, 331–332
as treatment for comorbidities of obesity in children, 460
weight management and, 272–276
Physical activity assessment, 68–73
analysis of physical activity data, 69
pedometers and accelerometers, 71–72
physical activity records, 69, 70
Physical activity index (PAI), 111
Physical Activity Readiness Questionnaire (PAR-Q), 290–291
Physical activity records, 69
interpretation of, 70
limitations of, 70
Physical examination, 59–60
Physical fitness, 256
Physical fitness assessment, 256–268
cardiorespiratory, 257–260
purpose of, 257
Physician referral form, 293
Physicians' Health Study, 30
Physiologic fitness, 256, 257
Phytochemicals, 163
Pica, 396–397
Planet Health, 450
Plant-based diets, 181–182
legumes and soy, 181–182
recommendations for, 181
risk of chronic diseases and, 181
Plasma, 236
Polycystic ovary syndrome, 64
Polyphenols, 199
Polysaccharides, 157, 158
Polyunsaturated fatty acids, 167
Portion size, 335
Portion sizes, 66, 67, 73, 441
Positron emission tomography/computed tomography (PET/CT), 104
Potassium, 222

PPA. *See* Phenylpropanolamine
Prader-Willi syndrome, 143
Precontemplation stage, 311
Precontemplators
consciousness-raising interventions, 313
determining perceptions of problem, 312–313
strategies for, 312–313
Prediction equations
from bioelectrical impedance analysis, 57
for calculating resting metabolic rate, 99
from circumference measures, 50
from skinfold measures, 53–54
Pregnancy
complications of, 27
exercise benefits during, 252–254
Preparation stage, 314
characteristics of, 315
Prevalence, 419
Prevention
of childhood obesity, 425–468
eating disorders, 418–420
nutrition policies for obesity, 443
obesity, 427, 428
primary, 418–420, 427
secondary, 419, 427
tertiary, 419, 427
Prevention concepts, 426–432
Primary prevention, 418–420, 427, 434–444
Primary pulmonary hypertension, 349
Problem-solving skills, 326–328
Proteins
fat replacers, 177
fish, 182–183
function of, 179–181
overview of, 180
plant-based diets and, 181–182
promotion of health and weight management and, 181–184
respiratory quotient of, 92
Prozac. *See* Fluoxetine (Prozac)
Pulmonary circulation, 237
Pulmonary embolism, 382
Push-up tests, 264–265
Pyruvate, 86, 87, 363
PYY. *See* Peptide YY

Q

Q. *See* Cardiac output
Qnexa, 352
Quality of life, optimal weight and, 32–33

R

RAND Corporation, 16
Rating of perceived exertion (RPE), 281–282
RDAs. *See* Recommended dietary allowances
Reactive oxygen species, 202
Reciprocal determinism, 317
Recommended dietary allowances (RDAs), 196
Reductil. *See* Sibutramine
REE. *See* Resting energy expenditure
Reinforcement, 317, 319